W0080258

Lecture Notes in Physics

For information about Vols. 1–99, please contact your bookseller or Springer-Verlag.

Vol. 100: Einstein Symposion Berlin. Proceedings 1979. Edited by H. Nelkowski et al. VIII, 550 pages. 1979.

Vol. 101: A. Martin-Löf, Statistical Mechanics and the Foundations of Thermodynamics. V, 120 pages. 1979.

Vol. 102: H. Hora, Nonlinear Plasma Dynamics at Laser Irradiation. VIII, 242 pages. 1979.

Vol. 103: P. A. Martin, Modèles en Mécanique Statistique des Processus Irréversibles. IV, 134 pages. 1979.

Vol. 104: Dynamical Critical Phenomena and Related Topics. Proceedings, 1979. Edited by Ch. P. Enz. XII, 390 pages. 1979.

Vol. 105: Dynamics and Instability of Fluid Interfaces. Proceedings, 1978. Edited by T. S. Sørensen. V, 315 pages. 1979.

Vol. 106: Feynman Path Integrals, Proceedings, 1978. Edited by S. Albeverio et al. XI, 451 pages. 1979.

Vol. 107: J. Kijowski, W. M. Tulczyjew, A Symplectic Framework for Field Theories. IV, 257 pages. 1979.

Vol. 108: Nuclear Physics with Electromagnetic Interactions. Proceedings, 1979. Edited by H. Arenhövel and D. Drechsel. IX, 509 pages. 1979.

Vol. 109: Physics of the Expanding Universe. Proceedings, 1978. Edited by M. Demiański. V, 210 pages. 1979.

Vol. 110: D. A. Park, Classical Dynamics and Its Quantum Analogues. VIII, 339 pages. 1979.

Vol. 111: H.-J. Schmidt, Axiomatic Characterization of Physical Geometry. V, 163 pages. 1979.

Vol. 112: Imaging Processes and Coherence in Physics. Proceedings, 1979. Edited by M. Schlenker et al. XIX, 577 pages. 1980.

Vol. 113: Recent Advances in the Quantum Theory of Polymers. Proceedings 1979. Edited by J.-M. André et al. V, 306 pages. 1980.

Vol. 114: Stellar Turbulence. Proceedings, 1979. Edited by D. F. Gray and J. L. Linsky. IX, 308 pages. 1980.

Vol. 115: Modern Trends in the Theory of Condensed Matter. Proceedings, 1979. Edited by A. Pekalski and J. A. Przystawa. IX, 597 pages. 1980.

Vol. 116: Mathematical Problems in Theoretical Physics. Proceedings, 1979. Edited by K. Osterwalder. VIII, 412 pages. 1980.

Vol. 117: Deep-Inelastic and Fusion Reactions with Heavy Ions. Proceedings, 1979. Edited by W. von Oertzen. XIII, 394 pages. 1980.

Vol. 118: Quantum Chromodynamics. Proceedings, 1979. Edited by J. L. Alonso and R. Tarrach. IX, 424 pages. 1980.

Vol. 119: Nuclear Spectroscopy. Proceedings, 1979. Edited by G. F. Bertsch and D. Kurath. VII, 250 pages. 1980.

Vol. 120: Nonlinear Evolution Equations and Dynamical Systems. Proceedings, 1979. Edited by M. Boiti, F. Pempinelli and G. Soliani. VI, 368 pages. 1980.

Vol. 121: F. W. Wiegel, Fluid Flow Through Porous Macromolecular Systems. V, 102 pages. 1980.

Vol. 122: New Developments in Semiconductor Physics. Proceedings, 1979. Edited by F. Beleznay et al. V, 276 pages. 1980.

Vol. 123: D. H. Mayer, The Ruelle-Araki Transfer Operator in Classical Statistical Mechanics. VIII, 154 pages. 1980.

Vol. 124: Gravitational Radiation, Collapsed Objects and Exact Solutions. Proceedings, 1979. Edited by C. Edwards. VI, 487 pages. 1980.

Vol. 125: Nonradial and Nonlinear Stellar Pulsation. Proceedings, 1980. Edited by H. A. Hill and W. A. Dziembowski. VIII, 497 pages. 1980.

Vol. 126: Complex Analysis, Microlocal Calculus and Relativistic Quantum Theory. Proceedings, 1979. Edited by D. Iagolnitzer. VIII, 502 pages. 1980.

Vol. 127: E. Sanchez-Palencia, Non-Homogeneous Media and Vibration Theory. IX, 398 pages. 1980.

Vol. 128: Neutron Spin Echo. Proceedings, 1979. Edited by F. Mezei. VI, 253 pages. 1980.

Vol. 129: Geometrical and Topological Methods in Gauge Theories. Proceedings, 1979. Edited by J. Harnad and S. Shnider. VIII, 155 pages. 1980.

Vol. 130: Mathematical Methods and Applications of Scattering Theory. Proceedings, 1979. Edited by J. A. DeSanto, A. W. Sáenz and W. W. Zachary. XIII, 331 pages. 1980.

Vol. 131: H. C. Fogedby, Theoretical Aspects of Mainly Low Dimensional Magnetic Systems. XI, 163 pages. 1980.

Vol. 132: Systems Far from Equilibrium. Proceedings, 1980. Edited by L. Garrido. XV, 403 pages. 1980.

Vol. 133: Narrow Gap Semiconductors Physics and Applications. Proceedings, 1979. Edited by W. Zawadzki. X, 572 pages. 1980.

Vol. 134: γγ Collisions. Proceedings, 1980. Edited by G. Cochard and P. Kessler. XIII, 400 pages. 1980.

Vol. 135: Group Theoretical Methods in Physics. Proceedings, 1980. Edited by K. B. Wolf. XXVI, 629 pages. 1980.

Vol. 136: The Role of Coherent Structures in Modelling Turbulence and Mixing. Proceedings 1980. Edited by J. Jimenez. XIII, 393 pages. 1981.

Vol. 137: From Collective States to Quarks in Nuclei. Edited by H. Arenhövel and A. M. Saruis. VII, 414 pages. 1981.

Vol. 138: The Many-Body Problem. Proceedings 1980. Edited by R. Guardiola and J. Ros. V, 374 pages. 1981.

Vol. 139: H. D. Doebner, Differential Geometric Methods in Mathematical Physics. Proceedings 1981. VII, 329 pages. 1981.

Vol. 140: P. Kramer, M. Saraceno, Geometry of the Time-Dependent Variational Principle in Quantum Mechanics. IV, 98 pages. 1981.

Vol. 141: Seventh International Conference on Numerical Methods in Fluid Dynamics. Proceedings. Edited by W. C. Reynolds and R. W. MacCormack. VIII, 485 pages. 1981.

Vol. 142: Recent Progress in Many-Body Theories. Proceedings. Edited by J. G. Zabolitzky, M. de Llano, M. Fortes and J. W. Clark. VIII, 479 pages. 1981.

Vol. 143: Present Status and Aims of Quantum Electrodynamics. Proceedings, 1980. Edited by G. Gräff, E. Klempt and G. Werth. VI, 302 pages. 1981.

Lecture Notes in Physics

Edited by H. Araki, Kyoto, J. Ehlers, München, K. Hepp, Zürich
R. Kippenhahn, München, H. A. Weidenmüller, Heidelberg
and J. Zittartz, Köln

181

Gauge Theories of the Eighties

Proceedings of the Arctic School of Physics 1982
Held in Äkäslompolo, Finland, August 1–13, 1982

Edited by R. Raitio and J. Lindfors

Springer-Verlag Berlin Heidelberg GmbH 1983

Editors

Risto Raitio
Juha Lindfors
University of Helsinki, Department of Theoretical Physics
Siltavuorenpenger 20 C, SF-00170 Helsinki 17

ISBN 978-3-540-12301-9 ISBN 978-3-540-39888-2 (eBook)
DOI 10.1007/978-3-540-39888-2

Originally published by Springer-Verlag Berlin Heidelberg New York in 1983

2153/3140-543210

PREFACE

The Arctic School of Physics 1982 was held in Äkäslompolo, Lapland from August 1 to 13, 1982. This was the second time that this school took place there. Its aim was to review and discuss gauge theories from the point of view of future developments. The lectures covered lattice theories, non-perturbative methods, supersymmetry, neutrino physics, quark plasma and thermodynamics. More speculative topics included random dynamics, compositeness and superunification. On the experimental side there were lectures on lepton scattering and antiproton-proton collisions.

The school was attended by about 60 students with various levels of experience, and about 20 lecturers. Their stay of 12 days - with no night between - offered excellent opportunities to discuss physics both in the lecture hall and on the nearby fells.

The school received financial support from Nordita, the Particle Physics Committee of the Academy of Finland, the Finnish Ministry of Education, and the Research Institute for Theoretical Physics of the University of Helsinki. The organizing committee wishes to thank these institutions for their kind support. We also thank all the speakers for sharing their knowledge and giving much inspiration to all of us.

Helsinki, February 1983

Risto Raitio Juha Lindfors

TABLE OF CONTENTS

The contribution by C. Rubbia was not submitted for publication

INTRODUCTION TO LATTICE GAUGE THEORIES AND NUMERICAL METHODS

Carleton DeTar
Research Institute for Theoretical Physics
University of Helsinki
Siltavuorenpenger 20 C
SF-00170 Helsinki 17
FINLAND
and
Department of Physics
University of Utah
Salt Lake City UT 84112
USA

1. Introduction

The lattice approximation has proved to be an important technique for studying properties of continuum space-time gauge theories - particularly for providing information about the ground state, the low-lying excitations, and thermodynamic behavior. Numerical lattice calculations have given the most convincing evidence to date that quarks are confined in SU(2) and SU(3) Yang-Mills theories, and they promise to provide values for the masses of the low-lying mesons, baryons, and "glueballs" and to give a thermodynamic description of the theory, without the need for introducing extraneous assumptions about the vacuum state. In principle all we should want to know about the structure of the QCD vacuum and the low-lying states is given by the numerical lattice calculations. We simply need a very large and fast computer. However, most of us find it unsatisfying to characterize the physical vacuum by a set of numbers that only modern mass storage devices can record. Moreover, with present computer facilities it is inconceivable that lattice gauge theories can give detailed information about hadron wave functions, such as, e.g. the charge radius of the neutron. Our ultimate goal should be to obtain approximate analytic models. Nevertheless, the numerical calculations now provide a high standard of comparison according to which any approximate description must be judged.

There is a distinguished history of numerical and analytical studies of lattice field theories in condensed matter physics. Many of these techniques have been adapted to high energy physics applications. Wilson and Polyakov [1] and Kogut and Susskind [2] proposed lattice versions of quantum chromodynamics in 1974. Creutz [3] demonstrated that it was feasible to do numerical calculations with Wilson's lattice version of the SU(2) Yang-Mills theory and gave evidence that the long range confinement of the strong coupling lattice methods persists even in the weak coupling regime, thereby "demonstrating" that the theory is confining. His work opened the door for a host of subsequent numerical studies. We now know a great deal about lattice gauge theories with practically any gauge group. Of particular interest for QCD, the string tension, a few glueball masses, and the deconfinement temperature have been calculated for the pure Yang-Mills SU(2) and SU(3) gauge theories. With fermions included, we have some preliminary evaluations of several low-lying hadron masses in an approximation that neglects Feynman diagrams with quark loops [4,5]. Efforts to include quark loops are beginning to succeed [6,7] .

Because of the considerable current interest in fermions and the computational technology my lectures will dwell more heavily on these topics. Lattice gauge theories have already been treated in several reviews and summer school lectures [8,9,10] . Therefore, I refer the interested student to these sources for a more historically balanced presentation. Nevertheless, I shall try to keep the discussion reasonably self-contained for the benefit of the non-experts in this field.

In these lectures I emphasize the connection between the hamiltonian and Euclidean action formulation of the gauge theories, since I believe this gives more insight into the meaning of finite temperature correlation functions. The massive Schwinger model on a finite lattice is used to motivate the generalization to quantum chromodynamics in four space-time dimensions. I begin by introducing the hamiltonian version of the Schwinger model and proceed to derive the functional integral expression for the partition function of this model. In the process a number of related topics are treated, including the determination of glueball and meson masses, the Grassmann calculus, and a brief discussion of various popular fermion actions. I conclude with a discussion of numerical methods.

2. Hamiltonian Formulation of a Simple Lattice Gauge Theory: Massive Schwinger Model

Let us begin by discussing a finite lattice version of the massive Schwinger model (quantum electrodynamics in one dimension). First, we focus on the fermions and introduce the electromagnetic field later. In one dimension Schwinger described relativistic fermions with two-component spinors. We introduce

$$\psi_x = \begin{pmatrix} \psi_{1x} \\ \psi_{2x} \end{pmatrix} \; ; \; \gamma_0 = \begin{pmatrix} 1 & 0 \\ 0 & -1 \end{pmatrix} \; ; \; \gamma_1 = \begin{pmatrix} 0 & 1 \\ -1 & 0 \end{pmatrix} , \tag{2.1}$$

where $x = 0,1...,N-1$ labels the lattice points. The lattice Hamiltonian is [2]

$$H_F = \frac{-i}{2a} \sum_{x=0}^{N-1} \bar{\psi}_x \gamma_1 (\psi_{x+1} - \psi_{x-1}) + \sum_{x=0}^{N-1} m \bar{\psi}_x \psi_x . \tag{2.2}$$

where a is the lattice spacing, and we use periodic boundary conditions $\psi_0 = \psi_N$. With the anticommutation relations

$$\{\psi_{x\alpha}, \psi^{\dagger}_{x'\alpha'}\} = \delta_{\alpha\alpha'} \delta_{xx'} \tag{2.3}$$

it is easily verified that the lattice version of the Dirac equation

$$i\dot{\psi}_x = [\psi_x, H_F] = \frac{-i\gamma_0\gamma_1}{2a} (\psi_{x+1} - \psi_{x-1}) + m\gamma_0\psi_x \tag{2.4}$$

is obtained. Notice that the hermiticity of H is associated with the appearance of the central difference in (2.4). The equation of motion is brought to 2×2 block diagonal form in momentum space:

$$\psi_k = \frac{1}{\sqrt{N}} \sum_{x=0}^{N-1} e^{ikx2\pi/N} \psi_x \qquad k = 0,1,...,N-1 .$$

The eigenfrequencies are the zeros of the determinant

$$\begin{vmatrix} m-\omega & \frac{1}{a} \sin\frac{2\pi k}{N} \\ \frac{1}{a} \sin\frac{2\pi k}{N} & -m-\omega \end{vmatrix} , \tag{2.5}$$

i.e.

$$\omega = \pm \sqrt{m_2 + \frac{1}{a^2} (\sin\frac{2\pi k}{N})^2} . \tag{2.6}$$

The continuum limit of the theory is defined by $a \to 0$ with Na fixed. For ω to remain finite, it is necessary that $k \approx 0$ or N or $k \approx N/2$.

$$\omega \approx \pm\sqrt{m^2+p^2}$$

$$p = \begin{cases} \dfrac{2\pi k}{aN} \quad k \gtrsim 0; \ \dfrac{2\pi(k-N)}{aN} \quad k \lesssim N \ . \\ \dfrac{2\pi(k-N/2)}{aN} \quad k \approx N/2 \ . \end{cases} \tag{2.7}$$

Therefore, there are two orthogonal sets of low-lying excitations with the physical characteristics of two types of free particles of mass m and their antiparticles. This species doubling is a persistent difficulty for lattice fermion theories. There is a simple remedy in the one dimensional case. Let N be even and define

$$\psi_x = (\gamma_0\gamma_1)^x \, \psi'_x \ . \tag{2.8}$$

Then

$$H_F = \frac{-i}{2a} \sum_{x=0}^{N-1} \psi'^{\dagger}_x \, (\psi'_{x+1} - \psi'_{x-1}) + \sum_{x=0}^{N-1} m\psi'^{\dagger}_x\psi'_x \, (-)^x, \tag{2.9}$$

which is the sum of two identical hamiltonians, one from the upper and one from the lower components. If we write $\phi_x = \psi'_{x\,1}$, the upper component, then we have the one-component hamiltonian

$$H'_F = \frac{-i}{2a} \sum_{x=0}^{N-1} \phi^*_x \, (\phi_{x+1} - \phi_{x-1}) + \sum_{x=0}^{N-1} m\phi^*_x \, \phi_x (-)^x. \tag{2.10}$$

The Fourier transform

$$\phi_k = \frac{1}{N} \sum_{x=0}^{N-1} e^{ikx2\pi/N} \, \phi_x \tag{2.11}$$

gives

$$H'_F = \sum_{k=0}^{N/2-1} (\phi^\dagger_k \ \phi^\dagger_{k+N/2}) \begin{pmatrix} \frac{1}{a}\sin\frac{2\pi k}{N} & m \\ m & \frac{-1}{a}\sin\frac{2\pi k}{N} \end{pmatrix} \begin{pmatrix} \phi_k \\ \phi_{k+N/2} \end{pmatrix} . \tag{2.12}$$

This expression is easily diagonalized and the dispersion relation (2.6) is the same. However, k now covers only half the range $0 \leq k < N/2-1$, giving a single fermion species in the continuum limit. In Susskind's language, the transformation from ψ to ϕ corresponds to keeping only the upper component of ψ_x on the even sites and lower on the odd sites [2].

Although the species doubling problem can be solved by this device the continuous chiral symmetry in the massless theory is then lost [11]. When m=0 the Hamiltonian is invariant under the global transformations

$$\begin{aligned} \psi_x &\to e^{i\theta/2}\psi_x \\ \psi_x &\to e^{i\gamma_5\theta/2}\psi_x \end{aligned} \tag{2.13}$$

where $\gamma_5 = \gamma_0\gamma_1$. The first symmetry is associated with charge conservation. It is valid also for the massive theory, and is unaffected by the reduction $\psi \to \phi$. The second is obviously lost in the reduction $\psi \to \phi$, since it involves both upper and lower components.

An alternative solution to the species doubling problem involves introducing a non-local difference operator. See [12]. The non-locality has discouraged widespread use of this form.

b. Hamiltonian with Gauge Fields

Kogut and Susskind [2] showed how to construct a lattice hamiltonian invariant under local gauge transformations. For a U(1) theory in $A_0 = 0$ gauge, it is

$$H = \frac{1}{2} g^2 a \sum_{x=0}^{N-1} \ell_x^2 - \frac{i}{2a} \sum_{x=0}^{N-1} \left[\bar{\psi}_x e^{i\theta_x}\gamma_1\psi_{x+1} - \bar{\psi}_{x+1} e^{-i\theta_x}\gamma_1\psi_x\right] + \sum_{x=0}^{N-1} m\bar{\psi}_x\psi_x \tag{2.14}$$

where the gauge link angle $\theta_x \in [0,2\pi)$, the conjugate flux variable ℓ_x, satisfying

$$[\theta_x, \ell_{x'}] = i\delta_{xx'}, \tag{2.15}$$

and the gauge coupling constant g have been introduced. The familiar continuum limit is obtained, if

$$\begin{aligned} \theta_x &= g\, a\, A_1\, (ax) \\ \ell_x &= E_1\, (ax)/g \\ \psi_x &= \sqrt{a}\, \chi\, (ax) \end{aligned} \tag{2.16}$$

and the limit $ga \to 0$ is taken. Then

$$H \sim \int dx \; \{ \tfrac{1}{2} E_1{}^2(x) + \bar{\chi}(x)\gamma_1\left[-i\partial_x + gA_1(x)\right]\chi(x) + m\bar{\chi}(x)\chi(x)\} . \tag{2.17}$$

Notice that finite values of the vector potential and electric field correspond respectively to small θ_x and large ℓ_x.

The Hamiltonian is invariant under the time independent gauge transformations

$$\begin{aligned} &\psi_x \to e^{i\lambda_x}\psi_x \\ &\theta_x \to \theta_x - \lambda_{x+1} + \lambda_x . \end{aligned} \tag{2.18}$$

The specification of the model is completed by requiring that the physical states satisfy Gauss' law:

$$\ell_x - \ell_{x-1} = \rho_x \equiv :\psi_x^\dagger\psi_x: . \tag{2.19}$$

where : : is a suitably chosen normal ordering discussed below. Because of the assumed periodicity of θ_x and ℓ_x under $x \to x+N$, Gauss' law requires that the physical states be neutral:

$$0 = \ell_{N-1} - \ell_{-1} = \sum_{x=0}^{N-1} \rho_x , \tag{2.20}$$

Usually one arranges for the vacuum to be neutral and studies neutral excitations thereof. In perturbation theory the vacuum in lowest order has all free negative energy fermion states filled. From (2.6) we see that there are N such states in the two-component theory. Thus we obtain

$$:\psi_x^\dagger\,\psi_x: = \psi_x^\dagger\,\psi_x - 1 , \tag{2.21}$$

which corresponds to the introduction on each site of a "heavy ion" with charge opposite to the fermion's charge. On the other hand, the one-component theory has 1/2 particle per site in the vacuum state. The charge must be quantized in whole units of g since ℓ_x is quantized. At the expense of changing the lattice theory, but presumably not the continuum limit, one can define

$$\rho_x = \begin{cases} \phi_x^\dagger\phi_x - 1 & x \text{ even} \\ \phi_x^\dagger\phi_x & x \text{ odd} , \end{cases} \tag{2.22}$$

which corresponds to the introduction of heavy ions on the even sites [13]. Another possibility involves doubling the species, but this time adding a particle of negative charge. Then

$$\rho_x = \phi_x^\dagger\phi_x + \phi_x^{c\dagger}\phi_x^c$$

is a suitable choice. This solution is typical of fermion actions with symmetric time derivatives and does not require the introduction of heavy ions.

3. The Pure Gauge Theory

a. Feynman Path Integral for the Partition Function

From the hamiltonian

$$H_G = \frac{1}{2} g^2 a \sum_{k=0}^{N-1} \ell_x^2 \tag{3.1}$$

with $\ell_x - \ell_{x-1} = 0$ for physical states and $\ell_0 = \ell_N$, we shall calculate the partition function

$$Z_G(\beta) = \mathrm{Tr}\, \exp(-\beta H_G) \ . \tag{3.2}$$

The problem is a trivial one with the solution

$$Z_G(\beta) = \sum_{\ell=-\infty}^{\infty} \exp(-\tfrac{1}{2} g^2 a \beta \ell^2 N), \tag{3.3}$$

but we shall take a devious approach to obtain the functional integral representation on the Wilson lattice, which is useful for non-trivial cases in higher dimensions and with non-Abelian gauge groups. To this end end we write

$$Z_G(\beta) = \mathrm{Tr}\ T^M \tag{3.4}$$

where

$$T = \exp(-H_G \Delta\tau) \tag{3.5}$$

and $\beta = M\Delta\tau$. T is the transfer matrix, an important operator in its own right. It is the evolution operator for an imaginary time interval. The functional integral is usually written on the group manifold - in this case, over the variables θ_x, conjugate to ℓ_x. We therefore use the basis in which θ_x is diagonal to calculate the trace. However the general state vector $|\{\theta_x\}\rangle$ does not satisfy Coulomb's law, so it is necessary to introduce the projection operator

$$P = \prod_{x=0}^{N-1} \int_0^{2\pi} \frac{d\omega_x}{2\pi} \exp[i(\ell_x - \ell_{x-1})\omega_x] \ , \tag{3.6}$$

whereupon the trace over physical states may be obtained by integrating over all θ_x as follows:

$$Z(\beta) = \mathrm{Tr}\,\exp(-\beta H_G) = \int \prod_{x=0}^{N-1} d\theta_x \,<\{\theta_x\}|\exp(-\beta H_G)\,P|\{\theta_x\}> . \qquad (3.7)$$

Now the projection P commutes with H, so

$$Z(\beta) = \mathrm{Tr}\,(TP)^M \qquad (3.8)$$

and on the θ_x basis the matrix product is

$$Z(\beta) = \int \prod_x d\theta_{x,M-1} \prod_x d\theta_{x,M-2} \cdots \prod_x d\theta_{x,0} <\{\theta_{x0}\}|TP|\{\theta_{x,M-1}\}><\{\theta_{x,M-1}\}|$$

$$TP|\{\theta_{x,M-2}\}> \;\ldots\; <\{\theta_{x1}\}|TP|\{\theta_{x0}\}> . \qquad (3.9)$$

To evaluate the transfer matrix element, we first find the matrix element

$$J = <\{\theta'_x\}|\,\exp\,(-\frac{\varepsilon}{2}\sum_{x=0}^{N-1} \ell_x^2)\,\exp[\sum_{x=0}^{N-1} i\omega_x(\ell_x - \ell_{x-1})]\,|\{\theta_x\}> \qquad (3.10)$$

with $\varepsilon = g^2 a\Delta\tau$ and then integrate over all ω_x. Since

$$e^{i\omega\ell_u}|\{\theta_x\}> = |\{\theta_0,\ldots,\theta_{u-1},\theta_u-\omega,\theta_{u+1},\ldots\theta_{N-1}\}> \qquad (3.11)$$

we find

$$J = <\{\theta'_x\}|\,\exp(-\frac{\varepsilon}{2}\sum_{x=0}^{N-1} \ell_x^{\,2})\,|\{\theta''_x\}> \qquad (3.12)$$

where

$$\theta''_x = \theta_x - \omega_x + \omega_{x+1} \qquad . \qquad (3.13)$$

The expression (3.12) is readily evaluated for small ε as follows:

$$<\theta'_x\,|\,\exp(-\frac{\varepsilon\ell x^2}{2})\,|\theta''_x> = \sqrt{4\pi\varepsilon}\,<\theta'_x\,|\int_{-\infty}^{\infty} d\theta e^{i\ell_x\theta-\theta^2/2\varepsilon}|\theta''_x>$$

$$= \sqrt{4\pi\varepsilon}\sum_{n=-\infty}^{\infty} \exp[-(\theta'_x - \theta''_x + 2\pi n)^2/2\varepsilon] \qquad (3.14)$$

$$\sim \sqrt{4\pi\varepsilon}\;\exp\{\frac{1}{\varepsilon}[\cos(\theta'_x - \theta''_x) - 1]\}$$

In the second step we used (3.11) and

$$\langle \theta'_x | \theta''_x + \theta \rangle = \sum_{n=-\infty}^{\infty} \delta(\theta'_x - \theta''_x - \theta + 2\pi n) \tag{3.15}$$

Putting everything together, we find that up to an irrelevant constant factor the matrix element is

$$J \sim \exp\{ \frac{1}{g^2 a\Delta\tau} \sum_{x=0}^{N-1} [\cos(\theta'_x - \theta_x + \omega_x - \omega_{x+1}) - 1]\} \tag{3.16}$$

for small $g^2 a\Delta\tau$. Since the projection operator P appears M times in (3.8) it is necessary to integrate over M sets of ω_x. Thus

$$Z(\beta) \sim \int \prod_{\tau=0}^{M-1} \prod_{x=0}^{N-1} d\theta_{x\tau} d\omega_{x\tau} \exp(-S_G) \tag{3.17}$$

$$S_G = \frac{1}{g^2 a\Delta\tau} \sum_{x=0}^{N-1} \sum_{\tau=0}^{M-1} [1-\cos(\theta_{x\tau+1} - \theta_{x\tau} + \omega_{x\tau} - \omega_{x+1,\tau})] \tag{3,18}$$

where $\theta_{xM} \equiv \theta_{x0}$. This is the Feynman path integral with the Wilson action for the U(1) gauge theory. The action can be described by constructing a space-time lattice shown in Fig.3.1. The angles $\theta_{x\tau}$ are the space-like link variables and $\omega_{x\tau}$, the time-like variables. The argument of the cosine is the plaquette angle θ_p formed from the "vectorial" sum of the angles around a square. Gauge invariance on the space-time lattice takes the following form: For an arbitrary set of angles $\lambda_{x\tau}$, let

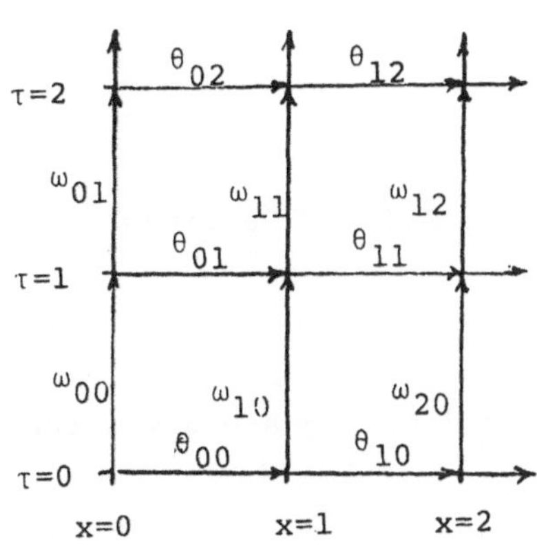

Fig.3.1

$$\theta_{x\tau} \rightarrow \theta_{x\tau} - \lambda_{x+1,\tau} + \lambda_{x\tau} \tag{3.19}$$

$$\omega_{x\tau} \rightarrow \omega_{x\tau} + \lambda_{x\tau} - \lambda_{x,\tau+1} . \tag{3.20}$$

The action S_G is invariant under this transformation.

The continuum limit of the action S_G is expressed in the familiar form by putting

$$\theta_{x\tau} = gaA_1(ax, \tau\Delta\tau) \tag{3.21}$$

$$\omega_{x\tau} = g\Delta\tau A_0(ax, \tau\Delta\tau). \tag{3.22}$$

Then in the limit $\Delta\tau g \to 0$ and $ag \to 0$ with A_1 and A_0 fixed,

$$S_G \sim \int dx \int_0^\beta d\tau \, \frac{1}{4} F_{\mu\nu}^2 \tag{3.23}$$

with $F_{\mu\nu} = \partial_\mu A_\nu - \partial_\nu A_\mu$.

The generalization of the Wilson action to higher dimensions and other gauge groups is straightforward. For SU(3) in four space-time dimensions on a symmetric hypercubical lattice of lattice constant a, each site is labeled by a four vector x and has eight links to its nearest neighbors in the directions $\pm e_\mu$, where e_μ is a unit lattice vector. To each of the four positive directions an SU(3) link matrix $U_{x\mu}$ is assigned. Each site x also has 24 associated plaquettes, $p_{x\mu\nu}$, six with $\mu>\nu>0$. The plaquette product around the border $\partial p_{x\mu\nu}$,

$$U(\partial p_{x\mu\nu}) = U_{x\mu} \, U_{x+e\mu,\nu} U^\dagger_{x+e\nu,\mu} \, U^\dagger_{x,\nu} \, , \tag{3.24}$$

appears in Wilson's Yang-Mills action:

$$S_G = \frac{1}{6g^2} \sum_x \sum_{\mu>\nu>0} \mathrm{Tr} \, [\, 2-U(\partial p_{x\mu\nu}) - U^\dagger(\partial p_{x\mu\nu}) \,] \tag{3.25}$$

The partition function is

$$Z(\beta) = \int \prod_{x,\mu} dU_{x\mu} \, \exp(-S_G), \tag{3.26}$$

where $dU_{x\mu}$ is the invariant Haar measure for SU(3). Putting

$$U_{x\mu} = \exp(iga \, A_\mu{}^c \lambda^c) \tag{3.27}$$

where λ^c are the eight generators of SU(3) and taking $a \to 0$ gives the Euclidean SU(3) Yang-Mills action

$$S_G^{YM} \sim \int_0^\beta d\tau \int d^3x \, \frac{1}{4} \, (\, F^c_{\mu\nu} \,)^2 \tag{3.28}$$

where

$$F^c_{\mu\nu} = \partial_\mu A^c{}_\nu - \partial_\nu A^c_\mu - g f_{abc} A^a{}_\mu A^b_\nu . \tag{3.29}$$

b. Measurements

The thermal expectation value of an operator is defined in general as

$$\langle \mathcal{O} \rangle = \mathrm{Tr}\,[\mathcal{O} \exp(-\beta H)] / \mathrm{Tr} \exp(-\beta H) . \tag{3.30}$$

At zero temperature the expectation value is

$$\langle \mathcal{O} \rangle \underset{\beta\to\infty}{\sim} \sum_d \langle 0,d | \mathcal{O} | 0,d \rangle \tag{3.31}$$

where $|0,d\rangle$ is the normalized ground state of the theory with a label d for a possible degeneracy. It is often quite easy to express the numerator of (3.30) as a functional integral just as we have done the denominator. In this way the functional integral provides information about the ground state of the theory.

Other quantities of interest involve correlations in Euclidean time. For example, we define the Euclidean Heisenberg operator

$$\mathcal{O}_\tau = \exp(\tau H)\, \mathcal{O} \exp(-\tau H) \tag{3.32a}$$

and consider the correlation

$$\langle \mathcal{O}_{\tau_2} \mathcal{O}_{\tau_1} \rangle = \mathrm{Tr}\,\{\exp[(-\beta+\tau_2)H]\, \mathcal{O} \exp[(-\tau_2+\tau_1)H]\, \mathcal{O} \exp(-\tau_1 H)\} / \mathrm{Tr} \exp(-\beta H) \tag{3.32b}$$

At low temperature and large $\tau_2 - \tau_1$

$$\langle \mathcal{O}_{\tau_2} \mathcal{O}_{\tau_1} \rangle \underset{\substack{\beta\to\infty \\ \tau_2-\tau_1\to\infty}}{\sim} |\langle 0 | \mathcal{O} | 1 \rangle|^2 \exp[-E_1(\tau_2 - \tau_1)] \tag{3.33}$$

where $|1\rangle$ is the lowest lying state with energy E_1 reached from the vacuum by applying the operator $\mathcal{O}$. (The degeneracy label has been suppressed.)

In the pure gauge theory correlations in time can be used to measure the potential energy of two sources of opposite charge in the presence of the gauge field. They also provide information about bound states (glueballs) of the gauge field. These applications are next discussed briefly.

Consider the simple one dimensional U(1) gauge theory again.

Suppose we add a fixed charge of $+g$ at site x_1 and $-g$ at site $x_2>x_1$. The projection operator P_s onto physical states in the presence of these sources is [(see (3.6)]

$$P_s = \int_0^{2\pi} \prod_{x=0}^{N-1} \frac{d\omega_x}{2\pi} \exp[\, i(\ell_x - \ell_{x-1})\omega_x] \exp[-i\omega_{x_1} + i\omega_{x_2}] \tag{3.34}$$

since Gauss' law now reads

$$\ell_x - \ell_{x-1} = \delta_{x,x_1} - \delta_{x,x_2} \ . \tag{3.35}$$

Between these sources is one extra unit of electric flux at each link. This added flux appears at one time when the sources are introduced and disappears when the sources are removed. The operator that must accompany the appearance of the sources is therefore

$$L = \exp(i \sum_{x=x_1}^{x_2-1} \theta_x) \tag{3.36}$$

since

$$e^{i\theta_x} \,|\ell_x\rangle = |\ell_x+1\rangle \tag{3.37}$$

increases the flux by one unit. The operator L^* accompanies the removal of the sources. Notice that $LP = P_s L$. A particularly interesting correlation introduced by Wilson is the Wilson loop, given by

$$W = \langle\, L^*_{\tau_2} \;\; L_{\tau_1} P \,\rangle \ . \tag{3.38}$$

It is written in functional integral form as

$$W = \int [d\theta\; d\omega] \exp(-S_G) \exp(i\theta_W) / \int [d\theta\; d\omega] \exp(-S_G) , \tag{3.39}$$

where θ_W is the gauge invariant expression

$$\theta_W = \sum_{x=x_1}^{x_2-1} (\theta_{x\tau_1} - \theta_{x\tau_2}) + \sum_{\tau=\tau_1}^{\tau_2-1} (\omega_{x_1\tau} - \omega_{x_2\tau}) \ . \tag{3.40}$$

This expression is the vectorial sum of the angles around the rectangular Wilson loop bounded by $x=x_1,x_2$ and $\tau=\tau_1,\tau_2$ [1]. For the Yang-Mills action (3.25) a similar expectation value is defined with $\exp(i\theta_W)$ replaced by the product of representation matrices U around the loop. Non-rectangular loops are also considered. At low temperature and large $\tau_2 - \tau_1$ the correlation W becomes

$$W \sim C \exp[-V(x_2-x_1)(\tau_2-\tau_1)] \tag{3.41}$$

where V gives the energy difference between the vacuum in the presence of the sources and in the absence of the sources. If the potential grows linearly with $|x_2-x_1|$ as in a confining theory, then

$$W \sim C \exp(-\sigma A), \tag{3.42}$$

where A is the area of the loop and σ is the string tension. Thus the color confining properties of SU(2) and SU(3) gauge theories were studied [1,3].

To measure the mass of a low lying glueball, it is necessary to consider operators $\mathcal{O}_\gamma$ that excite the state from the vacuum. A zero momentum state can be constructed from a local operator $\mathcal{O}_\gamma(x)$ by summing over x. At large β but any $\tau_2-\tau_1$, one may calculate the time correlation (3.33) for the general operator

$$\mathcal{O}_g = \sum_\gamma c_\gamma \mathcal{O}_\gamma \tag{3.43}$$

where c_γ are constants. The quantity of interest at large β is

$$C_{\tau_2\tau_1} = \langle \mathcal{O}_{g\tau_2} \mathcal{O}_{g\tau_1} \rangle - \langle \mathcal{O}_g \rangle^2 \doteq \sum_{\gamma,\gamma'} \sum_{n=1}^{\infty} \langle 0| \mathcal{O}_\gamma |n\rangle\langle n| \mathcal{O}_{\gamma'} |0\rangle c_\gamma c^*_{\gamma'} \exp[-E_n(\tau_2-\tau_1)] \tag{3.44}$$

where $|n\rangle$ for n=1,2,... is the nth glueball state of energy E_n excited from the vacuum $|0\rangle$ with energy $E_0=0$ by $\mathcal{O}_\gamma$. By letting $\tau_2-\tau_1 \to \infty$ the state with lowest E_n is selected [14]. This method may be converted into a variational calculation. The time correlation (3.44) is just

$$C_{\tau_2\tau_1} = \langle g| e^{-(\tau_2-\tau_1)HG} |g\rangle - |\langle g|0\rangle|^2 \tag{3.45}$$

where

$$|g\rangle = \sum_\gamma c_\gamma \mathcal{O}_\gamma |0\rangle \tag{3.46}$$

is the variational ansatz for the glueball state. By maximizing, for example,

$$C_{\tau_2\tau_1}/C_{\tau_2-1,\tau_1} = \exp(-F_g) \tag{3.47}$$

with respect to C_γ for a fixed $\tau_2-\tau_1$ one obtains $E_g < E_1$. In this spirit the masses of glueballs in SU(2) and SU(3) were measured [15]

4. The Free Fermion Theory

a. Grassmann Calculus

We now turn to a discussion of the fermion sector without the gauge field. We show how to derive the stadard functional integral representation from the hamiltonian theory. It is traditional to express the functional integration over fermion fields in terms of Grassmann variables. Therefore, we begin with a brief review of the Grassmann calculus [16]. Let $\eta_1,\ldots,\eta_n$ be a set of independent Grassmann numbers. They have the property that

$$\eta_i\eta_k+\eta_k\eta_i \equiv \{\eta_i,\eta_k\} = 0 \tag{4.1}$$

so that $\eta_i^2=0$. A function of these variables is usually expressed as a "Taylor" series, and has only a finite number of terms because of (4.1):

$$f(\eta) = a + \sum_i b_i\eta_i + \sum_{i>j} c_{ij}\eta_i\eta_j + \sum_{i>j>k} d_{ijk}\,\eta_i\eta_j\eta_k + \ldots + z\eta_1\ldots\eta_n \tag{4.2}$$

where a, c_{ij} etc. are ordinary numbers and b_i, d_{ijk} etc. are independent Grassmann numbers. Left partial differentiation with respect to the Grassmann numbers is defined as follows:

$$\frac{\partial}{\partial\eta_k}\eta_i = \delta_{ik}\ ;\quad \frac{\partial}{\partial\eta_k} a = 0 \tag{4.3}$$

$$\frac{\partial}{\partial\eta_k}\eta_i\eta_j = \delta_{ik}\eta_j - \delta_{kj}\eta_i . \tag{4.4}$$

Notice that the derivative and the number anticommute. Integration is the same as differentiation, but for historical and notational reasons it has been given a separate symbol:

$$\int d\eta = 0 \; ; \; \int d\eta_i \, \eta_j = \delta_{ij} \; ; \tag{4.5}$$

$$\int d\eta_k \eta_i \eta_j = \delta_{ik}\eta_j - \delta_{kj}\eta_i \; . \tag{4.6}$$

The Grassmann δ-function is easily shown to be

$$\delta(\eta-\eta') = \eta-\eta' \tag{4.7}$$

so that

$$\int d\eta \quad f(\eta)\delta(\eta-\eta') = f(\eta') \tag{4.8}$$

where f is a c-number function as in (4.2).

An important property of Grassmann integration is that if $\eta_1,\ldots,\eta_n$ and $\eta_1^* \ldots \eta_n^*$ are 2n <u>independent</u> Grassmann numbers,(the notation η^*does not mean the "complex conjugate" of η) and if M is an n×n matrix, then

$$\int [d\eta^* d\eta] \exp(-\eta^\dagger M\eta) = \det M, \tag{4.9}$$

where

$$[d\eta^* d\eta] = [d\eta_1^* d\eta_1 \ldots d\eta_n^* d\eta_n]. \tag{4.10}$$

This is easily derived from the definitions above. Also readily derived are the following useful expressions:

$$\int [d\eta^* d\eta] \eta_j \eta_k^* \exp(-\eta^\dagger M\eta) = (M^{-1})_{jk} \det M \tag{4.11}$$

$$\int [d\eta^* d\eta] \exp (-\eta^\dagger M\eta) \exp(\chi^\dagger\eta+\eta^\dagger\chi) = \exp(\chi^\dagger M^{-1}\chi) \det M \tag{4.12}$$

where χ^* and χ are 2n additional independent Grassmann numbers.

It will be convenient to define a correspondence between Grassmann numbers and Fock space vectors for fermions through the "Fermi-coherent" states [17], defined as follows. Let η be a Grassmann number and ϕ

and $\phi^\dagger$ be fermion annihilation and creation operators. Let $|0\rangle$ be the vacuum state defined so that

$$\phi|0\rangle = 0 , \tag{4.13}$$

Then define

$$|\eta\rangle = \exp(\phi^\dagger\eta)|0\rangle = (1+ \phi^\dagger\eta)|0\rangle , \tag{4.14}$$

This is a vector in an enlarged Hilbert space in which the vector components can be Grassmann numbers. The vector is "coherent" in the sense that it is a eigenstate of ϕ:

$$\phi|\eta\rangle = \eta|\eta\rangle , \tag{4.15}$$

which follows trivially from the definition . Now for the dual space, we define

$$\langle\eta| = \langle 0|\exp (\eta^*\phi) = \langle 0|(1+\eta^*\phi) \tag{4.16}$$

where η^* is independent of η. Introducing yet a third independent Grassmann variable η', we can compute the inner product

$$\langle\eta|\eta'\rangle = 1+ \eta^*\eta' = \exp(\eta^*\eta') \tag{4.17}$$

Finally, we have the easily derived completeness relation

$$1 = \int d\eta^* d\eta \; |\eta\rangle\langle\eta| \; \exp(-\eta^*\eta) \tag{4.18}$$

and the trace relation for an operator $\mathcal{O}$ on the ordinary Fock space

$$\mathrm{Tr}\,\mathcal{O} = -\int d\eta^* d\eta \; \langle\eta|\mathcal{O}|\eta\rangle \exp(\eta^*\eta) \tag{4.19}$$

b. Feynman Path Integral for the Partition Function

Armed with the technical preliminaries of Sec.4a, we derive a functional integral representation for the partition function for a trivial problem, namely, a single fermion level with the hamiltonian

$$H_F = m\phi^\dagger\phi \tag{4.20}$$

where $\phi^\dagger$ and ϕ are fermion creation and annihilation operators. The partition function is explicitly

$$Z(\beta) = \mathrm{Tr}\,\exp(-\beta H_F) = 1 + e^{-m\beta} ;$$

however, we want to obtain the Grassmann functional integral representation, useful in more difficult applications. We write as before (Sec.3a),

$$Z(\beta) = \mathrm{Tr}\, T^M , \tag{4.21}$$

where

$$T = \exp(-H_F \Delta\tau) \tag{4.22}$$

and $M\Delta\tau = \beta$. T is the transfer matrix. Using the trace and completeness relations for the fermion coherent states (4.18) and (4.19), we obtain

$$Z(\beta) = -\int d\eta_M^* d\eta_0 \; d\eta_{M-1}^* d\eta_{M-1} d\eta_{M-2}^* \ldots d\eta_1^* d\eta_1 \quad \exp(\eta_M^* \eta_0)$$
$$\langle\eta_M|T|\eta_{M-1}\rangle \exp(-\eta_{M-1}^*\eta_{M-1})\langle\eta_{M-1}|\ldots|\eta_1\rangle \exp(-\eta_1^*\eta_1)\langle\eta_1|T|\eta_0\rangle. \tag{4.23}$$

It is natural to define $\eta_M \equiv -\eta_0$, which result in the famous anti-periodic boundary condition for fermions. Now the matrix element of T on this basis is just

$$\begin{aligned}\langle\eta'|T|\eta\rangle &= \langle\eta'|\sum_{n=0}^{\infty} \frac{(-m\Delta\tau\phi^\dagger\phi)^n}{n!} |\eta\rangle \\ &= \langle\eta'|\,[\,1 + \sum_{n=1}^{\infty}\frac{(-m\Delta\tau)^n}{n!} \phi^\dagger\phi]|\eta\rangle \\ &= 1 + \eta'^*\eta e^{-m\Delta\tau} = \exp(\eta'^*\eta e^{-m\Delta\tau}) .\end{aligned} \tag{4.24}$$

Therefore

$$Z(\beta) = \int [\, d\eta^* d\eta\,] \exp(S_F) \tag{4.25}$$

where

$$d\eta^* d\eta = d\eta_0^* d\eta_0 \ldots d\eta_{M-1}^* d\eta_{M-1} \tag{4.26}$$

$$S_F = \sum_{\tau,\tau'=0}^{M-1} \eta_{\tau'}^* M_{F\tau'\tau}\eta_\tau \tag{4.27}$$

and the fermion action is

$$M_{F\tau'\tau} = \delta_{\tau',\tau} - \delta^F_{\tau',\tau+1} e^{-m\Delta\tau} , \tag{4.28}$$

where

$$\delta^F_{0,M} = \delta^F_{M,0} = -1, \tag{4.29}$$

but otherwise $\delta^F_{\tau'\tau}$ is the usual Kronecker symbol. For small $\Delta\tau$ the action S_F becomes the continuum action for this problem

$$S_F \sim {}_0\int^\beta d\tau\, \eta^*(\tau) \left[\frac{\partial}{\partial\tau} + m \right] \eta(\tau) \tag{4.30}$$

with $\eta(\beta) \equiv -\eta(0)$ and $\eta^*(\beta) \equiv -\eta^*(0)$. In this example the time derivative has been discretized as

$$d\tau \frac{\partial}{\partial\tau} \to \delta_{\tau'\tau} - \delta^F_{\tau'\tau+1} . \tag{4.31}$$

However, one encounters more frequently

$$d\tau \frac{\partial}{\partial\tau} \to \frac{1}{2}(\delta^F_{\tau',\tau-1} - \delta^F_{\tau',\tau+1}) , \tag{4.32}$$

the central difference operator. For small $\Delta\tau$ a symmetric form of the action is

$$\hat{M}_{F\tau'\tau} = \frac{1}{2} (\delta^F_{\tau',\tau-1} - \delta^F_{\tau',\tau+1}) + m\Delta\tau\, \delta_{\tau'\tau} , \tag{4.33}$$

which has the same continuum limit as $M_{F\tau'\tau}$. However, there is a profound difference between the lattice theories in the two forms. As we shall see, M_F describes the propagation of a single species, whereas $\hat{M}_F$ describes the propagation of two species. Thus we encounter a species doubling problem at the level of the action formulation. As before, doubling is associated with the central difference operator. However, unlike the case of the space variable, there is no hermiticity requirement that demands central differences in the time variable. They are introduced to make the action symmetrical under the interchange of space and time coordinates, the lattice version of Euclidean Lorentz invariance. However, this choice is only a matter of convenience.

To illustrate the species doubling phenomenon, we calculate the Euclidean fermion propagator. In the hamiltonian formulation, the finite temperature propagation function is

$$G(\tau_2-\tau_1) = < \phi(\tau_2)\phi^\dagger(\tau_1) > \tag{4.34}$$

where

$$\phi(\tau) = e^{\tau H}\phi e^{-\tau H} \tag{4.35}$$

is the Heisenberg annihilation operator for imaginary time, and $< \ >$ denotes the quantum statistical ensemble average, i.e.

$$G(\tau_2-\tau_1) = \mathrm{Tr}\ e^{-\beta H}\ \phi(\tau_2)\phi^\dagger(\tau_1)\ /\ \mathrm{Tr}\ e^{-\beta H}\ . \tag{4.36}$$

Therefore

$$G(\tau_2-\tau_1) = \mathrm{Tr}\ e^{-(\beta-\tau_2)H}\ \phi\ e^{-(\tau_2-\tau_1)H}\phi^\dagger\ e^{-\tau_1 H}/\ \mathrm{Tr}\ e^{-\beta H} \tag{4.37}$$

This expression has a functional integral form, which is readily found by following the same steps as before. The only difference is the appearance of

$$<\eta'|\phi^\dagger T|\eta> \ = \eta'^{*}\ \exp(\eta'^{*}\eta e^{-m\Delta\tau}) \tag{4.38}$$

$$<\eta'|T\phi|\eta> \ = \exp(\eta'^{*}\eta e^{-m\Delta\tau})\eta\ . \tag{4.39}$$

Therefore

$$G(\tau_2-\tau_1) = \int [d\eta^* d\eta\ ]\ \eta_{\tau_2}\ \eta^*_{\tau_1}\ \exp(-S_F)/Z(\beta)\ , \tag{4.40}$$

which is, from the identities of Sec. 4a

$$G(\tau_2-\tau_1) = (M^{-1}_{F\ \tau_2\tau_1}\det M_F)/\det M_F = M^{-1}_{F\tau_2\tau_1}. \tag{4.41}$$

Of course we could have written the result immediately from (4.36) for this trivial example:

$$G(\tau_2-\tau_1) = \exp[-m(\tau_2-\tau_1)\Delta\tau]/\{1 + \exp[-\beta(\tau_2-\tau_1)]\}. \tag{4.42}$$

The expression (4.40) is the general functional integral definition of the thermal Euclidean Green function. Therefore the Green function for the modified action with a central difference is

$$\hat{G}\;(\tau_2-\tau_1) = -(\hat{M}_F^{\;-1})_{\tau_2\,\tau_1} \;. \tag{4.43}$$

It is trivial to invert the matrices using a Fourier decomposition since with

$$\eta_\omega \equiv \frac{1}{\sqrt{M}}\sum_{\tau=0}^{M-1} e^{i(\omega+\frac{1}{2})\tau 2\pi/M}\,\eta_\tau \quad \text{for } \omega=0,1,\ldots,M-1, \tag{4.44}$$

the actions corresponding to M_F and $\hat{M}_F$ are diagonal:

$$S_F = \sum_{\omega=0}^{M-1} \eta_\omega^* \,(1 - e^{-m\Delta\tau + i(\omega+\frac{1}{2})2\pi/M})\,\eta_\omega \tag{4.45}$$

$$\hat{S}_F = \sum_{\omega=0}^{M-1} \eta_\omega^* \,[-i\,\sin(\omega+\tfrac{1}{2})2\pi/M + m\Delta\tau]\eta_\omega \;. \tag{4.46}$$

The additional $\frac{1}{2}$ is associated with the antiperiodicity. The inverse matrices are given by

$$-(M_F^{\;-1})_{\tau'\tau} = \frac{1}{M}\sum_{\omega=0}^{M-1} \frac{e^{-i(\omega+\frac{1}{2})(\tau'-\tau)2\pi/M}}{1-\exp[-m\Delta\tau+i(\omega+\frac{1}{2})2\pi/M]} \tag{4.47}$$

$$-(\hat{M}_F^{\;-1})_{\tau'\tau} = \frac{1}{M}\sum_{\omega=0}^{M-1} \frac{e^{-i(\omega+\frac{1}{2})(\tau'-\tau)2\pi/M}}{-i\sin(\omega+\frac{1}{2})2\pi/M + m\Delta\tau} \;. \tag{4.48}$$

With a little algebra the first expression can be simplified to (4.42). Let us consider, instead, the zero temperature limit $\beta \to \infty$ with $\Delta\tau$ fixed so that $M \to \infty$. Then

$$-(M_F^{\;-1})_{\tau'\tau} \sim \frac{1}{2\pi}\int_0^{2\pi} d\omega' \,\frac{e^{-i\omega'(\tau'-\tau)}}{1-\exp(-m\Delta\tau+i\omega')} \tag{4.49}$$

$$-(\hat{M}_F^{\;-1})_{\tau'\tau} \sim \frac{1}{2\pi}\int_0^{2\pi} d\omega' \,\frac{e^{-i\omega'(\tau'-\tau)}}{-i\,\sin\omega' + m\Delta\tau} \tag{4.50}$$

where $(\omega+\frac{1}{2})2\pi/M \sim \omega'$. Both integrals can be done by contour integration in $z=e^{i\omega'}$ with the result

$$(M_F^{\;-1})_{\tau'\tau} \sim e^{-m\Delta\tau(\tau'-\tau)}\theta(\tau'-\tau) \tag{4.51}$$

and

$$(\hat{M}_F^{\;-1})_{\tau'\tau} \sim \begin{cases} e^{-m\Delta\tau(\tau'-\tau)} & \tau'>\tau \\ -e^{-m\Delta\tau|\tau'-\tau|}(-)^{\tau'-\tau} & \tau'<\tau \end{cases} \;. \tag{4.52}$$

The first integral gets contributions from one pole at $e^{i\omega'} = e^{m\Delta\tau}$ whereas the second gets contributions from a pole at $e^{i\omega'} \approx -1-m\Delta\tau$ when $\tau'>\tau$ and a pole at $e^{i\omega'} \approx 1-m\Delta\tau$ when $\tau'<\tau$. The first propagator is a causal Euclidean Green function for a particle with a single energy. The second is analogous to a Euclidean Feynman propagator for a fermion that has an associated antiparticle. Therefore the action $\hat{S}_F$ does not correspond to the original hamiltonian, which had no antiparticle. It will take us a bit far afield to answer the question, what hamiltonian corresponds to $\hat{S}_F$. Suffice it to say that in order to construct a Fock space for the transfer matrix, it is necessary to use four basis vectors, corresponding to the doubling of the fermion species.

5. The Interacting Theory

a. Functional Integral Representation for the Lattice Schwinger Model

The technique described in Secs.3 and 4 enable us to construct the functional integral representation for the hamiltonian (2.14). We leave the details as an exercise for the reader and simply present the results. The new features are as follows: The projection operator P in (3.6) must be replaced by

$$P = \prod_{x=0}^{N-1} \int \frac{d\omega_x}{2\pi} \exp i(\ell_x - \ell_{x-1} - \rho_x)\omega_x . \tag{5.1}$$

where ρ_x is given by (2.21). A pair of two-component fermion Grassman variables $\eta_{x,\tau}$ and $\bar{\eta}_{x,\tau}$ is introduced for each spatial coordinate and time slice. The result is, for large M at fixed β

$$Z(\beta) \sim \int [d\omega d\theta][d\eta^* d\eta] \exp(-S) \tag{5.2}$$

where

$$S = S_G + \sum_{x',\tau',x,\tau} \bar{\eta}_{x'\tau'} M_{x'\tau' x\tau} \eta_{x\tau} + i\omega_I , \tag{5.3}$$

S_G is given by (3.18), and

$$M_{x'\tau',x\tau} = \gamma_0\delta_{xx'}\delta_{\tau\tau'} - \gamma_0\delta_{xx'}\delta^F_{\tau'\tau+1}\, e^{-i\omega_{x,\tau}} + \Delta\tau h_{x'\tau',x\tau} \tag{5.4a}$$

$$h_{x'\tau',x\tau} = [\frac{-i}{2a}\gamma_1\delta_{x,x'+1}\, e^{i(\theta_{x,\tau}-\omega_{x,\tau})} + \frac{i}{2a}\,\delta_{x'x+1}\, \gamma_1\, e^{-i(\theta_{x\tau}+\omega_{x+1,\tau})} + m\delta_{x'x}\, e^{-i\omega_{x\tau}}]\delta_{\tau'\tau+1} \tag{5.4b}$$

and

$$\omega_I = \sum_{x=0}^{N-1}\sum_{\tau=0}^{M-1} \omega_{x,\tau} \ . \tag{5.5}$$

The angle ω_I comes from the "heavy ions" required for a neutral vacuum in (2.21). This action describes two fermion species, since the hamiltonian (2.2) describes two. A corresponding action based on the reduction (2.10) would have only one species. The fermion action has the peculiarity that in the lattice of Fig.3.1., fermions always move forward in Euclidean time. The action is invariant under the gauge transformation (3.19) together with

$$\eta_{x,\tau} \to e^{i\lambda_{x,\tau}}\,\eta_{x,\tau}$$
$$\bar{\eta}_{x,\tau} \to e^{-i\lambda_{x,\tau}}\,\bar{\eta}_{x,\tau} \ . \tag{5.7}$$

The continuum limit of the action in (5.2) is found by making the replacements (3.21,3.22) together with

$$\eta_{x,\tau} = \sqrt{a}\,\chi(ax,\tau\Delta\tau) \ . \tag{5.8}$$

To obtain the desired form, it is necessary to take $\Delta\tau/a\to 0$ and $m\Delta\tau\to 0$ first, and then $ga\to 0$, with A_μ and χ fixed. Then

$$S \sim \int dx \int_0^\beta d\tau \{\frac{1}{4}\,F_{\mu\nu}{}^2 + \bar{\chi}(\hat{\gamma}_\mu\partial_\mu + igA_\mu\hat{\gamma}_\mu + m)\chi\} \tag{5.9}$$

where $\hat{\gamma}_0 = \gamma_0$ and $\hat{\gamma}_1 = -i\gamma$, with γ_0 and γ_1 given by (2.1). The sum over repeated indices has a positive metric. To my knowledge the action (5.2-4) is not used, since there are other, more elegant and convenient forms. Other choices are permitted, provided they have the same formal continuum limit. It may be a delicate question, however, whether two given choices have the same practical continuum limit. Species doubling may be only one of a host of artifacts.

The more symmetric gauge invariant lattice action (the "naive Wilson action" is given by

$$S = \frac{1}{g^2 a\Delta\tau} \sum_{x=0}^{N-1} \sum_{\tau=0}^{M-1} (1-\cos\theta_p) + \sum_{x,\tau,x',\tau'} \bar{\eta}_{x'\tau'} \hat{M}_{x'\tau',x\tau} \eta_{x\tau} \tag{5.10}$$

where

$$-\hat{M}_{x'\tau',x,\tau} = \frac{1}{2}(\delta^F_{\tau',\tau+1} e^{-i\omega_{x,\tau}} \gamma_0 - \delta^F_{\tau,\tau'+1} e^{i\omega_{x,\tau'}} \gamma_0)\delta_{xx'} + \frac{\Delta\tau}{2a}(\delta^F_{x',x+1} e^{-i\theta_{x,\tau}} \gamma_1 - \delta^F_{x,x'+1} e^{i\theta_{x',\tau}} \gamma_1)\delta_{\tau\tau'} - m\Delta\tau\, \delta_{xx'}\delta_{\tau\tau'} \quad , \tag{5.11}$$

which has the same formal continuum limit. However, it has four fermion species, compared with two for the action (5.2-4). The reason is that the gauge covariant time derivative is replaced by the one-sided form

$$(\frac{\partial}{\partial\tau} + igA_0)\delta\tau \rightarrow \delta_{\tau\tau'} - \delta_{\tau'\tau+1} e^{-i\omega_{x,\tau}} \tag{5.12}$$

in (5.2-4) and the symmetric form

$$(\frac{\partial}{\partial\tau} + igA_0)\delta\tau \rightarrow \frac{1}{2}(\delta^F_{\tau',\tau+1} e^{i\omega_{x,\tau}} - \delta^F_{\tau,\tau'+1} e^{-i\omega_{x,\tau'}})$$

in (5.10-11). Therefore each component in the hamiltonian is represented twice, just as in the elementary example of Sec.4b. Since the extra species have the opposite charge, it is not necessary to introduce the "heavy ions" to obtain a neutral half-filled band.

b. Measurements for Fermions

The typical fermion action for quantum chromodynamics in four space-time dimensions is obtained by introducing the spinors $\eta_{x\alpha c}$ and $\eta^*_{x\alpha c}$

with four Dirac components α and three color components c as Grassmann numbers on each site of the lattice of (3.26). The partition function has the generic form

$$Z(\beta) = \int [dU] \int d\eta^* d\eta \exp [-S_G(U)-S_F(\eta^*,\eta,U)] \tag{5.13}$$

where

$$S_F(\eta^*,\eta,U) = \eta^\dagger M_F(U)\eta \tag{5.14}$$

and $M_F(U)$ is a matrix in coordinate, spin, and color space. Before listing specific choices for M_F, let us consider the calculation of fermion and meson propagators.

One might guess that the fermion propagator can be constructed as in Sec.4b by calculating

$$S_{\alpha'c'\alpha c}(\vec{x}',\tau';\vec{x},\tau) = \langle\psi_{\vec{x}'\alpha'c'}(\tau')\ \psi^*_{\vec{x}\alpha c}(\tau)P\rangle \tag{5.15}$$

where P projects onto states satisfying Coulomb's law and α, c labels spinor and color components. The operator ψ^* creates a non-singlet intermediate state from the color singlet initial state. Gauge invariance requires that the intermediate state be color singlet. Therefore, to define the propagator correctly, we introduce a heavy ion that is moved along a path C from $\vec{x},\tau$ to $\vec{x}'\tau'$ to correct the charge imbalance. In the functional integral form we then have

$$S^C_{\alpha'\alpha}(\vec{x}',\tau',\vec{x},\tau) = \int[dU][d\eta^*d\eta]\eta_{\vec{x}'\tau'\alpha'c'}\ U(C)\eta^*_{\vec{x},\tau,\alpha,c}$$
$$\times \exp[-S_G(U)-S_F]/Z(\beta)\ , \tag{5.16}$$

where U(C) is the product of the gauge link matrices along C. From (4.11) we see that this propagator is just

$$S^C_{\alpha'\alpha}(\vec{x}'\tau',\vec{x}\tau) = \frac{\int[dU]\ \exp[-S_{eff}(U)]\ \mathrm{Tr}_c[M^{-1}_{\vec{x}'\tau'\alpha',\vec{x}\tau\alpha}U(C)]}{\int[dU]\ \exp[-S_{eff}(U)]} \tag{5.17}$$

where

$$\exp[-S_{eff}(U)] = \exp[-S_G(U)]\det M(U) \tag{5.18}$$

and Tr_c is a trace over color indices. In conventional language the choice of C corresponds to the choice of gauge in which the propagator is defined. The operator

$$\mathcal{O}_m = \sum_{c=1}^{3} \psi^{*}_{\vec{x}\alpha' c}\, \psi_{\vec{x}\alpha c} \tag{5.19}$$

creates a color singlet intermediate fermion-antifermion state. No string C is needed here. The mass of the lowest lying meson state thus created can be found using the same method as was used for the glueball states (3.44). If $\mathcal{O}_m$ has no vacuum expectation value, the propagator has the functional integral form

$$D(\vec{x}'\tau',\vec{x}\tau) \underset{\beta\to\infty}{\sim} \int[dU]\exp[-S_{eff}(U)]\{-\mathrm{Tr}_c(M^{-1}_{\vec{x}'\tau'\bar{\alpha}',\vec{x}\tau\bar{\alpha}}M^{-1}_{\vec{x}\tau\alpha,\vec{x}'\tau'\alpha'}) + \mathrm{Tr}_c M^{-1}_{\vec{x}'\tau'\bar{\alpha}',\vec{x}'\tau'\alpha'}\mathrm{Tr}_c M^{-1}_{\vec{x}\tau\bar{\alpha},\vec{x}\tau\alpha}\}/\int[dU]\exp[-S_{eff}(U)] \tag{5.20}$$

Otherwise it is necessary to subtract $\langle\mathcal{O}_m\rangle^2$ as in the glueball case.

c. Popular Fermion Actions

The naive QCD action is a generalization of (5.10) :

$$S_F^{NW} = \frac{1}{2}\sum_{x,\mu} (\bar{\eta}_{x+e_\mu}\hat{\gamma}_\mu U^{\dagger}_{x\mu}\eta_x - \bar{\eta}_x U_{x\mu}\hat{\gamma}_\mu\eta_{x+e_\mu}) + \sum_x ma\bar{\eta}_x\eta_x \,, \tag{5.21}$$

with antiperiodic boundary conditions in the $\tau=x_4$ direction [1]. The hermitian gamma matricies are $\hat{\gamma}_0=\gamma_0$ and $\hat{\gamma}_j=-i\gamma_j$ for j=1,2,3; they satisfy $\{\gamma_\mu,\gamma_\nu\}=\delta_{\mu\nu}$. This lattice action describes 2^4 fermion species, not counting the three colors and four Dirac components - an unacceptably large number. The change of variables [18,19]

$$\eta(x) = \hat{\gamma}_1^{x_1}\hat{\gamma}_2^{x_2}\hat{\gamma}_3^{x_3}\hat{\gamma}_4^{x_4}\,\eta'(x) \tag{5.22}$$

leads to a diagonalization in spinor components:

$$S_F^{NW} = \frac{1}{2}\sum_{x,\mu} (\bar{\eta}'_{x+e_\mu}U^{\dagger}_{x,\mu}\eta'_x - \bar{\eta}'_x U_{x,\mu}\eta'_{x+e_\mu})(-)^{n_{x\mu}} + \sum_x ma\bar{\eta}'_x\eta'_x \,, \tag{5.23}$$

where the sign is specified by the integers

$$\begin{aligned} n_{x1} &= 0 \\ n_{x2} &= x_1 \\ n_{x3} &= x_1+x_2 \\ n_{x4} &= x_1+x_2+x_3 \ . \end{aligned} \tag{5.24}$$

Throwing away all but one of the spinor components of η' in analogy with the reduction of Sec.1a leads to the "Euclidean Kogut-Susskind" action

$$S_F^{EKS} = -\frac{1}{2} \sum_{x,\mu} (\phi^*_{x+e_\mu} U^\dagger_{x,\mu} \phi_x - \phi^*_x U_{x\mu} \phi_{x+e_\mu})(-)^{n_{x\mu}} + \sum_x ma \phi^*_x \phi_x \ , \tag{5.25}$$

which still has a four-fold multiplication of fermion species. This residual degeneracy is associated with a discrete symmetry that may become a continuous global SU(4) flavor symmetry in the continuum limit. However, there is no flavor conservation at finite lattice spacing. A further reduction to two species has been suggested [19]. The chief potential drawbacks of the theory, apart from the unphysical multiplication of species, are that it does not have a continuous chiral symmetry for m=0, and that the transition from a discrete to a continuous flavor symmetry in the continuum limit may introduce undesirable lattice artifacts.

Wilson removed the degeneracy of the fermion multiplets by writing

$$S_F^W = \sum_x \bar{\eta}_x \eta_x (4+ma) - \frac{1}{2} \sum_{x,\mu} [\bar{\eta}_{x+e_\mu} (1+\hat{\gamma}_\mu) U^\dagger_{x\mu} \eta_x + \bar{\eta}_x U_{x\mu} (1-\hat{\gamma}_\mu) \eta_{x+e_\mu}] \ . \tag{5.26}$$

The unwanted fermion species acquire infinite masses in the continuum limit [8]. This action lacks even a discrete chiral symmetry for m=0. The formal continuum limit of the theory is, as with the others listed here, the QCD Lagrangian, which is chirally symmetric for m=0. For zero gauge field it is therefore expected that $\langle \bar{\psi} \psi \rangle_{m=0} \underset{a\to 0}{\sim} 0$. This is the case for S^{EKS} but not for S^W. Although the counting of physical states is correct for S^W, one may wonder whether the pion is described correctly by the theory - the usual chiral symmetry relationship between a vanishing renormalized quark mass and a vanishing pion mass is obscure, and soft pion theorems may fail.

The Wilson action is sometimes written with an unimportant factor (4+ma) removed to give

$$\hat{S}_F^W = \sum_x \bar{\eta}_x \eta_x - K \sum_{x,\mu} [\bar{\eta}_{x+e_\mu} (1+\hat{\gamma}_\mu) U^\dagger_{x\mu} \eta_x + \bar{\eta}_x U_x (1-\hat{\gamma}_\mu) \eta_{x+e_\mu}] \tag{5.27}$$

where $K = (8+2ma)^{-1}$ is the "hopping parameter". An expansion of (5.20) in powers of K corresponds approximately to an expansion in inverse powers of the fermion mass [8] and has been used to estimate hadron masses [6].

To complete the list we mention another approach due to Banks and Casher [20]. They introduce a separate latticization for the operators

$$\not{D}+m = \gamma_\mu (\partial_\mu + igA^c_\mu \lambda^c) + m \tag{5.28}$$

and

$$-D^2+m^2 = -(\partial_\mu + igA^c_\mu \lambda^c)^2 - \frac{ig}{2} \sigma_{\mu\nu} F^c_{\mu\nu} \lambda^c + m^2 \tag{5.29}$$

where

$$\sigma_{\mu\nu} = \frac{1}{2} [\hat{\gamma}_\mu, \hat{\gamma}_\nu] . \tag{5.30}$$

In the previous examples the operator $\not{D}$+m was replaced consistently by one of the lattice matrices M_F. The matrix appears in Green's functions (5.17) and (5.20) as the inverse M_F^{-1} and determinant det M_F. Banks and Casher observe that formally

$$\det (\not{D}+m) = \det \gamma_5 (\not{D}+m) \gamma_5 = \det (-\not{D}+m) . \tag{5.31}$$

Therefore, assuming det ($\not{D}$+m) is positive,

$$\det (\not{D}+m) = \det^{1/2} (-D^2+m^2). \tag{5.32}$$

The inverse is written

$$(\not{D}+m)^{-1} = (-\not{D}+m)/(-D^2+m^2) \ . \tag{5.33}$$

The last two expressions are latticized by writing $-\not{D}+m$ in the naive form $M_F^\dagger$, but by writing $-D^2+m^2$ in the form

$$-D^2+m^2 \rightarrow \sum_{\mu>0} (2\delta_{x'x} - \delta_{x',x+e_\mu} U_{x,\mu} - \delta_{x'+e_\mu,x} U^\dagger_{x,\mu}) + \frac{1}{8} \sum_{x,\mu\neq\nu} \sigma_{\mu\nu} U(\partial p_{x\mu\nu}) \tag{5.34}$$

rather than as a straight forward multiplication $M_F M_F^\dagger$. The result is a lattice prescription for calculating Green's functions without apparent species multiplication in the continuum limit and with a continuous chiral symmetry at m=0. The lattice Green functions do not correspond to a particular local lattice action S_F, although they are well-defined. The method deserves further study.

At present there is no general agreement about the best procedure for putting fermions on the lattice in QCD. The problems of avoiding species multiplication and preserving chiral symmetry plague most formulations as we have seen, and they may remain the most serious outstanding problem after the problems of numerical method have been solved.

6. Monte Carlo Techniques with Fermions

a. Review of the Monte Carlo procedure

Consider an elementary quantum mechanical system: namely a simple harmonic oscillator, described by the hamiltonian

$$H = \frac{1}{2} p^2 + \frac{1}{2} x^2 \ . \tag{6.1}$$

To calculate the partition function we may follow steps analogous to those of Sec.3a [21] and obtain

$$Z(\beta) = \mathrm{Tr}\, e^{-\beta H} = \int dx_0 \ldots dx_{M-1}\, e^{-S} \tag{6.2}$$

where

$$S = \sum_{\tau=0}^{M-1} \frac{(x_{\tau+1} - x_\tau)^2}{2(\Delta\tau)^2} + \frac{1}{2} x_\tau^2 \tag{6.3}$$

and where $x_M \equiv x_0$ and $\beta = M\Delta\tau$. Similarly, we can get Euclidean correlation functions at finite temperature. For example

$$< x(\tau') \; x(\tau) >_\beta = \int dx_0 \ldots dx_{M-1} \; x_{\tau'} x_\tau \; e^{-S} / \; Z(\beta). \qquad (6.4)$$

The idea of the Monte Carlo method is to calculate averages of operators like $x_{\tau'} x_\tau$ by generating a sequence of configurations

$$x_\tau^{(n)} \qquad n=1,2,\ldots L; \; \tau=0,1,\ldots,M-1 \qquad (6.5)$$

for which the probability distribution is proportional to e^{-S}, i.e.

$$P(\{x_\tau\}) \; dx_0 \ldots dx_{M-1} \; \alpha \; e^{-S} \; dx_0 \ldots dx_{M-1} \; . \qquad (6.6)$$

Then the average of $x_{\tau'} x_\tau$ in (6.4) is just a simple average over the sequence of configurations:

$$< x_{\tau'} x_\tau > = \lim_{L \to \infty} \frac{1}{L} \sum_{n=1}^{L} x_{\tau'}^{(n)} x_\tau^{(n)} . \qquad (6.7)$$

To obtain the set $x_\tau^{(n)}$ there are two popular algorithms, outlined below:

(i) Heat Bath Algorithm [22]

(a) Choose a starting configuration $x_\tau^{(1)}$.

(b) Fix all x_τ's but x_0. Select a new value of x_0 at random according to the probability distribution

$$e^{-S(x_0 | x_1 \ldots x_{M-1})} = e^{-\alpha x_0^2 + \beta x_0 + \gamma} \qquad (6.8)$$

where α, β, γ depend on $x_1, \ldots, x_{M-1}$. Then repeat, singling out x_1, etc. until new values for each x_τ have been chosen. This completes a "sweep" of the lattice. The sweep can also be done in random order.

(c) Repeat the sweep until "equilibrium" is reached. A test for equilibrium is that all quantities of interest such as $x_j x_0$ have stable values when the average is taken over the last several lattice sweeps.

(ii) Metropolis, Rosenbluth, Teller and Teller algorithm [23]

This method is used when it is inconvenient to generate x_τ directly according to the probability e^{-S}. It replaces step (b) with

(b') Choose, according to a present random method, independent of the other x_τ's, a value x_0' , a candidate for replacing x_0. Compute

$$\lambda = e^{-S(x_0'|x_1,\ldots,x_{M-1})} / e^{-S(x_0|x_1\ldots x_{M-1})} . \qquad (6.9)$$

Then choose a random number r with uniform probability on [0,1]. If $\lambda<r$, reject the change and keep the old value. If $\lambda>r$, accept the change in x_0. This treatment of x_0 is usually repeated a few times before proceeding to x_1, etc. Continue for each x_τ in turn to complete the sweep. It can be shown that the Metropolis et al. algorithm gives the same result as the heat bath method.

The set $\{ x_\tau^{(n)} \}$ at equilibrium contains a wealth of information about the system. For example, the square of the wavefunction averaged over the thermal ensemble is given by the distribution of values of x occuring in $\{x_\tau^{(n)}\}$

$$P(x \,\varepsilon\, \{x_\tau{}^{(n)}\}) = \sum_m e^{-\beta E_m} |\psi_m(x)|^2 \sum_m e^{-\beta E_m} . \qquad (6.10)$$

To see this, notice that

$$|\psi_m(x)|^2 = < m|x><x|m >, \qquad (6.11)$$

so the quantity of interest is

$$\mathrm{Tr}\, e^{-\beta H} |x><x| / \mathrm{Tr}\, e^{-\beta H} . \qquad (6.12)$$

It is easily shown that this expectation value is just

$$\int dx_0 \ldots dx_{M-1}\, \delta(x-x_0)\, e^{-S} \Big/ \int dx_0 \ldots dx_{M-1}\, e^{-S} \qquad (6.13)$$

which is with the Monte Carlo method $P(x \,\varepsilon\, \{x_\tau^{(n)}\})$.

b. Confronting the Fermion Determinant

For numerical calculations with fermions it is necessary to complete the Grassmann functional integration analytically before integrating over the gauge field variables by the Monte Carlo method. For typical Green functions as in (5.17) and (5.20) the ideal Monte Carlo weight for the gauge field configurations is

$$P(U^{(n)}) = |\det M_F(U)| \exp[-S_G(U)] . \qquad (6.14)$$

(If the determinant changes sign, it is necessary of course to include the sign as a factor in all expectation values.) The complexity of det $M_F(U)$ prohibits the use of a heat-bath algorithm. Indeed, the fermion determinant is sufficiently cumbersome that it is often omitted altogether, leading to the "quenched approximation" [4]. This approximation corresponds in the language of perturbation theory to omitting Feynman graphs with closed loops. The Metropolis et al algorithm requires an evaluation of

$$\det M_F(U') \exp\,[-S_G(U')]/ \det M_F(U) \exp\,[-S_G(U)] \tag{6.15}$$

where U' results from changing one link variable in the set U. The ratio of the pure Yang-Mills factors is easily evaluated because a local change in U affects only a small number of terms in S_G. The ratio of fermion determinants may be simplified [24,25] by noting that

$$M_F(U') = M_F(U) + \delta M_F \,, \tag{6.16}$$

where δM is zero except for a finite set of matrix elements corresponding to sites at the ends of the link where the change in U was made. Then

$$\frac{\det M_F(U')}{\det M_F(U)} = \det(1 + M_F^{-1}(U)\,\delta M_F) \,. \tag{6.17}$$

Owing to the near triviality of δM, the determinant can be reduced to that of a matrix of low rank. However, it is still necessary to compute at least a few matrix elements of M_F^{-1} , a difficult task. One method involves calculating M_F^{-1} for a starting configuration and updating M_F^{-1} at each step. This method requires storing the entire inverse matrix and for moderately large lattices, therefore, a very large computer [7].

Other methods require keeping only a single column of the inverse matrix:

(i) Relaxation method. The standard Jacobi or Gauss-Seidel methods with relaxation are popular [26].

(ii) Pseudofermion method [27]. Monte Carlo techniques can be used to evaluate M^{-1} as follows: If M is a hermitian positive definite matrix, then

$$(M^{-1})_{ij} = \frac{\int [dx] \exp[-\frac{1}{2}x^{\dagger}M\,x]\; x_i^{*}x_j}{\int [dx] \exp[-\frac{1}{2}x^{\dagger}\,M\,x]} \tag{6.18}$$

where x is an auxilliary c-number field introduced only to calculate M^{-1}. As in (i) if M is not hermitian and positive definite, then $MM^\dagger$ can be inverted instead, since $M^{-1} = (MM^\dagger)^{-1} M^\dagger$.

(iii) Stochastic Neumann Series

An old method of Neumann and Ulam, revived recently by Kuti [27], works particularly well for matrices with a band structure such as M_F, and promises to give convergence to M_F^{-1} at a rate independent of the size of the lattice. This is important, since methods (i) and (ii) take increasing amounts of computer time as the size of the lattice increases.

c. The Santa Barbara Chessboard

An entirely different approach to lattice fermion calculations was developed in Santa Barbara [28]. So far the only gauge theory application has been with the massive Schwinger model [29]. The method involves calculating the partition function

$$Z(\beta) = \mathrm{Tr}\,[\exp(-\beta H)] = \mathrm{Tr}\; T^M \tag{6.19}$$

and related quantities in the occupation basis directly, rather than computing the fermion determinant. The method breaks the fermion lattice hamiltonian into two parts

$$H = H_e + H_o, \tag{6.20}$$

where H_e allows propagation only between even sites and the next site to the right, and H_o, only between odd sites and the next site to the right. Corresponding to H_e and H_o are the transfer matrices T_e and T_o. In the limit of large M at fixed β, the partition function becomes

$$Z(\beta) \sim \mathrm{Tr}(T_e T_o)^M \tag{6.21}$$

The pattern of propagation suggests a chess board with propagation only across squares of one color. The advantage of this approximation is that the transfer matrices T_e and T_o are quickly computed products of elementary transfer matrices for the propagation between a pair of adjacent lattice sites. In the occupation basis $Z(\beta)$ is evaluated by summing over each sequence of 2M occupations $\{n_x^\tau\}$ for $\tau=0,1,\dots,2M-1$:

$$Z(\beta) = \sum_{n_x^\tau} \langle n_x{}^0|T_e|n_x^{2M-1}\rangle\langle n_x^{2M-1}|T_o|n_x^{2M-2}\rangle \ldots \langle n_x^1|T_o|n_x{}^0\rangle \quad . \tag{6.22}$$

The sum is carried out with Monte Carlo techniques. The non-zero contributions to $Z(\beta)$ can be represented graphically by fermion world lines passing through the lattice, joining the occupied sites. A pattern of world lines corresponds to a particular easily calculated contribution to $Z(\beta)$, which is treated as a Monte Carlo weight. The world lines are moved stochastically according to this weight.

The method gives a local algorithm for the fermion problem. In more than one space dimension there may be problems with sign fluctuations in the contributions to the trace, but the method is promising, nevertheless. Because the fermion occupation is controlled in the transfer matrix approach, this method is uniquely suited to studying properties of theories at large particle density.

7. Conclusion

In these lectures I have tried to give a brief introduction to lattice gauge theories with a particular emphasis on the construction of the lattice theories with fermions and their relationship to hamiltonian theories. I have not discussed the physical behavior of the theories, the details of the hadron spectroscopy, the manifestation of various lattice artifacts, the renormalization group scaling, or the phase structure as a function of coupling constants and temperature. I leave the exploration of these topics to my colleagues Jacobs, McLerran, Mack, Makeenko, and Satz and to the interested student.

Aknowledgement

The preparation of these lectures was supported in part by grant PHY-8008249 from the U.S. National Science Foundation and by the Research Institute for Theoretical Physics, University of Helsinki. I thank Tom DeGrand for teaching me about projections onto physical states in the U(1) gauge theory. To the organisers of the Arctic School for their kind hospitality, kiitoksia paljon!

References

[1] K.G. Wilson, Phys.Rev. D10, 2445 (1976)
A.M. Polyakov, Phys.Lett. 59B, 82 (1975).

[2] John Kogut and Leonard Susskind, Phys.Rev. D11,395,3594 (1975), L. Susskind, Phys.Rev. D16,3031 (1977).

[3] M. Creutz, Phys.Rev. D21,2308 (1980).

[4] H. Hamber and G. Parisi, Phys.Rev.Lett. 47,1792 (1981) ; E. Marinari, G. Parisi and C. Rebbi, ibid., 1975 ; H. Hamber, E. Marinari, G. Parisi and C. Rebbi, Phys.Lett. 108B,314 (1982) ; D. Weingarten, Phys.Lett. 109B,57 (1982).

[5] F. Fucito, G. Martinelli, C. Omero, G. Parisi, R. Petronzio, and F. Rapuano, Ref.TH. 3288-CERN (1982).

[6] A. Hasenfratz, P. Hasenfratz, Z. Kunszt, and C.B. Lang, Phys.Lett. 110B,289 (1982), Ref.TH. 3313-CERN (1982).

[7] A. Duncan, R. Roskies, and H. Vaidya, University of Pittsburgh report PITT-82-6 (1982).

[8] K.G. Wilson in New Phenomena in Subnuclear Physics, ed A. Zichichi (Erice 1975) (Plenum, N.Y., 1977).

[9] John B. Kogut, Rev.Mod.Phys. 51, No. 4 (1979).

[10] L. Susskind, Bonn Summer School, 1974 (unpublished)
M. Creutz Erice, 1981, BNL 29840 (1981)
C. Rebbi 1981 GIFT School, ICTP preprint (1981).

[11] H.B. Nielsen and M. Ninomiya, Nucl.Phys. B 185,20 (1981); Phys.Lett. 105B,219 (1981).

[12] S.D. Drell, M. Weinstein, and S. Yankielowicz, Phys.Rev. D14, 1627 (1976).

[13] T. Banks, L. Susskind, and J. Kogut, Phys.Rev. D13,1043 (1976).

[14] B. Berg, Phys.Lett. 97B,401 (1980); G. Bhanot and C. Rebbi, Nucl.Phys. D180 [FS2],469 (1981).

[15] B. Berg, A. Billoire and C. Rebbi, Ann.Phys. (N.Y.)
B. Berg and A. Billoire, Ref.TH.3230-CERN (1982); K. Ishikawa, M. Teper, and G. Schierholz, Phys.Letters 110B,399 (1982); DESY 82-24 (1982).

[16] F.A. Berezin, The Method of Second Quantization, Pure and Appl.Phys. vol.24 (Academic, N.Y., 1966).

[17] M. Creutz, Phys.Rev. D15,1128 (1977). See also D. Soper, Phys.Rev. D18,4590 (1978).

[18] N. Kawamoto and J. Smit, Nucl.Phys. B192,100 (1981).

[19] H.S. Scharatchandra, H.J. Thun, and P. Weisz, Nucl.Phys. B192, 205 (1981).

[20] T. Banks and A. Casher, Nucl.Phys. B169,103 (1980).

[21] R.P. Feynman and A.R. Hibbs, Quantum Mechanics and Path Integrals, (McGraw-Hill, N.Y.,1965).

[22] K. Binder, in Phase Transitions and Critical Phenomena ed. C. Domb and M.S. Green (Academic, N.Y., 1976), vol 5B.

[23] N. Metropolis, A.W. Rosenbluth, A.H. Teller, and E. Teller, J. Chem.Phys. 21 ,1087 (1953).

[24] D.J. Scalapino and R.L. Sugar, Phys.Rev.Lett. 46,519 (1981).

[25] F. Fucito, E. Marinari, G. Parisi, C. Rebbi, Nucl.Phys. B180, [FS2] 369 (1981).

[26] See for example B. Carnahan, H.A. Luther, and J.O. Wilkes Applied Numerical Methods (Wiley, N.Y., 1969).

[27] J. Kuti, NSF-ITP-81-151 (1981).

[28] J. Hirsch, D. Scalapino, R. Sugar, R. Blankenbecler , Phys.Rev. Lett. 47,1628 (1981).

[29] Olivier Martin and Steve Otto, CALT-68-901 (1982).

NONPERTURBATIVE METHODS

Gerhard Mack
II. Institut für Theoretische Physik der Universität Hamburg

0. INTRODUCTION

During the last ten years, quantum field theory and classical statistical mechanics have merged into a single subject and the same methods are used in both fields. Accordingly, speaking of nonperturbative methods in quantum field theory one usually means methods of classical statistical mechanics (other than standard perturbation theory). This includes the Monte Carlo method. Since this method is covered by other lectures at the school, I will concentrate on analytical methods, chiefly expansion methods. For illustration, applications to some models - ferromagnets and pure Yang Mills theory on a lattice - will be discussed. Presentation of this material covered the first 3 lectures at the school. For further reading I recommend E. Seilers book [1]. The 4-th lecture dealt with the effective Z(N) theory (vortex condensation theory) of quark confinement. It is omitted in these notes, see refs. 2a. Tests of this theory by Monte Carlo computations were performed by Pietarinen and the author [2b].

In classical statistical mechanics and Euclidean quantum field theory one wants to compute partition functions (free energies) and correlation functions. This involves computation of ∞-dimensional integrals (in the ∞-volume limit). Expansion methods to achieve that fall into two categories

a) <u>simple</u>. By a suitable expansion, the whole problem is reduced to the computation of finite dimensional integrals.

b) <u>sophisticated</u>: The integration variables are divided into groups. (Often this step is preceded by variable transformations, and sometimes by use of integral representations for some of the factors in the integrand, such as a Kramers Wannier duality transformation). Then the groups of variables are treated individually (one after the other) by suitable expansion methods of type a) (high temperature expansions, low temperature expansions, Mayer expansions, cluster expansions of constructive field theory [3], to name the most important ones). Renormalization group calculations [4] fall into this category, they involve many identical steps of type a). More generally, after some of the integrations are done, the integrand $\mathfrak{Z}$ still depends on the remaining groups of variables. $\ln \mathfrak{Z}$ is then called an effective action. The procedure amounts to compute a sequence of effective actions by suitable expansion methods.

Quite complicated systems have already been analyzed by this method, and significant progress is still being made. Here are some examples. Glimm, Jaffe and Spencer [5] have developed a method to deal with field theories with spontaneously broken discrete symmetry, it uses a combination of low temperature - and cluster expansions. Brydges and Federbush have established Debye screening in very dilute 3-dimensional Coulomb gases [6]. Fröhlich and Spencer were able to analyze the Kosterlitz Thouless phase of the two-dimensionale plane rotator model [7]. Göpfert and the author have proven confinement of static quarks in 3-dimensional U(1) lattice gauge theory for all values of the coupling constant [8]. (The results of this work will be described in lecture 3.) Finally, Gawedzki and Kupiainen have announced a rigorous renormalization group treatment of the dipole gas and anharmonic crystal [9]. This was a particularly difficult problem because it requires an infinite number of renormalization group steps to determine the long distance behavior of correlation functions. All of this was achieved by combination of standard expansion techniques of classical statistical mechanics.

1. EXPANSION METHODS

The derivation of any expansion for a free energy or correlation functions may be divided into two steps: i) transformation of the model into a 'polymer system', and ii) application of expansion formulae for polymer systems [10]. Different expansions (for instance high and low temperature expansions) are based on different transformations into polymer systems, while the second step ii) is always essentially the same. I will illustrate the method first at the example of high temperature expansions for pure Yang Mills theory on a lattice. In the course of this discussion I will also review the proof of confinement of static quarks for strong coupling [11].

1.1. High temperature expansions,

and reasons to hope that their intrinsic limitations are not more stringent than those of the Monte Carlo Method if suitable "partially summed" expansions are used.

Sites, links, plaquettes, cubes of the lattice will be denoted by x, ℓ, p, c respectively. In pure Yang Mills theory on a lattice, the basic variables U_ℓ are attached to the links ℓ of the lattice, they are unitary matrices in the gauge group .

Let C be a closed loop which consists of links $\ell_1 \ldots \ell_n$. Then the parallel transporter around C is defined by $U_C = U_{\ell_1} \ldots U_{\ell_n}$. In particular, the boundary ∂p of a plaquette consists of four links $\ell_1 \ldots \ell_4$ and $U_{\partial p} = U_{\ell_1} \ldots U_{\ell_4}$.

Let D^k be some representation of the gauge group, and $\chi_k = \mathrm{tr}\, D^k$ the corresponding character. According to Wilson, static quarks which transform according to representation D^k of the gauge group will be confined by a linearly rising potential αr if the Wilson loop expectation value $\langle \chi_k(U_C) \rangle$ obeys an area law [12],

$$\langle \chi_k(U_C) \rangle \sim \exp[-\alpha A]$$

where A is the minimal area of a surface whose boundary is C. α is called the string tension.

It will be usefull to consider partition functions Z(X) that are associated with subsets X of the lattice Λ. The action for such a subset X in a typical pure lattice gauge theory model is

$$L_X(U) = \sum_{p \in X} \mathcal{L}_p(U) \tag{1.1}$$

Summation extends only over those plaquettes p in Λ whose corners are all in X.

$$\mathcal{L}_p(U) = \frac{\beta}{2}\,\mathrm{tr}\,U_{\partial p} + \mathrm{const.} \qquad \text{for the SU(2) Wilson action} \tag{1.2}$$

The partition function Z(X) for an arbitrary sublattice $X \subseteq \Lambda$ is defined by

$$Z(X) = \int \prod_{\ell \in X} dU_\ell \, e^{L_X(U)} \tag{1.3}$$

The Wilson loop expectation values on Λ read in this notation

$$\langle \chi_k(U_C) \rangle = \frac{1}{Z(\Lambda)} \int \prod_{\ell \in \Lambda} dU_\ell \; \chi_\ell(U_C) \, e^{L_\Lambda(U)} \tag{1.4}$$

Reformulation as a polymer system.

One writes

$$e^{\mathcal{L}_p(U)} = 1 + f_p(U) \, , \tag{1.5}$$

considers f_p as "small", and expands in products of f_p's

$$e^{L_X(U)} = \prod_{p \in X} [1 + f_p(U)] = 1 + \sum_{\mathcal{B} \subseteq X} \prod_{p \in \mathcal{B}} f_p(U) \tag{1.6}$$

Summation is over all nonempty sets B of plaquettes p on X; $B \subseteq X$ is a somewhat imprecise short hand notation to keep track of this restriction. The partition functions become

$$Z(X) = 1 + \sum_{\mathcal{B} \subseteq X} \int \prod_{\ell \in X} dU_\ell \prod_{p \in X} f_p(U) \tag{1.7}$$

Next one decomposes B into connected pieces P, they will be called polymers. A set B is said to decompose into two disjoint subsets B_1 and B_2 if no plaquette p in B_1 shares a link in its boundary with a plaquette p' in B_2. A nonempty set P of plaquettes is called connected, or a polymer, if it is not a union of two nonempty disjoint subsets B_1 and B_2. Examples are shown in figure 1.

I will use the symbol Σ for union of disjoint subsets.

For polymers P one defines "activities" A(P) by

Figure 1

$$A(P) = \int \prod_{\ell \in P} dU_\ell \prod_{p \in P} f_p(U) \qquad (1.8)$$

"$\ell \in P$" is shorthand for "ℓ is a link in the boundary of a plaquette p in P".

If B decompoes into disjoint pieces P_i, then the integral in eq. (1.7) factorizes. (If ℓ is not in the boundary of any plaquette $p \in B$ then the integration over U_ℓ is trivial and gives a factor 1 because the Haar measure dU_ℓ is normalized.) Therefore

$$Z(X) = 1 + \sum_{n \geq 1} \sum_{\substack{(P_1,\ldots,P_n) \\ \Sigma P_i \subseteq X}} \prod_{i=1}^{n} A(P_i) \qquad (1.9)$$

Let us temporarily write $X_\square$ for the set of all plaquettes on X. Summation in eq. (1.9) extends over all partitions of $X_\square$ into disjoint polymers and empty plaquettes. The reader is invited to think of a chessboard as $X_\square$ and of an ample supply of polymers that are cut out of card board. Each polymer can cover a certain number of plaquettes (squares) on the chessboard. Any union of squares that can be cut out of card board without falling apart is a polymer. A certain weight $\mu(P)$ can be given to every such polymer by glueing pieces of lead on top of them; polymers of the same shape are not distinguished and should have the same weight. Now the chessboard can be covered or partly covered by polymers. The rule is that no squares may be covered by more than one polymer, and no two polymers may touch along a line. (They are allowed to touch at corners.) Every such covering adds a contribution exp(total weight of all polymers on the chessboard) to the partition function. The activities are $A(P\ ) = e^{\mu(P)}$ in this example.

An expression of the form (1.9) is called <u>partition function of a polymer system</u>. It is sometimes useful to admit activities which are not necessarily positive, but it is always required that Z(X) is positive for all X. In our applications to lattice gauge theory this requirement is fulfilled by definition (1.3) of Z(X).

<u>Extension to Wilson loops.</u>

Suppose the loop C consists of links $\ell_1 \ldots \ell_m$ so that $U_C = U_{\ell_1} U_{\ell_2} \ldots U_{\ell_m}$. The same steps that lead to eq. (1.7) give

$$Z(\Lambda)\langle\chi_k(U_C)\rangle = \int \prod_{\ell\in\Lambda} dU_\ell \, \chi_k(U_{\ell_1}\cdots U_{\ell_m})\Big\{1+\sum_{B\subseteq\Lambda}\prod_{p\in B} f_p(U)\Big\} \tag{1.10}$$

We want to exploit the factorization properties of the integral again, to do so we introduce a suitable new definition of polymer. Every set B of plaquettes on Λ specifies a set of $n+1 \geq 1$ polymers. We decompose B into connected pieces as before. Polymers P_o consists of all (possibly none) those connected pieces of B that touch the Wilson loop C along a link. An example is shown in figure 2. It is convenient to consider the links ℓ in the loop C as part of P_o also. (It can happen that P_o consists only of C.)

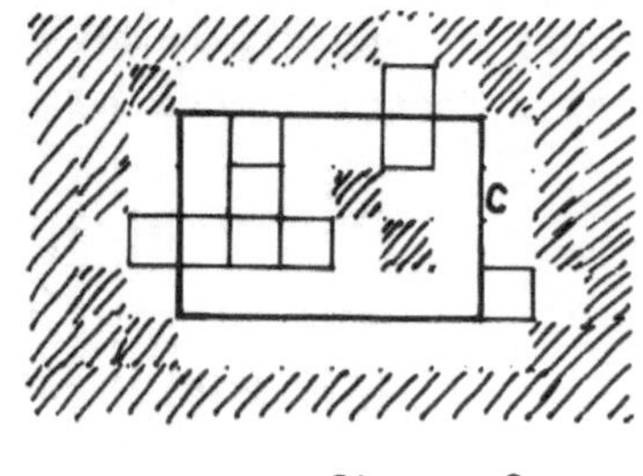

figure 2

$P_1, \ldots, P_n$ are the other connected pieces of B. They do not touch the loop C along any link. An activity $K_C(P_o)$ is defined by

$$K_C(P_o) = \int \prod_{\ell\in P_o} dU_\ell \, \chi_k(U_C) \prod_{p\in P_o} f_p(U) \tag{1.11}$$

The activities of the other polymers which do not touch C are defined by eq. (1.8) as before.

Proceeding as in the derivation of eq. (1.9) one obtains

$$Z(\Lambda)\langle\chi_k(U_C)\rangle = \sum_{n\geq 0}\ \sum_{(P_o,\ldots,P_n)} K_C(P_o)\prod_{i=1}^{n} A(P_i) \tag{1.12}$$

An empty product, which arises when n = 0, is read as 1. Summation is over partitions of Λ into disjoint polymers $P_o, \ldots, P_n$. The loop C is considered as part of P_o, and disjoint means that different polymers may not overlap or touch along a link.

The next step is a partial resummation [11] : One sums over all those partitions $(P_o, P_1, \ldots, P_n)$ with a fixed P_o. They are in one to one correspondence with partitions $(P_1, \ldots, P_n)$ of $\Lambda - \bar{P}_o$. $\Lambda - \bar{P}_o$ consists of all those plaquettes on Λ which do not touch P_o along a link (see figure 2). Making use of eq. (1.9) the result of this partial resummation takes the form

$$\langle \chi_k(U_C)\rangle = \sum_{P_o} \frac{Z(\Lambda-\bar{P}_o)}{Z(\Lambda)} K_C(P_o) \qquad (1.13)$$

Summation is over polymers P_o as described above; they contain the path C.

Proof of confinement of static quarks (with nontrivial transformation law under the center of the gauge group) for small β (strong coupling) [11].

To be specific, let us consider a theory with gauge group SU(2) and Wilson action. I choose the additive constant in expression (1.2) for $\mathcal{L}_p$ so that

$$\mathcal{L}_p(U) = \frac{\beta}{2}[\mathrm{tr}\, U_{\partial p} + 2] \geq 0 \qquad (1.14)$$

Then

$$0 \leq f_p(U) \leq \varkappa\beta \qquad (1.15)$$

for a constant $\varkappa \gtrsim 1$ and sufficiently small β. Since integrations over U_ℓ for $\ell \notin X$ are trivial, expression (1.3) for Z(X) is equivalent to

$$Z(X) = \int \prod_{\ell\in\Lambda} dU_\ell \prod_{p\in\Lambda} [1+\tilde{f}_p(U)]$$

with

$$\tilde{f}_p(U) = \begin{cases} f_p & \text{if } p\in X \\ 0 & \text{otherwise} \end{cases}$$

Since $f_p \geq 0$ it follows that

$$0 < Z(X) \leq Z(\Lambda) \quad \text{for } X \subseteq \Lambda$$

Consequently, eq. (1.13) tells us that

$$|\langle \chi_k(U_C)\rangle| \leq \sum_{P_o} |K_C(P_o)| \qquad (1.16)$$

The crucial question is now: For which P_o is $K_C(P_o) \neq 0$? Here the center of the gauge group comes in.

The center $\mathbb{Z}_2$ of SU(2) consists of the two matrices ± 1, and the character χ_k of the (2k+1)-dimensional representation of SU(2) obeys

$$\chi_k(-U_C) = (-1)^{2k}\chi_k(U_C) \tag{1.17}$$

The static quarks transform nontrivially under the center of the gauge group SU(2) if they have fractional colour-isospin $k = \frac{1}{2}, \frac{3}{2}, \ldots$. It follows from eq. (1.17) that

$$K_C(P_o) = 0 \tag{1.18}$$

unless P_o contains all plaquettes in a surface with boundary C.

Idea of the proof of assertion (1.18) [13]. Suppose P_o leaves a hole. Then I can find a coclosed set **T** of plaquettes which winds around C and shares no plaquette with P_o. "Coclosed" means that every 3-dimensional cube in Λ has an even number of plaquettes in P_o in its boundary. An example is shown in fig. 3.

Because **T** is coclosed, there exists a variable transformation

$$U_\ell \to U'_\ell = U_\ell \gamma_\ell \quad , \gamma_\ell \in Z_2 \tag{1.19}$$

with the property that

$$U_{\partial p} \to \begin{cases} -U_{\partial p} & \text{if } p \in \mathbf{T} \\ U_{\partial p} & \text{otherwise} \end{cases} \tag{1.20}$$

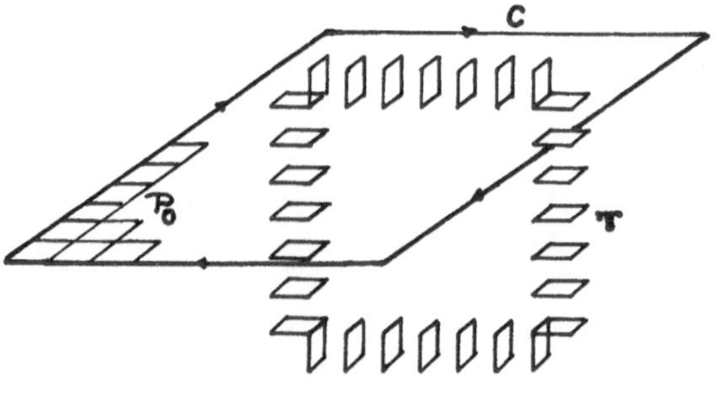

figure 3

For instance, if **T** is as in figure 4, one chooses $\gamma_\ell = -1$ for $\ell \in S$, and $\gamma_\ell = +1$ otherwise. Moreover, since **T** winds around C, the transformation takes

$$U_C \to U'_C = -U_C \tag{1.21}$$

Finally, because of invariance of Haar measure

$$dU_\ell = dU'_\ell \tag{1.22}$$

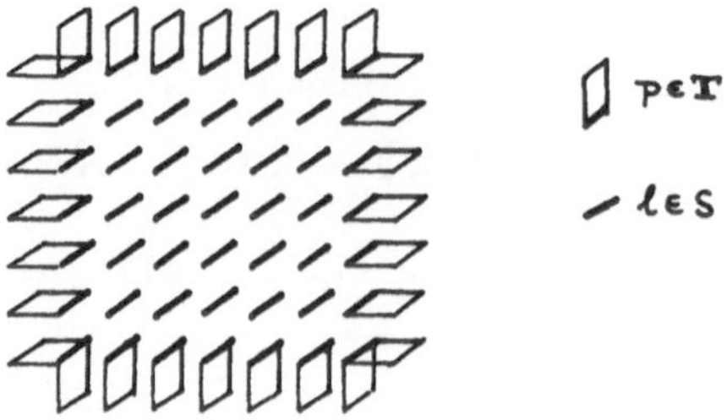

figure 4

Since **T** does not intersect P_o, $f_p(U) = f_p(U')$ for $p \in P_o$.

On the other hand, because of eq. (1.17), $\chi_k(U'_C) = -\chi_k(U_C)$ if $k = \frac{1}{2}, \frac{3}{2}, \ldots$. Therefore the integrand of expression (1.11) for $K_C(P_o)$ changes sign under the above variable transformation. In conclusion, $K_C(P_o) = 0$ if P_o leaves a hole, q.e.d.

If the static quarks carry integral isospin $k = 1, 2, \ldots$ then assertion (1.17) does not hold. In this case $K_C(P_o) \neq 0$ for P_o as shown in figure 5b.

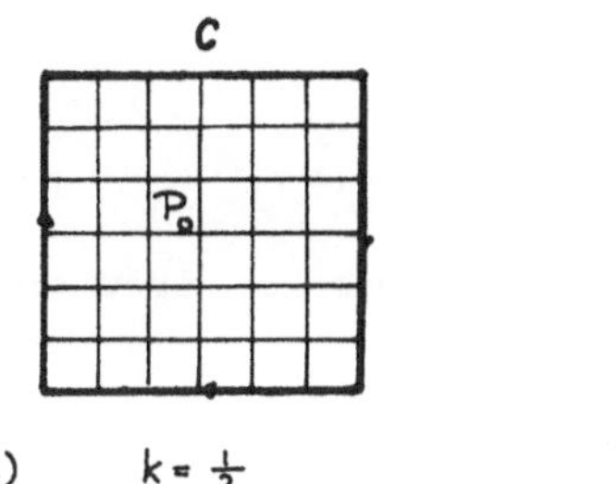

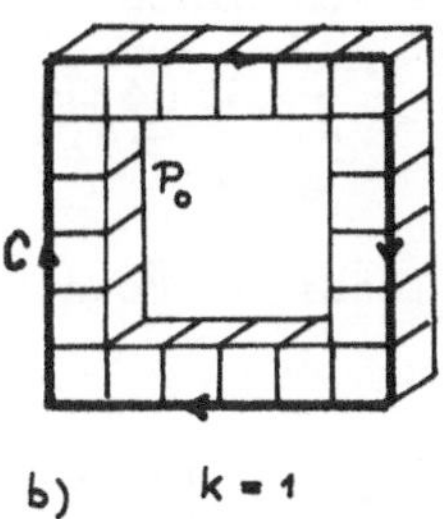

fig. 5 leading contributions to (1.13)

Now we return to inequality (1.16). Suppose that C bounds a rectangle if size $L \cdot T$. Then a surface with boundary C contains at last $L \cdot T$ plaquettes. Therefore (1.18) implies that

$$|\langle \chi_k(U_C)\rangle| \leq \sum_{\substack{P_o \\ |P_o| \geq L\cdot T}} |K_C(P_o)|$$

if $k = \frac{1}{2}, \frac{3}{2}, \ldots$. I write $|P_o|$ for the number of plaquettes in P_o.

Since $\int dU_\ell = 1$ and $0 \leq f_p \leq \varkappa\beta$, if follows from definition (1.11) that

$$|K_C(P_o)| \leq \chi_k(1)\,(\varkappa\beta)^{|P_o|} \tag{1.23}$$

Finally we need a combinatorial estimate. It is a corollary of Eulers solution of the Königsberg bridge problem [14] and says that the number of polymers P_o with a given number $n = |P_o|$ of plaquettes is bounded by $\varkappa_1^{|C|+n}$, where $|C|$ is the length of C (= 2(L+T)) and $\varkappa_1$ is some constant. It follows that

$$\begin{aligned} |\langle \chi_k(U_C)\rangle| &\leq \sum_{n \geq L\cdot T} \varkappa_1^{|C|}\,(\varkappa\varkappa_1\beta)^n\,\chi_k(1) \\ &= \varkappa_1^{|C|}\,(\varkappa_1\varkappa\beta)^{L\cdot T}\,(1-\varkappa\varkappa_1\beta)^{-1}\,\chi_k(1) \\ &\leq \text{const}\cdot \varkappa_1^{|C|}\,e^{-\alpha_o L\cdot T} \end{aligned} \tag{1.24}$$

if β is small, and $k = \frac{1}{2}, \frac{3}{2}, \ldots$, with $\alpha_o = -\ln \varkappa\varkappa_1\beta > 0$. This is the promised area law. If k is integer, the contribution of the polymer P_o of figure 5b produces a perimeter law.

Möbius inversion [15]

So far we have obtained only an upper bound on the Wilson loop expectation value. If one wants to actually calculate $\langle\chi_k(U_C)\rangle$ one needs $Z(X)/Z(\Lambda)$ for subsets $X\subset\Lambda$ since

$$\langle\chi_k(U_C)\rangle = \sum_{P_o} \frac{Z(\Lambda-\bar{P}_o)}{Z(\Lambda)} K_C(P_o)$$

by eq. (1.13). This leads us to ask for expansions for free energies $-\ln Z(X)$, since

$$Z(X)/Z(\Lambda) = \exp[\ln Z(X) - \ln Z(\Lambda)] \tag{1.25}$$

$Z(X)$ has been exhibited as partition function of a polymer system in eq. (1.9), for $X\subseteq\Lambda$. From now on we will regard X and Λ as sets of plaquettes. We look for a "Lagrangean" $\mathbf{L}(Y)$ of our polymer system which is defined for all subsets $Y\subseteq\Lambda$ and has the property that

$$\ln Z(X) = \sum_{Y\subseteq X} \mathbf{L}(Y) \tag{1.26}$$

$\mathbf{L}(Y)$ is uniquely determined by this equation, since it can be inductively computed from it. The explicit inversion formula is given in eq. (1.27) below. Eq. (1.26) provides an inductive specification of localization of free energy . $-L(Y)$ is the part of the free energy of a system in Y that is spread out throughout Y, i.e. equal to the free energy in Y minus whatever part of it is already localized in some proper subsets of Y.

The inversion formula for $\mathbf{L}(Y)$ reads

$$\mathbf{L}(Y) = \sum_{W\subseteq Y} \vartheta_{YW} \ln Z(W) \quad \text{with} \quad \vartheta_{YW} = (-1)^{|Y|-|W|} \tag{1.27}$$

$|Y| \equiv$ number of plaquettes in Y. $L(Y)$ has the important property that

$$\mathbf{L}(Y) = 0 \text{ if Y is not polymer-connected.} \tag{1.28}$$

Y is "not polymer connected" if it is union of two nonempty subsets Y_1, Y_2 such that there exists no polymer which is a subset of Y and intersects both Y_1 and Y_2. We say that Y_1, Y_2 are "polymer disjoint" in this case.

The Möbius inversion formulae (1.26), (1.27) have been known for a long time, but they appear to be not as generally known as they ought to be. We pause to give the proof of (1.27) and (1.28).

The proof of eq. (1.27) proceeds in two steps. First one notes that

$$\sum_{\substack{Y \\ X \subseteq Y \subseteq X'}} \vartheta_{YX} = \begin{cases} 1 & \text{if } X = X' \\ 0 & \text{otherwise} \end{cases} \tag{1.29}$$

Then one uses this to show that expression (1.27) statisfies (1.26). By uniqueness of $\mathbf{L}(Y)$ it must therefore be true. Inserting (1.27) into (1.26) we obtain

$$\text{r.h.s. of } (1.26) = \sum_{\substack{Y \\ W \subseteq Y \subseteq X}} \sum_{W} \vartheta_{YW} \ln Z(W) = \ln Z(X)$$

by eq. (1.29). The proof of eq. (1.29) follows from the binomial theorem. Let $s = |X|$, $t = |X'|$, $n = |Y|$ be the number of plaquettes in X, X', Y, respectively. Then $\vartheta_{YX} = (-1)^{n-s}$. Y is fixed by selecting n-s elements of X'-X. This can be done in $\binom{t-s}{n-s}$ ways. So

$$\begin{aligned} \text{l.h.s. of } (1.29) &= \sum_{\substack{n \\ s \leq n \leq t}} (-1)^{n-s} \binom{t-s}{n-s} = \sum_{k=0}^{t-s} (-1)^k \binom{t-s}{k} \\ &= (1-1)^{t-s} = \begin{cases} 1 & \text{if } t = s \\ 0 & \text{otherwise} \end{cases} \end{aligned}$$

Since $X \subseteq X'$, $t = s$ implies $X = X'$. □

Next, I give the proof of (1.28). Suppose that Y is not polymer connected, so that it is union of two polymer disjoint subsets Y_1, Y_2. Let X be any subset of Y. Then $X_1 = Y_1 \cap X$ and $X_2 = Y_2 \cap X$ are also polymer disjoint if they are both nonempty. It follows from formula (1.9) for the partition function that in this case

$$Z(X) = Z(X_1) Z(X_2)$$

If X_1 or X_2 is empty, the same is also true because $Z(\emptyset) = 1$. From its definition it follows that ϑ_{YX} also factors: $\vartheta_{YX} = \vartheta_{Y_1X_1}\vartheta_{Y_2X_2}$. Therefore

$$\mathbf{L}(Y) = \sum_{X_1 \subseteq Y_1} \sum_{X_2 \subseteq Y_2} \vartheta_{Y_1X_1} \vartheta_{Y_2X_2} [\ln Z(X_1) + \ln Z(X_2)] \tag{1.30}$$

Y_1 and Y_2 are nonempty by hypothesis. It follows from eq. (1.29) with X = empty set that both terms in (1.30) are zero. Thus $\mathbf{L}(Y) = 0$. □

Use as an expansion formula

I propose to use eq. (1.26) as an expansion formula for the free energy

$$\ln Z(X) = \sum_{\substack{Y \\ (Y \subseteq X)}} \mathbf{L}(Y) \tag{1.26}$$

To obtain a useful approximation for the free energy of an arbitrarily large system X, one should truncate the sum over Y. For instance, one may consider omitting all subsets Y with a diameter bigger than some d. To use a truncated expansion (1.26) for computational purposes, one has to compute $\mathbf{L}(Y)$ for small sets Y from eq. (1.27) combined with eq. (1.9) for Z(W), see eq. (1.32) below. Both equations involve finite sums.

Expansion (1.26) has the essential property that it is a finite sum for a finite lattice X. It can therefore be expected to converge fast for an infinite system if the polymer system has no long range correlations. This is in sharp contrast with the standard high temperature expansions (s. below). They involve infinite sums on finite lattices X already, and they may diverge on a finite lattice. Their divergence is therefore not in general indicative of any kind of long range correlations in the polymer systems. This is important for instance in the SU(2) lattice gauge theory model (1.2). This model is thought (or hoped) not to have infinite range correlations for any finite value of the coupling parameter β. Nevertheless the standard high temperature expansions appear to diverge outside the strong coupling regime β < 2.2. It is hoped that use of the expansion (1.26) offers a way to overcome this limitation. As we shall see in the next subsection, (1.26) may be regarded as a partially summed form of the standard high temperature expansions.

In the expansions (1.26) we expect "divergence" (or very slow convergence) only if the polymer system has long range correlations, for instance

$$\frac{Z(\Lambda - A - B)}{Z(\Lambda)} - \frac{Z(\Lambda - A)}{Z(\Lambda)}\,\frac{Z(\Lambda - B)}{Z(\Lambda)} \;\underset{\text{fast}}{\not\to}\; 0 \tag{1.31a}$$

when distance (A,B) → ∞

that is

$$\ln \frac{Z(\Lambda - A - B)\,Z(\Lambda)}{Z(\Lambda - A)\,Z(\Lambda - B)} = -\sum_{\substack{Y \\ A \cup B \subset Y \subseteq \Lambda}} \mathbf{L}(Y) \;\underset{\text{fast}}{\not\to}\; 0 \tag{1.31b}$$

The r.h.s. of eq. (1.31b) involves the sum over (large) polymer-connected sets Y that contain both the widely separated subsets A and B. Expressions of the form on the r.h.s. of (1.31a) are called reduced correlation functions in the theory of polymer systems [10] .

The relation between convergence and long range correlations in expansions (1.26) for infinite systems deserves further study.

It is amusing to see how the expansions react when the additive constant in the Lagrangean $\mathcal{L}_p$ of our lattice gauge theory model is changed: $\mathcal{L}_p \to \mathcal{L}_p + c$. If $f_p = e^{\mathcal{L}_p} - 1$ is small for some choice, it is not small for another choice. As a result, the convergence of the standard high temperature expansions is affected by such a change. Not so for the expansions (1.26). The change $\mathcal{L}_p \to \mathcal{L}_p + c$ changes $Z(X) \to Z(X)e^{c|X|}$. As a result $\mathbf{L}(Y) \to \mathbf{L}(Y)+c$ if Y is a single plaquette, while $\mathbf{L}(Y)$ for all other sets Y are unchanged. (To see this it suffices to note that eq. (1.26) remains valid for any finite X after these substitutions, if it was valid before.) Therefore, convergence properties of (1.26) are unaffected.

The standard high temperature expansions

They can be obtained from eq. (1.26) by further expansion. One starts from expansion (1.26) with eqs. (1.27), (1.9) for $\mathbf{L}(Y)$ inserted, viz.

$$\ln Z(\Lambda) = \sum_Y \mathbf{L}(Y) \tag{1.26}$$

with

$$\mathbf{L}(Y) = \sum_{X \subseteq Y} \vartheta_{YX} \ln \Big(1 + \sum_{\substack{(P_1,\ldots,P_n) \\ \Sigma P_i \subseteq X}} \prod_i A(P_i)\Big) \tag{1.32}$$

Now one expands $\mathbf{L}(Y)$ in powers of the activities, using $\ln(1+x) = x - \frac{x^2}{2} + \ldots$ and the multinominal theorem. As a result one obtains an expression of the form [10]

$$\mathbf{L}(Y) = \sum_Q a(Q)A(Q) \tag{1.33}$$

Summation is over sets Q of not necessarily distinct polymers, polymer P_i may occur with multiplicity n_i. Every plaquette in Y must be contained in at last one polymer P_i in Q. a(Q) are combinatorial coefficients independent of the activities, and

$$A(Q) \equiv A(P_1)^{n_1} \ldots A(P_k)^{n_k} \qquad \text{if } Q = (P_1^{n_1}, \ldots, P_k^{n_k})$$

The set Y of plaquettes which are contained in at least one polymer in Q is called the shadow of Q.

Inserting (1.33) into (1.26) one obtains the final result

$$\ln Z(\Lambda) = \sum_{Q} a(Q) A(Q) \tag{1.34}$$

Summation is over all Q whose shadow is polymer connected and contained in Λ.

Expansion (1.34) may diverge on a finite lattice.

Example: Λ = a single plaquette, a single polymer $P = \Lambda$ with activity A.

$$\ln Z(\Lambda) = \ln(1+A) = \sum \frac{(-1)^{n+1}}{n} A^n \qquad \text{diverges for } |A| > 1$$

Sufficient conditions for convergence of expansions (1.34) are known. (They are derived by use of Kirkwood Salsburg equations, cp ref. 10.) Convergence is assured if there exists a number $\xi > 1$ such that

$$\frac{1}{\xi}\left[1 + \sup_{p} \sum_{\substack{P \\ p \in P}} \xi^{|P|} A(P)\right] < 1 \tag{1.35}$$

Summation is over all polymers which contain a given plaquette (in case the polymers are made of plaquettes).

The problem of the roughening transition

Let us now return to the expansion for Wilson loop expectation values, eq. (1.13). In the high temperature regime (β small enough) only polymers P_o (with boundary C = Wilson loop) which differ from the minimal surface with boundary C by small deformations need be taken into account. An example is given in figure 6.

One can make use of this to obtain an expansion for $\ln\langle\chi_k(U_C)\rangle$ in the limit of large loops, that is for the string tension α. This is done by performing another high temperature expansion in which the deformations in P_o (and the clusters which come from the expansion of $\ln Z(\Lambda-\bar{P}_o)/Z(\Lambda)$) are treated as the polymers. Series expansions for the string tension α were computed in this way by Münster [16].

This series for the string tension α itself must be expected to diverge when $\beta > \beta_R$, where β_R is the roughening transition point [17]. A roughening transition is expected to take place on the basis of general arguments [18]. It means the following. One may look at fluctuations of the surface P_o, i.e. the average deflection d squared, see figure 7. It is determined by the relative size of the contribution from surfaces P_o in expansion (1.13). For $\beta < \beta_R$, d is bounded

figure 6 figure 7

independent of the side length T of the Wilson loop. In contrast, for $\beta > \beta_R$ is grows logarithmically with T: $d \propto \ln T$. (*) This means that increasingly larger deformations in P_o become important for increasing T and therefore the series expansion for α, which is an expansion in increasing size of the deformation for infinitely large loops C, cannot converge for $\beta > \beta_R$. In SU(2) lattice gauge theory will Wilson action, $\beta_R \approx 2.$according to ref. 17. On the other hand, the roughening transition is a change in asymptotic behavior for large Wilson loop, it is not expected to correspond to nonanalytic behavior of $\langle \chi_k(U(C)) \rangle$ for fixed finite C.

Conclusion: If one wants to use series expansions beyond $\beta = \beta_R$, one must be content with expansions for finite Wilson loops, and the "partially summed" high temperature expansions which I have described earlier in this lecture should be used. The larger the loop, the more terms in the expansion will be needed.

Let us note that the Monte Carlo Method is also limited to finite Wilson loops (3 x 3 or so).

1.2 Other expansions

I will very briefly mention two other expansions.

1) Low temeprature expansions, e.g. for lattice gauge theories with discrete gauge group G 19

example: $G = \mathbb{Z}_2 \, , \quad U_\ell = \pm 1$

$$Z = \sum_U \exp \sum_p \beta [U_{\partial p} - 1]$$

so that

(*) The roughening transition can also be defined without recourse to expansions in terms of the behavior of expectaion values $\langle \mathrm{tr}\, U_C \, \mathrm{tr}\, U_{\partial p} \rangle$.

$$Z = 1 + \sum_{\mathbb{B}} e^{-2\beta|\mathbb{B}|} \qquad (1.36)$$

B is the set of plaquettes where $U_{\partial p} = -1$. It is coclosed, i.e. every 3-dimensional cube c in Λ contains an even number of plaquettes in B in its boundary. $|B|$ is the number of plaquettes in B. B can be decomposed into connected components P. In this way the partition function is found to equal the partition function of a polymer system. The polymers P are connected coclosed sets of plaquettes. An example is shown in figure 8. Their activities are

$$A(P) = e^{-2\beta|P|} \qquad (1.37)$$

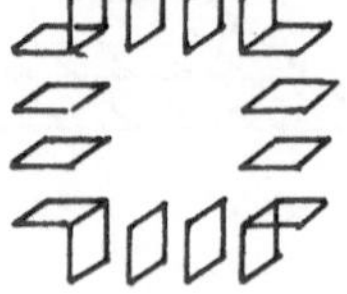

figure 8

2. Mayer expansion for dilute gases [20]

example: A gas of particles with "charges" $q_j = \pm 1$ which interact through a potential $q_i q_j v(x_i - x_j)$ of finite range.

The canoncial partition function of N particles is

$$Z_N = \sum_{\{q_j\}} \int dx_1 \dots dx_n \exp\left[-\beta \sum_{(ij)} q_i q_j v(x_i - x_j)\right] \qquad (1.38)$$

Polymers P are subsets of the N particles $\{1 \dots N\}$, their activities will only depend on the number of particles that are selected. We may label them by $1 \dots n = |P|$

$$A(P) = \sum_{\mathcal{G}} \int dx_1 \dots dx_n \sum_{\{q_j\}} \prod_{(ij)\in\mathcal{G}} f_{ij} \qquad (1.39)$$

with

$$f_{ij} = -1 + \exp\left[-\beta q_i q_j v(x_i - x_j)\right] \qquad (1.40)$$

A graph $\mathcal{G}$ on n vertices $\{1 \dots n\}$ is specified by prescribing the pairs (ij) of vertices which are joined by a line. These pairs are considered as the elements $(ij)\in\mathcal{G}$ of $\mathcal{G}$. Summation in (1.39) is over all connected graphs on vertices $\{1 \dots n\}$. There exists a useful formula (the "tree formula") with which the sum over graphs can be reexpressed, see ref. 21.

A simplification occurs when one considers the grand canoncial partition function with fugacity $z = e^{\beta\mu}$ (μ = chem. potential)

$$Z = \sum \frac{z^N}{N!} Z_N$$

It is expressed in terms of the activities by

$$\ln Z = \sum_P z^{|P|} A(P)$$

provided the sum is absolutely convergent.

2. THE NOTION OF AN EFFECTIVE ACTION (WITH ILLUSTRATION USING A SIMPLE APPROXIMATION)

I will consider lattice models (ferromagnets) with n-component spin variables $\varphi(x) = (\varphi^1(x), \ldots \varphi^n(x))$, and with partition function of the form

$$Z = \int \prod_x d^n\varphi(x)\, e^{L_0(\varphi)} \prod_x f(\varphi(x)) \tag{2.1a}$$

with $f(\varphi(x)) \geqslant 0$ and

$$L_0(\varphi) = -\frac{n}{2f_0} \sum_x \nabla_\mu \varphi^i(x) \nabla_\mu \varphi^i(x) + \text{const.} \tag{2.1b}$$

<u>Example</u> 1. 0(n)-symmetric Heisenberg ferromagnet:

$$f(\varphi(x)) = \delta\left(|\varphi(x)|^2 - 1\right) \tag{2.2a}$$

<u>Example</u> 2. Discrete Gaussian Model = Z-ferromagnet: n = 1, and

$$f(\varphi(x)) = \sum_{k=0,\pm1,\pm2,\ldots} \delta(\varphi(x) - 2\pi k) \tag{2.2b}$$

Let us write $\tilde{\varphi}(k)$ for the Fourier transform of $\varphi(x)$. Instead over $\varphi(x)$ we may integrate $\tilde{\varphi}(k)$. Suppose that the integrations over $\varphi(k)$ with $|k| > M$ have been done. Let $\Phi(x)$ be the sum of the Fourier components $\varphi(k)e^{ikx}$ with $|k| < M$, i.e. equal to $\varphi(x)$ minus its high frequency components. Then the result of the integrations will be of the form

$$Z = \int \mathcal{D}\Phi\, e^{L_{eff}(\Phi)} \tag{2.3}$$

This defines the effective action L_{eff}. It is convenient to take out a kinetic term from $L_{eff}(\Phi)$ and to write

$$Z = \int d\mu_C(\Phi)\, e^{-V_{eff}(\Phi)} \tag{2.4}$$

$d\mu_C$ is a free field measure with UV-cutoff M. (This means that it contains δ-functions which constrain the Fourier components $\tilde{\varphi}(k)$ with $|k| > M$ to zero.)

It is of importance to study the localization properties of L_{eff}. (Because of the integrations that have been performed, L_{eff} will include nonlocal terms.) [4].By using the technique of Möbius inversion of section 1 one can exhibit L_{eff} as a sum of terms which are localized in regions Y of space (Euclidean spacetime if we consider QFT). As an example, the complete effective action for the Z-ferromagnet with UV-cutoff M is exhibited in eq. (3.2). (For technical reasons, a Pauli Villars UV-cutoff is used there instead of a sharp momentum cutoff $|k| < M$, and Φ is rescaled by a factor $\beta^{\frac{1}{2}}$.)

In this section I will use a simple approximation to illustrate how (approximate) computations of effective actions can be used to determine the phase structure of a model.

2.1 Normal ordering.

A free field theory is uniquely determined by its <u>propagator</u>

$$C = \beta(-\Delta + m^2)^{-1} \quad \text{without UV-cutoff, or on a lattice} \tag{2.5a}$$

or

$$C = \beta[(-\Delta+m^2)^{-1} - (-\Delta+M^2)^{-1}] \quad \text{with a Pauli-Villars cutoff M} \tag{2.5b}$$

etc. . $-\Delta$ should be read as k^2 (continuum) or $-2\sum_\mu[\cos k_\mu a - 1]$ (lattice) to obtain the propagator in momentum space. The factor β could be absorbed by rescaling the field.

The corresponding Euclidean free field measure is

$$d\mu_C(\varphi) = \frac{1}{Z_0} \prod_x d^n\varphi(x) \exp[-\tfrac{1}{2}(\varphi, C^{-1}\varphi)] \tag{2.6}$$

C is also called the <u>covariance</u> of this <u>Gaussian measure</u>.

Normal ordering with respect to a Gaussian measure with covariance C is defined by

$$\begin{aligned} e^{ik\cdot\varphi(x)} &= :e^{ik\cdot\varphi(x)}: \int d\mu_C(\varphi')\, e^{ik\cdot\varphi'(x)} \\ &= e^{-k^2C(0)/2} :e^{ik\cdot\varphi(x)}: \end{aligned} \tag{2.7}$$

C(0) = propagator at distance 0.

To write an arbitrary function $f(\varphi(x))$ in normal ordered form, one expands it in a Fourier series or - integral and uses (2.7). It follows that, if f is any <u>tempered distribution</u> then

$$f(\varphi(x)) = :F(\varphi(x)): \tag{2.8a}$$

with

$$F(\xi) = (2\pi C(0))^{-\frac{n}{2}} \int d^n\eta \, f(\eta) \, e^{-\frac{1}{2C(0)}(\xi-\eta)^2} \tag{2.8b}$$

and F is an entire analytic function of ξ, hence posesses an everywhere convergent power series (for any finite number n of components of φ - in the $n \to \infty$ limit it need not be so [30]). Note in oarticular that also δ-functions can be written in normal ordered form, and then expanded in normal ordered products.

Generalization to functions which depend on the field at several points is obvious.

example: $\varphi(x)^2 = :\varphi(x)^2: + nC(0)$

2.2 A simple approximation

To carry out the integrations of high frequency components of the fields, we will use the approximation (for C a propagator which propagates the high frequency components)

$$\int d\mu_C(\varphi) \prod_x f_x(\varphi(x)) \equiv \int d\mu_C(\varphi) \prod_x :F_x(\varphi(x)): \approx \prod_x F_x(0) \tag{2.9}$$

In this formula, :: is normal ordering will respect to Gaussian measure with covariance C; it depends on C because C(0) enters into eq. (2.8b).

Motivation for the approximation

Write the integral as a sum of Feynman vacuum-diagrams by use of Wicks theorem. Neglect all nontrivial diagrams (with > 1 vertex. Diagrams ο + ◎ + ◎ +...are taken into account by the above normal ordering). For instance, let n=1. Expand

$$F_x(\varphi(x)) = F_x(0) + F_x'(0)\varphi(x) + \cdots$$

The leading term in $\prod_x :F_x(\varphi(x)):$ is $\prod_x F_x(0)$. Other terms include for instance something proportional to

$$:\varphi^2(x_1)::\varphi^2(x_2): \overset{\text{Wick}}{=} :\varphi^2(x_1)\varphi^2(x_2): + C(x_1-x_2):\varphi(x_1)\varphi(x_2): + 4C(x_1-x_2)^2$$

The last term is graphically represented by a graph ⊙ with two vertices, which we neglect. The first two terms integrate to zero.

2.3 Application

I use this approximation to carry out the integration of the high frequency parts of the field $\varphi(x)$, viz. $\tilde{\varphi}(k)$ with $|k| > M$. The aim is to obtain in this way an effective action with UV-cutoff M, with M of the order of the ultimate physical mass. The (approximate) true vacuum etc. is then obtained by a semiclassical treatment of this effective action. This reveals then the phase structure of the model.

NB: In very nonlinear theories (with δ-functions etc.), semiclassical approximations are only trustworthy once the cutoff M has been brought down to O(phys.mass). To obtain a rigorous justification of a semiclassical approximation when M is low enough one relies on Glimm-Jaffe-Spencer expansions (compare introduction) if $n = 1$.

Let us split the original propagator ($\beta = f_o/n$)

$$C = \beta(v+u) \qquad , \quad \beta\tilde{u}(k) = \tilde{C}(k)\,\theta(\pi M - |k|)$$

Then

$$\begin{aligned} d\mu_C(\varphi) &= d\mu_{\beta u}(\Phi)\, d\mu_{\beta v}(\chi) \\ \varphi &= \Phi + \chi \end{aligned} \tag{2.10}$$

Φ has only (propagating) components $\Phi(k)$ with $|k| < \pi M$, the other components are called χ. (Note that $d\mu_{u=0}(\Phi)$ = Dirac measure concentrated at $\Phi = 0$.) We obtain

$$\begin{aligned} Z &= \int d\mu_C(\varphi) \prod_x f(\varphi(x)) = \int d\mu_{\beta u}(\Phi)\, d\mu_{\beta v}(\chi) \prod_x f(\varphi(x)+\chi(x)) \\ &= \int d\mu_{\beta u}(\Phi)\, d\mu_{\beta v}(\chi) \prod_x :F(\varphi(x)+\chi(x)): \end{aligned}$$

where :: is normal ordering of the function $F_x(\chi(x)) = F(\Phi(x) + \chi(x))$ with respect to covariance βv. Using our simple approximation, the above expression becomes

$$\approx \int d\mu_{\beta u}(\Phi) \prod_x F(\Phi(x)) \equiv \int d\mu_{\beta u}(\Phi)\, e^{-V_{eff}(\Phi)}$$

with

$$V_{eff}(\Phi) = -\sum_x \ln F(\Phi(x))$$

The effective action consists of $-V_{eff}$ plus the "kinetic term" from the Gaussian measure.

Thus we obtain the following Rules for an approximate determination of L_{eff}:

Suppose the partition function to be evaluated is of the form

$$Z = \int d\mu_C(\varphi) \prod_x f(\varphi(x))$$

Then proceed as follows. Normal order f with respect to a covariance βv that equals C except for the presence of an infrared cutoff at M , viz. $f(\chi(x)) = :F(\chi(x)):$. Drop the dots ::, modify the kinetic term by an UV-cutoff M, set

$$L_{eff}(\phi) = -\frac{1}{2\beta}(\phi, u^{-1}\phi) - V_{eff}(\phi) \quad ; \quad V_{eff}(\phi) = -\sum_x \ln F(\phi(x)). \tag{2.11}$$

Application to the Z-ferromagnet (example 1, eq. (2.2b)).

In 3 dimensions this model is the (Kramers Wannier) dual transform of the U(1) lattice gauge theory with Villain action [22].

In 2 dimensions it is the dual transform of the plane rotator (XY-model) with Villain action.

The partition function is ($\beta = 1/f_o$, $v_{Cb} = (-\Delta)^{-1}$)

$$Z = \int d\mu_{\beta v_{Cb}}(\varphi) \sum_k \prod_x \delta(\varphi(x) - 2\pi k(x))$$

We split

$$v_{Cb} = v + u$$

Instead of a sharp momentum cutoff one could also use a Pauli Villars-cutoff. In this case

$$v = (-\Delta + M^2)^{-1} \quad , \quad u = (-\Delta)^{-1} - (-\Delta + M^2)^{-1}$$

Alternatively, one may use a lattice cutoff $M = L^{-1}$. In this case the propagator $u(x,y)$ is obtained by averaging $v_{Cb}(x,y)$ over blocks of side length L in each argument. We want M of the order of the ultimate physical mass m. According to our rules, we have to write the periodized δ-function in normal ordered form. The result is immediate from eq. (2.8b): it is equal to a normal ordered periodized Gaussian. As a result of the application of our rules,

$$Z = \int d\mu_{\beta u}(\phi) \sum_k \prod_x e^{-\frac{1}{2\beta v(0)}(\phi(x) - 2\pi k(x))^2} \tag{2.12}$$

For $\beta v(0)$ large this becomes

$$\approx \int d\mu_{\beta u}(\phi) \exp\left[z \sum_x \cos\phi(x) + const \right]$$

with

$$z = \exp\left[-\beta v(0)/2\right] \tag{2.13}$$

The effective action in the presence of a cutoff M is thus given by

$$V_{eff}(\phi) = -z \sum_x \cos \phi(x) \tag{2.14}$$

We now proceed to the further analysis for the <u>3-dimensional case</u> (U(1) lattice gauge theory).

Suppose that $\beta \to \infty$ in units of lattice spacing, and $M = \lambda m$, m = phys.mass. For large β, m will be small and $v(0) \approx v_{Cb}(0)$. Assume for simplicity that we use a lattice cutoff $M = L^{-1}$. Then the field $\Phi(x)$ is constant on blocks of lattice spacing L, and the effective action becomes

$$V_{eff}(\phi) \approx -z L^3 \sum_{\text{blocks}} \cos \phi(x) \tag{2.15}$$

We look at this as an effective potential for a theory an a block lattice of lattice spacing L. Because of the large factor zL^3 (s. below), V_{eff} has minima which are separated by <u>very high</u> maxima. Expanding the cos around $\Phi = 0$ we obtain the mass m,

$$m^2 \approx 2\beta e^{-\beta v_{Cb}(0)/2} \tag{2.16}$$

Therefore $zL^3 = \lambda^{-3}(m\beta)^{-1} \to \infty$ exponentially as $\beta \to \infty$. Here $\lambda = (M/m) = O(1)$. An approximate expression for the <u>surface tension</u> (see figure 10) can be obtained by a (semi)classical approximation also. The result is

$$\alpha = 8m\beta^{-1} \tag{2.17}$$

Duality transformation shows that the surface tension in the Z-ferromagnet equals the string tension in the U(1) lattice gauge theory.

Let me briefly discuss the <u>2-dimensional case</u> for contrast. In this case the factor zL^3 in eq. (2.15) is replaced by zL^2. This makes a profound difference - we are now lacking the one factor $L = O(m^{-1})$ which made $zL^3 \to \infty$ as $\beta \to \infty$.

Consider the situation with small or zero physical mass $0 \leq m \ll 1$ in units of (lattice spacing)$^{-1}$. The mass m is determined by the curvature of V_{eff} at its minima, in the limit of low enough cutoff.

$$m^2 = 2\beta z \quad \text{with} \quad z = \exp[-\beta v(0)/2] \tag{2.18}$$

z depends on M, and therefore indirectly on m because M is constraint to values $\geq O(m)$. For small M

$$v(0) \approx -\frac{1}{2\pi} \ln Ma' \qquad \text{with } a' = 0 \text{ (lattice spacing)} \tag{2.19}$$

Suppose that $\beta > \beta_c = 8\pi$. Then z is exceedingly small for moderately small M, and $m^2 = 2\beta z$ tends to 0 faster than M^2. Therefore $m \to 0$ as I lower M, maintaining $M \gg O(m)$. Thus

$$m = 0 \qquad \text{for} \quad \beta > \beta_c \approx 8\pi \tag{2.20}$$

This is the Kosterlitz-Thouless phase.

Application to the nonlinear σ-model (Heisenberg ferromagnet)

This is example 1, eq. (2.2a). According to our result, we must find the normal ordered form (with respect to covariance βv) of the δ-function that is concentrated on the unit sphere S_{n-1}.

According to eq. (2.8b)

$$\delta(\varphi(x)^2 - 1) = :F(\varphi(x)):$$

with ($\beta = f_o/n$, $\xi = (\xi^1, \ldots, \xi^n)$)

$$F(\xi) = [2\pi\beta v(0)]^{-\frac{n}{2}} \int d^n\eta \; \delta(\eta^2 - 1) \, e^{-\frac{1}{2\beta v(0)}(\xi-\eta)^2} \tag{2.21}$$

We split the propagator as before

$$C = \beta(v + u)$$

Application of our rules gives

$$\begin{aligned} V_{eff}(\Phi) &= \sum_x \vartheta_{eff}(\Phi(x)) \\ \vartheta_{eff}(\xi) &= -\ln F(\xi) \; . \end{aligned} \tag{2.22}$$

To find the behavior V_{eff} at $\xi = 0$ (maximum or minimum), one computes V_{eff}'' from eqs. (2.21), (2.22). Qualitatively, V_{eff} looks as follows as a function of $|\xi|$

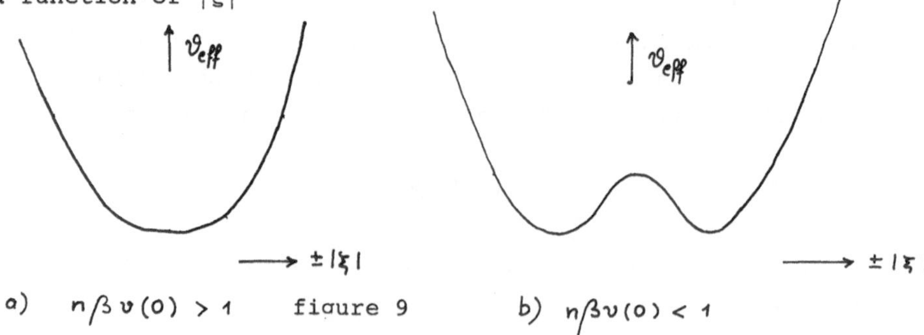

a) $n\beta v(0) > 1$ figure 9 b) $n\beta v(0) < 1$

We are interested in the behavior when the cutoff M is small compared to the (lattice spacing)$^{-1}$. $M \to 0$ gives $\upsilon(0) \to (-\Delta)^{-1}(0)$.

In 2 dimensions, $\upsilon(0) \to \infty$ as $(-\Delta+m^2)^{-1}(0) \to \infty$ when $m \to 0$ in 2 dimensions. As a result, $\mathcal{V}_{eff}$ alway looks as in figure 9a) for sufficiently low cutoff M. Therefore there will be a finite mass and no spontaneous symmetry-breakdown.

In more than 2 dimensions, there will be spontaneous symmetry-breaking if $n\beta\upsilon_{Cb}(0) < 1$, i.e. if $f_o^{-1} > (-\Delta)^{-1}(0)$. The potential $\mathcal{V}_{eff}$ looks like figure 9b) for this range of f_o and low cutoff M.

These results agree with the results of the $1/n$ expansion.

3. RIGOROUS RESULTS FOR THE 3-DIMENSIONAL U(1) LATTICE GAUGE THEORY

In this lecture I will report results of a study by M. Göpfert and myself [8]. It dealt with lattice gauge theory in 3 space time dimensions without matter fields and with gauge group U(1). It is known from the work of A. Guth [23], that such a model shows a deconfining phase transition in 4 dimensions, so that its weak coupling continuum limit will not show confinement (if it exists at all). In contrast, in 3 dimensions, the work of Polyakov [24], Banks,Myerson and Kogut [22], Drell et al. [25] and DeGrand and Toussaint [26] lead to the belief that confinement will be true for all values of the coupling constant. We proved that this is indeed the case, for the model with Villain action. It turned out, however, that the ratio α/m_D^2 of the string tension to asymptotic physical mass squared becomes infinite in the weak coupling limit (continuum limit).

The model lives on a 3-dimensional cubic lattice of lattice spacing a. It is a classical statistical mechanical system whose random variables are attached to the links b = (x,y) of the lattice

$$U_\ell = e^{ia\theta_\mu(x)} \quad \text{for} \quad \ell = (x, x+e_\mu) \quad , \ -\frac{\pi}{a} \leq \theta_\mu \leq \frac{\pi}{a} \tag{3.1a}$$

(e_μ = lattice vector in μ direction). The action is of the form

$$L(U) = \sum_p \mathcal{L}(U_{\partial p}) \tag{3.1b}$$

with $U_{\partial p} = U_{\ell_1} \dots U_{\ell_4}$ if p is the plaquette whose boundary consists of links $\ell_1 \dots \ell_4$, and

$$\mathcal{L}(e^{i\varphi}) = \ln \sum_{m=0,\pm 1,\pm 2,\dots} \exp\left[-\frac{1}{2ag^2}(\varphi - 2\pi m)^2\right] \tag{3.1c}$$

g^2 is the unrenormalized electric charge squared. It has dimension of a mass in 3 dimensions. The Boltzmann factor is exp L(U).

Since the gauge group U(1) of this model is abelian, the model can be subject to a Kramers-Wannier duality transformation. As a result one obtains a ferromagnet with a global symmetry group $\mathbb{Z}$. Its random variables n(x) are attached to the sites of a 3-dimensional cubic lattice Λ (the dual of the origianl one) and assume values which are integer multiples of 2π. The new action is

$$\hat{L}(n) = -\frac{1}{2\beta}\int_x [\nabla_\mu n(x)]^2 \qquad \text{with } \beta = 4\pi^2/g^2$$

We use the standard notations

$$\int_x = a^3 \sum_x \quad , \quad \nabla_\mu f(x) = a^{-1}[f(x+e_\mu) - f(x)]$$

a is the lattice spacing, I prefer not to set a = 1 in this section. Expectation values are computed with the help of the Boltzmann factor exp $\hat{L}(n)$. This model is known as the "discrete Gaussian model", for obvious reasons. We call it the " $\mathbb{Z}$-ferromagnet" in order to emphasize its symmetry properties. I have already considered this model as an example in section 2.

The global $\mathbb{Z}$-symmetry of this model is always spontaneously broken (if $\langle n(x)\rangle$ exists at all) since the equation

$$\langle n(x)\rangle = \langle n(x)\rangle + 2\pi I \qquad , I \in \mathbb{Z}$$

has no solution if $I \neq 0$. The surface tension α of the model is defined as the cost of free energy per unit area of a domain-wall which separates two domains whose spontaneous magnetization $\langle n(x)\rangle$ differs by 2π, see figure 10. The duality transformation shows that α equals the string tension of the U(1) gauge model.

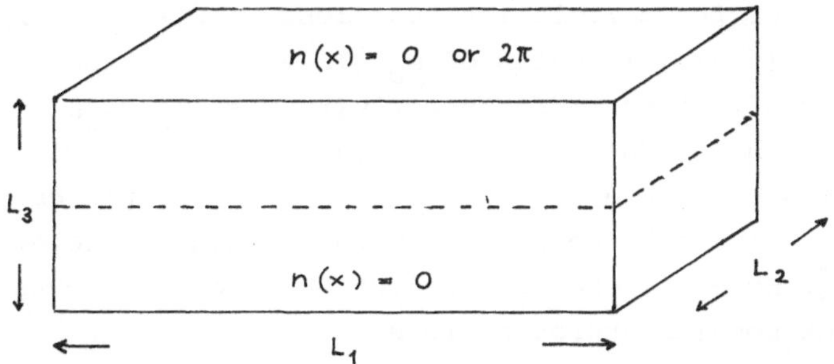

figure 10: Definition of the surface tension α. One considers the free energy of a box for the two choices of boundary conditions i) $n(x) = 0$ everywhere and ii) $n(x) = 2\pi$ above the dashed line and $n(x) = 0$ below. α equals the difference of the free energy for the two boundary conditions, divided by L_1L_2, in the limit $L_i \to \infty$ $(i = 1,2)$, followed by $L_3 \to \infty$.

It is convenient to introduce the following quantity with the dimension of a mass squared

$$m_D^2 = (2\beta/a^3)\exp[-\beta v_{Cb}(0)/2]$$

where v_{Cb} is the lattice Coulomb potential. As was shown by Banks et al., the model can also be transformed into a (special) Coulomb system. m_D^{-1} is the prediction of a Debye Hückel approximation for the sceening length of that system, for large β/a.

Our main results are as follows.

Theorem 1 There is a dimensionless constant $C > 0$ such that

$$\alpha \geq C \cdot m_D \beta^{-1} \qquad \text{for sufficiently large } \beta/a.$$

Since αa^2 is a monotone decreasing function of β/a by Guth's inequality [23], it follows that $\alpha > 0$ for all values of the coupling constant $g^2 > 0$. We believe that the r.h.s. of the inequality of theorem 1 represents the true asymptotic behavior of α. It is amusing to compare with the leading term of the high temperature expansion, which is valid when β/a is small. It reads $\alpha = 2\pi^2 a^{-1}\beta^{-1} + \ldots$.

The meaning of m_D as an asymptotic mass is clarified by the second result. The effective action L_{eff} mentioned below depends on a real field $\Phi(x)$. It is obtained by integrating out the high frequency components of $\beta^{-\frac{1}{2}}n(x)$. Symbolically we may write

$$\phi(x) = \beta^{-1/2}\, n(x) \qquad \text{with Pauli-Villars cutoff } M.$$

Theorem 2 Consider the correlation functions $\langle\Phi(x_1) \ldots \Phi(x_n)\rangle$ for fixed distances $m_D|x_i - x_j|$ in units of m_D. They tend to the correlation functions of a massive free field theory with mass m_D as $\beta/a \to \infty$ and $M/m_D \to \infty$ (proportional $(\beta/a)^{1/12}$, for instance).

These results were obtained by a rigorous block spin calculation, and are therefore perfectly consistent with the general renormalization group theory [4]. However, they contradict what would be obtained by making simple but popular approximations.

Suppose that one could set up a renormalization group procedure (block spin calculation) for the U(1) gauge theory (3.1) in such a way that the effective action at each step of this interative procedure is still (approximately) of the same form (3.1), except for the replacement of g^2 by a running coupling constant $g^2_{eff}(a')$ and a new value $a' > a$ of the lattice spacing. Suppose moreover that $a' g^2_{eff}(a')$ reaches values in the domain of validity of high temperature expansions after sufficiently many iteration steps (depending on ag^2), no matter how small the bare coupling constant $g^2 a$ is. Then it would follow immediately that the string tension α should be proportional to the physical mass (= mass gap) squared.

In contrast, theorem 1 tells us that

$$\alpha/m_D^2 \geq C\,(\beta m_D)^{-1} = C\,(2\beta^3/a^3)^{-1/2} \exp \tfrac{1}{4}\beta v_{Cb}(0) \to \infty$$

as $\beta/a = 4\pi^2/g^2 a \to \infty$. (The numerical values of $v_{Cb}(0)$ is found in the literature to be $0.2527\ a^{-1}$).

I will now briefly describe the main steps of the analysis. The first step is to integrate out the high frequency components of the field $\varphi(x) = \beta^{-\frac{1}{2}} n(x)$. This produces an effective action $L_{eff}(\Phi)$ for a real field $\Phi(x)$ (on the original lattice) with Pauli-Villars cutoff M. $L_{eff}(\Phi)$ is obtained in the form of an infinite series of the following form, with real coefficients $\rho_s(\ldots)$.

$$L_{eff}(\Phi) = -\tfrac{1}{2}\int_x \Phi(x)\left\{-\Delta\left(1-\tfrac{\Delta}{M^2}\right)\right\}\Phi(x) + \sum_{s\geq 1}\frac{1}{s!}\sum_{m_1\ldots m_s}\int_{x_1\ldots x_s}\rho_s(m_1,x_1,\ldots,m_s,x_s)\left[e^{im_1\beta^{1/2}\Phi(x_1)}-1\right]\ldots\left[e^{im_s\beta^{1/2}\Phi(x_s)}-1\right] \tag{3.2}$$

m-summations are over $m_j = \pm 1, \pm 2, \ldots$. The first term is the usual kinetic term for a real field with a Pauli-Villars cutoff M [27].

The main problem was to establish convergence of the expansion, for $M > \lambda m_D$ (λ independent of β) and large β/a, and bounds on the individual terms. In particular, it was shown that the coefficients ρ_s for $s > 2$ decay exponentially with distances $|x_i - x_j|$ so that they become negligibly small for distances $|x_i - x_j| \gg M^{-1}$. This means that no interactions of range much larger than the cutoff length M^{-1} have been generated in the process of integrating out the high frequency components of the field $\varphi(x)$. This is a basic requirement in a block spin calculation.

Moreover, the bounds show also that the dominant terms are the terms with $s = 1$ and $m_1 = \pm 1$, for large β/a, and that $\rho_1(m,x) \approx \frac{1}{2}\beta^{-1} m_D$. Thus

$$L_{eff}(\Phi) \approx \text{kinetic term} - m_D^2\beta^{-1}\int_x \left[1-\cos\beta^{1/2}\Phi(x)\right] + \ldots \tag{3.3}$$

This reproduces the result (2.13), (2.14) of the simple approximation that I used in section 2. The rigorous treatment of the model justifies this approximation, for large β/a, by producing bounds on the correction terms ... in the effective action. If the resulting theory with effective action (3.3)(without the correction terms ...)is treated by a classical approximation, one obtains the result

$$\alpha = 8 m_D \beta^{-1}$$

It is possible that this is exact for $\beta/a \to \infty$, but our bounds are not sharp enough to prove it. There is, however, an upper bound on α due to Ito which complements the lower bound of theorem 1 and is very close to it[28].

The second step is the analysis of a theory with action (3.2). Such an analysis was already performed by Brydges and Federbush [6] and we could use their result. The Glimm-Jaffe-Spencer expansion of constructive field theory [5] (see introduction) in the basis tool in this step.

Finally I would like to explain briefly how the effective action (3.2) is obtained. It is known from the work of Banks, Myerson and Kogut [22] that the model can be transformed into a Coulomb system with partition function

$$Z = \sum_{m \in \mathbb{Z}^\Lambda} e^{-\beta (m, v_{Cb} m)/2}$$

Here we set the lattice spacing equal to one, $m(x) = 0, \pm 1, \pm 2, \ldots$ is the charge at site x of Λ, $v_{Cb} = (-\Delta)$ is the lattice Coulomb potential, and

$$(m, v_{Cb} m) = \sum_x \sum_y m(x) v_{Cb}(x-y) m(y)$$

A self-interaction term $x = y$ is included. Following Fröhlich [29] one splits the Coulomb potential into a Pauli-Villars cutoff Coulomb potential $u = (-\Delta)^{-1} - (-\Delta+M^2)^{-1}$, and a Yukawa potential $v = (-\Delta+M^2)^{-1}$ of range M^{-1}.

$$v_{Cb} = v + u$$

One inserts this and uses the formula for the characteristic function of a Gaussian measure with covariance u,

$$\int d\mu_u(\phi) e^{i(f,\phi)} = e^{-\frac{1}{2}(f,uf)}$$

to rewrite the partition function as

$$Z = \int d\mu_u(\phi) \mathcal{Z}(\phi)$$

with

$$\mathcal{Z}(\phi) = \sum_{m \in \mathbb{Z}^\Lambda} e^{i\beta^{1/2}(m,\phi)} e^{-\beta(m,vm)/2}$$

$\mathcal{Z}(\Phi)$ is the partition function of a Yukawa gas with complex space dependent fugacity $z = \exp i\beta^{\frac{1}{2}} m(x)\Phi(x)$. Its logarithm is the desired effective action

$$L_{eff}(\Phi) = \text{kinetic term} + \ln \mathcal{Z}(\Phi)$$

It is natural to try to use a Mayer expansion to compute $\ell n\Xi$ (compare section 1). The leading term in such a Mayer expansion comes from clusters with only one particle and gives (3.3). Unfortunately, known methods to prove convergence of such a Mayer expansion were not nearly good enough to cover the values of parameters (fugacity z, inverse temperature β) of interest here. We have therefore developed a refined version of such Mayer expansions. It is based on splitting the Yukawa potential v into a sum of interactions of decreasing strength and increasing range, and then treating one after the other of these by Mayer expansions as usual. Recursive bounds are established, and these combine to prove convergence of the complete expansion and produce bounds on the individual terms.

References

1. E. Seiler, Gauge theories as a problem of constructive quantum field theory and Statistical Mechanics, Lecture Notes in Physics, Vol. 159, Heidelberg: Springer 1982

2a. G. Mack, Phys. Rev. Letters 45 (1980) 1378, Acta Austriaca Physica, Suppl. XXII (1980) 509

2b. G. Mack and E. Pietarinen, Phys. Letters 94B (1980) 397, Nucl. Phys. B205 [FS5] (1982) 141

3. J. Glimm, A. Jaffe, T. Spencer: The particle structure of weakly coupled $\mathbf{P}(\varphi)_2$ model and other applications of high temperature expansions. Part II: The cluster expansion. In: Constructive quantum field theory. G. Velo, A. Wightman (eds.) Lecture Notes in Physics, Vol. 25, Berlin, Heidelberg, New York, Springer 1973

4. K. Wilson, Phys. Rev. D2 (1970) 1473

5. J. Glimm, A. Jaffe, T. Spencer, Ann. Phys. 101 (1975) 610, 631

6. D. Brydges, P. Federbush, Commun. Math. Phys. 73 (1980) 197

7. J. Fröhlich and T. Spencer, Phys. Rev. Letters 46 (1980) 1006, Commun. Math. Phys. 81 (1981) 527

8. M. Göpfert and G. Mack, Commun. Math. Phys. 82 (1982) 545

9. A. Kupiainen and K. Gawedzki (in preparation)

10. Ch. Gruber and A. Kunz, Commun. Math. Phys. 22 (1971) 133
D. Ruelle, Statistical Mechanics: Rigorous results, Benjamin, New York 1966

11. K. Osterwalder and E. Seiler, Ann. Phys. 110 (1978) 440

12. K. Wilson, Phys. Rev. D10 (1974) 2445

13. G. Mack and V. Petkova, Ann. Phys. 123 (1979) 442, p. 464

14. L. Euler, Solutio problematis ad geometriam situs pertinentis, Commentarii Academiae Petropolitanae 8 (1736) 128, engl. transl. Sci. Amer. 7 (1953) 66

15. G.S. Rushbrooke, G.A. Baker, P.J. Woods, in: Phase transitions and critical phenomena, C. Domb and M. Green, eds., Academic Press New York (1974) Vol. 3
C. Domb, ibid, p. 77
G.S. Rushbrooke, J. Math. Phys. 5 (1964) 1106

16. G. Münster, Nucl. Phys. B180 [FS2] (1981) 23

17. A., E. and P. Hasenfratz: Nucl. Phys. B180 [FS2] (1981) 353
M. Lüscher, G. Münster, P. Weisz, Nucl. Phys. B180 [FS2] (1981)1
C. Itzykson, M. Peskin, I. Zuber, Phys. Letters 95B (1980) 259

18. M. Lüscher, Nucl. Phys. B180 [FS2] (1981) 317

19. R. Marra and S. Miracle Solé, Commun. Math. Phys. 67 (1978) 233

20. see e.g. D. Ruelle, ref. 10, or standard text books.
21. D. Brydges and P. Federbush, J. Math. Phys. 19 (1978) 2064
22. T. Banks, R. Myerson, J. Kogut, Nucl. Phys. B129 (1977) 493
23. A. Guth, Phys. Rev. D21 (1980) 2291
24. A.M. Polyakov, Nucl. Phys. B120 (1977) 429
25. S.D. Drell, H.R. Quinn, B. Svetitsky, M. Weinstein, Phys. Rev. D19 (1979) 619
26. T.A. DeGrand, D. Toussaint, Phys. Rev. D22 (1980) 2478
27. N.N. Bogolubov and D.V. Shirkov, Introduction to the theory of quantized fields. London, Interscience 1959
28. K.R. Ito, Nucl. Phys. B205 [FS5] (1982) 440
29. J. Fröhlich, Commun. Math. Phys. 47 (1976) 233
30. A. Holtkamp, Ph.D.Thesis, Hamburg 1982 (DESY-T-82/02)

LARGE N

Yu. M. Makeenko

Institute for Theoretical and
Experimental Physics

117259 Moscow, USSR

Contents:

Introduction

The major dynamical problem we have met in QCD is that of strong coupling in the infrared domain. An effective charge grows in the infrared domain so that perturbative expansions cannot be applied. In the Lectures by Professors C.DeTar, L.Jacobs, G.Mack, L.McLerran and H.Satz given at this School, the lattice formulation of QCD is discussed. Among the numerous non-perturbative methods employed for the lattice QCD, a special role is played by the numerical Monte Carlo method which allows us to probe the continuum.

At the same time, two non-perturbative approaches were employed directly for the continuum QCD. The first one suggests a possibility to exactly solve QCD (or may be only gluodynamics) analytically by using the higher conservation laws (which are not yet found for QCD) and/or some hidden symmetry. The second approach implies a less ambitious task - to find an approximate solution of QCD by expanding in the inverse number of colors, 1/N. Expansions of such a type (in the inverse number of field components) are widely applicable in quantum mechanics, statistical physics and so on. There expansion in 1/N works as a semiclassical expansion while the leading order in 1/N represents the Hartree-Fock approximation. The name "manycolor QCD" is also used for this approximation in case of QCD.

The topic of the present lectures is the dynamics of manycolor QCD. The main attention is paid to the semiclassical nature of the large-N limit of QCD. It is discussed how the factorization property is satisfied diagrammatically at $N = \infty$. The problem of reformulating QCD in term of colorless composite field, the Wilson loop, which becomes classical as $N \to \infty$ is considered. It is shown that loop equations are an adequate tool for studying the large-N limit of QCD. The reason why the factorization property holds at $N = \infty$ is discussed. Some comments concerning the concept of "master field" in manycolor QCD are included. Solutions of the loop equation, which are known at present, as well as a connection between manycolor QCD and the relativistic string are briefly discussed. The loop equation on the lattice is considered. A special attention is paid to a solution found by Eguchi and Kawai.

1. Useful exercise: sigma model in 1+1 dimensions

Our goal in the present lectures is to consider the large-N limit of 3+1 dimensional QCD. In order to illustrate the general approach, we shall first consider more simple case of the non-linear sigma model in 1+1 dimensions, which is loosely taken as a toy model for QCD. The O(N) sigma model in 1+1 dimensions is exactly solvable so that 1/N expansion can be compared with the exact solution.

The partition function for the non-linear sigma model reads

$$Z \propto \int \mathcal{D}\vec{n}\, \delta(\vec{n}^2 - 1/g)\, e^{\frac{i}{2}\int d^2x\, (\partial_\mu \vec{n})^2} . \tag{1.1}$$

Here $\vec{n}$ is N-component field

$$\vec{n} = (n_1, \ldots, n_N) \tag{1.2}$$

which is restricted by

$$\vec{n}^2(x) = 1/g \quad . \tag{1.3}$$

It follows from the definition (1.1) that the action of the sigma model is of order N at large N since the action involves a sum over N components. However the entropy (i.e. contribution to an effective action coming from the measure in the path integral (1.1)) is of the same order, so that a straightforward WKB method is not applicable.

In case of the sigma model it is well-known how to overcome this difficulty. Let us introduce an auxiliary field $u(x)$ (in our normalization $u \sim 1$):

$$Z \propto \int \mathcal{D}u(x) \int \mathcal{D}\vec{n}(x) \exp\left\{\frac{i}{2}\int d^2x \left[(\partial_\mu \vec{n})^2 + u(\vec{n}^2 - \frac{1}{g})\right]\right\}. \tag{1.4}$$

Therefore we have reduced integration over $\vec{n}$ to a Gaussian form.

Now integration over $\vec{n}$ yields

$$Z \propto \int \mathcal{D}u(x) \exp\left\{-\frac{N}{2} \operatorname{tr} \ln(\partial_\mu^2 - u(x)) + \frac{i}{2g}\int d^2x \, u^2(x)\right\}. \tag{1.5}$$

This path integral can already be calculated by a saddle-point method because the action $\sim$ N (both terms in the exponent $\sim$ N since $g \sim 1/N$ from eq. (1.3)) while the entropy $\sim$ 1. Thus the semiclassical expansion can be applied to the sigma model only after reformulating in terms of <u>singlet</u> composite field.

Another significant property of the large-N limit, that

of factorization, can be seen for the sigma model. The factorization property holds for the average of singlet operators, with the averaging being defined by the path integral as in eq. (1.3), say

$$\langle u(x_1)\ldots u(x_n)\rangle \equiv \frac{\int \mathcal{D}u \int \mathcal{D}\vec{n}\, e^{\frac{i}{2}\int d^2x\left[(\partial_\mu \vec{n}^2)+u(\vec{n}^2-\frac{1}{g})\right]} u(x_1)\ldots u(x_n)}{\int \mathcal{D}u \int \mathcal{D}\vec{n}\, e^{\frac{i}{2}\int d^2x\left[(\partial_\mu \vec{n})^2+u(\vec{n}^2-\frac{1}{g})\right]}} \quad (1.6)$$

In the large-N limit, the path integral over u has a saddle point at

$$u(x) = u_{SP}(x) \quad (1.7)$$

which is an extremum of the action entering the exponent in eq. (1.5). Therefore to the leading order in 1/N the path integral (1.6) is equal to the saddle-point value

$$\langle u(x_1)\ldots u(x_n)\rangle = u_{SP}(x_1)\ldots u_{SP}(x_n)\,. \quad (1.8)$$

In other words the factorization property holds

$$\langle u(x_1)\ldots u(x_n)\rangle = \langle u(x_1)\rangle\ldots\langle u(x_n)\rangle \quad (1.9)$$

which means that $u(x)$ becomes classical as $N \to \infty$ and its quantum fluctuations are suppressed.

2. Factorization property in perturbation theory

Now we are going to apply the same strategy to QCD.

The Euclidean partition function of QCD, with the gauge group being SU(N), is defined by

$$Z \propto \int \underbrace{\mathcal{D}A_\mu^a}_{N^2-1} \exp\left\{-\int d^4x \left[\frac{1}{4} F_{\mu\nu}^a F_{\mu\nu}^a + \bar{\Psi}(\nabla\!\!\!/ + m)\Psi\right]\right\}. \tag{2.1}$$

Here $F_{\mu\nu}^a$ stands for the non-Abelian field strength of gluon field, while $\nabla\!\!\!/ = \gamma_\mu \nabla_\mu$ is the covariant derivative. The action $\sim N^2$ as $N \to \infty$ since $a = 1, \ldots, N^2-1$ but the entropy is of the same order, so that the path integral (2.1) has no saddle point.

The lesson we have learned from the above example of the sigma model says that the lack of saddle point is due to the fact that original field, $A_\mu^a(x)$, is colorful. Therefore we might expect a saddle-point method to be applied at large N after reformulation in terms of colorless composite field.

Some argument in favor of existence of such a saddle point can be presented before QCD is reformulated in terms of colorless composite field. That is based on the factorization property which in turn can easily be established by analyzing a topology of planar diagrams dominating, as was pointed out by 't Hooft[1], in the large-N limit.

To analyze the planar diagrams let us introduce convenient graphic notations due to 't Hooft. Represent the propagator of the matrix gluon field

$$(A_\mu)^{ij} = A_\mu^a \, (t^a)^{ij} \tag{2.2}$$

by the double line in index space

$$G_{\mu\nu}^{ij,kl}(x-y) \equiv \langle A_\mu^{ij}(x) \, A_\nu^{kl}(y) \rangle = \begin{array}{c} i \qquad\qquad l \\ \overline{\overline{\qquad\qquad\quad}} \\ j \qquad\qquad k \end{array} \; . \tag{2.3}$$

Since

$$\langle A_\mu^a(x) \, A_\nu^b(y) \rangle \propto \delta^{ab} \tag{2.4}$$

one gets

$$G_{\mu\nu}^{ij,kl}(x-y) \propto (t^a)^{ij} (t^a)^{kl} \; ; \tag{2.5}$$

$$(t^a)^{ij} (t^a)^{kl} = \delta^{il} \delta^{kj} - \frac{1}{N} \delta^{ij} \delta^{kl} \tag{2.6}$$

where the latter equation is nothing but the completeness condition for matirces of the fundamental representation of the SU(N).

As $N \to \infty$ the second term on the r.h.s. of eq. (2.6) becomes insignificant, so that we arrive at the following "Feynman rules" in index space. The Kroneker symbol corresponds to each open line

$$i \underline{\qquad\qquad} j \quad \Longleftrightarrow \quad \delta^{ij} \; ; \tag{2.7}$$

the trace $\delta^{ii} = N$ corresponds to each closed line

$$\bigcirc \quad \Longleftrightarrow \quad \delta^{ii} = N \; ; \tag{2.8}$$

the coupling constants correspond to each vertex

$$\begin{aligned} &\succ \quad \Longleftrightarrow \quad g \, , \\ &\times \quad \Longleftrightarrow \quad g^2 \, . \end{aligned} \tag{2.9}$$

Let us consider in some detail an example of the simplest gauge-invariant quantity

$$G(x) = \frac{1}{N^2}\, tr\, F_{\mu\nu}^2(x) \tag{2.10}$$

The normalization factor, $1/N^2$, guarantees that $G(x) \sim 1$, since

$$tr\, F_{\mu\nu}^2 = \frac{1}{2} F_{\mu\nu}^a F_{\mu\nu}^a \, . \tag{2.11}$$

The average of G's is defined by the path integral

$$\langle G(x_1) \ldots G(x_n) \rangle = \frac{\int \mathcal{D}A \exp\left\{-\int d^4x \frac{1}{2} tr\, F_{\mu\nu}^2\right\} G(x_1) \ldots G(x_n)}{\int \mathcal{D}A \exp\left\{-\int d^4x \frac{1}{2} tr\, F_{\mu\nu}^2\right\}} \, . \tag{2.12}$$

Here the quark term is omitted because, as was shown by 't Hooft [1], quark loops are insignificant as $N \to \infty$.

Feynman diagrams can be obtained from (2.12) by a standard Faddeev-Popov trick of gauge fixing. Then (infinite) volume of the gauge group appears both in the numerator and in the denominator since $G(x)$ is gauge-invariant. Finally we arrive at a path integral similar to (2.2) except gauge is fixed as well as ghosts are present.

We are now in a position to show how the factorization property is fulfilled in $N = \infty$ QCD diagrammatically. Let us consider the index-space diagrams for the average of two colorless quantities, $G(x_1)$ and $G(x_2)$, which are depicted in fig. 1. The graph a represents the zeroth order of perturbation theory. The quantity $G(x)$ is depicted by two lines as is prescribed by the general recipe (the completeness condition (2.6) is used).

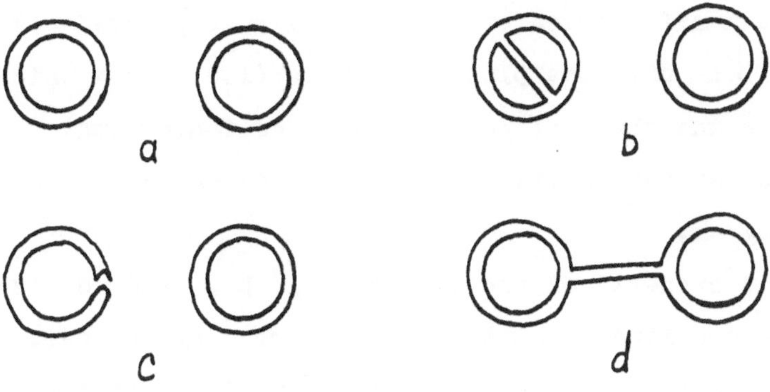

Fig. 1.

The graph a involves four closed lines (factor N^4). The factor $1/N^2$ comes from the normalization so that the contribution of the graph a is of order 1:

$$\text{graph } a \sim \frac{1}{N^2} \cdot N^2 \cdot \frac{1}{N^2} \cdot N^2 \sim 1 \,. \qquad (2.13)$$

Analogously for the graph b :

$$graph\ b \sim g^2 \frac{1}{N^2} \cdot N^3 \cdot \frac{1}{N^2} \cdot N^2 \sim 1 \ . \tag{2.14}$$

The graphs b has five closed lines (N^5), N^{-4} coming from the normalization, and g^2 due to two three-gluon vertices. If $g^2 N \sim 1$ *), the contribution of the graph b is of order 1.

On the contrary the graph c , which is of the same order in the coupling constant as the graph b , involves only three closed lines (N^3) and is of order $1/N^2$:

$$graph\ c \sim g^2 \frac{1}{N^2} N \frac{1}{N^2} N^2 \sim 1/N^2 \ . \tag{2.15}$$

This fact is a manifestation of the general rule pointed out by 't Hooft [1] : only the planar diagrams (i.e. those which can be drawn on the sheet of paper without self-crossing) survive in the large-N limit. The graph c is an example of non-planar diagram.

The common property of the graphs a , b , and c is that virtual gluon line is emitted and absorbed by the same operator. This is not the case for the graph d where the gluon line is emitted by one and absorbed by another operator. This graph involves only three closed lines and is of order $1/N^2$:

---

*) This fact follows, say, from the well-known formula of asymptotic freedom

$$g^2 = 24\pi^2 / 11 N \log \Lambda$$

relating the bare charge to the cut-off.

$$\text{graph } d \sim g^2 \frac{1}{N^4} N^3 \sim 1/N^2 . \tag{2.16}$$

In a similar way one can verify to any order in g^2 that only (planar) diagrams with the gluon lines emitted and absorbed by the same operators survive as $N \to \infty$. Therefore correlations between the colorless operators $G(x_1)$ and $G(x_2)$ are of order $1/N^2$ so that the factorization property holds as $N \to \infty$

$$\langle G(x_1)\, G(x_2) \rangle = \langle G(x_1) \rangle \langle G(x_2) \rangle + O\left(\frac{1}{N^2}\right) \tag{2.17}$$

A.A.Migdal (published in refs. 2,3) and E.Witten [4] were the first who advocated the factorization property in QCD at $N = \infty$. This property looks like the factorization for the sigma model discussed in the previous section. Moreover it is shown below that QCD can be reformulated (like the sigma model) in terms of colorless composite field (loop variables) which becomes classical as $N \to \infty$. As a consequence, the factorization property holds for the average of quantities (like $G(x)$) which can be expressed via this field. From this point of view the factorization property indicates that some kind of such a field exists. A less formal explanation of the factorization property in $N = \infty$ QCD, which treats the large-N limit as an additional statistical averaging, is given in sect. 6.

3. Loop-space approach in QCD

QCD can be entirely reformulated in terms of the colorless composite field

$$\Phi(C) = \frac{1}{N} \operatorname{tr} P \exp ig \oint_C d\xi_\mu A_\mu(\xi) , \quad (3.1)$$

which is called the Wilson loop operator. Here P orders the matrices along the loop C. The fact that QCD can be reformulated via $\Phi(C)$ involves two main points:

i) All the observables are represented via $\Phi(C)$.

ii) Dynamics is entirely reformulated in terms of $\Phi(C)$.

In the given context, i) was first advocated by Wilson[5]. The appropriate formulas for the continuum theory were derived by Makeenko and Migdal [6] . At finite N, observables are expressed via the n-loop averages

$$W_n(C_1, \ldots, C_n) = \langle \Phi(C_1) \ldots \Phi(C_n) \rangle . \quad (3.2)$$

Formulas are presented in ref. 6. Great simplifications occur in these formulas at $N = \infty$, when all the observables are expressed via $W_1(C)$ only. For example the expression for the average of the product of two colorless quark currents is given by

$$\langle \bar{\Psi}(x_1) \Psi(x_1) \bar{\Psi}(x_2) \Psi(x_2) \rangle = \sum_C J(C) W_1(C) . \quad (3.3)$$

Here the sum over paths on the r.h.s. goes over all the paths, C, connecting the points x_1 and x_2, as shown in fig.2.

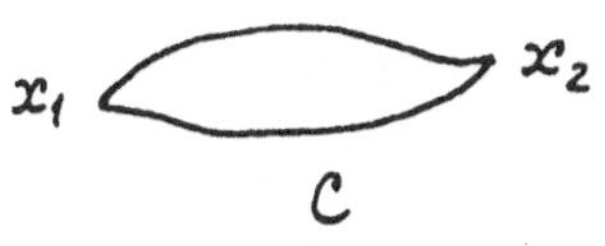

Fig. 2.

The measure $J(C)$ depends on quark quantum numbers only and can be calculated in free theory. For the case of spinor quarks, the measure $J(C)$ reads

$$J(C) = \int \mathcal{D}K \, tr \, P exp \int d\tau \left[K_\mu (\dot{x}_\mu - i\gamma_\mu) - m \right] \qquad (3.4)$$

where the path integration goes over the four-momentum $K_\mu(\tau)$; while τ is a parametrization of the contour $C = \{ x_\mu(\tau) \}$.

Formulas for the averages of the product of several currents look like (3.3). For the three-point Green function $\langle \bar{\Psi}(x_1)\Psi(x_1)\bar{\Psi}(x_2)\Psi(x_2)\bar{\Psi}(x_3)\Psi(x_3) \rangle$ the sum goes over all the paths passing the points x_1, x_2, and x_3 as shown in fig. 3. For the n-point function the sum goes over all the contours passing these n points. Therefore, given $W_1(C)$, any observable can be obtained to the leading order of the 1/N expansion by summing $W_1(C)$ over paths. So the problem is how to find $W_1(C)$.

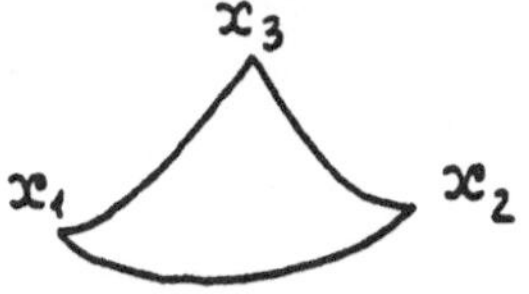

Fig. 3.

In the original formulation, $W_1(C)$ is defined by

$$W_1(C) = \langle \Phi(C) \rangle \qquad (3.5)$$

where the average is given by the path integral (2.12). As mentioned above, $W_1(C)$ can be calculated perturbatively after gauge fixing ($\Phi(C)$ is gauge-invariant).

Despite of the fact that only the planar diagrams survive in the large -N limit, the problem of summing them is as hopeless as that of summing all the diagrams, albeit the number of planar diagrams is much smaller in high orders of perturbation theory. As was shown by Koplik, Neveu, and Nussinov [7], the number of planar diagrams of given order grows only exponentially, while the total number grows factorially. Nevertheless there is no further simplification in ordinary approach as $N \to \infty$, so that a semiclassical nature of the large-N limit is disguised. It is a remarkable fact that great simplifications occure at large N, if dynamics is reformulated in loop space.

Two ways of reformulating dynamics entirely in loop space are to rewrite via $\Phi(C)$:

i) the original path integral;

ii) the original Schwinger equation of motion.

The way i) was advocated by Jevicki and Sakita [8,9], and in a different form by Banks and Yoneya [10]. Jevicki and Sakita tempt to introduce the collective variables, $\Phi(C)$, into the original Yang-Mills path integral in order to entirely re-express it via $\Phi(C)$ (like the path integral for the sigma model can be rewritten via $\mathcal{U}(x)$).

Unfortunately this turned out to be difficult at finite N. The reason is that the collective variables $\Phi(C)$ are not independent at finite N. There are nontrivial relations, say, the Mandelstam relations [11] among them. Only in the large-N limit, these relations are insignificant so that the variables $\Phi(C)$ become independent, and the original path integral can be rewritten as path integral over $\Phi(C)$.

On the contrary the Schwinger-Dyson equations can be rewritten in terms of $\Phi(C)$ at any N. This results in the loop equations briefly discussed in sect. 5. Thus at finite we have an interesting situation when the equations of motion for $\Phi(C)$ in loop space are known but their solution in terms of path integral over $\Phi(C)$ is still not found. Only at $N=\infty$, as showed by Jevicki and Sakita [9], the stationary points, $\Phi_{sp}(C)$, of some effective action for $\Phi(C)$ (which can be calculated by the method of collective variables) [8,9] satisfy the loop equation (5.16).

4. Schwinger-Dyson equation: example of φ^3- theory

Before discussing the loop equations in QCD, I would like to establish the connection between the Schwinger-Dyson equations and the path intergal by a simple example of φ^3-theory. We shall also see how the semiclassical expansion can be obtained by using the Schwinger-Dyson equations.

In the Euclidean formulation of φ^3-theory, the n-point Green functions are defined by

$$G_n(x_1,\dots,x_n) = \langle \varphi(x_1)\dots\varphi(x_n)\rangle . \tag{4.1}$$

The averaging is given by the path integral

$$\langle Q[\varphi]\rangle = \frac{\int \mathcal{D}\varphi\, e^{-\frac{S}{\hbar}} Q[\varphi]}{\int \mathcal{D}\varphi\, e^{-\frac{S}{\hbar}}} \tag{4.2}$$

while the action, S , reads

$$S = \int d^4x \left[\frac{1}{2}(\partial_\mu \varphi)^2 + \frac{\lambda}{6}\varphi^3\right] . \tag{4.3}$$

In eq. (4.2) we explicitly gave a dependence upon Planck's constant, $\hbar$, in order to trace an analogy between $\hbar$ in φ^3-theory and $1/N^2$ in QCD.

Given path integral (4.2), the corresponding equations of motion can be derived as follows. Let us shift the variable of the path integration in the numerator:

$$\varphi(x) \rightarrow \varphi(x) + \delta\varphi(x) . \tag{4.4}$$

Since the integral goes over all the fields (providing those decrease at the Euclidean infinity), the variation of the path integral vanishes. This yields the equation of motion

$$\left\langle Q[\varphi]\frac{\delta S}{\delta\varphi(x)}\right\rangle = \hbar \left\langle \frac{\delta Q[\varphi]}{\delta\varphi(x)}\right\rangle . \tag{4.5}$$

The r.h.s. of eq. (4.5) is called the commutator term because in the operator approach it is due to the equal time commutator of $\varphi(x)$ and $\partial\varphi(x)/\partial x_0$.

The Schwinger-Dyson equations are obtained from (4.5) by substituting

$$Q[\varphi] = \varphi(x_1)\dots\varphi(x_n) . \tag{4.6}$$

Computing the functional derivative by means of the formula

$$\frac{\delta\varphi(x_i)}{\delta\varphi(x)} = \delta^{(4)}(x-x_i) , \tag{4.7}$$

one gets the following Schwinger-Dyson equations

$$\begin{aligned}\Box_{x_1} G_n(x_1,\dots,x_n) &= \frac{\lambda}{2} G_{n+1}(x_1,x_1,\dots,x_n) + \\ &+ \hbar \sum_{i=2}^{n} \delta(x_1-x_i)\, G_{n-2}(x_2,\dots,\not{x}_i,\dots,x_n) .\end{aligned} \tag{4.8}$$

The notation $\not{x}_i$ means that the argument x_i is omitted.

Thus we arrived at the Schwinger-Dyson equations, starting from the definition of the Green functions via the path integral. Alternatively one might start from eq. (4.8) and obtain the Feynman path integral as its solution.

In the semiclassical limit, $\hbar \to 0$, the terms on the r.h.s. of eq. (4.8) which are proportional to $\hbar$ vanish. In this limit one gets the equation

$$\Box_{x_1} G_n(x_1,\dots,x_n) = \frac{\lambda}{2} G_{n+1}(x_1,x_1,\dots,x_n) \tag{4.9}$$

whose solution is of the form

$$G_n(x_1,\dots,x_n) = \varphi_{cl}(x_1)\dots\varphi_{cl}(x_n) \tag{4.10}$$

provided that $\varphi_{cl}(x)$ satisfies

$$\Box\varphi_{cl}(x) + \frac{\lambda}{2}\varphi_{cl}^2(x) = 0 . \tag{4.11}$$

It is easy to understand eq. (4.11) to be nothing else the condition of extremum of the action entering the path integral (4.2). Thus we have reproduced, using the Schwinger-Dyson equations, the well-known fact that path integral is dominated by a classical trajectory as $\hbar \to 0$.

5. The loop equations

As already mentioned, the most important part of the loop-space approach in QCD consists in reformulating dynamics in terms of $\Phi(C)$. More generally, we completely translate dynamics in the language of loop space. Loop space consists of functionals defined on closed loops. The functionals we deal with are of the type of the Wilson loop (3.1). Motion in loop space corresponds to variation of the form of the loop.

An idea to write down a functional equation of motion satisfied by $\Phi(C)$ for infinitesimal variation of the loop was first proposed by Gervais and Neveu, Nambu, Polyakov [12] . For the lattice formulation, loop equations were advocated by Eguchi, Foerster, Weingarten [13] , and Migdal. The continuum version of the loop equations, described in this section, was derived by Makeenko and Migdal [2,3,6,14] .

The loop equations is nothing but the Schwinger-Dyson equations for gluodynamics written in loop space. Those can be derived by virtue of the trick similar to one used in the previous section for derivation of the Schwinger-Dyson equations for the φ^3 -theory, i.e. by shifting the vector-potential, $A_\mu^a(x)$, in the path integral (2.12). An analogue of eq. (4.5) reads (for $\hbar = 1$)

$$\left\langle Q[A] \nabla_\mu F^a_{\mu\nu}(x) \right\rangle = \left\langle \frac{\delta Q[A]}{\delta A^a_\nu(x)} \right\rangle \tag{5.1}$$

where ∇_μ is the covariant derivative in the adjoint representation.

Let us take

$$Q[A] = \frac{1}{N} tr \left[t^a \mathcal{U}(C_{xx}) \right] \tag{5.2}$$

where the matrix $\mathcal{U}(C_{xx})$ equals

$$\mathcal{U}^{ij}(C_{xx}) = \left[P \exp ig \int_{C_{xx}} d\xi_\mu A_\mu \right]. \tag{5.3}$$

Computation of the functional derivative on the r.h.s. of eq. (5.1), with $Q[A]$ given by (5.2), can be performed by the formula

$$\frac{\delta A^b_\mu(y)}{\delta A^a_\nu(x)} = \delta^{ab} \delta_{\mu\nu} \delta(x-y) . \tag{5.4}$$

Using the chain rule as well as the completeness condition (2.6). eq. (5.1) can be rewritten as

$$\left\langle \frac{1}{N} tr P\left[\nabla_\mu F_{\mu\nu}(x) \mathcal{U}(C_{xx}) \right) \right\rangle = \frac{gN}{i} \oint_C dy_\nu \delta(x-y) \left[W_2(C_{xy}, C_{yx}) - \frac{1}{N^2} W_1(C) \right]. \tag{5.5}$$

Here the contours C_{xy} and C_{yx} are two parts of C from x to y and from y to x, respectively (see fig. 4). These contours are closed due to the δ function.

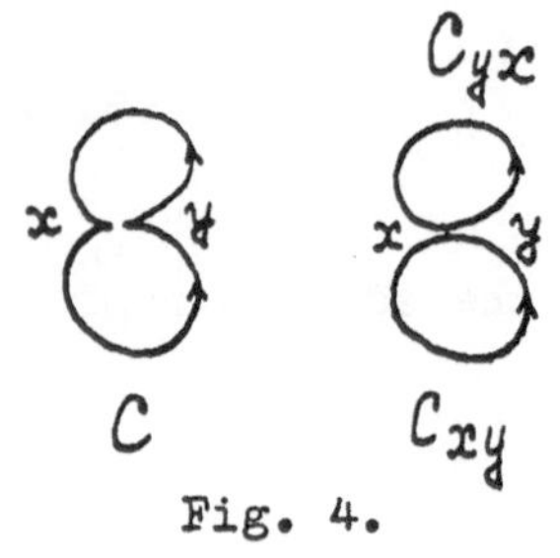

Fig. 4.

Both terms on the r.h.s. of eq. (5.5) are functionals of the type (3.2). However another quantity enters the l.h.s. The crucial point is that the l.h.s. can be rewritten via $W_1(C)$, if operations of differentiating with respect to the path and the area are defined in loop space.

The <u>area derivative</u>, $\delta / \delta \sigma_{\mu\nu}(x)$, is defined by the difference

$$\frac{\delta \Phi(C)}{\delta \sigma_{\mu\nu}(x)} = \frac{1}{\delta \sigma_{\mu\nu}} \left[\Phi\left(\text{(loop with little loop at } x \text{ in } \mu,\nu \text{ plane)} \right) - \Phi\left(\bigcirc \right) \right] . \tag{5.6}$$

Here a little loop, lying in the μ,ν – plane, encloses the area $\delta \sigma_{\mu\nu}$. By definition the area derivative is finite only if the increment on the r.h.s. of eq. (5.6) is proportional to $\delta \sigma_{\mu\nu}$. The finiteness of the area derivative is in fact a nontrivial condition for $\Phi(C)$. The functionals whose area derivative is finite are called those of the Stokes type. An alternative (but equivalent) definition of the functional of the Stokes type is based on the "back-forth" condition

$$\Phi\left(\text{(loop with appendage)} \right) = \Phi\left(\bigcirc \right) . \tag{5.7}$$

Here the loop on the r.h.s. differs from that on the l.h.s. by the insertion of a curve traversed back and forth.

As follows from the representation (3.1), the functional

$\Phi(C)$ satisfies eq. (5.7) and, therefore, $\Phi(C)$ belongs to the Stokes type. The area derivative, $\delta\Phi(C)/\delta\sigma_{\mu\nu}$, can be calculated by choosing the little loop to be the square lying in the μ,ν -plane. The result reads

$$\frac{\delta\Phi(C)}{\delta\sigma_{\mu\nu}(x)} = \frac{ig}{N}\, tr\left[F_{\mu\nu}(x)\,\mathcal{U}(C_{xx})\right]. \tag{5.8}$$

This formula is called the Mandelstam formula.

We need one more operation in loop space, the path derivative. The <u>path derivative</u> is defined for the Stokes functional with a marked point (like the functional on the r.h.s. of eq. (5.8)). The path derivative is defined by the difference

$$\partial^x_\mu\, \Phi(C_{xx}) = \frac{1}{\delta x_\mu}\left[\Phi\left(\bigcirc\!\!\!\!\overset{x}{\longrightarrow}_{\mu}\right) - \Phi\left(\bigcirc\!\!\!\bullet\right)\right] \tag{5.9}$$

where δx_μ is the length of the infinitesimal path added to the point x in the μ -direction. For the Stokes functionals, the path derivative is a well-defined operation since the increment on the r.h.s. of (5.9) does not depend on the form of the path.

Let us now apply the operator ∂^x_μ to eq. (5.8). Due to the definition (5.9), ∂^x_μ acts both on $F_{\mu\nu}(x)$ and on the contour integral (5.3) so that one gets

$$\partial^x_\mu \frac{\delta\Phi(C)}{\delta\sigma_{\mu\nu}(x)} = \frac{ig}{N}\, tr\left[\nabla_\mu F_{\mu\nu}(x)\,\mathcal{U}(C_{xx})\right]. \tag{5.10}$$

Thus we have written the l.h.s. of eq. (5.5) completely in loop space.

Finally eq. (5.5) reads

$$\partial_\mu^x \frac{\delta W_1(C)}{\delta \sigma_{\mu\nu}(x)} = g^2 N \oint_C dy_\nu \delta(x-y)\left[W_2(C_{xy}, C_{yx}) - \frac{1}{N^2} W_1(C)\right]. \quad (5.11)$$

This equation is not closed. Starting from W_1, we obtain another quantity, W_2, so that equation of motion connects the 1-loop average with a 2-loop one. We have met such a situation for φ^3-theory whose chain of the Schwinger-Dyson equations is given by eq. (4.8). It is clear that a similar chain should be derived for the n-loop averages in gluodynamics.

To derive the loop equations satisfied by $W_n(C_1, ..., C_n)$, one has to replace $Q[A]$ given by (5.2) with

$$Q[A] = \frac{1}{N} tr\left[t^a U(C_{1xx})\right] \Phi(C_2) \ldots \Phi(C_n). \quad (5.12)$$

Proceeding as before, we get the following chain of the loop equations

$$\frac{1}{g^2 N} \partial_\mu^x \frac{\delta W_n(C_1, \ldots)}{\delta \sigma_{\mu\nu}(x)} = \oint_{C_1} dy_\nu \delta(x-y)\left[W_{n+1}(C_{xy}, C_{yx}, \ldots) - \right.$$

$$\left. - N^{-2} W_n(C_1, \ldots)\right] + N^{-2} \sum_{j \geq 2} \oint_{C_j} dy_\nu \delta(x-y)\left[W_{n-1}(C_1 + C_j, \ldots, \not{C_j}, \ldots) - W_n(C_1, \ldots)\right] \quad (5.13)$$

Here x belongs to C_1 ; $C_1 + C_j$ stands for the joining of C_1 and C_j ; $\not{C_j}$ means that C_j is omitted.

Eqs. (5.13) look like eqs. (4.8) for φ^3-theory. Moreover the number of colors, N, enters eqs. (5.13) simply as the scalar factor, N^{-2}, similarly to how the Planck's constant, $\hbar$, enters eqs. (4.8). It is the major advantage of the use of loop space. A "semiclassical" expansion of the loop equations (5.13) in $1/N^2$ is straightforward. We will show now that loop equations are greatly simplified to the leading order in $1/N^2$.

For $N=\infty$ ($g^2N\sim 1$) eq. (5.13) is simplified as

$$\partial_\mu^x \frac{\delta W_n(C_1,\ldots)}{\delta\sigma_{\mu\nu}(x)} = g^2N\oint_{C_1} dy_\nu\,\delta(x-y)\,W_{n+1}(C_{xy},C_{yx},\ldots)\,. \tag{5.14}$$

This equation possesses a factorized solution

$$W_n(C_1,\ldots,C_n) = W_1(C_1)\ldots W_1(C_n) \tag{5.15}$$

provided $W_1(C)$ satisfies the "classical" equation

$$\partial_\mu^x \frac{\delta W_1(C)}{\delta\sigma_{\mu\nu}(x)} = g^2N\oint_C dy_\nu\,\delta(x-y)\,W_1(C_{xy})\,W_1(C_{yx})\,. \tag{5.16}$$

An analogy of the formulas (5.14), (5.15), (5.16) with those (4.9), (4.10), (4.11) is evident.

The factorization property (5.15) holds to the leading order in N^{-2} , same as eq. (5.13) (we have seen already in sect. 2 that nonfactorized diagrams are of order $1/N^2$). The factorization means the field $\Phi(C)$ becomes "classical" as $N\to\infty$, i.e. its quantum fluctuations (entropy) are N^2 times smaller than the effective action. The word "classical" is understood here in the sence of semiclassical expansion in $1/N^2$. From the point of view of the original Yang-Mills theory, eq. (5.16) is, of course, quantum. The r.h.s. of eq. (5.16) represents as discussed above the commutator term and, in fact, is proportional to the Plank's constant. It is very nontrivial that $N=\infty$ quantum Yang-Mills theory becomes "classical" after reformulation in loop space.

6. Some comment on the large-N limit

There is another, pure statistical, explanation [6,16] why the large-N limit is a "semiclassical" limit for the collective variables $\Phi(C)$. For this purpose let us consider index space, introduced in sect. 2, where the gauge field $A^a_\mu(x)$ possesses N^2-1 degrees of freedom. The matrix $U^{ij}(C_{xx})$ can be reduced by a gauge transformation to

$$U(C_{xx}) = \Omega(C_{xx})\, diag\left(e^{ig\alpha_1(C)},\dots,e^{ig\alpha_N(C)}\right)\Omega^{+}(C_{xx}). \tag{6.1}$$

Then $\Phi(C)$ reads

$$\Phi(C) = \frac{1}{N}\sum_{j=1}^{N} e^{ig\alpha_j(C)} \quad . \tag{6.2}$$

The phases $\alpha_j(C)$ are gauge-invariant and normalized so that $\alpha_j(C)\sim 1$ as $N\to\infty$. For simplicity we omit below all the indices (including space ones) exept color.

The commutator of Φ's can be estimated using the representation (6.2). Since

$$[\alpha_i,\alpha_j] \propto \delta_{ij} , \tag{6.3}$$

one gets

$$\left[\Phi(C), \Phi(C') \right] \propto g^2 \frac{1}{N} \propto \frac{1}{N^2}, \quad (6.4)$$

i.e. the commutator can be neglected as $N \to \infty$ and the field $\Phi(C)$ becomes classical.

Note the commutator (6.4) is of order $1/N^2$. One factor, $1/N$, is associated with the fact that g enters the definition (6.2) of $\Phi(C)$ while another one has a deep reason. Image the summation over j in eq. (6.2) as some statistical averaging. It is well-known in statistics that such averages weakly fluctuate as $N \to \infty$, so that the dispersion is of order $1/N$. It is the factor that enters the commutator (6.4).

So, we see the factorization to be valid only for the gauge--invariant quantities which involve the averaging of the type (6.3). There is no reason to expect factorization for the gauge invariants which do not involve this averaging, say, for the phases, $\alpha_i(C)$. Moreover the commutator (6.3) ~ 1 and, therefore, $\alpha_i(C)$ fluctuate strongly even at $N = \infty$. Haan [16] was the first who explicitly constructed some example of strongly fluctuating gauge-invariant quantities for the case of the two-matrix model.

Note that $G(x)$ of sect. 2 belongs to the class (6.2). $G(x)$ is related to $\Phi(C)$ by

$$G(x) = \frac{1}{g^2 N} \sum_{\mu,\nu} \left(1 - \Phi(\delta C_{\mu\nu}) \right) / (\delta \sigma_{\mu\nu})^2 \quad (6.5)$$

where $\delta C_{\mu\nu}$ is an infinitesimal contour in the μ,ν-plane (enclosing the area $\delta\sigma_{\mu\nu}$) which passes the point x.

We see that to explane factorization there is no need to assume that $A_\mu^a(x)$ becomes classical as $N \to \infty$, i.e. the path integral over $A_\mu^a(x)$ is saturated by a single field configuration, $A_\mu^a{}^{cl}(x)$ (or by a single gauge orbit to preserve the gauge invariance). Such a field is called the "master field in the strong sence". This type of master field exsits only for the simplest matrix models (one-plaquette, one-matrix etc.). However even in a little bit more complicated model of two matrices there is no master field in the strong sence [16,18]. In the two-dimensional QCD, difficulties associated with such a master field also appear for the self-intesecting contours [19].

Despite there is no arguments in favor of the master field in the strong sence to exist for the four-dimensional QCD, it would be very nice if that exists nonetheless, say approximately, due to some dynamical reasons. The point is that solving eq. (5.16) in loop space is an extremely formidable task (some obtained solutions are discussed in the next sections). It would be much more simple to solve the ordinary classical Yang-Mills equation for $A_\mu^a{}^{cl}(x)$ instead of eq. (5.16) (see ref. 20) as is implied in the approach based on the master field in the strong sence. Attempts [21] to introduce other collective variables instead of $\Phi(C)$ seem to be also due to the complicatedness of the loop-space approach.

Recently the name "master field" has been used basically in a weak sence. This means that "saddle-point" field, $A_\mu^a{}^{cl}(x)$, is allowed to be an operator in some space rather than a function. For the case of this "internal-symmetry" space being loop space, the loop equation (5.16) represents [22] the saddle-point master equation. One more example of the master field in the weak sence in briefly discussed in sect. 8.

7. Solutions of the loop equation: connection with string theory

In this section we shall review some solutions of the continuum loop equation (5.16) which are known at present.

First of all, eq. (5.16) can be solved iteratively in the coupling constant, g^2N , imposing the initial condition

$$W(0) = 1 . \tag{7.1}$$

Here 0 stands for the loop shrank to a point. The iterative solution of eq. (5.16) is unique provided that Euclidean boundary conditions at infinity are imposed. It is shown in refs. 6,14 how the set of diagrams of Faddeev-Popov perturbation theory, including the ghost loops, is reproduced in this way. To perturbatively solve eq. (5.16), it is convenient to utilize the manifestly gauge-invariant diagram technique in loop space [6] .

In two dimensions Kazakov and Kostov [23] have found the exact solution of the loop equation in case of loops with arbitrary self-intersections. An interesting property of this solution (which is significant for the string theory) is the polynomial preexponential factor which arises for self-intersecting loops.

Starting from the loop equation (5.16), Durhuus and Olesen [24] have derived an equation for the spectral density, $\rho_C(\alpha)$, which describes the eigenvalues of the matrix $U^{ij}(C_{xx})$ as $N \to \infty$. The two-dimensional solution [25] of this equation turned out to be similar to $\rho_C(\alpha)$ found by Gross and Witten [26] in the lattice case: for small loops $\rho_C(\alpha)$ has support only in the interval $|\alpha| \leqslant \alpha_C < \pi$ while for large loops the distribution is almost uniform in the interval $[-\pi, \pi]$.

In four dimensions Makeenko and Migdal [6,27] showed that the loop equation (5.6), reduced to a slightly different (bootstrap) form, possesses for large smooth loops a self-consistent solution

$$g^2 N W(C) \underset{C\to\infty}{\longrightarrow} Z\, e^{-\rho\cdot \mathrm{Perimeter}(C)} \sum_{S:\partial S = C} e^{-K\cdot \mathrm{Area}(S)} \tag{7.2}$$

where the sum goes over the surfaces enclosed by C. With the exponential accuracy, (7.2) reproduces the area law

$$W(C) \propto e^{-K\cdot \mathrm{Area}(S_{min})} \tag{7.3}$$

where S_{min} is the minimal surface. On the preexponential level there arise two relations between K, ρ and Z entering (7.2). The unknown constant should be expressed via $g^2 N$ by joining the self-consistent solution (7.2) with the perturbation theory for intermediate loops. This procedure resembles the bootstrap approach in the theory of phase transitions.

One of the most interesting results obtained with the aid of the loop equations is a connection between manycolor QCD and the relativistic string theory. 't Hooft [1] was the first who showed that only diagrams planar in index space dominate as $N \to \infty$, which have exactly the topology of some free string. The type of the string (Nambu-Goto or Ramond-Neveu-Schwarz string, etc.) should be determined by the dynamics of manycolor QCD.

The idea of refs. 12,28 was to show that $W_1(C)$ approximately satisfies an equation similar to the equation of the Nambu string. For large N this question can be studied using the loop equation. Eq. (5.16) itself does not resemble the equation of free Nambu string. Eq. (5.6) is nonlinear and should be interpreted, therefore, as equation of an interacting string.

Substituting the Ansatz (7.2), corresponding to the Nambu string, into the loop equation shows that (7.2) is a solution only for the asymptotically large loops. For the intermediate loops, the Ansatz (7.2) does not satisfy the loop equation. The reason consists, roughly speaking, in the fact that application of $\partial_\mu^x \delta/\delta\sigma_{\mu\nu}(x)$ to the Ansatz (7.2) does not yield the factorized expression entering the r.h.s. of eq.(5.16) The factorization holds only asymptotically (e.g. for the Ansatz (7.3) it is valid due to the additiveness of the minimal area).

It is not surprising, in fact, that manycolor QCD is not reduced to the free Nambu string. Even within the strong coupling expansion of the lattice formulation of QCD, the free Nambu string arises only to lowest orders. Weingarten [29] explicitly constructed the non-planar surfaces which contribute to the strong coupling expansion of $W_1(C)$ at $N = \infty$.

Thus, in order that string Ansatz of the type (7.2) would

satisfy the loop equation for the loops of arbitrary size, that should be modified. The only such modification known at present is the fermion string discovered by Migdal [30]. The fermion string differs from the Nambu string by the fact that two dimensional elementary fermions (elves) live on the surface S. The Ansatz for $W_1(C)$ reads

$$W_1(C) = \sum_{S:\partial S = C} \int \mathcal{D}\psi \, e^{-\int d^2\xi \left[\bar{\psi}\sigma_\kappa \partial_\kappa \psi + \bar{\psi}\psi m^4 \sqrt{g}\right]} \tag{7.4}$$

where the world sheet of the string is parametrized by the coordinates ξ_1 and ξ_2 for which the metric, $g_{\kappa m}$, is conformal, and m stands for the elf mass.

Elves were introduced in order to guarantee the factorization which holds now due to some remarkable properties of two-dimensional fermions. For large loops, the internal fermionic structure becomes frozen so that we recover the empty string (7.2). However, for small loops, elves are necessary for asymptotic freedom.

I have asked recently A.A.Migdal about the present status of the elfin theory, and he said me that Ansatz (7.4) satisfies the loop equation only for the Majorana field ψ (but not for two complex fields as in ref. 30). There are two advantages of the Majorana elves. The first one is that boundary conditions for ψ can be deduced from the Lagrangian. The second one is that (7.4) is the most general action in case of real ψ. For instance the term responsible for the four-fermion interaction, which is possible for the complex ψ, cannot be written for the real ψ . Therefore one might expect the theory of the Majorana elves to be renormalizable.

After some string solution of the loop equation would be found, there arises the problem of solving the string theory, say, that of calculating the spectrum, physical amplitudes etc. Unfortunately, the quantum string theory is not constructed at present. Some recent advance inspired by Polyakov [31] is discussed in the Lectures by Professor H.B.Nielsen given at this School.

The main attention in constructing quantum string theory is naturally paid to the simplest version of the string, the Nambu string, albeit that does not realize in QCD. To my mind such a situation looks like the situation in quantum field theory of fifties when the methods we use now in QCD were developed for more simple (unphysical) φ^4 -theory before QCD had been invented. Maybe the Nambu string will play a role with respect to the elfin theory, same as φ^4 -theory played with respect to QCD.

8. Loop equations on the lattice: Eguchi-Kawai solution

Many interesting results are obtained for the lattice analogue of the continuum loop equation (5.16). The lattice provides, as usual, the proper regularization . The lattice loop equation were first advocated in ref. 13.

Loop equations on the lattice can be derived just as in the continuum (see sect. 5). The lattice analogue of the vector-potential is the matrix $\mathcal{U}_{x,\mu}$, assigned to links, which is an element of the group SU(N). The loop product, $\mathcal{U}(C_{xx})$, is defined as the product of $U_{x,\mu}$ along the contour C_{xx} composed from the links of the lattice. The lattice action is built up from the matrices

$$U(\partial p_{\mu\nu}) = U_{x,\mu} U_{x+\hat{\mu},\nu} U^{+}_{x+\hat{\nu},\mu} U^{+}_{x,\nu} \tag{8.1}$$

where $\partial p_{\mu\nu}$ stands for the boundary of the plaquette $p_{\mu\nu}$ which lies in the μ, ν -plane. The Wilson action reads

$$S = \sum_{p} \left(1 - \mathrm{Re}\, \frac{1}{N} \mathrm{tr}\, U(\partial p) \right) . \tag{8.2}$$

The quantum averaging is defined by the path integral over $U_{x,\mu}$

$$\langle Q[U] \rangle = \frac{\int \prod_{x,\mu} dU_{x,\mu}\, e^{-\beta S}\, Q[U]}{\int \prod_{x,\mu} dU_{x,\mu}\, e^{-\beta S}} \tag{8.3}$$

where $dU_{x,\mu}$ stands for the Haar measure on the SU(N).

To derive equation of motion, let us shift $U_{x,\mu}$ in the path integral (8.3) by

$$U_{x,\mu} \rightarrow (1 + i\,\mathcal{E}_{x,\mu})\, U_{x,\mu} . \tag{8.4}$$

Here $\mathcal{E}_{x,\mu}$ is an infinitesimal Hermitian matrix. Performing the variation (8.4) on each of four links coming from the point in the direction ν = 1,...,4 and repeating the substitutions of sect. 5 for the lattice case, we get in the large-N limit the following lattice loop equation

$$\frac{\beta}{2N^2}\sum_{\mu}\left[W_1(C+\partial p_{\mu\nu})-W_1(C-\partial p_{\mu\nu})\right]=\sum_{\ell\in C}\tau_\nu(\ell)\,\delta_{xy}\,W_1(C_{xy})W_1(C_{yx}). \quad (8.5)$$

Here the contours $C+\partial p_{\mu\nu}$ and $C-\partial p_{\mu\nu}$ are obtained from C_{xx} by adding $\partial p_{\mu\nu}$ at the point x ($-\partial p_{\mu\nu}$ means that orientation of the boundary is opposite). These contours are displayed in fig.5. The sum on the r.h.s. of eq. (8.5) goes over all the links belonging to the contour C. $\tau_\nu(\ell)$ = 0; ±1 stands for the projection of the (oriented) link $\ell\in C$ on the axis ν ($\tau_\nu(\ell)=+1$ for the parallel directions, $\tau_\nu(\ell)=-1$ for the antiparallel ones, and $\tau_\nu(\ell)=0$ for the perpendicular case). The point y is defined as the beginning of the link ℓ , if that has a positive direction, or as the end, if the direction of the link ℓ is negative. Such an asymmetry is due to the fact that we have derived the loop equation resulting from the left-shift (8.4) of $U_{x,\mu}$. The Kroneker symbol guarantees that C_{yx} and C_{xy} are always closed (see fig. 4).

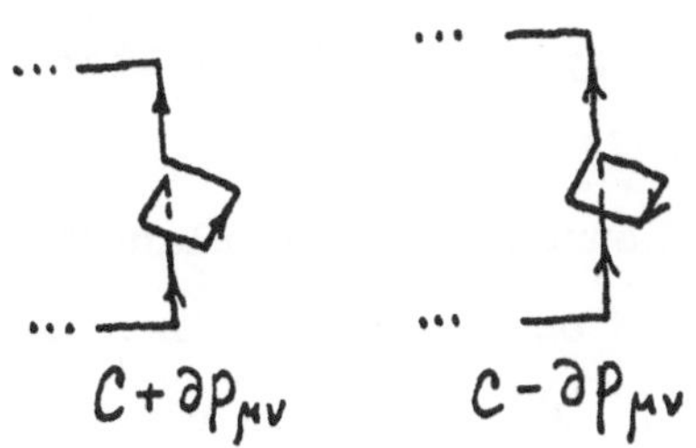

Fig. 5.

Just as in sect. 5, one can derive the chain of lattice loop equations for W_n and show the factorization property (5.15) to be valid on the lattice with the accuracy $O(1/N^2)$ not only to each order of the strong or weak coupling expansions but exactly.

The lattice loop equation can be solved by iterations in N^2/β *) imposing the initial condition (7.1) as well as the Euclidean boundary conditions at infinity. As a result the weak coupling expansion for the loop averages is recovered in this way. On the contrary the strong coupling expansion can be reproduced by solving eq. (8.5) iteratively in β/N^2. This can be done as follows.

Let us substitute in eq. (8.5) the boundary of the plaquette, ∂p, instead of C. Then δ_{xy} on the r.h.s. in non-vanishing only for y coinsiding with x. Rewrite eq. (8.5) as

$$W(\partial p) = \frac{\beta}{2N^2} \sum_{\mu} \left[W(\partial p + \partial p_{\mu\nu}) - W(\partial p - \partial p_{\mu\nu}) \right] . \tag{8.6}$$

One of the terms on the r.h.s. reads

$$W\left[\, \square \, \right] = 1 . \tag{8.7}$$

(Eqs. (5.7) and (7.1) are used.). Thus to the leading order in β/N^2 we get

$$W(\partial p) = \beta/2N^2 . \tag{8.8}$$

Other terms of the r.h.s. of eq. (8.6) are of the next order in β/N^2.

The lattice loop equation (8.5) was solved exactly in two dimensions. For the one-plaquette quantities this was done by

---

*) Let us recall that β is related to g^2 by $g^2 = 2N/\beta$

Paffuti and Rossi, and by Friedan [32], who reproduced the results by Gross and Witten [26] in a more simple way. For contours with arbitrary self-intersections, the solution of two-dimensional lattice loop equation was found by Kazakov and Kostov [33]. For some models, lattice loop equations were solved in ref. 34.

Among the other known solutions of the loop equations, a special role is played by that of Eguchi and Kawai [35] valid in any dimensions. For this solution, all the matrices $U_{x,\mu}$ depend only on the direction μ rather than on the space-point x. In other words, the following equality holds

$$\langle Q[U_{x,\mu}]\rangle = \langle Q[U_\mu]\rangle_{EK} \tag{8.9}$$

where the average (8.3) for the Wilson theory enters the l.h.s. while the r.h.s. is the corresponding average for the Eguchi-Kawai (EK) model which is defined by

$$\langle Q[U_\mu]\rangle_{EK} = \frac{\int \prod_{\nu=1}^{d} dU_\nu \, e^{-\frac{\beta}{2}\sum_{\mu\neq\nu}\left(1-\frac{1}{N}\operatorname{tr} U_\mu U_\nu U_\mu^+ U_\nu^+\right)} Q[U_\mu]}{\int \prod_{\nu=1}^{d} dU_\nu \, e^{-\frac{\beta}{2}\sum_{\mu\neq\nu}\left(1-\frac{1}{N}\operatorname{tr} U_\mu U_\nu U_\mu^+ U_\nu^+\right)}} \tag{8.10}$$

To prove let us derive the loop equation for the EK model, performing in (8.10) the shift

$$U_\nu \rightarrow (1 + i\,\varepsilon_\nu)\, U_\nu \,. \tag{8.11}$$

Proceeding as before we find that the l.h.s. of the loop equation for the EK model exactly coinsides with that of eq. (8.5)

However a differences arises, generally speaking, on the r.h.s.'s, since the shift (8.11) varies the $\mathcal{U}$ -matrices on all links of C_{xx} which have the ν -direction while for the shift (8.4) those are varied on the link coming from the point x only. Owing to this fact, δ_{xy} does not appear on the r.h.s. of the loop equations for the EK model which, therefore, involve superfluous terms corresponding to open contours C_{xy} .

The key point in proving the equivalence between the EK and Wilson theories is that the EK action (see (8.10)) possesses $[U(1)]^4$ invariance

$$\mathcal{U}_\mu \to e^{i\alpha_\mu} \mathcal{U}_\mu \tag{8.12}$$

where α_μ is a number rather than a matrix. If this invariance was not broken spontaneously, then those superfluous terms, which are not invariant under (8.12), would vanish so that the loop equation for the EK model would coinside with eq. (8.5).

The invariance (8.12) is not broken, indeed, in the strong coupling expansion. However as was first shown by Bhanot, Heller and Neuberger [36], $[U(1)]^4$ symmetry is spontaneously broken at large β/N^2 . Therefore the equality (8.9) is not valid at large β/N^2 (and, consequently, in the continuum). In ref. 36 it was also discovered the quenched version of the EK model which is equivalent to the Wilson lattice gauge theory in the weak coupling regime as well [37] .

The quenched EK model can be studied by the Monte Carlo technique. The obtain results is refferred to the manycolor

QCD. The present methods allow to perform the Monte Carlo simulations up to the group $U(20)$ [38].

Another interesting method to numerically study the quenched EK model was advocated by Greensite and Halpern [39] and by Migdal. It is based on (numerical) solving the Langevin equation. Its solution is, in general, a functional of the noise (i.e. an operator in corresponding Hilbert space). The averaging with the Gaussian weight should then be performed over the noise. However the noise is quenched at $N=\infty$ as well, so that master field can be obtained by simply substituting this quenched noise into one of the (equilibrium) solutions of the Langevin equation. Thus the quenched master field is in fact the master field of $N=\infty$ QCD in the weak sence as defined in sect. 6.

Acknowledgements

I would like to thank the Organizers of this School and especially Claus Montonen and Risto Raitio for their warm hospitality.

R e f e r e n c e s

1. G.'t Hooft Nucl.Phys. B72(1974) 461
2. Yu.M.Makeenko, A.A.Migdal Phys.Lett. 88B(1979) 135; 89B(1980) 437(E)
3. A.A.Migdal Ann.Phys 126(1980) 279
4. E.Witten 1979 Cargese Lecture, in Recent developments in gauge theories, ed. G.'t Hooft (Plenum 1980)
5. K.Wilson Phys.Rev. D10(1974) 2445
6. Yu.M.Makeenko, A.A.Migdal Nucl.Phys. B188(1981)269
7. J.Koplik, A.Neveu, S.Nussinov Nucl.Phys. B123(1977) 109
8. B.Sakita Phys.Rev. D21(1980)1067
   A.Jevicki, B.Sakita Nucl.Phys. B165(1980) 511; Phys.Rev. D22(1980) 467
9. A.Jevicki, B.Sakita Nucl.Phys. B185(1981) 89
10. T.Banks Phys.Lett. 89B(1980) 369
    T.Yoneya Nucl.Phys. B183(1981) 471
11. S.Mandelstam Phys.Rev. D19(1979) 2391
    R.Giles Phys.Rev. D24(1981) 2160
    G.A.Christos Phys.Lett. 110B(1982) 471
12. J.L.Gervais, A.Neveu Phys.Lett. 80B(1979) 255
    Y.Nambu Phys.Lett. 80B(1979) 372
    A.M.Polyakov Phys.Lett. 82B(1979) 247
13. T.Eguchi Phys.Lett. 87B(1979) 91
    D.Foerster Phys.Lett. 87B(1979) 87; Nucl.Phys. B170(1980) 107
    D.Weingarten Phys.Lett. 87B(1979) 97; 90B(1980) 277
14. Yu.M.Makeenko Yad.Fiz. 33(1981) 526

15. A.M.Polyakov Nucl.Phys. B164(1980) 171

16. O.Haan Phys.Lett. 106B(1981) 207

17. S.Coleman "1/N", Erice lecture notes (1979)

18. C.Itzykson "The planar approximation", in Proc. of the Bad-Honnef meeting, Bonn(1980)

19. R.Crewther in Proc. of V-th Seminar on High Energy Physics, Serpukhov (1982)

20. T.Banks, A.Casher Nucl.Phys. B167(1980) 215

21. A.A.Slavnov Phys.Lett. 112B(1982) 154
E.A.Ivanov, S.O.Krivonos JINR Preprint E2-81-824(1981)

22. J.L.Gervais, A.Neveu Nucl.Phys. B192(1981) 463

23. V.A.Kazakov, I.K.Kostov Nucl.Phys. B176 (1980) 199

24. B.Durhuus, P.Olesen Nucl.Phys. B184(1981) 406

25. B.Durhuus, P.Olesen Nucl.Phys. B184(1981) 461

26. D.Gross, E.Witten Phys.Rev. D21(1980) 446
S.Wadia Chicago Preprint EFI 79/44 (1979)

27. Yu.M.Makeenko, A.A.Migdal Phys.Lett. 97B(1980) 253

28. J.L.Gervais, A.Neveu Nucl.Phys. B153 (1979) 445
E.Corrigan, B.Hasslacher Phys.Lett. 81B (1979) 181
N.V.Borisov, M.V.Ioffe, M.I.Eides Yad.Fiz. 29(1979) 1421
I.Ya.Aref'yeva Lett.Math.Phys. 4(1979) 270
L.Durand, E.Mendel Phys.Lett. 85B(1979) 241

29. D.Weingarten Phys.Lett. 90B(1980) 285

30. A.A.Migdal Nucl.Phys. B189(1981) 253

31. A.M.Polyakov Phys.Lett. 103B(1981) 207; 211

32. G.Paffuti, P.Rossi Phys.Lett. 92B(1980) 321
D.Friedan Comm.Math.Phya. 78(1981) 353

33. V.A.Kazakov, I.K.Kostov Phys.Lett. 105B(1981) 453

34. R.Brower, P.Rossi, C.Tan Phys.Rev. D23(1981) 953
T.Chen, C.Tan Providence preprint, BROWN-HET-452(1981)

35. T.Eguchi, H.Kawai Phys.Rev.Lett. 48(1982) 1063

36. G.Bhanot, U.Heller, H.Neuberger Phys.Lett. 113B(1982) 47

37. G.Parisi Phys.Lett. 112B(1982) 463
G.Parisi, Zhang Yi-cheng Phys.Lett. 114B(1982) 319
D.Gross, Y.Kitazawa Princeton Univ.Preprint, PRE 2564 (1982)
S.Das, S.Wadia Chicago Preprint EFI 82/15(1982)
V.A.Kazakov, A.A.Migdal Chernogolovka Preprint (1982)

38. G.Bhanot, U.Heller, H.Neuberger Phys.Lett. 115B(1982) 237
M.Okawa Phys.Rev.Lett. 49(1982) 353; 705

39. J.Greensite, M.B.Halpern Berkeley Preprint UCB-PTH-82/14 (1982)

EXPERIMENTING WITH QCD

Laurence Jacobs

Institute for Theoretical Physics
University of California
Santa Barbara, CA 93106

ABSTRACT

A survey of some of the current predictions of QCD obtained by Monte Carlo methods is presented.

Contents:

I. Introduction

Most of our current understanding of the physics of elementary particles is based on the belief that the fundamental interactions are described by a local gauge theory. This is a rather well founded belief considering the remarkable success of theoretical physics in the description of weak and electromagnetic interactions. We believe that the strong interaction is also described by a local gauge theory, Quantum Chromodynamics. However, the predictive power of QCD has been limited due to the essential role of non-perturbative aspects of the theory. Nonetheless, because of asymptotic freedom [1], there are instances, like deep inelastic scattering, where the models agree extremely well with experiment. It is therefore clear that the confrontation of theory with experiment requires the development of techniques which do not rely on a perturbative expansion. The development of such techniques has been hampered by the fact that the severe ultraviolet divergences of quantum field theory require the introduction of cutoffs which generally require a perturbative implementation. A major advance in this direction was accomplished by Wilson [2] and Polyakov [3] with the invention of a gauge-invariant, non-perturbative regulator. In this scheme, space-time is replaced by a discrete lattice of points in four dimensions, thus introducing an ultraviolet cutoff. Apart from the fact that this regularization serves to define the theory correctly, an added bonus comes from the fact that a field theory defined on a lattice is formally equivalent to a spin system for which the powerful methods of statistical mechanics can be fruitfully applied. One of these techniques, the time-honored method of Monte Carlo simulations [4] has yielded invaluable information in condensed-matter physics. This method can be applied in the study of gauge theories [5,6] and over the past three years has proven to be a very valuable tool.

In the couse of these lectures I will present a survey of some of the main results which have emerged from the application of Monte Carlo methods in non-Abelian gauge theories. A more complete and detailed description of these and other results will soon appear in a review written in collaboration with Mike Creutz and Claudio Rebbi [7] which will include a fairly extensive guide to the existing literature on this subject. Also, a number of reviews on lattice technology exist in the literature [8] which can be consulted by the interested reader. These lectures are not self-contained, but should become reasonably so when complemented by some of the other lectures in this school [9].

It is clear that the application of these methods in quantum field theory is still in its early stages of development and many areas are either only roughly understood or not understood at all, but the results obtained thus far are most encouraging.

Although some insight into the complex nature of high-energy physics has emerged from these numerical methods, a truly satisfactory state of affairs will have to await the development of new analytic tools. In a few cases, though, rigorous analytical confirmation of Monte Carlo results has already been achieved. Notable is the case of the existence of a U(1) deconfining transition in four dimensions and its absence in three; first seen numerically [10] and later rigorously established [11].

One of the thorniest areas in the field, about which I will have something to say here, concerns the existence and properties of a continuum limit: the removal of the lattice cutoff. The continuum limit of a lattice theory cannot be investigated by simply taking the lattice spacing to zero, since reducing the lattice spacing corresponds to a scale transformation which requires the simultaneous readjustment of the couplings. Indeed, for physical correlation lengths (and other dimensionful observables) to remain finite as the lattice spacing approaches zero, these should be defined over an increasing number of lattice spacings. In the continuum limit, physical correlation lengths will thus have to extend over an infinity of lattice spacings. This is the definition of a continuous phase transition. Therefore, a necessary requirement for a continuum limit to exist is that the bare lattice couplings renormalize to a critical value as the spacing is taken to zero. The existence and properties of the possible critical points of a lattice theory is therefore of central importance. Even when a lattice model exhibits the right critical properties, the possibility of extracting continuum physics from the model requires a non-trivial assumption about the validity of universality. Recall that the only input available for deciding what particular lattice model to use is that the lattice action be gauge-invariant and that it reduce in the classical continuum limit to the usual Yang-Mills action. What the assumption of universality entails is, basically, that any model with the above two properties should lead to the same values of physical quantities in the continuum as long as the lattice spacing is small enough for the given model to be in its scaling regime. This is a highly non-trivial assumption and should be the focus of intensive future research. Universality is also, at present, a necessary assumption, since, without it, we would have no way of knowing what the correct lattice action might be. Fortunately, all tests of universality performed thus far seem to support its validity and all apparent violations [12] have been resolved [13] (at least in principle).

At present, all analyses seem to support the coexistence of confinement and asymptotic freedom [14]; measurements of the string tension have produced results which are not inconsistent with present-day phenomenology [15]; the conjectured deconfinement temperature [16] has been measured [17]; masses of glueball states have been estimated [18]; and preliminary results for the masses of light hadrons have been obtained [19]. This is unarguably a very impressive track record for the Monte Carlo technique. Although we cannot say that we now understand the fundamental interactions, what these methods have allowed is the possibility of comparing experiments carried out with the theory with experiments performed on physical systems. So far the theory seems to have fared well.

After a brief overview of the theory behind the extraction of continuum physics from lattice calculations, given in Section II, I present the basics of the Monte Carlo technique in Section III. Section IV contains a discussion of some of the important observables which have been obtained for the SU(2) and SU(3) pure gauge theories. Some further conclusions are given in Section V.

II. Probing the continuum

Any field theory can be written on a lattice by simply substituting partial derivatives by finite-difference operators. The result of this discretization, when applied to gauge theories has two major disadvantages: the resulting model is only gauge-invariant in the continuum limit and the integral over gauges is not compact. These unattractive features are solved in a natural way [2,3] by associating a finite element of the gauge group, U_{ij}, to each <u>link</u> of the lattice. The gauge potentials emerge as the infinitesimal group generators when the lattice spacing is taken to zero. A further advantage of this formulation comes from the fact that it allows the definition of a gauge theory with a discrete symmetry group. For technical reasons, which I discuss later, this possibility is quite relevant.

The classical continuum limit is obtained by writing the $U_{ij}=(U_{ji})^{-1}$ in terms of the gauge potential $A\mu(x)$:

$$U_{ij} = \exp\left[-ig_0(x^i-x^j)_\mu \; A_\mu\left(\frac{x^i+x^j}{2}\right)\right] \tag{2.1}$$

where g_0 is a (bare) coupling constant and x^i, x^j are the coordinates of two neighboring sites in the lattice. A gauge-invariant action can be constructed in terms of closed loops. The simplest case is when these loops are the elementary squares (plaquettes) of the lattice. The Wilson action is then written as

$$S(U) = \sum_p S_p \tag{2.2}$$

with

$$S_p = 1 - \frac{1}{N} \mathrm{Tr}\;(U_{ij}U_{jk}U_{kl}U_{li}) \tag{2.3}$$

where p stands for a plaquette, N is the dimension of the U matrices and i,j,k,l label the corners of p. If A_μ is smooth, it can be expanded about the center of p, leading to

$$S_p = \frac{g_0^2}{4N}\, a^4 \;\mathrm{Tr}\;(F_{\mu\nu}F_{\mu\nu} + F_{\nu\mu}F_{\nu\mu}) \tag{2.4}$$

(no sum over μ,ν) where a is the lattice spacing and $F_{\mu\nu}$ is the usual

Yang-Mills field strength. In the limit $a\to 0$, $a^4 \sum_p \to \int d^4x$ and $S(U)\to \beta\int d^4x\frac{1}{2}\mathrm{Tr}(F_{\mu\nu}F_{\mu\nu})$, $\beta = 2N/g_0^2$. So the naive, classical continuum limit of the quantum theory defined by the path integral

$$Z = \int \prod_{ij} dU_{ij}\, e^{-\beta S(U)} \tag{2.5}$$

with $S(U)$ given by (2.2) is the usual classical Yang-Mills theory. As explained in the Introduction, though, the quantum continuum limit of (2.5) is more subtle. Observables in the quantum theory are defined as functional averages with the measure of (2.5). Thus, the expectation value of an operator $\mathcal{O}(A_\mu)$ has the (finite) lattice representation

$$\langle \mathcal{O} \rangle = \sum_{\{U_{ij}\}} \mathcal{O}(U_{ij})\, e^{-S(U_{ij})}/Z \tag{2.6}$$

where $\sum$ is an ordinary sum for a discrete group or a multiple invariant integral for a Lie group.

If $\langle \mathcal{O} \rangle$ is a dimensionful observable, say, a correlation length ℓ, the arguments of the preceeding section imply that it is given by

$$\ell = a\lambda(g_0) \tag{2.7}$$

where the complexities of the quantum theory are embodied in the dimensionless function $\lambda(g_0)$. For a continuum limit to exist, it must be possible to define $g_0 = g_0(a)$ in such a way that (2.7) remains constant as $a\to 0$. The critical point at which $\lambda(g_0)$ must diverge to satisfy this requirement must be a scaling critical point, since it must be true that a single relationship $g_0 = g_0(a)$ suffices to define all observables.

Given a particular renormalization scheme (the choice of physical parameter to be held fixed as $a\to 0$), the functional relationship between g_0 and a defines a Gell-Mann-Low function $\gamma(g_0)$ as

$$\gamma(g_0) = a\,\frac{\partial g_0(a)}{\partial a}. \tag{2.8}$$

In a non-Abelian gauge theory, it can be shown that $g_0=0$ is a scaling critical point. In fact, $g_0(a)$ satisfies

$$\gamma(g_0) = \gamma_0 g_0^3 + \gamma_1 g_0^5 + \mathcal{O}(g^7) \tag{2.9}$$

where, for SU(N), the first two coefficients are given by [20]

$$\gamma_0 = \frac{11}{3}\left(\frac{N}{16\pi^2}\right)$$
$$\gamma_1 = \frac{34}{3}\left(\frac{N}{16\pi^2}\right)^2 \qquad (2.10)$$

and are independent of the renormalization scheme. Eq. (2.9) can be integrated, leading to

$$\frac{1}{g_0^2(a)} = \gamma_0 \ln\left(\frac{1}{a^2\Lambda^2}\right) + \frac{\gamma_1}{\gamma_0} \ln\left[\ln\left(\frac{1}{a^2\Lambda^2}\right)\right] + O(g_0^2) \qquad (2.11)$$

where Λ is an undetermined integration constant which fixes the scale. The above equation is the statement of asymptotic freedom: the bare coupling g_0 vanishes logarithmically with the lattice spacing. For small a, we can write

$$a\Lambda \equiv \eta(g_0) = (g_0^2\gamma_0)^{\gamma_1/2\gamma_0^2} \exp\left(-\frac{1}{2\gamma_0 g_0^2}\right) \qquad (2.12)$$

If $g_0=0$ is a scaling critical point, the value of an observable O with mass dimension m^D must be given by the expression $a^{-D}\Omega(g_0)$ where, as $g_0 \to 0$, $\Omega(g_0) \to K\, \eta^D(g_0)$ with constant K. If this behavior is satisfied, the observable is given in terms of Λ by the expression $K\Lambda^D$. Because $\eta(g_0)$ given by (2.12) has an essential singularity at $g_0=0$, the value of any such observable cannot be seen in any order of perturbation theory.

III. Review of the method

Even for a very small lattice with a small discrete gauge group, the number of configurations entering an expression like (2.6) is exorbitant, making a direct numerical evaluation impossible. However, because configurations are weighted exponentially, it is true, in practice, that only a relatively small subset of states contributes significantly to the sums. The Monte Carlo method [21] is a stochastic, importance sampling technique for constructing such sets of states.

If a sequence of configurations $\{C_i\}$ is constructed such that the probability density of finding any one configuration C_0 in the sequence is proportional to exp $[-S(C_0)]$, the value of an observable, given by (2.6) can be approximated by an average over states in this sequence,

$$\langle O \rangle \approx \frac{1}{M} \sum_{n=1}^{M} O(C_n). \tag{3.1}$$

Given an arbitrary initial configuration C_i, the above equilibrium distribution can be obtained by the application of a transition operator $T(C,C')$ which satisfies

$$T(C,C') > 0 \quad \forall\ C,C' \tag{3.2}$$

$$\sum_{C'} T(C,C') = 1 \tag{3.3}$$

$$e^{-S(C)}T(C,C') = e^{-S(C')}T(C',C) \tag{3.4}$$

It is easy to see that, if the above conditions are met, an iteration of this transformation will eventually produce the desired sequence; that is, given any C_i, repeated application of T will eventually produce configurations C distributed according to exp $[-S(C)]$. To see this, first note that exp $[-S(C)]$ is an eigenvector of T:

$$\sum_C e^{-S(C)}T(C,C') \overset{(3.4)}{=} \sum_C e^{-S(C')}T(C',C) \overset{(3.3)}{=} e^{-S(C')} \tag{3.5}$$

To see that any initial configuration approaches this eigenvector define first the distance $\mu(\varepsilon,\varepsilon')$ between two ensembles of configurations ε and ε' such that the probability of C in $\varepsilon(\varepsilon')$ is $P(C)$ $[P'(C)]$,

$$\mu(\varepsilon,\varepsilon') = \sum_{C} |P(C) - P'(C)| \tag{3.6}$$

If ε' is obtained from ε by the application of T, that is, if

$$P'(C) = \sum_{C'} T(C,C')P(C') \tag{3.7}$$

and calling ε_{eq} an ensemble of configurations distributed according to $\exp[-S(C)]$, then using (3.7)

$$\begin{aligned} \mu(\varepsilon',\varepsilon_{eq}) &= \sum_{C} \left| \sum_{C} T(C,C') \left[P(C') - P_{eq}(C')\right] \right| \\ &\leq \sum_{C'} |P(C') - P_{eq}(C')| \\ &= \mu(\varepsilon,\varepsilon_{eq}) \end{aligned} \tag{3.8}$$

because of the positivity and conservation properties of T, (3.2) and (3.3).

The conditions above do not determine T uniquely and many choices are possible. Two popular algorithms are the following:

(a) The Metropolis Method [21]

There are two steps involved in the transition from U to U'. One first selects a new $\overline{U} \neq U$ randomly according to some distribution $T_0(U,\overline{U}) = T_0(\overline{U},U)$. Then one computes the action for U replaced by $\overline{U}$ and all other links fixed, S(U). If $\Delta S = S(\overline{U}) - S(U) \leq 0$, $U' = U$ with unit probability. Otherwise the change is accepted or rejected according to whether $\exp(-\Delta S)$ is larger or smaller than a random number R uniformly distributed in the unit interval. It is trivial to see that this algorighm satisfies (3.2-4).

(b) The Heat-Bath Method [22]

The trial value U' is selected from all possible values with a probability distribution proportional to $\exp[-S(U')]$, satisfying (3.4) by definition.

It can be shown that the heat-bath algorithm is equivalent to an iterated application of the Metropolis algorithm for each link holding the others fixed. It is therefore apparent that the heat-bath method converges faster than the Metropolis algorithm. In a number of special cases this is true. However, in other cases, the generation of trial variables according to the distribution exp [-S] is so time-consuming that it offsets the gain in relaxation time. In most cases, though, improved Metropolis algorithms exist which are almost as efficient as the heat-bath method but are computationally simpler. One such improvement is achieved by performing several Metropolis hits on a given link before proceeding to the next one. This procedure increases the probability of finding an acceptable upgrade for the link in question. The main advantage of this modification comes from the fact that an important fraction of the time needed to compute ΔS involves only neighboring links and this computation does not have to be repeated for every hit. This modified Metropolis algorithm interpolates between (a) and (b) above; it corresponds to (a) for a single hit per link, and tends to (b) in the large hit/link limit. It is usually possible to find the number of hits per link which optimizes this procedure.

IV. Applications

I will restrict my discussion on the application of the Monte Carlo method to the pure SU(2) and SU(3) gauge theories in four dimensions. I will describe the technology required and the results obtained for, (i), the string tension, (ii) the deconfinement temperature, and (iii), the mass of the glueball states. The interested reader will find the discussion of other topics in [7] and the literature which is cited there, as well as in the other lectures on lattice gauge theory presented at this School [9].

(i) The String Tension

Consider the product of M group elements arranged along some closed path Γ on the lattice, $U_1U_2\ldots U_M \equiv U_\Gamma$. A very important quantity in the study of lattice gauge theory is the Wilson loop [2]. It is defined in terms of U_Γ as

$$W_\Gamma = \frac{1}{N}\,\mathrm{Tr}\,U_\Gamma \tag{4.1}$$

for the group SU(N). The statistical average of W_Γ in the sense of (2.6) provides information about the propagation sources in the fundamental representation. If all the U_{ij} along Γ are set to δ_{ij}, the group identity (or to a pure gauge), $\langle W_\Gamma\rangle = 1$. On the other hand, if even a single link along Γ is randomized, $\langle W_\Gamma\rangle = 0$. Therefore, as correlations between the U_{ij} along Γ drop, so does the value of $\langle W_\Gamma\rangle$. For an equilibrium configuration this implies that $\langle W_\Gamma\rangle$ decays with the size of the loop. However, the detailed rate of decay of $\langle W_\Gamma\rangle$ with the size of the loop is dictated by the details of the dynamics of the theory, since the insertion of W_Γ into the functional integral corresponds to the inclusion of a current loop along Γ. This is seen very clearly in the formal continuum limit, where W_Γ is just the trace of a path-ordered exponential of the integral along Γ of A_μ coupled to a loop of external (static) charge.

Consider for Γ a rectangle of lattice dimensions $n_x\cdot n_t$ (and, therefore, of physical size $a^2 n_x\cdot n_t$). Let $\{|\alpha\rangle\}$ be a complete set of states in the presence of the current loop. Then $\langle W_{n_x n_t}\rangle$ is the ratio of the partition function with the sources inserted to the partition function without sources. It can be written as

$$\langle W_{n_x n_t}\rangle = \sum_\alpha |\langle 0|\alpha\rangle|^2 e^{-E_\alpha(x)t} \tag{4.2}$$

where $\langle 0|\alpha\rangle$ is the amplitude for a transition between the vacuum and a state with two static sources a distance x apart, with energy $E_\alpha(x)$. As $t\to\infty$, only the lowest energy state will contribute to (4.2), so that

$$\langle W_{n_x n_t}\rangle \underset{t\to\infty}{\to} w_0\, e^{-E(x)t} \qquad (4.3)$$

with w_0 constant. On the lattice, therefore, it should be true that

$$\langle W_{n_x n_t}\rangle \underset{n_t\to\infty}{\sim} e^{-E(g_0,n_x)n_t}. \qquad (4.4)$$

However, since $E(g_0,n_x)$ contains the self energies of the sources, it diverges in the continuum limit unless a subtraction is performed, so a direct comparison between (4.3) and (4.4) would be fruitless. Nonetheless, the extraction of an asymptotic force between the charges is still possible if one compares expressions like (4.4) for loops of different sizes. Notice that an asymptotic constant attractive force between the charges would manifest itself by the behavior

$$E(g_0,n_x) \to K(g_0)n_x \qquad (4.5)$$

This force, called the string tension, would be given by $\sigma = K(g_0)/a^2$ in physical units. However, unlike E, σ is a renormalization group invariant and, in fact, serves to define the renormalization scheme. This is done by demanding that σ remain fixed as $a\to0$ (and hence, because of (2.12), $g_0\to0$) by adjusting $K(g_0)$ according to

$$K(g_0) = \frac{\sigma}{\Lambda^2}\,\eta^2(g\ ) \qquad (4.6)$$

with $\eta(g_0)$ defined in (2.12). Notice that, because of (4.4) and (4.5), a non-vanishing $K(g_0)$ would be signaled by an area-law decay for $\langle W_{n_x n_t}\rangle$. (This is of course true for loops of arbitrary shape and orientation. For general loops, the decay rate is controlled by the minimal area defined by the loop).

The following should be noted at this point. In perturbation theory about $g_0=0$,

$$K(g_0) = a_0 g_0^2 + O(g_0^4) \qquad (4.7)$$

a behavior clearly distinguishable from (4.6) which has an essential singularity at $g_0=0$. As $g_0\to0$, the physical size of any loop of finite lattice extent approaches zero. Therefore, the coexistence of confinement and asymptotic freedom should manifest itself by a smooth matching of the strong coupling (logarithmic) behavior of $K(g_0)$ with that predicted by (4.6) [14]. However, once the physical size of a loop becomes smaller than the confinement scale, $K(g_0)$ should show the power behavior given by (4.7). All of this is confirmed dramatically by Monte Carlo results.

The actual computation of $K(g_0)$ (or σ) requires care. The decay rate of $\langle W \rangle$ is not given by a simple area law but instead includes subdominant terms (constant, perimeter, etc.) and a precise determination of σ cannot be made by fitting the measurement of $\langle W \rangle$ to any simple form. It has been suggested [23] that a way to extract a pure area law is to measure the ratios

$$K'(g_0,n) = -\ln \frac{\langle W_{nn} \rangle \langle W_{n-1,n-1} \rangle}{\langle W_{n,n-1} \rangle^2} \tag{4.8}$$

since it can be shown that in this way the unwanted pieces drop out. Fitting the envelope of curves $K'(g_0,n)$ for several n to the scaling form (4.6) leads to

$$\begin{aligned} \sigma &= (5.9 \pm 1.8) \times 10^3 \, \Lambda^2 \quad [SU(2)] \\ \sigma &= (2.8 \pm 0.9) \times 10^4 \, \Lambda^2 \quad [SU(3)]. \end{aligned} \tag{4.9}$$

As mentioned in Section II holding σ fixed as $a \to 0$ defines a particular renormalization scheme. If one wishes to compare the values of observables obtained in this scheme with the corresponding values from a different scheme, one must determine the relationship between the renormalization scales. The scale Λ present in (4.9) is related to the momentum cutoff Λ_{mom} by [24].

$$\begin{aligned} \Lambda_{mom} &= 57.47 \, \Lambda \quad [SU(2)] \\ \Lambda_{mom} &= 83.33 \, \Lambda \quad [SU(3)] \end{aligned} \tag{4.10}$$

leading to

$$\begin{aligned} \Lambda_{mom} &= (0.75 \pm 0.12) \sqrt{\sigma} \quad [SU(2)] \\ \Lambda_{mom} &= (0.5 \pm 0.1) \sqrt{\sigma} \quad [SU(3)] \end{aligned} \tag{4.11}$$

The original freedom to adjust g_0 in the classical theory reflects itself by the freedom of adjusting σ in the quantum theory (or whatever other dimensionful parameter used to renormalize). This phenomenon has been called Dimensional Transmutation [25]. The value of all observables predicted by the theory is therefore fixed once σ is given as input. One can extract a phenomenological value for σ using its string model [26] connection with the Regge slope $\sigma = 1/2\pi\alpha'$ with $\alpha' = 1.0\ (\mathrm{GeV})^{-2}$, giving $\sigma = 0.16\ (\mathrm{GeV})^2$ and, hence, $\Lambda_{mom}^{SU(3)} \approx 200$ MeV. This number is not inconsistent with phenomenology. Apart from the fact that experimental estimates for Λ_{mom} are not too precise, a serious comparision between theory and experiment at this level would still require including the effects of light quarks in calculating Λ.

A word about universality: Extraction of numbers from lattice calculations is done by performing simulations at couplings near the crossover region as soon as scaling seems to have set in. This is a valid procedure if universality holds. However, when comparing theoretical estimates with Monte Carlo for various different forms of the lattice action it should be noted that the scaling formulas have power corrections (in g_0) which are usually ignored, when in fact, comparison with MC is done at $g_0 = \mathcal{O}(1)$.

(ii) The Deconfinement Temperature

It has been argued by several authors [16] that at some very high temperature, T_d, hadronic matter should undergo a phase transition to a plasma state. Thus, for $T>T_c$, if this conjecture were correct, the confinement mechanism seen for small T should disappear. An impressive demonstration of this effect in QCD has been obtained by Monte Carlo methods [17]. I will briefly review these simulations here.

A gauge theory defined on a lattice of infinite spatial extent and n_τ sites along the time direction with periodic boundary conditions represents a system at a finite physical temperature T given by

$$T = (a\cdot n_\tau)^{-1} \tag{4.12}$$

where a is the lattice spacing. The continuum limit is taken by keeping T fixed as $a\to 0$.

Consider a product of links along the time direction at some spatial point x. Because of the periodic boundary conditions a path with n_τ links is a closed loop, $U_x = U_{1n_\tau}\cdots U_{32}U_{21}$. The time propagation of a static source can be investigated by studying the expectation values of Wilson loops

$$W_x = \frac{1}{N} \text{Tr } U_x \tag{4.13}$$

The ratio of the partition functions for the system with a static source present and absent defines the shift in free energy, ΔF, induced by the presence of the source.

$$\Delta F = -T \ln \langle W_x \rangle . \tag{4.14}$$

If the theory confines, ΔF diverges and $\langle W_x \rangle = 0$. If, however, gauge field fluctuations which are excited thermally screen the source, the infrared effects responsible for confinement disappear, making ΔF finite and, hence, $\langle W_x \rangle \neq 0$. The numerical analysis [17] clearly shows this effect, with $T_d \sim 200$ MeV. In accordance with the expectation that deconfinement is a physical effect, the constancy of $n_\tau \, a(g_c(n_\tau))$, where $g_c(n_\tau)$ is the value of the bare coupling at the point where $\langle W_x \rangle$ becomes non-vanishing, was verified.

The above described transition can be understood as arising from the spontaneous breakdown of a global symmetry which is explicit in the classical action and realized for $T<T_d$. To see this, first note that W_x given in (4.13) is a gauge-invariant function of time-like links, so that one cannot set $U_0=1$ ($A_0=0$) everywhere as a gauge condition. However, a partial gauge fixing is accomplished by setting $U_0(\vec{x},t)=1$ everywhere except at one value of t, say t_0. Thus $W_x = \text{Tr } U_0(\vec{x},t_0)$. But the gauge-field action is invariant under the global transformation $U_0 \to -U_0$, and therefore, $\langle W_x \rangle = 0$.

The transition between the confined and Coulomb phases can be pictured by studying the two-point correlation $\Gamma_R = W_0 W_R^\dagger = \exp[-F(T,R)/T]$, where $F(T,R)$ measures the potential between two heavy sources a distance R apart. In the confined phase one expects $\Gamma_R \sim \exp[-\sigma(T)R/T]$ with $\sigma(T)$ the string tension at temperature T. The Debye-screened Coulomb phase, on the other hand, would be characterized by $\Gamma_R \sim 1 + (3/16\pi T) \times (g_0^2/R) \exp(-KR)$, where K^{-1} is the Debye screening length. The behavior of Γ_R in the two phases is beautifully reproduced by the Monte Carlo analysis [27].

(iii) The Glueball Mass

Because of confinement, non-Abelian gauge theories have no long-range forces and, hence, no massless states. In these theories, therefore, the mass spectrum must start at some finite point m_g. This is usually referred to as the mass gap. In QCD without quarks, m_g corres- to the mass of a stable particle called a glueball. Since the real world contains light quarks, an experimental detection of glueball states may be quite complicated.

The determination of m_g by Monte Carlo analysis is considerably more difficult than measuring the string tension.

In principle, the idea is simple. Consider the connected correlation function

$$G(t) = P(t)P(0) - P(0)^{2} \qquad (4.15)$$

where $P = \mathrm{Tr}\, U_p$ and U_p is the product of four group elements around a plaquette. G(t) depends on the relative location and orientation of the two plaquettes, but, if these are taken at the same spatial location and with the same orientation, one can introduce a complete set of energy eigenstates $|n\rangle$ in (4.15) to obtain

$$G(t) = \sum_{n \neq 0} |\langle n|P(t)|0\rangle|^{2} e^{-E_n t} \qquad (4.16)$$

For large t, one could therefore extract the lowest value of E_n and in this way determine m_g. Initial attempts to measure m_g in this way [28] were only successful in giving a rough estimate. The main reason for this is that, in the region where the simulations can be done, the correlation length is still only of the order of a few lattice spacings. At three or four lattice spacings, the numerical value of G is already buried in the noise. At such short separations, moreover, the summation over momenta introduces power-law corrections which mask the exponential decay.

An improvement of this method involves considering correlations between plaquettes at different positions and with arbitrary relative orientation. If one then uses for P in (4.15) a sum over all locations and orientations of the plaquette operator, higher-spin states as well as states with nonvanishing momentum contribute nothing to the expectation values, leading to an expression like (4.16) but with m_n instead of E_n [29]. Thus, for large t, one expects $G(t) \sim \exp[-m_g t]$.

In the numerical analysis, one measures effective masses (t-dependent) by comparing G at two values of t=na. In lattice units, a convenient ratio would be

$$U(n,g_0) = \ln \frac{G(n)}{G(n+1)}. \qquad (4.17)$$

In the scaling limit $g_0 \to 0$ the dimensionless gap μ should go to zero with the lattice spacing $a(g_0)$ in such a way that $m_g = \mu a^{-1}(g_0)$ approaches a finite limit. As $g_0 \to 0$, and n large, then, one expects to see $\mu(g_0)$ scale like $m_g a(g_0)$. For any finite n and g_0 small enough, our arguments for the case of the string tension apply, and the curves $\mu(n, g_0)$ will start to follow the perturbative prediction, and m_g will be extracted from the envelope of those curves.

Further improvement of the above procedure is possible by considering a variational problem in which one uses for P some linear combination of operators, $P = \sum_i C_i O_i$, and then minimizes m_g with respect to the C_i [30]. The idea behind this is that if one could find an operator which excited only the glueball state, then two-point correlations would fall exponentially with the glueball mass at all separations and not just asymptotically. Although one cannot, in general, construct such an operator, the variational procedure above leads to better upper bounds in the determination of m_g. The result for SU(2) is [31].

$$m_g = (2.4 \pm 0.6)\sqrt{\sigma} \tag{4.18}$$

or $m_g \approx (960 \pm 240)$ MeV. This kind of analysis has been also done for the SU(3) system with the result [32] $m_g = (920 \pm 310)$ MeV.

V. Conclusions

I have presented in these lectures a survey of some of the applications of the Monte Carlo technique in quarkless QCD. The field is clearly still in its developing stages and the results which I have presented above should not be taken as final values. For one thing, the effects of light quarks have been completely ignored in my discussion. This is generally not justified. However, much progress has been made in extending these techniques to include the effects of quarks [33]. Indeed, preliminary results exist for the masses of low-lying hadrons which are in quite reasonable agreement with experimental observation [33].

The ultimate goal of the application of these techniques in the study of QCD is to provide stringent tests of the theory. In view of the rapid progress in this field, such an exciting possibility may not be too far away in the future.

Acknowledgements

I thank the organizers of this School for their kind hospitality and support. This work was partially supported by the NSF under grant PHY-77-27084.

References

1. D. Gross and F. Wilczek, Phys. Rev. Lett. 30, 1346 (1973) and Phys. Rev. D8, 3633 (1973); H.D. Politzer, Phys. Rev. Lett. 30, 1343 (1973).
2. K. Wilson, Phys. Rev. D10, 2445 (1974).
3. A.M. Polyakov, Phys. Let. 59B, 82 (1975).
4. See, for example; K. Binder in "Phase Transitions and Critical Phenomena" ed. C. Domb and M.S. Green (Academic Press, NY, 1976).
5. M. Creutz, L. Jacobs and C. Rebbi, Phys. Rev. Lett. 42, 1390 (1979) and Phys. Rev. D20, 1915 (1979).
6. K. Wilson, Cornell preprint 79-1032 (1979) and Cargese lectures (1979).
7. M. Creutz, L. Jacobs and C. Rebbi, Phys. Rep. C (to appear).
8. See, for example, K. Wilson in New Phenomena in Subnuclear Physics, ed. A. Zichichi (Plenum Press, N.Y., 1977); J. Kogut, Rev. Mod. Phys. 51, 659 (1979); M. Creutz, (Cambridge Univ. Press, to appear).
9. See the Lectures by C. DeTar, G. Mack, Yu. Makeenko, L. McLerran and H. Satz in these proceedings.
10. For U(1) in d=4, M. Creutz, L. Jacobs and C. Rebbi, Phys. Rev. D20, 1915 (1979) and M. Creutz, Phys. Rev. Lett 43, 553 (1979); for d=3, G. Bhanot and M. Creutz, Phys. Rev. D21, 2892 (1980).
11. For d=4; A. Guth, Phys. Rev. D21, 2291 (1980); for d=3, M. Gopfert and G. Mack, Commun. Math. Phys. 82, 545 (1981).
12. C.B. Lang, C. Rebbi, P. Salomonson and B.S. Skagerstam, Phys. Lett. 101B, 173 (1981) and Goteborg preprint 81-22 (1981); G. Bhanot and R. Dashen, Phys. Lett. 113B, 299 (1982).
13. Yu M. Makeenko and M.I. Polikarpov. Nucl. Phys. B205, 386 (1982). C.B. Lang, C. Rebbi, P. Salomonson and B.S. Skagerstam, Ref. 12; B. Grossman and S. Samuel, Columbia preprint CU-TP-230 (1982).
14. See, for example, M. Creutz, Erice Lectures (1981) and C. Rebbi, Lectures for the 1981 GIFT School, Brookhaven preprints (1981) and references therein.
15. M. Creutz, Phys. Rev. D21, 2308 (1980); C. Rebbi, ibid, 3350 (1980).
16. J.C. Collins and M.J. Perry, Phys. Rev. Lett. 34, 1353 (1975); A.M. Polyakov, Phys. Lett. 72B, 477 (1978); L. Susskind, Phys. Rev. D20, 2610 (1979).
17. L. McLerran and B. Svetitsky, Phys. Lett. 98B, 195 (1981) and Phys. Rev. D24, 450 (1981); J. Kuti, J. Polonyi and K. Szlachnayi, Phys. Lett. 98B, 199 (1981); J. Engels, F. Karsch, I. Montvay and H. Satz, Phys. Lett. 101B, 89 (1981).
18. B. Berg, A. Billoire and C. Rebbi, Brookhaven preprint BNL-30826 (1981) and references therein; B. Berg and A. Billoire, Phys. Lett. 113B, 65 (1982).

19. See, for example, C. Rebbi, talks presented at the Orbis Scientiae meeting (1982) and BNL preprint and references therein.

20. W.E. Caswell, Phys. Rev. Lett. 33, 244 (1974); D.R.T. Jones, Nucl. Phys. B75, 531 (1974).

21. N. Metropolis, A.W. Rosenbluth, M.N. Rosenbluth and A. Teller, J. Chem. Phys. 21, 1087 (1953).

22. M. Creutz, L. Jacobs and C. Rebbi, reference 5.

23. M. Creutz, Phys. Rev. Lett. 45, 313 (1980); C. Rebbi, Phys. Rev. D21, 3350 (1980).

24. A. Hasenfratz and P. Hasenfratz, Phys. Lett. 93B, 165 (1980); R. Dashen and D. Gross, Phys. Rev. D23, 2340 (1981).

25. S. Coleman and E.J. Weinberg, Phys. Rev. D7, 1888 (1973).

26. P. Goddard, J. Goldstone, C. Rebbi and C.B. Thorn, Nucl. Phys. B56, 109 (1973).

27. J. Kuti, J. Polonyi and K. Szlachnayi, reference 17.

28. G. Bhanot and C. Rebbi, Nucl. Phys. B180, [FS2], 469 (1981); B. Berg and J. Stehr, Z. Phys. C9, 333 (1981).

29. G. Münster, Nucl. Phys. B190, [FS3], 454 (1981).

30. K. Wilson, talk given at the Abingdon Meeting on Lattice Gauge Theories (1981).

31. B. Berg, A. Billoire and C. Rebbi, reference 18.

32. B. Berg and A. Billoire, reference 18.

33. C. Rebbi (ref. 19) and references therein. See also J. Kuti, Phys. Rev. Lett. 49, 183 (1982).

THE U_1 PROBLEM AND INSTANTONS

D. I. Dyakonov

Leningrad Nuclear Physics Institute
188350 Gatchina, Leningrad
U.S.S.R.

These lectures are devoted to the so-called U_1 problem |1| and related topics, topics that seem to me of key importance to the theory of QCD. The point is that the U_1 problem cannot be solved within perturbation theory. It is my feeling that it also is not settled in the non-perturbative lattice approach to QCD. If this is indeed so, it means that some very important field configurations are not, and probably cannot be, taken into account in Monte Carlo simulations. I shall try to explain what type of configurations are missing.

In the first lecture I will review the U_1 problem and explain how it is solved at the present stage of our knowledge. It is remarkable that even though we do not possess a full theory, many important quantities appear to be interrelated, and these relations can be compared with experimental data.

In my second lecture I shall discuss the instanton field configurations (which appear crucial for the U_1 problem in particular), and show how to build a non-trivial self-consistent theory of an interacting instanton medium manifesting many desirable features.

Lecture 1.

Let me first review the U_1 problem. The characteristic mass scale of ordinary hadrons is several hundred MeV (one can reasonably take the mass of a typical hadron, $m_\rho = 770$ MeV, as the scale). The "current" masses of three light quarks u, d, s are extremely small: $m_u \sim 4$ MeV, $m_d \sim 7$ MeV, $m_s \sim 150$ MeV. Therefore, what one calls a chiral limit, $m_{u,d,s} \to 0$, is considered to be a good approximation to the real world. Meanwhile, in this limit the QCD Lagrangian has high symmetry, called chiral symmetry, with respect to independent transformations of left-handed and right-handed components (helicities) of the u,d,s quarks into one another. Combining these two transformations, one can get a 9 parameter γ_5 rotation of the u,d,s quarks. The latter transformation mixes states with opposite parity.

Therefore, if the chiral symmetry is preserved, hadrons with the same quantum numbers except for parity should be degenerate. Meanwhile, the actual splitting (say, $m_{A_1} - m_\rho = 1200\text{-}770 \simeq 400$ MeV) is very large, and it cannot be attributed to the very small quark masses neglected in these considerations. Chiral symmetry breaking due to the spontaneous condensates $\langle\bar{u}u\rangle$, $\langle\bar{d}d\rangle$, $\langle\bar{s}s\rangle$ is by far a more likely possibility in the chiral limit $m_{u,d,s} = 0$. If so, then according to the Goldstone theorem 9 massless bosons should emerge, owing to the spontaneous violation of a 9 parameter continuous symmetry. Indeed, 8 pseudoscalar mesons (π, K, η) have masses much lower than the masses of other hadrons, and their non-zero masses can be ascribed to the small but non-zero masses of quarks that violate somewhat the chiral symmetry from the very beginning. The ninth pseudoscalar meson η' (958 MeV), however, is too heavy. This paradox was called the U_1 problem |1|.

In order to give a more systematic introduction to the subject, I shall use the apparatus of Ward identities |2|. To this end, let us introduce 8 axial currents

$$J^a_{\mu 5}(x) = \bar{q}(x)\gamma_\mu\gamma_5 t^a q(x)$$, t^a are SU_3 flavour generators (1)

which are conserved in the chiral limit, since

$$\partial_\mu J^a_{\mu 5}(x) = i\bar{q}\gamma_5 \{m, t^a\}_+ q \equiv P^a(x)$$

(2)

$$m = \begin{pmatrix} m_u & 0 & 0 \\ 0 & m_d & 0 \\ 0 & 0 & m_s \end{pmatrix}$$

Consider a 2-point correlation function

$$\int d^4x \; e^{ipx} \; \partial_{\mu x} \, i \langle T \, J^a_{\mu 5}(x), P^b(0) \rangle \, .$$

Since we know that there are no massless hadrons in the spectrum, in the limit of zero momentum this quantity is zero. On the other hand, this zero can be rewritten as |2,3|

$$0 = \int d^4x \; i \langle T \, P^a(x), P^b(0) \rangle + \langle \bar{q} \{t^a, \{t^b, m\}\} q \rangle \qquad (3)$$

To derive this, one applies the ∂_μ derivative to the Green function. The second term in (3) arises from the differentiation of the θ function in the T product. The arising equal time commutator is calculated according to the canonical commutation relations for Heisenberg operators.

Let us analyze eq. (3). The second term is linear in the quark masses since we assume that the $\langle\bar{q}q\rangle$ condensate which violates the chiral symmetry is formed even in the chiral limit $m_q = 0$. The first term is naively quadratic in quark masses (see the definition of the pseudoscalar density (2)). Therefore, one has to conclude that there is a large contribution to the first term in eq. (3) that is also linear in quark masses.

To be more precise, let us first notice that both terms in eq. (3) diverge quadratically in perturbation theory. Let us subtract perturbative contributions to these quantities. The second term is then just the non-perturbative part of the condensate. The first term is now rapidly convergent, and one can saturate the two-point correlation function by the lowest intermediate states.

To be specific, let us consider the pseudoscalar density with the π^0 quantum numbers:

$$J_{\mu 5}^{\pi^0} = \frac{1}{\sqrt{2}} \left(\bar{u} \gamma_\mu \gamma_5 u - \bar{d} \gamma_\mu \gamma_5 \bar{d} \right),$$

$$\partial_\mu J_{\mu 5}^{\pi^0} = \mathcal{P}^{\pi^0} = \sqrt{2}\, i \left(m_u \bar{u} \gamma_5 u - m_d \bar{d} \gamma_5 d \right).$$

Introducing the conventional pion decay constant f_π as

$$\langle 0 | J_{\mu 5}^{\pi^0} | \pi^0 \rangle = i p_\mu f_\pi \; ; \; \langle 0 | \mathcal{P}^{\pi^0} | \pi^0 \rangle = f_\pi m_\pi^2 ,$$

and inserting the π^0 as an intermediate state in the two-point correlation function at zero momentum, one gets from eq. (3)

$$0 = \left. \frac{(f_\pi m_\pi^2)^2}{m_\pi^2 - p^2} \right|_{p=0} + 2 m_u \langle \bar{u} u \rangle + 2 m_d \langle \bar{d} d \rangle$$

or

$$f_\pi^2 m_\pi^2 = - 2 m_u \langle \bar{u} u \rangle - 2 m_d \langle \bar{d} d \rangle . \tag{4}$$

One can see that the pion's mass squared is indeed linear in the quark masses.

Problem. Derive similar relations for the other members of the pseudoscalar octet (π, K, η). Assuming that $\langle \bar{u} u \rangle = \langle \bar{d} d \rangle = \langle \bar{s} s \rangle$ and that $f_\pi = f_K = f_\eta$, obtain the values of quark masses mentioned at the beginning of the lecture. Note that the difference $m_s - m_{u,d} \sim 150$ MeV is known from the mass splittings of the hyperons.

Now let us turn to the SU_3 singlet pseudoscalar meson, η'. If the same technique is applied to this state, one gets $m_{\eta'}^2 \sim 0.16\ \mathrm{GeV}^2$ (experimentally, $0.917\ \mathrm{GeV}^2$). This paradox comprises the U_1 problem |1|. What is going on?

Let us consider the appropriate SU_3 singlet axial current

$$J_{\mu 5}(x) = \bar{q}\gamma_\mu\gamma_5 I q \quad , \quad I = \frac{1}{3}\begin{pmatrix} 1 & 0 & 0 \\ 0 & 1 & 0 \\ 0 & 0 & 1 \end{pmatrix}. \tag{5}$$

Its divergence, similar to the octet case, has a contribution linear in the quarks masses, a contribution which can be obtained by using the equations of motion of QCD. However in the singlet case there is also the famous axial anomaly |4|. Thus,

$$\partial_\mu J_{\mu 5}(x) = i\bar{q}\gamma_5\{m, I\}q + \frac{g^2}{32\pi^2}\epsilon_{\alpha\beta\mu\nu}F^a_{\alpha\beta}F^a_{\mu\nu} \equiv P(x) + 2Q(x), \tag{6}$$

where $F^a_{\alpha\beta}$ is the gluon field strength.

In this case Crewther |2| has derived two independent Ward identities. (Note that they can be derived both from canonical quantization rules and by means of changing the variables in the functional integral |5|. A detailed discussion of certain subtleties connected with these Ward identities can be found in ref. |6|). One has

$$\begin{aligned} 0 &= \int d^4x\, \partial_\mu\, i\langle T\, J_{\mu 5}(x), P(0)\rangle = \\ &= \int d^4x\, i\,\langle T\, P(x) + 2Q(x), P(0)\rangle + \langle \bar{q}\,\{I,\{I,m\}\}\,q\rangle, \end{aligned} \tag{7}$$

$$0 = \int d^4x\, \partial_\mu\, i\langle T\, J_{\mu 5}(x), Q(0)\rangle = \int d^4x\, i\langle T\, P(x) + 2Q(x), Q(0)\rangle. \tag{8}$$

In the last identity there is no contact term because in this case one is to calculate an equal-time commutator of quark operators with the gluon gluon ones. This commutator is zero. Combining identities (7) and (8) we get

$$\int d^4x\, i\langle T\, P(x), P(0)\rangle - 4\int d^4x\, i\langle T\, Q(x), Q(0)\rangle + \frac{4}{9}\langle \bar{q}\, m\, q\rangle = 0. \tag{9}$$

This identity is similar to the non-singlet case (3), with the difference that there is now an extra term which is the zero-momentum correlation function of gluon operators. Had it been zero, we would repeat the same argument as in the non-singlet case, and would obtain $m^2_{\eta'} = 0.16\ \mathrm{GeV}^2 = O(m_q)$ - a disaster! Meanwhile, it can be easily seen that in any order of perturbation theory the Q-Q correlation function at zero momentum is identically zero, due to the antisymmetric $\varepsilon_{\alpha\beta\gamma\delta}$. Therefore, the large mass of η' is a clear indication that the perturbation theory of QCD misses some important features of the true theory; the Q-Q correlation function at zero momentum should not be zero.

This requirement to the true theory of QCD becomes even more disturbing if one recalls that the pseudoscalar gluon field density is, in its turn, a total divergence of a certain gluon current:

$$Q = \frac{g^2}{32\pi^2} F^a_{\mu\nu} \tilde{F}^a_{\mu\nu} = \partial_\mu K_\mu ,$$
$$K_\mu = \frac{g^2}{16\pi^2} \epsilon_{\mu\alpha\beta\gamma} A^a_\alpha \left(\partial_\beta A^a_\gamma + \frac{g}{3} f^{abc} A^b_\beta A^c_\gamma \right). \tag{10}$$

This means that the Q-Q correlation function (hereafter we use the notation $\int d^4x\, i \langle TQ(x)Q(o)\rangle \equiv \langle QQ\rangle$) can be written as

$$\langle QQ \rangle = \lim_{p\to 0} p_\mu p_\nu \int d^4x\, e^{ip\cdot x}\, i \langle T K_\mu(x), K_\nu(0)\rangle \neq 0. \tag{11}$$

(Again there are some subtleties here, connected with the proper definition of the T product and with the Schwinger terms. What one can show (see ref. |6|) is that the quantity encountered in the Ward identity (9) should actually be understood as given by eq. (11)).

It can be seen from eq. (11) that the Fourier transform of the $\langle K_\mu K_\nu \rangle$ correlation function should have a ghost pole at zero momentum |7|:

$$\langle K_\mu K_\nu \rangle \sim -\,\mathrm{const.} \frac{p_\mu p_\nu}{p^4} , \text{ or } -\,\mathrm{const} \frac{\delta_{\mu\nu}}{p^2} , \text{ etc.} \tag{12}$$

I would like to emphasize that a massless pole in a theory is always a reflection of its most profound features, and it would be very important to understand its origin in the case under consideration. Before we discuss the physical meaning of this ghost pole necessary to solve the U_1 problem in QCD, I would like to digress to two cases already known in particle physics where a similar phenomenon occurs: a would-be Goldstone boson mixes with a vector ghost state, and the resulting physical state is massive. This phenomenon is what we now want for the η' meson.

Digression 1. Higgs phenomenon

Let us consider a simple case of scalar electrodynamics. There is a charged scalar field which is described by a complex field $\phi = \rho \exp i\theta$, there is a transverse photon with a propagator $i\langle A_\mu A_\nu \rangle^T = (g_{\mu\nu} - p_\mu p_\nu/p^2)/p^2$, and one can say that there is also a longitudinal (unphysical) photon with a propagator $i\langle A_\mu A_\nu\rangle^L = -p_\mu p_\nu/p^4$.

Let the effective potential of the theory be such that a spontaneous condensate occurs: $\langle\phi\rangle = \langle\rho\rangle \neq 0$. There is then a would-be Goldstone boson associated with the phase θ of the scalar field. However, via the scattering on the condensate, there is a transition between the longitudinal degree of freedom of the photon field and the Goldstone boson. As a result, the mass operator for the θ field is non-zero:

$$\Sigma = (ie\langle\rho\rangle p_\mu)\left(-\frac{p_\mu p_\nu}{p^4}\right)(-ie\langle\rho\rangle p_\nu) = -e^2\langle\rho\rangle^2.$$

$\langle\rho\rangle$ $\langle\rho\rangle$

$-iep_\nu$ $\quad -\frac{p_\mu p_\nu}{p^4}$ $\quad iep_\mu$

Therefore, the propagator of the would-be massless field θ is now

$$\frac{1}{0-\Sigma-p^2} = \frac{1}{e^2\langle\rho\rangle^2 - p^2} .$$

A physical state has the mass $m^2 = e^2\langle\rho\rangle^2$, the transition amplitude squared. This is the mass of the third polarization state of the massive vector boson. The other two (transverse) polarization states acquire the same mass, as a matter of fact, but they are irrelevant here.

Digression 2. Two-dimensional spinor electrodynamics

This theory, known as the Schwinger model, is exactly solvable. The charge screening takes place, and there are no charged fermions left in the spectrum. Instead, there is a neutral "hadron" which may be called the η' meson of the model.

There is an axial anomaly

$$\partial_\mu J_{\mu 5} = \frac{e}{2\pi} \epsilon_{\mu\nu} F_{\mu\nu} \equiv Q = \partial_\mu K_\mu \ , \quad K_\mu = \frac{e}{\pi} \epsilon_{\mu\nu} A_\nu \ .$$

The $\langle K_\mu K_\nu \rangle$ correlation function evidently has a pole corresponding to a scalar or a longitudinal photon, or to a Coulomb potential - depending on the gauge choice:

$$\langle K_\mu K_\nu \rangle = \left(\frac{e}{\pi}\right)^2 \epsilon_{\mu\alpha} \epsilon_{\nu\beta} \langle A_\alpha A_\beta \rangle_0 = -\left(\frac{e}{\pi}\right)^2 \begin{cases} \dfrac{p_\mu p_\nu}{p^4} \ , & \text{Landau gauge} \\ \dfrac{g_{\mu\nu}}{p^2} \ , & \text{Feynman gauge} \\ \dfrac{\delta_{\mu 0} \delta_{\nu 0}}{p_0^2} \ , & A_0 = 0 \text{ gauge} \end{cases}$$

Note that the quantity $p_\mu p_\nu \langle K_\mu K_\nu \rangle = -(e/\pi)^2$ is gauge invariant, as it should be.

We described the "pure gluodynamics" of the model, with fermions switched off. Let us now switch in the massless fermions. We know (see refs. |8,9|) that a $\langle \bar{q} q \rangle$ condensate develops, and there is a point-like quark-antiquark massless state manifesting itself, as a pole in the exact photon polarization operator:

$$\Pi_{\mu\nu} = \frac{e^2}{\pi} \left(g_{\mu\nu} - \frac{p_\mu p_\nu}{p^2} \right) .$$

One can say that there is a non-zero transition amplitude between the ghost (scalar photon, for example) and a quark-antiquark massless state

|10, 11|. Diagonalizing the two states, one obtains a real physical state "η'" with mass equal to the transition amplitude e^2/π. This state manifests itself in the exact photon propagator which takes the form (say, in the Landau gauge)

$$\langle A_\alpha A_\beta \rangle = \frac{g_{\alpha\beta} - p_\alpha p_\beta / p^2}{p^2 - e^2/\pi} .$$

These two examples demonstrate how a would-be Goldstone state may mix with a massless vector ghost state, so that the resulting physical mass is equal to the transition amplitude. Fortunately, in both examples the nature of the ghosts is quite transparent; their presence can be seen just at the Lagrangian level.

In QCD we have a singlet (pseudo) Goldstone state. Can it be seen from the pure gluodynamics Lagrangian that there is a massless ghost state, so that these states mix?

To answer the question let us study the

Schrödinger equation for Quantum Gluodynamics

To set it up, it is necessary to use the Hamiltonian $A_0^a = 0$ gauge. The dynamical variables of the theory are then the space components $A_i^a(\vec{x})$. The Schrödinger equation for the vacuum state (as well as for excited states) is

$$\int d^3x \left[-\frac{1}{2} \left(\frac{\delta}{\delta A_i^a(\vec{x})} \right)^2 + \frac{H^2}{2} \right] \Psi[A_i^a] = \mathcal{E}\, \Psi[A_i^a] ,$$

$$H_i^a(\vec{x}) = \epsilon_{ijk} \left(\partial_j A_k^a + \frac{g}{2} f^{abc} A_j^b A_k^c \right) . \tag{13}$$

One should also take into account the local gauge invariance condition (the analog of div E = 0),

$$D_i^{ab} \frac{\delta}{\delta A_i^b(\vec{x})} \Psi = 0 , \tag{14}$$

meaning that the wave function for physical states should not be changed by infinitely small gauge transformations. This condition allows one to assume that the wave function ψ as well as the potential energy of the system

$$V = \frac{1}{2} \int d^3x \; H^2(\vec{x}) \tag{15}$$

depends on the "generalized coordinates" X,Y,Z,... that are functionals of $A_i^a(\vec{x})$ invariant under small gauge transformations. In particular, we are specially interested in a generalized coordinate

$$X = \int d^3x \; K_0(\vec{x}) \tag{16}$$

where K_0 is the zero-component of the K_μ vector (10):

$$K_0 = \frac{g^2}{16\pi^2} \epsilon_{ijk} A_i^a \left(\partial_j A_k^a + \frac{g}{3} f^{abc} A_j^b A_k^c \right). \tag{17}$$

Under the gauge transformation

$$A_i \longrightarrow S A_i S^+ - \frac{i}{g} (\partial_i S) S^+$$

the quantity X transforms as follows:

$$X \longrightarrow X - \frac{1}{24\pi^2} \int d^3x \; \epsilon_{ijk} \operatorname{Tr} \left(\partial_i S S^+ \partial_j S S^+ \partial_k S S^+ \right)$$

$$= X + \text{integer}.$$

This integer is equal to the "topological charge" of the gauge transformation S (see, e.g. ref. |12|). At the same time, the potential energy V is not changed by any gauge transformation. Thus the potential energy V is a <u>periodic</u> function of the generalized coordinate X, with period unity |13| (see Fig. 1).

The situation thus turns out to be similar to that of a particle in a periodic field (other generalized coordinates Y,Z,.. are of no interest to us now). In this case we know that there is a band spectrum with levels labelled by quasi-momentum k, and that, due to the quantum tunnelling through the barriers, the particle behaves practically as a free one. Thus at the bottom of the 1st band its energy E is just $E = k^2/2m^*$, m^* being the effective mass.

Another manifestation of the free motion is the Green function associated with the coordinates X:

$$i\langle T\, X(t)\, X(0)\rangle = |t|\cdot \mathrm{const}\;, \quad t\to\infty,$$

implying that the system cannot be localized in X. Making the Fourier transformation, we get

$$\int dt\; e^{i\omega t}\, i\langle T\, X(t)\, X(0)\rangle = -\frac{1}{\omega^2}\,\mathrm{const}\;, \quad \omega\to 0\,. \tag{18}$$

But this pole is just what is needed for saving the η' mass!

Indeed, recalling the definition of the generalized coordinate X (16) let us rewrite eq. (18) as follows:

$$\begin{aligned} -\,\mathrm{const} &= \omega^2 \int dt\; e^{i\omega t}\; i\left\langle T \int d^3x\, K_0(\vec{x},t),\, \int d^3y\, K_0(\vec{y},0)\right\rangle \\ &= V^{(3)}\, p_0^2 \int d^4x\; e^{ip_0x_0}\; i\langle T\, K_0(x),\, K_0(0)\rangle \\ &= V^{(3)}\, p_\mu p_\nu \int d^4x\; e^{ip\cdot x}\; i\langle T\, K_\mu(x),\, K_\nu(0)\rangle\Big|_{\vec{p}=0} \end{aligned} \tag{19}$$

-cf. eq. (12).

We thus conclude that the ghost pole needed to solve the U_1 problem is there because one cannot localize the system in the X space. Moreover, the development along X is free.

One should note that a possible weak point in this argument was that we assumed that the potential barriers in X were penetrable. That such is the case can be seen from the existence of instantons |14| - classical tunneling trajectories in imaginary time connecting the neighbour minima in X, with finite action. The finiteness of action implies the penetrability of the barriers.

The fact that instantons have a role of saving from the U_1 paradox was first noted by 't Hooft |15|. Here we present a more general argument |6| which is not necessarily based on quasiclassics. However, it seems quite plausible that the barriers in X are thick (since the classical tunnelling action, equal to $8\pi^2/g^2$, appears to be large). If so, the quasiclassics (instantons) must be taken into account. Unfortunately, up to now there has been no self-consistent theory of instantons. An attempt to build such a theory comprises the next lecture.

To end this section, I would like to mention that an exact though formal solution to the Schrödinger eq. (13) with an additional condition (14) was found in ref. |6|. We note that from eqs. (16) and (17) it follows that

$$\frac{\delta X[A]}{\delta A_i^a(\vec{x})} = \frac{g^2}{8\pi^2} H_i^a(\vec{x}) .$$

Therefore, a functional

$$\Psi[A] = \exp \pm \frac{8\pi^2}{g^2} X[A]$$

satisfies eq. (13) with zero eigenenergy! Unfortunately, the solution is merely formal, since it has growing asymptotes in some directions of the A_i^a space. Nevertheless, a generalization of this solution to a lattice formulation of the theory, where the potentials are actually angle variables, may be instructive.

How to work with the ghost?

Though formally we learned only that the correlation function $\langle K_o K_o \rangle$ has a pole $1/p^2$ (at $\vec{p} = 0$), it is natural to assume by virtue of the Lorentz invariance that in an arbitrary reference frame the $\langle K_\mu K_\nu \rangle$

correlation function has a ghost pole at $p^2 \to 0$, so that

$$\langle Q\,Q\rangle = \lim_{p\to 0} p_\mu p_\nu \langle K_\mu K_\nu\rangle = -\lambda^4 \tag{20}$$

where λ is some yet unknown constant with the dimension of mass. We emphasize that all physical results which follow could be obtained also in the reference frame $\vec{p} = 0$; we prefer the Lorentz-invariant technique from aesthetic considerations.

It is convenient to introduce the ghost formally as a lowest intermediate state in the correlation function $<K_\mu K_\nu>$. We write

$$\langle K_\mu K_\nu\rangle = \sum_P \langle 0|K_\mu|g^P\rangle \frac{1}{-p^2} \langle g^P|K_\nu|0\rangle$$

where Σ_p denotes summation over the polarization of the ghost. Denoting the polarization vector by ϵ_μ^P we define

$$\langle 0|K_\mu|g^P\rangle = \lambda^2 \epsilon_\mu^P \;, \quad \langle g^P|K_\nu|0\rangle = \lambda^2 \epsilon_\nu^{P*}$$

Hence we have

$$\langle QQ\rangle = \lambda^4 p_\mu p_\nu \sum_P \frac{\epsilon_\mu^P \epsilon_\nu^{P*}}{-p^2} = -\lambda^4 . \tag{21}$$

It is natural to introduce the propagator of the ghost (similar to that of a photon)

$$\langle g_\mu g_\nu\rangle_0 = \sum_P \frac{\epsilon_\mu^P \epsilon_\nu^{P*}}{-p^2}$$

In fact, this propagator is not gauge invariant. It depends on the choice of the gauge for A_μ. The only thing we know is the gauge-invariant condition (21). It is convenient to work with a gauge

$$\langle g_\mu g_\nu\rangle_0 = -\frac{p_\mu p_\nu}{p^4} \tag{22}$$

(In the Schwinger model example this corresponded to the Landau gauge for A_μ .)

The above discussion concerns pure gluodynamics. We now switch on the three light quarks. Eight non-singlet pseudo-Goldstone bosons have nothing to do with the ghost, the latter being an SU_3 flavour singlet. However, the ninth (singlet) would-be pseudo-Goldstone boson couples to the ghost through the axial anomaly

$$\pi_g \sim \frac{\bar{u}\gamma_5 u + \bar{d}\gamma_5 d + \bar{s}\gamma_5 s}{\sqrt{3}} \quad \text{[diagram: triangle loop connecting to } g_\mu \text{ ghost]}$$

$$\langle \pi_g | g_\mu \rangle = -i p_\mu M \, , \quad M = \frac{2\lambda^2}{f_\pi} \, . \tag{23}$$

Indeed, according to the current algebra, the divergence of the singlet axial current, divided by f_π, acts as an interpolating field for the ninth boson. Meanwhile, the divergence of the axial current is proportional to $F\tilde{F}$. The latter operator is the source for the ghost, and couples to it with the amplitude λ^2. One thus derives the transition amplitude (23). Another way to derive eq. (23) is to use the Ward identity (9) and find the relation between the introduced transition amplitude M and $-\lambda^4$, the value of the $\langle QQ \rangle$ correlation function in a pure gluonic world. I leave this exercise for the student.

Without the mixing with the ghost, the pseudo-Goldstone boson has the propagator

$$\langle \pi_g \pi_g \rangle_0 = \frac{1}{m_g^2 - p^2} \, , \quad m_g^2 = -\frac{1}{f_\pi^2}\frac{4}{9}\langle \bar{q} m q \rangle \simeq 0.16 \text{ GeV}^2 = O(m_q) \, . \tag{24}$$

Combining (22, 23, 24) we get for the exact propagator of the SU_3 singlet state

$$\langle \pi_g \, \pi_g \rangle = \frac{1}{m_g^2 + M^2 - p^2}$$

The new pole gives the mass of the physical η' meson. Note that in the chiral limit (m = 0) the ' mass is equal to the transition amplitude.

A quantity of interest is also <QQ> for the real world with three light quarks. One finds

$$\langle QQ \rangle_{R.W.} = \lim_{p\to 0} \lambda^4 p_\mu p_\nu \langle g_\mu \, g_\nu \rangle_{R.W.} =$$

$$= \lambda^4 p_\mu p_\nu \left(- \frac{p_\mu p_\nu}{p^4} \right) \frac{m_g^2 - p^2}{m_g^2 + M^2 - p^2} \Bigg|_{p\to 0} = - \lambda^4 \frac{m_g^2}{m_g^2 + M^2} = \mathcal{O}(m_q). \quad (25)$$

As should be expected |16|, this quantity dies out with the quark masses.

Actually, we have presented here the construction of Veneziano |7| who, following Witten |16|, considered the problem from the viewpoint of the large N_c limit. It can be seen |16, 7| that in the limit $N_c \to \infty$ the mass of $m_{\eta'}$ goes down as $1/\sqrt{N_c}$. Therefore, one can argue that in this limit the whole construction is cleaner from the theoretical point of view. Indeed, since Ward identities at zero momentum are used one has not to continue the amplitudes from their mass shell too far. However, as we shall consider in the next lecture, the large N_c limit may have problems of its own. As for the continuation of amplitudes to zero momentum, the actual accuracy for such a procedure is determined by the position of the next-lying pole with a non-negligible residue in the Q-Q channel. According to the estimations of the ITEP group |17| the glueball mass in the 0^- channel may be rather large. Therefore, one can reasonably apply the "polology" technique without considering N_c to be large.

In fact there are rather strong effects violating SU_3 which are due to the relatively large mass of the s quark. This circumstance leads in particular to the octet-singlet mixing. Actually, one has to diagonalize 3 states: $\pi_1 \sim (\bar{u}\gamma_5 u + \bar{d}\gamma_5 d)/\sqrt{2}$, $\pi_2 \sim \bar{s}\gamma_5 s$, and the ghost. In this case one can use 5 independent Ward identities of the type (9) in order to relate various transition amplitudes to one another. The details can be found in ref. |6|. Using f_π, f_K, m_π, m_K, m_η as an input we calculate the mass of $m^2_{\eta'}$ = 0.912 (exp. 0.917 GeV^2), the singlet-octet mixing angle θ^o = -11 (exp. -10°), the partial width $\Gamma(\eta' \to 2\gamma)$ = 5.2 (exp. 5.8 ± 1.2) KeV, and many other physical quantities in fair agreement with the experiment. We also find the numerical value for λ^4, the residue of the ghost pole for the pure gluonic world (see eq. 20):

$$\lambda^4 = (0.188 \text{ MeV})^4 \tag{26}$$

At present this is the best-known quantity for that imaginary world, and a QCD theory should first of all explain this most important quantity.

Lecture 2.*)

It was mentioned in the previous lecture that instantons |14| are classical solutions of Yang-Mills equations in imaginary time. They can be understood as classical tunneling trajectories connecting the neighbour minima in X (see Fig. 1) (Gribov, 1976. See also ref. 13). Instantons correspond to the transitions $X \to X + 1$, antiinstantons to the $X \to X - 1$ transitions.

---

*) This lecture is based on work of V. Yu. Petrov and myself, submitted for publication.

In this lecture I shall slightly change the notations used previously. Namely, I shall absorb the coupling constant g into the definition of the vector potentials. I shall also work in Euclidian space-time, so that the action will be written as

$$\exp\left(-\beta \frac{1}{32\pi^2}\int d^4x \,(F^a_{\mu\nu})^2\right), \quad \beta = \frac{8\pi^2}{g^2} \quad (27)$$

This notation is convenient since the action for one (anti)instanton is

$$\int d^4x \; F^2_{\mu\nu} = 32\pi^2 \quad (28)$$

Though the physical picture for instantons is more transparent in the Hamiltonian formulation with the $A_0 = 0$ gauge, it is more convenient to work in a covariant so-called singular gauge. In this gauge the instanton field has the following explicit form:

$$A^a_\mu(x) = O^{ab}\,\bar{\eta}_{b\mu\kappa}\,\frac{2(x-x_0)_\kappa}{(x-x_0)^2}\,\frac{\rho^2}{(x-x_0)^2+\rho^2} \quad (28)$$

Here x_0 is the position, ρ is the size, O^{ab} is the orientation of the instanton in colour space given by an orthogonal 3x3 matrix (for SU_2). For SU_{N_c} one can use rectangular $3\times(N_c^2 - 1)$ matrices to describe the orientations of instantons. By $\eta_{b\mu\kappa}$ we denote 't Hooft's symbols, the properties of which can be found in the Appendix to ref. 15. For the antiinstanton one should substitute η for $\bar{\eta}$.

In order to maintain the property that instantons smear the system along the X axis one should see that the (anti)instanton tunneling transitions occur many times in different points of space-time. That is one has to consider N_+ instantons and N_- antiinstantons (hereafter I and $\bar{I}$) with various locations, sizes and orientations, where $N_\pm$ should be of the order of the space-time volume, V.

(Anti)instantons are also known to have a property of being (anti)-

self-dual fields:

$$F^a_{\mu\nu} = \tilde{F}^a_{\mu\nu} \quad \text{for } I, \qquad \tilde{F}^a_{\mu\nu} = \frac{1}{2}\epsilon_{\mu\nu\alpha\beta} F^a_{\alpha\beta} . \tag{29}$$
$$F^a_{\mu\nu} = -\tilde{F}^a_{\mu\nu} \quad \text{for } \bar{I},$$

Therefore, the $\langle QQ \rangle$ correlation function we are interested in can be written as

$$\lambda^4 V = V\langle Q Q\rangle = \left\langle \int d^4x \frac{F\tilde{F}(x)}{32\pi^2}, \int d^4y \frac{F\tilde{F}(y)}{32\pi^2} \right\rangle$$
$$= \langle (N_+ - N_-)^2 \rangle \sim \text{const}\cdot(N_+ + N_-) \sim \text{const}\cdot V \tag{30}$$

We see that the quantity in question is the mean square of the difference between the number of Is and $\bar{\text{I}}$s.

It is instructive to compare eq. (30) with another correlation function:

$$\left\langle \int d^4x \frac{F^2(x)}{32\pi^2}, \int d^4y \frac{F^2(y)}{32\pi^2} \right\rangle - \left(\left\langle \int d^4x \frac{F^2}{32\pi^2} \right\rangle \right)^2 =$$
$$= \langle (N_+ + N_-)^2 \rangle - \langle N_+ + N_- \rangle^2 . \tag{31}$$

This expression represents the dispersion of the total number of "particles" in a statistical ensemble. If they are completely independent, they satisfy the Poisson distribution law which gives

$$\langle (N_+ + N_-)^2 \rangle - \langle N_+ + N_- \rangle^2 = \langle N_+ + N_- \rangle .$$

Meanwhile, there is a remarkable theorem for the l.h.s. of eq. (31) which follows from the renormalizability |18|, saying that it is equal to

$$\frac{4}{b} \left\langle \int d^4x \frac{F^2}{32\pi^2} \right\rangle = \frac{4}{b} \langle N_+ + N_- \rangle , \quad b = \frac{11}{3} N_c . \tag{32}$$

It is thus absolutely necessary to take into account some sort of instanton interactions in order to get a self-consistent theory.

Let us start with the so-called dilute instanton gas approximation |19|. It suggests the following approximate way of writing down the functional integral of QGD (divided by the functional integral where only configurations obtained in perturbation theory are taken into account):

$$Z = \sum_{N_\pm} \frac{1}{N_+!\,N_-!} \underbrace{\int d^4x_1 \frac{d\rho_1}{\rho_1^5} dO_1 \left[C_{N_c} \beta^{2N_c}(\rho_1) e^{-\beta(\rho_1)} \right] \cdots}_{\text{repeated } N_+ + N_- \text{ times}} \tag{33}$$

Here the small quantum oscillations around each instanton is also taken into account. They manifest themselves in the preexponential factor and also in the fact that the running coupling constant, taken at the size of the instanton, appears:

$$\beta(\rho) = \frac{8\pi^2}{g^2(\rho)} = b \log \frac{1}{\rho\Lambda} \quad . \tag{34}$$

The constant C_{N_c}, which depends on the number of colours N_c, was calculated by 't Hooft |15| at $N_c=2$ and generalized by C. Bernard |20|:

$$C_{N_c} \simeq \frac{4.7 \exp(-1.7 N_c)}{\pi^2 (N_c-1)!\,(N_c-2)!} \tag{35}$$

(Eq. (35) is for the Pauli-Villars regularization scheme.)

Unfortunately, the dilute-gas expression (33) is senseless, since it produces highly divergent integrals over the sizes of instantons:

$$\int \frac{d\rho}{\rho^5} \rho^b$$

This means that the dilute gas does not, in fact, exist: the instantons tend to expand in size until they touch each other. However, when the instantons start to overlap, two types of corrections to the non-interacting formula (33) come into play:

i) "Classical" - the total action is not a sum of free actions of individual instantons;

ii) "Quantum" - the determinant is no longer a product of determinants in the background field of separated instantons.

Both "classical" and "quantum" interactions of instantons can be investigated, at least in the case when the distances between instantons are still larger than their sizes, i.e. when the interaction can be considered as a small correction. The main contribution to the "quantum" corrections comes from the diagonalization of the would-be zero modes.

However, any "quantum" interaction, being written in the exponential form of the "classical" interaction, does not contain the β factor in the exponent. Now, what is β ? The inverse effective charge or the inverse effective temperature (in the language of statistical mechanics), β is a slowly varying logarithmic function of the sizes of the instantons. If the system stabilizes, one can estimate β according to eq. (34) as $\beta \simeq b \log 1/\bar{\rho}\Lambda$, where $\bar{\rho}$ is the average size of instantons, to be determined self-consistently at the end of the calculations. It should be emphasized that there is no theoretically small (or large) parameter in the theory - there are only certain numbers. However, let us <u>assume</u> that $\beta(\bar{\rho})$ is a large number. (We justify this at the end of calculations; indeed it happens that β = 20-30 for SU_3 and β = 10-20 for SU_2.)

Therefore, one can safely neglect the "quantum" interactions as not containing a large factor β in the exponent, <u>provided</u> the system indeed stabilizes at some $\bar{\rho} \ll \Lambda^{-1}$ due solely to the "classical" interaction. We shall see that this will be the case.

Classical instanton interaction

Let us consider a field which is a sum of an I and an $\bar{I}$ with arbitrary sizes, $\rho_{1,2}$, positions $x_{1,2}$ and relative orientations O:

$$A_\mu^a(x) = A_\mu^{(1)a}(x) + A_\mu^{(2)a}(x) \quad , \tag{36}$$

$$A_\mu^{(1)a}(x) = 2\,\bar\eta_{a\mu\varkappa}\frac{y_\varkappa\,\rho_1^2}{y^2(y^2+\rho_1^2)}\,,\quad y = x - x_1\,,$$

$$A_\mu^{(2)a}(x) = 2\,O^{aa'}\,\eta_{a'\mu\lambda}\frac{z_\lambda\,\rho_2^2}{z^2(z^2+\rho_2^2)}\,,\quad z = x - x_2\,. \tag{36}$$

The corresponding field strength can be written as a sum of strengths of separated I and $\bar{I}$, plus a mixed term arising from the off-diagonal terms in the commutator:

$$F_{\mu\nu}(A^1 + A^2) = F_{\mu\nu}(A^1) + F_{\mu\nu}(A^2) + F_{\mu\nu}(A^1, A^2).$$

The classical interaction potential is then, by definition,

$$u_{int}(R,\rho_1,\rho_2,O_{12}) = \frac{1}{32\pi^2}\int d^4x\left[F_{\mu\nu}^2(A^1+A^2) - F_{\mu\nu}^2(A^1) - F_{\mu\nu}^2(A^2)\right]$$

$$= \frac{1}{32\pi^2}\int d^4x\left[2F(1)F(2) + 2F(1)F(12) + 2F(2)F(12) + F(12)^2\right]. \tag{37}$$

Being dimensionless, u_{int} depends actually only on ρ_1/R, ρ_2/R and O_{12}, R being the distance between I and $\bar{I}$.

One can expand u_{int} at large distances in powers of $\rho_{1,2}/R$. For the sake of simplicity we put $\rho_1 = \rho_2 = \rho$ and restrict ourselves to the terms ρ^4/R^4 and ρ^6/R^6. The ρ^8/R^8 terms have also been computed, but are of little significance in what follows. We get

$$u_{int}^{I\bar{I}} = (\bar\eta O\eta RR/R^2)\left(4\frac{\rho^4}{R^4} - 15\frac{\rho^6}{R^6}\right) + 9\frac{\rho^6}{R^6} + O\left(\frac{\rho^8}{R^8}\right) \tag{38}$$

$$(\bar\eta O\eta RR/R^2) \equiv \bar\eta_{a\mu\varkappa}\,O_{12}^{ab}\,\eta_{b\mu\lambda}\frac{R_\varkappa R_\lambda}{R^2}\,. \tag{39}$$

This expression is for the SU_2 group. For SU_{N_c} replace the number 9 in eq. (38) by $27N_c/2(N_c^2-1)$. We also note that, being averaged over the relative orientations of I and $\bar{I}$, u_{int} is positive definite, i.e. repulsive. If one integrates eq. (37) over all distances between I and $\bar{I}$, one also gets a repulsion:

$$\int d^4R\, u_{int}(\rho_1, \rho_2, R, O_{12}) = \frac{27}{4}\frac{N_c}{N_c^2-1}\pi^2\rho_1^2\rho_2^2 . \tag{39}$$

Let us analyze eq. (37). The R^{-4} term is the well-known "dipole-dipole" interaction first found by Callan, Dashen and Gross |19|. The orientation-dependent structure (38) can be of any sign:

$$-3 \le (\bar{\eta} O \eta RR/R^2) \le 3 .$$

Averaged over the relative orientation in colour and/or ordinary space, this term gives zero.

The R^{-6} terms which, to my knowledge, had not been calculated up till now, consist of orientation-dependent and universally repulsive pieces. Note that the orientation-dependent piece has a sign opposite that of the standard "dipole-dipole" term.

Let us fix the relative orientation of I and $\bar{I}$ so that at large distances we would have the maximal possible attraction ($\bar{\eta}O\eta RR/R^2 = -3$). Then it is easy to find the minimum of the u_{int}:

$$\left(\frac{R^2}{\rho^2}\right)_{min} = \begin{cases} 6.75, & N_c = 2 \\ 6.26, & N_c = 3 \\ 5.63, & N_c = \infty \end{cases}, \quad u_{min} = \begin{cases} -0.088 \\ -0.102 \\ -0.126 \end{cases} \tag{40}$$

We see that the minimum is situated at rather large distances, and that it is a shallow one. A schematic presentation of the u_{int} is given in Fig. 2 where the solid line corresponds to the maximal possible attraction, and the dashed one corresponds to u_{int} averaged over the orientations.

It can be understood that in a disordered phase the "dipole-dipole" interaction is effectively averaged out, since only very peculiar orientations and positions of instantons lead to attraction. Therefore, one should take seriously only the R^{-6} universal repulsion and treat the "dipole-dipole" interaction as a perturbation. However, a very intriguing question arises: since the $I\bar{I}$ interaction is of a molecular type (see Fig. 2) why do instantons not form a NaCl-type crystal with "particles" sitting in minima of the potential, given by eq. (40)? If such an ordered phase had a smaller free energy than a disordered phase (or, speaking more accurately, a smaller thermodynamical potential, since the total number of Is and $\bar{I}$s is not fixed but should be found self-consistently from the extremum condition), spontaneous violation of global SU_{N_c} and/or Lorentz symmetry would occur.*)

Why not a crystal?

The first problem to consider is whether one can actually build, say, a cubic NaCl-type crystal with a maximal possible gain in energy on every link, or is frustration inevitable? The answer is that such a crystal can be built - see Fig. 3, where we show the orientation matrices of Is and I's:

$$\Omega_1 = \begin{pmatrix} 1 & & \\ & -1 & \\ & & -1 \end{pmatrix}, \quad \Omega_2 = \begin{pmatrix} -1 & & \\ & 1 & \\ & & -1 \end{pmatrix}, \quad \Omega_3 = \begin{pmatrix} -1 & & \\ & -1 & \\ & & 1 \end{pmatrix}, \quad \Omega_4 = \begin{pmatrix} 1 & & \\ & 1 & \\ & & 1 \end{pmatrix}.$$

This construction is for the SU_2 group. For larger groups the symmetry breaking pattern may be fixed by the interaction of not nearest neigh-

*) In principle, an intermediate situation can perhaps take place: the medium is somewhat polarized, but not ordered in the configuration space. Only colour symmetry would then be violated.

bours. We did not investigate this difficult problem in detail; however, it seems that a $SU_{N_c-2} \times U_1$ subgroup of the SU_{N_c} group remains unbroken.

Returning to the SU_2 crystal, the partition function for N Is and $\bar{I}$s can be easily computed:

$$Z = \exp\left(-\frac{b\lambda}{4} N \log \frac{N}{V\Lambda^4} + N \log 0.044\right),$$
$$\lambda = 1 + 4u_{min} = 0.65 \tag{41}$$

From (41) it follows that the dispersion in the number of Is is

$$\langle N^2 \rangle - \langle N \rangle^2 = \frac{4}{b\lambda} \langle N \rangle . \tag{42}$$

The average action is

$$\left\langle \frac{1}{32\pi^2} \int d^4x\, F_{\mu\nu}^2 \right\rangle = (1 + 4u_{min}) \langle N \rangle - \lambda \langle N \rangle . \tag{43}$$

Substituting (43) into (42) we see that we did not violate the low-energy theorem (see eq. 31 and 32). Indeed, we get

$$\left\langle \int d^4x \frac{F^2(x)}{32\pi^2}, \int d^4y \frac{F^2(y)}{32\pi^2} \right\rangle_{con} = \frac{4}{b} \left\langle \int d^4x \frac{F^2}{32\pi^2} \right\rangle ,$$

The grand partition function is

$$Z = \exp\left(\frac{b\lambda}{4} \langle N \rangle\right) = \exp\left(\frac{b}{4} V \left\langle \frac{F^2}{32\pi^2} \right\rangle\right) .$$

All that looks very nice, but from eq. (41) it follows that the average number of Is is

$$\langle N \rangle = V \Lambda^4 \lambda\, 0.0027$$

hence the gluon condensate is

$$\left\langle \frac{F^2}{32\pi^2} \right\rangle = \Lambda^4\, 0.027$$

which is a very small number. (We get much larger values for this quantity assuming the disordered phase.)

We thus conclude that at least for small groups instantons prefer a disordered phase. However, it should be kept in mind that the effective repulsion decreases as $N_c \to 0$ (see eqs. 38, 39). Therefore, at the moment we do not exclude the possibility that large groups exhibit spontaneous breakdown of global colour and/or rotational symmetry via the crystallization of instantons.

Disordered phase (= liquid)*)

A theory of instanton liquid with the interaction given by eq. (37) is a rather difficult matter, but it can be managed. Actually, the liquid stabilizes at rather large distances between instantons, so that one can use for the u_{int} its expansion (38). I present here some characteristic features of our analysis.

1. The free instanton density $d(\rho) \sim (\rho)^{\frac{11}{3}N_c}$ is replaced by

$$D(\rho) = \mathrm{const}\ d(\rho) \exp\left(-\frac{b-4}{2}\frac{\rho^2}{\bar{\rho}^2}\right)$$

with $\bar{\rho}^2$ given by

$$(\bar{\rho}^2)^2 = \frac{b-4}{2\pi^2\beta\gamma\frac{N}{V}} \quad , \quad \gamma = \frac{27}{4}\frac{N_c}{N_c^2-1} \ .$$

Since the average distance between instantons is simply $R^4 = V/N$, this formula gives the ratio $\bar{\rho}/\bar{R} \sim 1/3$ ($SU_{2,3}$). This means that the packing parameter in the liquid is, in fact, small.

---

*) Quite recently Shuryak |21| strongly advocated the liquid phase for instantons on phenomenological grounds. I take the opportunity to thank E. V. Shuryak for enlightening discussions which triggered this work.

2. It should be noted that similar results were obtained recently by Ilgenfritz and Mueller-Preussker |22| who, following Jevicki |23|, have introduced "by hand" a restriction on the sizes of Is of type $\theta(R^4_{12} - a' \rho_1^2\rho_2^2)$. Comparing their formulae with ours we find that the dimensionless core parameter a' should be taken equal to $8\beta\gamma/b$.

3. Another manifestation of the relative diluteness of the liquid is the "correlation energy" $E_{cor} = (b - 4)/4\beta$. This quantity, being compared with unity, shows the deviation of the averaged action of the liquid from the sum of free actions, a deviation due to the interactions. This quantity is of the order of a few percent.

4. The effective $\beta(\bar{\rho})$ is found to be ~15 for SU_2 and ~27 for SU_3. As mentioned above, such a large value justifies the neglect of "quantum" corrections to the partition function of instantons.

The ratio of the $<QQ>$ correlation function to the gluon condensate $<F^2/32\pi^2>$ is found to be ~1/3. This result agrees perfectly with what one expects for these quantities |24|.

However, taken at face value, the numbers we obtained suggest that the absolute value of the condensate is an order of magnitude smaller than one would like to have. This may be and may not be a real problem. Actually, all dimensional quantities of the theory are proportional to Λ. An increase of Λ by a factor 1.5 increases the gluon condensate by an order of magnitude. Furthermore, we did not calculate the two-loop corrections. Therefore, there is an undetermined factor of order of unity in the argument of the logarithm for the effective $\beta(\bar{\rho})$. Such an undetermined factor also contributes to the uncertainty of our numerical results.

To sum up, we have observed a repulsion between instantons and anti-instantons which leads to their stabilization. For small groups the disordered phase (liquid) has a smaller thermodynamical potential than does an ordered phase (crystal). The liquid is relatively dilute one. This makes it possible to speak of individual instantons in the medium.

<QQ> on a lattice

Finally, I would like to mention that Monte Carlo simulations for this fundamental quantity have been performed on a lattice for the SU_2 group |25|. The numerical result seems to be extremely small:

$$\langle QQ \rangle = \lambda^4 \sim \left(\frac{1}{3} \Lambda_{P.V.} \right)^4$$

A possible explanation of this result may follow from the picture outlined above. If, indeed, instantons have sizes much smaller than the distances between them (according to our estimates $\bar{\rho} \sim (800\ \text{MeV})^{-1}$ for SU_3), and the actual lattice spacing used in calculations is of the same order of magnitude, then instanton field configurations would be simply missed in the lattice calculations. However, they are crucial if we are to obtain the correct value for the <QQ> correlation function.

Acknowledgements

These lectures are based on work of M. I. Eides and of V. Yu. Petrov and myself. The reader is asked to refer to our original papers.

I would like to thank the organizers of the Äkäslompolo school, especially Juha Lindfors, Claus Montonen and Risto Raitio for their kind hospitality. My special thanks are due to Alice McLerran for her patience in improving the English of these lectures.

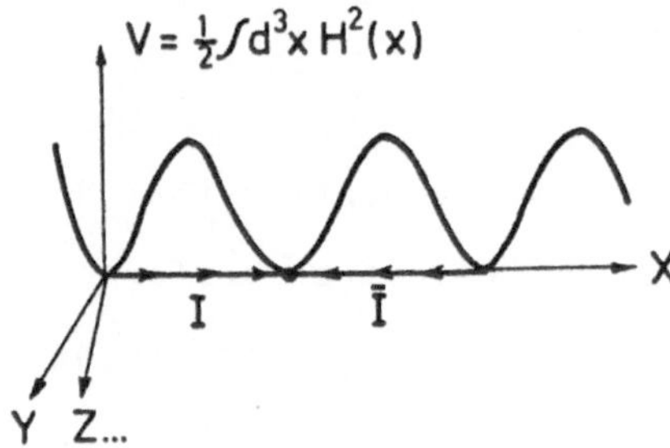

Fig. 1.

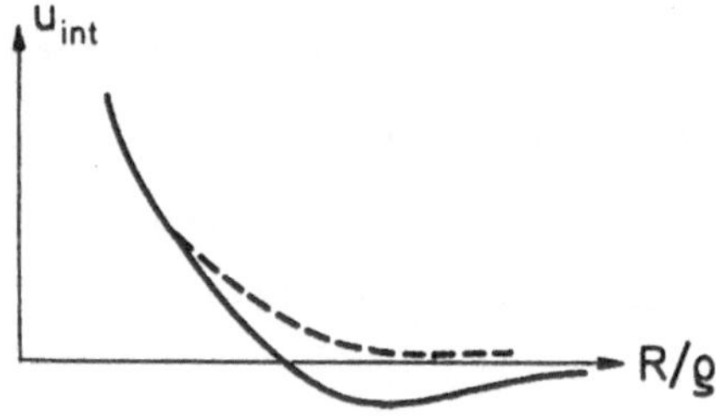

Fig. 2.

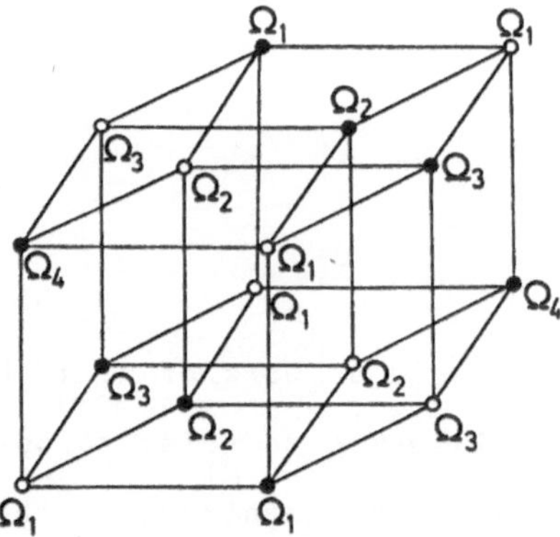

Fig. 3.

References

|1| S. Weinberg, in Proc. XVIII Intern. Conf. on High Energy Physics, London (1974) vol. 3, p. 59; Phys. Rev. D11, 3583 (1975)

|2| R.J. Crewther, Phys. Lett. 70B, 349 (1977); La Rivista del Nuovo Cimento 2, No 8, 63-117 (1979).

|3| S. Glashow, S. Weinberg, Phys. Rev. Lett. 20, 224 (1968).

|4| J. Schwinger, Phys. Rev. 82, 664 (1951), S.L. Adler, Phys. Rev. 177, 2426 (1969), J.S. Bell and R. Jackiw, Nuovo Cimento 60A, 47 (1969).

|5| K. Fujikawa, Phys. Rev. Lett. 44, 1733 (1980).

|6| D.I. D'yakonov and M.I. Eides, Sov. Phys. JETP 54, 232 (1981).

|7| G. Veneziano, Nucl. Phys. B159, 213 (1979).

|8| A. Casher, J. Kogut and L. Susskind, Phys. Rev. Lett. 31, 792 (1973).

|9| G.S. Danilov, I.T. Dyatlov and V. Yu. Petrov, Nucl. Phys. B174, 68 (1980); ZhETF 78, 1314 (1980).

|10| J. Kogut and L. Susskind, Phys. Rev. D11, 3594 (1975).

|11| D.I. Dyakonov and M.I. Eides, preprint LNPI-639 (1981).

|12| S. Coleman, The uses of instantons, preprint HUTP-78/A004 (1978)

|13| L.D. Faddeev, in Proc. of IV Int.Symp.on Non-Local Field Theories, April 20-28, 1976, Alushta, USSR, p.207; R. Jackiw and C. Rebbi, Phys.Rev.Lett. 37, 172 (1976); C. Callan, R. Dashen and D. Gross, Phys. Lett. 63B, 334 (1976).

|14| A.A. Belavin, A.M. Polyakov, A.S. Schwartz and Yu.S. Tyupkin, Phys. Lett. 59B, 85 (1975).

|15| G. 't Hooft, Phys. Rev. Lett. 37, 8 (1976); Phys. Rev. D14, 3432 (1976).

|16| E. Witten, Nucl. Phys. B156, 269 (1979).

|17| A.I. Vainshtein, V.I.Zakharov, V.A. Novikov and M.A. Shifman, in: Materials of XV Leningrad Nuclear Physics Institute Winter School, vol. 1, p. 5 (1980).

|18| V.A. Novikov, M.A. Shifman, A.I. Vainshtein and V.I. Zakharov,

Nucl. Phys. B165, 167 (1980).

|19| C. Callan, R. Dashen and D. Gross, Phys. Rev. D17, 2717 (1978).

|20| C. Bernard, Phys. Rev. D19, 3013 (1979).

|21| E.V. Shuryak, INP (Novosibirsk) preprint 81-118 (1981), to be published in Nucl. Phys. B.

|22| E.-M. Ilgenfritz, M. Mueller-Preussker, Nucl. Phys. B184, 443 (1981); Phys. Lett. 99b, 128 (1981).

|23| A Jevicki, Phys. Rev. D21, 992 (1979).

|24| D.I. Dyakonov and M.I. Eides, Leningrad Nuclear Physics Institute preprint -639 (1981).

|25| P. Di Vecchia, K. Fabricius, G.C. Rossi and G. Veneziano, Phys. Lett. 108B, 323 (1982).

GAUGE THEORIES

IN

THREE DIMENSIONS (= AT FINITE TEMPERATURE)

R. Jackiw
Center for Theoretical Physics
Massachusetts Institute of Technology
Cambridge, MA 02139

CONTENTS

I. INTRODUCTION

These lectures concern gauge theories in three dimensions[1], whose study is motivated by two considerations. Firstly, they provide interesting examples of non-trivial quantum field theories. Many have learned much from models in two dimensions; now it is the turn of three.[2] Indeed already one mechanism seen here has found application in four dimensions, and will be mentioned below.[3] Secondly, d-dimensional field theories [on a Euclidean space] govern the infinite temperature tail of d + 1 dimensional theories.[4] In this sense 3-dimensional models are physically interesting because they describe high-temperature behavior of the physical, 4-dimensional world. I shall not dwell here on details of this dimensional reduction, and I view the models under discussion as theories in their own right, without reference to the physical setting. Nevertheless, results relevant to the finite temperature application will be pointed out, and now I shall describe briefly some aspects of the temperature reduction.

Finite temperature formulations of field theory proceed in the well known Martin-Schwinger manner.[5] Amplitudes of interest are computed by the same "momentum-space" rules as at zero temperature, except that the energy variable is discrete, rather than continuous. Specifically a boson propagator becomes

$$D(p) = \frac{i}{p_0^2 - \vec{p}^2 - m^2}, \quad p_0 = i\pi(2n)T$$

while a fermion propagator is

$$S(p) = \frac{i}{\gamma_0 p_0 - \vec{\gamma}\cdot\vec{p} - m}, \quad p_0 = i\pi(2n+1)T$$

Here n is any integer and T is [proportional to] the temperature. Furthermore an integral over [virtual] 4-"momenta" becomes a sum over the integers n, and an integration over spatial 3-momenta.

$$\int \frac{d^4 p}{(2\pi)^4} \longrightarrow iT \sum_{n=-\infty}^{\infty} \int \frac{d^3 \vec{p}}{(2\pi)^3}$$

Consider now the contribution of a propagator to some perturbation theoretic diagram. With bosons, we have

$$iT \sum_{n=-\infty}^{\infty} \int \frac{d^3\vec{p}}{(2\pi)^3} \; e \; \frac{i}{-4n^2\pi^2T^2 - \vec{p}^{\,2} - m^2} \; e \cdots$$

Here e is the coupling constant governing the vertices that are connected by the propagator; the summation over the discrete "energy" is weighted by iT; the dots signify further terms. In the high T limit, all modes with $n \neq 0$ decouple, since our expression vanishes owing to the temperature dependence of the denominator -- the modes behave like very heavy particles. Only the $n = 0$ mode escapes this fate, and we conclude that, as $T \to \infty$, the above approaches

$$\int \frac{d^3\vec{p}}{(2\pi)^3} \; e\sqrt{T} \; \frac{1}{\vec{p}^{\,2} + m^2} \; e\sqrt{T}$$

This is exactly what one would find in a field theory on an Euclidean space of one dimension less, with an effective coupling constant which is $\sqrt{T}$ times the physical coupling. Moreover, it is further seen that the fermion contribution is subdominant, since the energy modes in the fermion propagator never vanish.

A position-space version of the above argument relies on the fact that finite temperature field theory is formulated over a finite temporal interval which is purely imaginary and extends from 0 to 1/iT. Fields are defined over this interval to be periodic for bosons and anti-periodic for fermions. As T becomes large, the interval shrinks to zero, and the temporal dimension disappears from the problem, as do the anti-periodic Fermi fields.

From this we conclude that the graphical perturbation series for a simple theory -- e.g. involving spin-1/2 fields and self-interacting spin-less fields -- coincides at sufficiently large temperature with the graphical series of a theory possessing the same structure of bosonic interactions, with all fermions deleted, and everything evaluated in a Euclidean space diminished by one dimension. Since 4-dimensional renormalizable theories become super-renormalizable in three dimensions, our effective field theory is less divergent than the full, finite temperature one. This reflects the fact that zero temperature renormalization is sufficient to remove ultraviolet infinities from a finite temperature theory.[5] Also 3-dimensional coupling constants have different dimension than the corresponding 4-dimensional ones. This comes about, as we have seen, from the circumstance that the effective couplings acquire powers of T in the high temperature domain.

The above is the most superficial inference that one can draw from perturbation theory. However as we move away from the simplest models, new questions arise. Upon considering a 4-dimensional theory with a 4-component Bose vector field, we realize that a 3-dimensional vector theory involves a 3-component field and we wonder how to take the temporal component of the 4-dimensional theory into account in a high temperature reduction. An answer requires detailed calculation. What is found in quantum electrodynamics (QED) and quantum chromodynamics (QCD) is that the propagator of the temporal [electric] component acquires, in lowest order perturbation theory, a contribution which acts as a mass term that increases linearly with temperature. This is just the Debye screening length arising from the charged particle plasma which is present at finite temperature. The effect is that the temporal component of the gauge potential decouples, and one is left in three dimensions with a truly 3-dimensional gauge theory. For QED, this is a non-interacting theory without dynamical interest [the fermions decouple!]; for QCD [or pure Yang-Mills theory] we are still facing a 3-dimensional non-Abelian gauge theory, with highly non-trivial dynamics.

However, perturbative analysis beyond lowest order encounters new difficulties. The problem is that we are dealing with massless gauge fields, whose propagator diverges at zero momentum. Thus in 4-dimensional, finite temperature perturbation theory we encounter integrals of the form $\int d^3p/p^2$ arising from the zero mode in the energy summation. Alternatively and equivalently we meet the same integral in the perturbative analysis of a 3-dimensional gauge theory at zero temperature. While in lowest order, the $p = 0$ singularity is integrable, it becomes enhanced in higher orders, integrals are infrared divergent, and a direct perturbative approach is vitiated.

This is the central problem with perturbation theory for gauge theories at finite temperature in four [or fewer] dimensions and at zero temperature in three [or fewer] dimensions. Similar difficulties afflict any zero temperature super-renormalizable, massless theory in any number of dimensions. While a complete understanding of these models is still lacking, it is generally agreed that the infrared divergences arise only in perturbation theory, but the complete theory is free from inconsistencies.

I shall report research by colleagues and me concerning 3-dimensional gauge theories.[1] Several topics are covered. Firstly, in the second Lecture, it is established that super-renormalizable, 3-dimensional gauge theories give rise to amplitudes that are not analytic in the coupling constant; they cannot be expanded in its powers; when an expansion is forced, divergences are encountered. We show how the non-analytic terms can be computed perturbatively, but certain contributions remain inaccessible in perturbation theory.[6] Secondly, in the third Lecture, I discuss how a 3-dimensional gauge theory may possess a mass term which does not violate gauge invariance. This fascinating structure makes use of a non-trivial topological structure -- the

Chern-Simons secondary class characteristic -- and leads to a quantization condition on the parameters of the theory.[7] In the fourth Lecture, the topological reasons for the quantization condition are explored and it is shown that the theory is inconsistent if quantization is not enforced.[7] A related quantization requirement has been recently obtained in a 4-dimensional theory,[3] and this result is also described. Finally, the fifth Lecture is devoted to concluding remarks and suggestions for further research.

II. HOW SUPER-RENORMALIZABLE GAUGE THEORIES HEAL THEIR INFRARED DIVERGENCES

Super-renormalizable interactions are governed by coupling constants with dimensions of positive powers of mass. When the fields are also massless, perturbative expansions in the super-renormalizable coupling lead to infrared divergences, even in off-mass-shell amplitudes, for the following reason. Computation of any Green's function to some [high] order yields a formula which must involve, for dimensional reasons, a [high] power of the coupling constant, divided by a [high] power of a momentum variable, characteristic of the process in question. Upon inserting this result into some further diagram and attempting a further momentum integration, infrared divergences will in general be encountered, as a consequence of the momenta in the denominator.

A. Preliminary Simple Example

It is evident that one cannot extract results from such theories by straightforward application of perturbation theory. How to proceed? Let us first look at a very simple example which in fact possess all the features of the 3-dimensional gauge theories with which we are concerned. The example is based on a 4-dimensional integral equation, which is modeled on a Schwinger-Dyson [integral] equation in a massless 4-dimensional field theory with cubic interactions.[8] In Minkowski space, the integral equation reads

$$\Gamma(p^2) = 1 - ig^2 \int \frac{d^4r}{(2\pi)^4} \frac{\Gamma(r^2)}{[(r-p)^2 + i\varepsilon][r^2 + i\varepsilon]^2} \tag{2.1}$$

Γ is the unknown function and g a coupling constant with dimensions of mass. A graphical represention is given in Fig. 1.

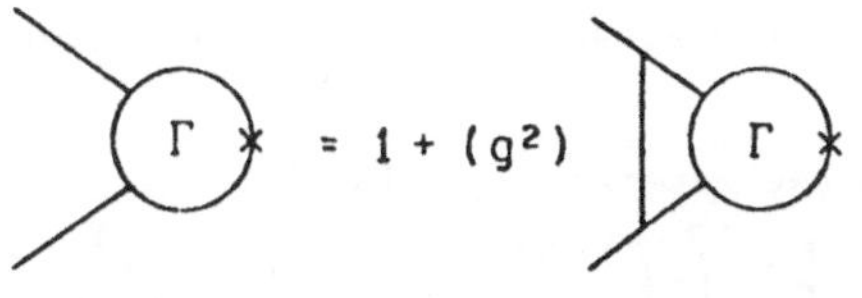

Γ = 1 + (g^2) Γ

Fig. 1. Graphical Representation of Eq. (2.1).

Eq. (2.1) cannot be solved perturbatively, because the first iteration is infrared divergent. With an infrared cutoff in the double propagator, one gets

$$\Gamma(p^2) = 1 - ig^2 \int \frac{d^4r}{(2\pi)^4} \frac{1}{[(r-p)^2 + i\varepsilon][r^2 - \mu^2 + i\varepsilon]^2} + O(g^4)$$
$$= 1 - \frac{g^2}{16\pi^2(-p^2 - i\varepsilon)} \ln\left(1 - \frac{p^2 + i\varepsilon}{\mu^2}\right) + O(g^4) \tag{2.2}$$

Nevertheless, a solution exists for (2.1), which is easily obtained by recognizing that the equation is of the Volterra type. Since $[(r - p)^2 + i\varepsilon]^{-1}$ is a Green's function for the d'Alembertian with respect to p, we may convert (2.1) to a differential equation by operating with $\Box_p$. Equivalently, we may rotate to Euclidean space, perform the angular integrations in (2.1), and find a one-variable equation.

$$f(x) = 1 - \int_0^x dy\, f(y) - x \int_x^\infty \frac{dy}{y} f(y)$$
$$\Gamma(-p^2) = f\left(\frac{g^2}{16\pi^2 p^2}\right) \tag{2.3}$$

Perturbation theory now corresponds to solving (2.3) by a power series in x; a procedure which again yields logarithmic divergences. But the differential equation which follows from (2.3),

$$f''(x) - \frac{1}{x} f(x) = 0 \tag{2.4a}$$

together with the boundary conditions implied by (2.3),

$$f(0) = 1, \quad f'(\infty) = 0 \tag{2.4b}$$

are solved by the modified Bessel function of the first kind.

Thus the solution to the integral equation (2.1) is

$$\Gamma(-p^2) = \left(\frac{g^2}{4\pi^2 p^2}\right)^{1/2} K_1\left(\left(\frac{g^2}{4\pi^2 p^2}\right)^{1/2}\right) = 1 + \frac{g^2}{16\pi^2 p^2} \ln \frac{g^2}{16\pi^2 p^2} + \frac{g^2}{16\pi^2 p^2}(2\gamma - 1)$$
$$+ O(g^4) \tag{2.5}$$

where γ is Euler's constant.

Note that the infrared divergence has disappeared and the infrared cutoff in (2.2) has been replaced by the coupling constant.

$$\mu^2 \to g^2/16\pi^2 \tag{2.6}$$

Indeed the coefficient of the non-analytic piece, viz., of the logarithm in (2.5), is exactly the same as that of the infrared divergent logarithm in (2.2). But the analytic, non-logarithmic $O(g^2)$ contribution to (2.5) cannot be found in (2.2)

The lessons to be drawn from the above exercise are the following. Infrared divergences arising from super-renormalizable interactions can be cured by considering the complete integral equations of the theory, and not attempting an expansion in powers of the coupling constant. In this way one finds coupling-constant logarithms in the perturbative expansion. The coefficients of these logarithms are determined by perturbation theory; however, there remain terms, not involving logarithms, that are not computable perturbatively.[8] All this also happens in the gauge theories to which we now turn.[9]

B. 3-Dimensional Spinor Electrodynamics

(i) Preliminaries

We consider a massless fermion field interacting with a massless Abelian gauge field A_μ, in three space-time dimensions (QED).[6, 10]

$$\mathcal{L} = -\frac{1}{4} F^{\mu\nu} F_{\mu\nu} + i \bar{\Psi} \gamma^\mu (\partial_\mu - ie A_\mu) \Psi$$

$$F_{\mu\nu} = \partial_\mu A_\nu - \partial_\nu A_\mu \tag{2.7}$$

The square of the coupling constant e has dimensions of mass; the interaction is super-renormalizable. In three space-time dimensions the Dirac matrices can be chosen to be the 2 × 2 Pauli matrices,

$$\gamma^0 = \sigma^3, \quad \gamma^1 = i\sigma^1, \quad \gamma^2 = i\sigma^2 \tag{2.8}$$

and ψ is a two-component spinor.

There does not appear a fermion mass $-m\,\bar{\psi}\psi$; in fact in three dimensions such a term violates P and T symmetries. Note there is no chiral symmetry for the massless fermions, since no matrix anti-commutes with all the Dirac [Pauli] matrices.

Discrete symmetries act in an unexpected way in 3-dimensional space-time; hence it is good to review them. Let us first recall that in two spatial dimensions, parity correpsonds to inverting one axis, say the x axis. [Inverting both would be a rotation.] One verifies that the theory (2.7) is invariant under the following parity transformation P.

$$\mathcal{P}\,\psi(t,\vec{x})\,\mathcal{P}^{-1} = \sigma_1\,\psi(t,\vec{x}')$$
$$\mathcal{P}A^0(t,\vec{x})\mathcal{P}^{-1} = A^0(t,\vec{x}'),\ \mathcal{P}A^1(t,\vec{x})\mathcal{P}^{-1} = -A^1(t,\vec{x}'),\ \mathcal{P}A^2(t,\vec{x})\mathcal{P}^{-1} = A^2(t,\vec{x}')$$
$$\vec{x} = (x,y)\ ,\qquad \vec{x}' = (-x,y) \tag{2.9a}$$

Also time inversion T is a symmetry.

$$\mathcal{T}\,\psi(t,\vec{x})\,\mathcal{T}^{-1} = \sigma_2\,\psi(-t,\vec{x})$$
$$\mathcal{T}A^0(t,\vec{x})\mathcal{T}^{-1} = A^0(-t,\vec{x}),\ \mathcal{T}\vec{A}(t,\vec{x})\mathcal{T}^{-1} = -\vec{A}(-t,\vec{x}) \tag{2.9b}$$

It is now easy to check that a fermion mass term is odd under P and T, hence we do not include it in $\mathcal{L}$. However, if $-m\bar{\psi}\psi$ were inserted in (2.7), then another reflection non-invariant term, involving only the electromagnetic fields, must be considered: $(1/4)\mu\,\varepsilon^{\alpha\beta\gamma}F_{\alpha\beta}A_{\gamma}$. We shall show in the next Lecture that this gives a gauge invariant mass to the vector field, but for the present both terms are ignored.

(ii) Infrared structure in conventional perturbation theory

To begin our study of the infrared structure, we compute first-order corrections to the fermion and gauge field propagators.

$$\mathcal{D}_{\mu\nu}(p) = \int d^3x\, e^{ipx} \langle 0|T A_\mu(x) A_\nu(0)|0\rangle \tag{2.10a}$$

$$\Delta(p) = \int d^3x\, e^{ipx} \langle 0|T \Psi(x) \bar{\Psi}(0)|0\rangle \tag{2.10b}$$

The self-energies are defined by

$$\mathcal{D}^{-1}_{\mu\nu}(p) = i P_{\mu\nu}(p)[p^2 - \Pi(p^2)] + \frac{i}{\alpha} p_\mu p_\nu \tag{2.11a}$$

$$P_{\mu\nu}(p) = g_{\mu\nu} - p_\mu p_\nu / p^2$$

$$\Delta^{-1}(p) = \frac{1}{i}[\not{p} - \Sigma(p)] \tag{2.11b}$$

We shall always work in a class of covariant gauges, parametrized by the constant α, and we shall describe our results as "gauge invariant" when they are α independent. The lowest order formulas for $\Pi_{\mu\nu}(p) = P_{\mu\nu}\Pi(p^2)$ and $\Sigma(p)$ are

$$\Pi_{\mu\nu}(p) = -ie^2 \int \frac{d^3k}{(2\pi)^3} \operatorname{tr} \gamma_\mu S(p+k) \gamma_\nu S(k) \tag{2.12a}$$

$$\Sigma(p) = -ie^2 \int \frac{d^3k}{(2\pi)^3} \gamma^\mu S(p+k) \gamma^\nu D_{\mu\nu}(k) \tag{2.12b}$$

where $D_{\mu\nu}$ and S are the free propagators.

$$D_{\mu\nu}(p) = \frac{-i}{p^2 + i\varepsilon} P_{\mu\nu}(p) - \frac{i\alpha}{(p^2 + i\varepsilon)^2} p_\mu p_\nu \tag{2.13a}$$

$$S(p) = \frac{i}{\not{p}} \tag{2.13b}$$

The graphical representation of (2.12a) and (2.12b) is contained in Fig. 2.

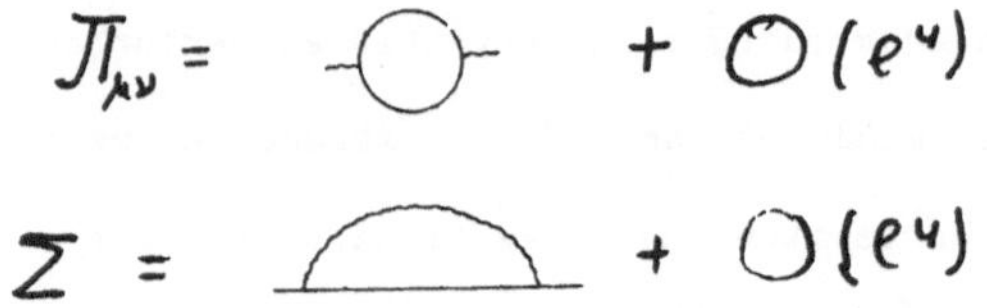

$$\Pi_{\mu\nu} = \qquad + O(e^4)$$

$$\Sigma = \qquad + O(e^4)$$

Fig. 2. Graphical representation for lowest order photon and fermion self-energies, Eqs. (2.12).

The integrals are elementary; no infrared divergences are encountered. In spite of superficial ultraviolet divergences, they too are absent, by symmetric integration in the fermion case and by gauge invariance in the gauge field case. [In this simple evaluation no regulators are needed, but if they are used, one must respect the masslessness of the fermion.]

The results are

$$\Pi(p^2) = \frac{e^2}{16}(-p^2 - i\varepsilon)^{1/2} + O(e^4) \tag{2.14a}$$

$$\Sigma(p) = -\frac{e^2\alpha}{16}\frac{\not{p}}{(-p^2 - i\varepsilon)^{1/2}} + O(e^4) \tag{2.14b}$$

The gauge dependent fermion correction vanishes in the Landau gauge ($\alpha = 0$); the gauge invariant vacuum polarization is positive for spacelike momenta, as it should be.

Next we attempt to calculate $O(e^4)$ terms. Of the several relevant graphs, the one depicted in Fig. 3 is infrared divergent.

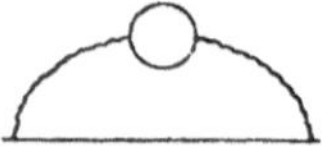

Fig. 3. Infrared divergent $O(e^4)$ two-loop contribution to fermion self-energy $\Sigma(p)$.

We conclude that to $O(e^4)$, the gauge field propagator remains finite, but that of the fermion propagator acquires a logarithmic divergence, which we now show is cured by a non-analytic $e^4 \ell n e^2$ term.

In order to avoid the logarithmic divergence, we must not attempt an expansion in powers of e^2; i.e. we must remain with the complete integral equations which determine the propagators $\mathcal{D}_{\mu\nu}$ and $\mathcal{S}$. The form of these equations, which is appropriate to the problem in hand, expresses $\mathcal{D}_{\mu\nu}$ and $\mathcal{S}$ as an infinite series in terms of themselves. To derive these, one writes the vacuum functional in terms of $\mathcal{D}_{\mu\nu}$ and $\mathcal{S}$, keeping only two-particle-irreducible graphs. The equations for the propagators follow by demanding that the variation of the functional with respect to the propagators vanishes[11].

Of course there is no hope of solving the equations exactly; we shall be content merely to determine the $O(e^4)$ logarithms. To this end we need only keep truncated equations.

$$\mathcal{D}^{-1}_{\mu\nu}(p) = i P_{\mu\nu}(p)\, p^2 - e^2 \int \frac{d^3k}{(2\pi)^3} \operatorname{tr} \gamma_\mu\, \mathcal{S}(p+k)\, \gamma_\nu\, \mathcal{S}(k) + \frac{i}{\alpha} p_\mu p_\nu + O(e^4)\Big|_{reg} \tag{2.15a}$$

$$\mathcal{S}^{-1}(p) = \frac{1}{i} \not{p} + e^2 \int \frac{d^3k}{(2\pi)^3} \gamma^\mu\, \mathcal{S}(p+k)\, \gamma^\nu\, \mathcal{D}_{\mu\nu}(k) + O(e^4)\Big|_{reg} \tag{2.15b}$$

Here $O(e^4)|_{reg}$ represents contributions that are regular to $O(e^4)$, which we shall not calculate. The omitted terms do give rise to logarithms, but only in terms $O(e^6)$ and higher. [Observe a significant difference from the toy model: There we find a single power of the logarithm; now because of non-linearities, the $O(e^4)$ logarithm fuels higher logarithms in higher orders. Thus in $O(e^8)$ there is an $\ln^2 e^2$ term as well as $\ln e^2$. All the leading logarithms can be explicitly calculated and summed; see below.] To $O(e^4)$, the insertions into fermion lines are innocuous; hence, on the right-handed side of (2.15) we may replace $\mathcal{S}$ by S, its free-field part. Thus are arrive at completely simplified equations.

$$\mathcal{D}^{-1}_{\mu\nu}(p) = ip^2 P_{\mu\nu}(p) - e^2 \int \frac{d^3k}{(2\pi)^3} \operatorname{tr} \gamma_\mu S(p+k) \gamma_\nu S(k) + \frac{i}{\alpha} p_\mu p_\nu + O(e^4)\Big|_{reg} \tag{2.16a}$$

$$\mathcal{S}^{-1}(p) = \frac{1}{i}\not{p} + e^2 \int \frac{d^3k}{(2\pi)^3} \gamma^\mu S(p+k) \gamma^\nu \mathcal{D}_{\mu\nu}(k) + O(e^4)\Big|_{reg} \tag{2.16b}$$

These are of course trivial to solve. From the lowest order result, we have an improved formula for the photon propagator.

$$\mathcal{D}_{\mu\nu}(p) = -i P_{\mu\nu}(p)\left[p^2 + i\varepsilon - \frac{e^2}{16}(-p^2 - i\varepsilon)^{1/2}\right]^{-1} - \frac{i\alpha}{(p^2+i\varepsilon)^2} p_\mu p_\nu + O(e^4)\Big|_{reg} \tag{2.17}$$

To evaluate $\mathcal{S}^{-1}$, we merely need to insert (2.17) into (2.16b). The expression for the fermion propagator which follows is

$$\mathcal{S}^{-1}(p) = \frac{1}{i}\not{p}\left[1 + \frac{e^2\alpha}{16(-p^2-i\varepsilon)^{1/2}} - \frac{e^4}{48\pi^2(p^2+i\varepsilon)} \ln \frac{e^2}{(-p^2-i\varepsilon)^{1/2}} + O(e^4)\Big|_{reg}\right] \tag{2.18}$$

The coefficient of the logarithm is gauge invariant [α independent].

It is important to appreciate that the regular $O(e^4)$ terms, which we do not calculate, cannot be obtained without first solving the theory completely. [This is analogous to what is seen in the toy model.] These analytic terms reflect the ambiguity in the normalization of $e^4 \ln e^2$; fixing them requires non-perturbative information about the amplitudes. [One may relate these unknowns to matrix elements of composite operators; but the ambiguity remains in the subtraction procedure needed to define these infinite quantities.][6, 9]

Although the logarithmic effect that we have exposed is gauge invariant [α independent], it occurs in a gauge variant quantity -- the fermion propagator. If one examines a gauge invariant amplitude, then to leading order the infrared divergences are absent. Consider for example, $\langle 0|T\bar{\psi}(x)\Gamma\psi(x)\ \bar{\psi}(y)\Gamma\psi(y)|0\rangle$, where Γ is any 2×2 matrix. Naively one would expect infrared divergences at the three-loop $O(e^4)$ level. There are two dangerous graphs, depicted in Fig. 4.

Fig. 4. Infrared divergent $O(e^4)$ three-loop contributions to a gauge invariant amplitude.

For a finite evaluation we may extract the $O(e^4)$ terms from the same graphs constructed with the improved photon propagator (2.17); see Fig. 5, where the double wavy line depicts the improved propagator $\mathcal{D}_{\mu\nu}$.

Fig. 5. Resummed expressions from which logarithmic $O(e^4)$ contributions to a gauge invariant amplitude may be extracted.

But an explicit calculation shows that $e^4 \ln e^2$ is absent from the sum, even though each individual graph contains it.

One may understand the cancellation of the infrared divergence on the basis of gauge invariance. The summed graphs of Fig. 5 can also be represented by

$$e^2 \int \frac{d^3k}{(2\pi)^3} \mathcal{D}_{\mu\nu}(k)\, T^{\mu\nu}(k,p)$$

where $T^{\mu\nu}(k,p)$ is a forward Compton amplitude for the scattering of photons on the "particles" $\bar{\psi}\Gamma\psi$. Since $T^{\mu\nu}$ is gauge invariant, viz., transverse to k^μ, the integral is simply

$$-ie^2 \int \frac{d^3k}{(2\pi)^3} \left[k^2 + i\varepsilon - \frac{e^2}{16}(-k^2 - i\varepsilon)^{1/2} \right]^{-1} T^{\mu}{}_{\mu}(k,p)$$

The $O(e^4)$ contribution

$$-\frac{ie^4}{16} \int \frac{d^3k}{(2\pi)^3} \frac{1}{(-k^2 - i\varepsilon)^{3/2}}\, T^{\mu}{}_{\mu}(k,p)$$

is not infrared divergent, in spite of the k^{-3} factor in the integrand, since by virtue of the transversality condition, $T^{\mu}{}_{\mu}$ vanishes at zero photon momentum. A higher order calculation, which is seen to involve at least four loops, must be done to exhibit any non-analytic and non-perturbative contributions to gauge invariant amplitudes.

Presumably logarithmic dependence on e^2 does occur even in gauge invariant quantities, in sufficiently high order. Indeed for scalar, massless electrodynamics a large-N analysis of the 3-dimensional theory has been performed, and non-analytic dependence on the coupling constant is found explicitly in the gauge invariant vacuum polarization tensor.[12] However, no such explicit result is as yet available in spinor electrodynamics.

While our calculation determines the $e^4 \ln e^2$ terms accurately and systematically, one may also go beyond this and sum [in a non-systematic approximation] all the leading coupling constant logarithms. Results for the fermion propagator and vacuum polarization tensor are as follows.[13]

The exact functional integrals for these two amplitudes [in Euclidean space], when evaluated in an approximation which correctly summarizes the leading coupling-constant logarithms, lead to a representation in terms of ordinary integrals [*viz*. position-dependent fields become constant fields in this study of the infrared region].

$$\Delta(p) = \int \frac{d^3A^\mu}{(2\pi)^{3/2}}\, e^{-\frac{1}{2}A^2} (\not{p} + i\lambda \not{A})^{-1} \Big/ \int \frac{d^3A^\mu}{(2\pi)^{3/2}}\, e^{-1/2 A^2} \tag{2.19a}$$

$$\Pi^{\mu\nu}(p) = e^2 \int \frac{d^3A^\mu}{(2\pi)^{3/2}}\, e^{-\frac{1}{2}A^2} \int \frac{d^3k}{(2\pi)^3}\, \mathrm{tr}\, \gamma^\mu (\not{p} + \not{k} + i\lambda\not{A})^{-1} \gamma^\nu (\not{k} + i\lambda\not{A})^{-1} \Big/ \int \frac{d^3A^\mu}{(2\pi)^{3/2}}\, e^{-\frac{1}{2}A^2} \tag{2.19b}$$

$$\lambda^2 = \frac{e^4}{48\pi^2} \ln P/e^2 > 0$$

A formal expansion of the above in powers of λ gives the $e^4 \ln e^2$ series encountered in perturbation theory. [In the vacuum polarization tensor (2.19b), all terms beyond the first integrate to zero, since the leading coupling-constant logarithms are absent.] Evaluation of the integrals yields

$$\Delta(p) = \frac{1}{\not p}\left[1 + \left(\frac{\pi}{2}\right)^{1/2}\frac{\lambda}{p}e^{p^2/2\lambda^2}\,\mathrm{erfc}\left(\frac{p}{\sqrt{2}\lambda}\right) - \left(\frac{\pi}{2}\right)^{1/2}\left(\frac{\lambda}{p}+\frac{p}{\lambda}\right)e^{-p^2/2\lambda^2}\right] \tag{2.20a}$$

$$\Pi^{\mu\nu}(p) = e^2 P^{\mu\nu}(p)\left[\frac{p}{16} + \frac{2\lambda}{(2\pi)^{3/2}} - \frac{p}{8\pi}\int_0^{\pi/2} d\theta\,\mathrm{erfc}\left(\frac{p\sin\theta}{2^{3/2}\lambda}\right)\right] \tag{2.20b}$$

The results exhibit unexpected features. For the fermion propagator the first two terms in the brackets, when expanded for small λ in an asymptotic series, reproduce the Borel-summable perturbative series involving even powers of λ. [The first two contributions to that series agree with (2.18).] However, the last term possesses an essential singularity at $\lambda^2 = 0$ and is not seen in perturbation theory. For the vacuum polarization tensor, only the first term in the brackets reproduces the perturbative result (2.19). The remainder is entirely non-perturbative; an asymptotic expansion for small λ yields a series in odd powers of $\lambda = (e^2/\sqrt{48\pi})\ln^{1/2}(p/e^2)$, which do not occur in perturbation theory. Of course the reduction of the exact functional integrals to ordinary integrals is justified by perturbation theory. Hence the significance of the non-perturbative contributions remains unclear.

In summary, 3-dimensional massless QED cures is perturbative, infrared divergences by giving rise to coupling-constant logarithms. The effect is gauge invariant in the class of covariant Lorentz gauges and first occurs in $O(e^4)$, but only to higher order [four-loop or higher] in gauge-invariant amplitudes. Subdominant terms include non-logarithmic $O(e^4)$ contributions, as well as double, triple, *etc.*, logarithms in $O(e^8)$, $O(e^{12})$, *etc*. The coefficients of the leading logarithms are computable in perturbation theory, but the normalization of the logarithm, which determines the non-logarithmic term, is not. Our results should be valid in the region where $e^2/(-p^2)^{1/2} \ll 1$ and $\left|\ln[e^2/(-p^2)^{1/2}]\right| \gg 1$, so

$e^2/(p^2)^{1/2}\ln[e^2/(-p^2)^{1/2}]$ also is small. In particular, the present method gives no information about the propagators at zero momenta; this involves $e^2/(-p^2)^{1/2} \gg 1$ and requires solving the theory completely. Thus we cannot illuminate the interesting question whether masses are generated spontaneously.

(iii) Infrared structure in dimensional regularization.

Our derivation of the coupling-constant non-analytic formula (2.18) was achieved by resumming perturbation theory, and by extracting from the resummed series the non-analytic $O(e^4)$ pieces. In this way the method and the result are similar to what is found in bound-state perturbation theory, e.g. $\alpha^2\ln\alpha$ contributions to the Lamb shift;[14] also they are similar to chiral perturbation theory, where logarithms of the chirality breaking parameter are encountered.[15] It is interesting to see how the same result emerges when dimensional regularization is used; it happens almost magically, since no resummation is required.[16]

In dimensional regularization, we use Feynman rules implied by (2.7), except the theory is considered in d dimensions, and the coupling constant e is set equal to $\mu^{2-d/2}$, where μ is a parameter with dimensions of mass. At the end of the calculation, answers are continued to d = 3, and evaluated there whenever possible. Evaluation to two-loop order [in the Landau gauge, $\alpha = 0$] results in an expression for Σ which has a pole at d = 3, corresponding to the infrared logarithmic divergence encountered previously.

$$\begin{aligned}\Sigma(p) &= -\frac{\not p}{96\pi^2}\,\frac{1}{3-d}\left(\frac{\mu^2}{-p^2-i\varepsilon}\right)^{4-d} + \cdots \\ &= \frac{1}{\not p}\,\frac{\mu^2}{96\pi^2}\left(\frac{1}{3-d} + \ln\frac{\mu^2}{-p^2-i\varepsilon}\right) + \cdots \\ &= \frac{1}{\not p}\,\frac{e^4}{96\pi^2}\left(\frac{1}{3-d} + \ln\frac{e^4}{-p^2-i\varepsilon}\right) + \cdots\end{aligned} \tag{2.21}$$

Here the dots stand for terms which vanish at d = 3, as well as for terms regular and higher order in μ^2. Observe that the logarithm is exactly the same as in (2.18), but (2.21) has an additional, unwanted divergent piece, proportional to 1/d-3.

It is remarkable that the divergent piece may be removed by an unconventional counter term. Let us observe that the usual action, together with the gauge fixing term,

$$I_\alpha = \int d^3x \left[-\frac{1}{4} F^{\mu\nu} F_{\mu\nu} - \frac{1}{2\alpha} (\partial_\mu A^\mu)^2 + i \bar{\Psi} \gamma^\mu (\partial_\mu - ie A_\mu) \Psi \right] \tag{2.22a}$$

may be supplemented by a further contribution, which formally does not affect dynamics: we add to (2.22a)

$$I_c = -\frac{1}{2} e^4 c^2 \int d^3x\, J^\mu(x) \int d^3y\, J_\mu(y)$$

$$J^\mu = \bar{\Psi} \gamma^\mu \Psi \tag{2.22b}$$

where c^2 is a dimensionless constant.

To see that (2.22b) may be added at will, we use a functional formulation. Consider the gauge-fixed quantity

$$Z = \int \mathcal{D}A^\mu \mathcal{D}\bar{\Psi} \mathcal{D}\Psi \exp i I_\alpha \tag{2.23a}$$

Obviously this may be multiplied by unity, represented as an ordinary integral over a constant vector B^μ.

$$1 = \frac{1}{Z_0} \int \frac{d^3B^\mu}{(2\pi)^{3/2}} \exp \frac{i}{2} B^2$$

Next shift the functional A^μ integration in (2.23a) by the constant amount ceB^μ, and perform the B^μ integral. Since the shift of A^μ in I_α results in

$$I_\alpha \to I_\alpha + ce^2 B^\mu \int d^3x\, j_\mu(x)$$

it follows that

$$\begin{aligned} Z &= \frac{1}{Z_0} \int \frac{d^3B^\mu}{(2\pi)^{3/2}} \exp \frac{i}{2} B^2 \int \mathcal{D}A^\mu \mathcal{D}\bar{\Psi} \mathcal{D}\Psi \exp i I_\alpha \\ &= \frac{1}{Z_0} \int \frac{d^3B^\mu}{(2\pi)^{3/2}} \exp \frac{i}{2} B^2 \int \mathcal{D}A^\mu \mathcal{D}\bar{\Psi} \mathcal{D}\Psi \exp i \left(I_\alpha + ce^2 B^\mu \int d^3x\, j_\mu(x) \right) \\ &= \int \mathcal{D}A^\mu \mathcal{D}\bar{\Psi} \mathcal{D}\Psi \exp i (I_\alpha + I_c) \end{aligned} \tag{2.23b}$$

[One may view this procedure as removing the gauge freedom which remains in covariant gauges, which fix only $\partial_\mu A^\mu$, but allow shifts of A^μ by a constant vector, corresponding to a gauge transformation with a gauge function linear in x. Moreover, because adding $\int d^3x j^\mu(x) \int d^3y j_\mu(y)$ to the action is equivalent to introducing additional zero-momentum photon vertices into the amplitudes, we see that gauge invariant Green's functions are not modified.][17]

I_c gives rise to new vertices. By chosing c^2 to be proportional to $1/3-d$, the singularity at $d = 3$ in (2.21) may be cancelled, however leaving an undetermined $O(e^4)$ finite part.

In this elegant and formal way the entire expression (2.21) is regained. By extending these arguments, all other results obtained by resummation methods may also be reproduced.[16]

C. 3-Dimensional Yang-Mills Theory

Next we consider non-Abelian SU(N) gauge theories [without fermions].

$$\mathcal{L} = -\frac{1}{4} F^{a\,\mu\nu} F^{a}_{\ \mu\nu}$$

$$F^a_{\mu\nu} = \partial_\mu A^a_\nu - \partial_\nu A^a_\mu + g f^{abc} A^b_\mu A^c_\nu \tag{2.24a}$$

$$f^{abc} f^{a'bc} = N\,\delta^{aa'} \tag{2.24b}$$

As in the Abelian case, it is possible to construct a gauge invariant mass term, but we postpone consideration of this until the next Lecture. The gauge-fixing and gauge-compensating terms to be added to the Lagrangian are

$$-\frac{1}{2\alpha}(\partial^\mu A^a_\mu)^2 + \bar{u}^a \Box u^a + \bar{u}^a \partial^\mu (g f^{abc} A^b_\mu u^c) \tag{2.24c}$$

where u is the Faddeev-Popov ghost field.

We shall study the two-point functions of the gauge potentials,

$$\delta^{ab} \mathcal{D}_{\mu\nu}(p) = \int d^3x\, e^{ipx} \langle 0 | T A^a_\mu(x) A^b_\nu(0) | 0 \rangle \tag{2.25a}$$

as well as of the ghosts.

$$\delta^{ab} \mathcal{G}(p) = \int d^3x\, e^{ipx} \langle 0 | T u^a(x) \bar{u}^b(0) | 0 \rangle \tag{2.25b}$$

The self-energies are defined by

$$\mathcal{D}^{-1}_{\mu\nu}(p) = i P_{\mu\nu}(p)\left[p^2 - \Pi(p^2)\right] + \frac{i}{\alpha} p_\mu p_\nu \tag{2.26a}$$

$$\mathcal{G}^{-1}(p) = i\left[p^2 - M(p^2)\right] \tag{2.26b}$$

The lowest order results are[18]

$$\Pi(p^2) = -g^2 \frac{N}{64}\left(10 + (1+\alpha)^2\right)(-p^2 - i\varepsilon)^{1/2} \tag{2.27a}$$

$$M(p^2) = -g^2 \frac{N}{16}(-p^2 - i\varepsilon)^{1/2} \tag{2.27b}$$

The vacuum polarization is negative for space-like momenta, in contrast to the Abelian theory. This circumstance is of course familiar from the 4-dimensional calculation. The gauge dependence [α dependence] is to be expected, since we are calculating a gauge variant quantity. The ghost self-mass (2.27b) also is negative for spacelike momenta; moreover it has the surprising feature that it is unexpectedly gauge invariant [α independent].

Observe that the negative self-masses produce a pole in the gauge and ghost propagators at spacelike momenta. Admittedly, the location of the pole is at $g^2/(-p^2)^{1/2} \sim O(1)$, where we can no longer rely on our approximate calculation. Nevertheless, it is puzzling to encounter this further infrared singularity, which is the 3-dimensional residue of asymptotic freedom, and presumably signals the infrared instability of the theory.

In the next order, infrared divergences are encountered. To avoid them, we proceed as in the Abelian case: the complete Schwinger-Dyson equations are truncated, so that they are exact up to, but not including $O(g^4)|_{reg}$. They are then solved, and the $O(g^4)$ coupling constant logarithms are found. The procedure is entirely similar to the Abelian calculation, but the calculation is much more lengthy. We record only the answers.

$$\mathcal{D}^{-1}_{\mu\nu}(p) = i\,P_{\mu\nu}(p)\left[p^2 + g^2 c(-p^2-i\varepsilon)^{1/2} + \frac{g^4 c N}{3\pi^2}(2+\alpha)\ln\frac{g^2}{(-p^2-i\varepsilon)^{1/2}}\right] + \frac{i}{\alpha}p_\mu p_\nu + O(g^4)\big|_{reg} \tag{2.28a}$$

$$\mathcal{G}^{-1}(p) = i\left[p^2 + \frac{g^2 N}{16}(-p^2-i\varepsilon)^{1/2} - \frac{g^4 c N}{3\pi^2}\ln\frac{g^2}{(-p^2-i\varepsilon)^{1/2}}\right] + O(g^4)\big|_{reg} \tag{2.28b}$$

$$c \equiv \frac{N}{64}\left(10 + (1+\alpha)^2\right) \tag{2.28c}$$

In arriving at this result it was important that only $O(g^4 \ln g^2)$ terms were kept; only they are insensitive to the [unphysical?] pole at $g^2/(-p^2)^{1/2} \sim O(1)$. Everything is gauge dependent, but note that for no real value of α can c vanish. Hence the ghost propagator always possess a non-analytic piece, even though the one in the gauge field propagator may be eliminated for $\alpha = -2$.

In summary, to the order here investigated, the perturbative non-Abelian theory behaves similarly to the Abelian one, except for a greater gauge dependence and a characteristic reversal of signs. As in the Abelian case, the validity of these results requires $g^2/(-p^2)^{1/2} \ll 1$, $|\ln g^2/(-p^2)^{1/2}| \gg 1$ and $g^2/(-p^2)^{1/2} \ln g^2/(-p^2)^{1/2} < 1$. Consequently we cannot probe the region $p^2 \to 0$, and cannot determine whether the gauge field acquires a mass.

III. TOPOLOGICALLY MASSIVE GAUGE THEORIES

While presenting 3-dimensional gauge theories, I mentioned that it is possible to add to the Lagrangian a mass term for the gauge field, without spoiling gauge invariance. Although this appears to be a peculiar feature of 3-dimensional theories, it is an interesting phenomenon, whose certain aspects have 4-dimensional analogs. Also there may be a direct physical [high temperature] significance to this mass. For all these reasons it is profitable to study the subject.[19]

A. Abelian Theory

(i) Non-interacting theory

Consider the following Lagrange density in 3-dimensional space-time.

$$\mathcal{L} = -\frac{1}{4} F^{\mu\nu} F_{\mu\nu} + \frac{\mu}{4} \varepsilon^{\mu\nu\alpha} F_{\mu\nu} A_\alpha$$
$$F_{\mu\nu} = \partial_\mu A_\nu - \partial_\nu A_\mu \tag{3.1}$$

Dimensional arguments show that μ has dimension of mass. Although the Lagrange density is not gauge invariant, the equation of motion which follows from (3.1) is.

$$\partial_\mu F^{\mu\nu} + \mu \, {}^*F^\nu = 0 \tag{3.2}$$

Here we have defined the dual field, which in three dimensions is a vector.

$${}^*F^\mu = \frac{1}{2} \varepsilon^{\mu\alpha\beta} F_{\alpha\beta}$$
$$F^{\alpha\beta} = \varepsilon^{\alpha\beta\mu}\, {}^*F_\mu \tag{3.3}$$

Note that the dual field is identically conserved.

$$\partial_\mu {}^*F^\mu = 0 \tag{3.4}$$

This Bianchi identity is a consequence of the definitions of $F^{\mu\nu}$ and ${}^*F^\mu$; alternatively, it follows from the equation of motion (3.2), owing to the antisymmetry of $F^{\mu\nu}$. Under a gauge transformation The Lagrange density changes by a total derivative.

$$A_\mu \to A_\mu + \partial_\mu \Theta \tag{3.5a}$$

$$\mathcal{L} \to \mathcal{L} + \frac{\mu}{2} \partial_\mu ({}^*F^\mu \Theta) \tag{3.5b}$$

This is why the equation of motion is gauge invariant.

While it is clear that μ has dimension of mass, it still remains to demonstrate that it is indeed a mass term for the field. This is most easily done by writing the field equation (3.2) in terms of the dual tensor (3.3). Eq. (3.2) is equivalent to

$$(\mu g^{\mu\alpha} + \varepsilon^{\mu\alpha\beta} \partial_\beta)\, {}^*F_\alpha = 0 \tag{3.6a}$$

Multiplying this with the differential operator $(\mu g^{\nu\mu} - \varepsilon^{\nu\mu\gamma}\partial_\gamma)$ yields

$$(\Box + \mu^2)\, {}^*F^\nu = 0 \tag{3.6b}$$

which demonstrates clearly that the gauge field excitations are massive.

It is gratifying that μ^2 occurs in (3.6b) with the correct sign for a propagating particle. Although we have no <u>a priori</u> control over this sign [$\mathcal{L}$ is linear in μ], we may understand that it must emerge the way it does by considering the energy-momentum tensor $\theta^{\mu\nu}$. When coupling our theory to an external metric, $(\mu/4)\int d^3x\ \varepsilon^{\mu\nu\alpha} F_{\mu\nu} A_\alpha$ is already a coordinate invariant world scalar, without additional metric factors. Hence the variation of the action $I = \int d^3x \mathcal{L}$ with respect to the metric [this variation defines the energy-momentum tensor] does not see the mass term. Consequently $\theta^{\mu\nu}$ has its conventional Maxwell form.

$$\Theta^{\mu\nu} = - F^{\mu\alpha} F^\nu{}_\alpha + \frac{1}{4} g^{\mu\nu} F^{\alpha\beta} F_{\alpha\beta} \tag{3.7}$$

In particular the energy $\mathcal{E}$ is a positive definite quantity,

$$\mathcal{E} = \frac{1}{2}\int d^2\vec{x}\,\left(\vec{E}^{\,2} + B^2\right)$$

(3.8a)

$$\vec{E} = -\vec{\nabla}A^0 - \dot{\vec{A}}, \quad B = -\frac{1}{2}\varepsilon^{ij}F_{ij} = \vec{\nabla}\times\vec{A}$$

(3.8b)

and the system's excitations cannot be tachyonic. Of course $\theta^{\mu\nu}$ remains conserved in our theory, as a consequence of the field equation (3.2).

The fact that the action associated with our mass term is a world scalar is evidence for its topological nature. This will have profound implication for the quantum theory of the non-Abelian generalization, which we shall discuss later. Here I want to record another curious topological property. Consider the time component of the field equation (3.2), in the presence of an external charge density ρ. This is the analog of Gauss' law; in our theory it reads

$$\vec{\nabla}\cdot\vec{E} - \mu B = \rho$$

(3.9a)

Upon integrating (3.9a) over all space, the first term vanishes, since the fields, being massive, decrease exponentially at large distances. One is left with

$$-\mu\int d^2\vec{x}\,B = \int d^2\vec{x}\,\rho = Q$$

(3.9b)

The magnetic flux passing out of our 2-dimensional space is proportional to the external charge Q. Correspondingly, the magnetic potential is long range, even though the magnetic field is short range.

$$\vec{A} \xrightarrow[r\to\infty]{} -\vec{\nabla}\frac{Q}{2\pi\mu}\tan^{-1} y/x$$

(3.10)

This is similar to the electromagnetic configuration supported by vortices in the Higgs model.[20]

Finally, let us note that apart from a total derivative, $\mathcal{L}$ may be written in a gauge invariant form, which, however, is spatially non-local, and Lorentz non-invariant. This follows from the identity

$$\frac{\mu}{4}\varepsilon^{\mu\nu\alpha}F_{\mu\nu}A_{\alpha}=\frac{\mu}{2}B\frac{\vec{\nabla}}{\nabla^2}\cdot\vec{E}-\frac{\mu}{2}\vec{E}\cdot\frac{\vec{\nabla}}{\nabla^2}B-\frac{\mu}{4}\varepsilon^{\mu\nu\alpha}\partial_{\alpha}\left(F_{\mu\nu}\frac{\vec{\nabla}}{\nabla^2}\cdot\vec{A}\right)$$

Such an explicitly gauge invariant formulation is not available for the non-Abelian generalization.

(ii) Interacting quantum theory

An interesting interacting generalization of the above model involves coupling [2-component] fermions to (3.1). Their gauge invariant Lagrange density is

$$\mathcal{L}_F=i\bar{\Psi}\gamma^{\mu}(\partial_{\mu}-ieA_{\mu})\Psi-m\bar{\Psi}\Psi \tag{3.11}$$

As already remarked above, the reflection non-invariant gauge field mass term should be accompanied by a Fermion mass term with the same quantum numbers. That is why we include such a term in (3.11). [If one were omitted, it would be induced by the other through radiative corrections.]

An analysis of the kinematics of this theory shows that both the massive photon and fermion carry non-vanishing spin: $\mu/|\mu|$ and $m/|2m|$, respectively. In three dimensions, "spin" is a pseudoscalar. The existence of a single excitation with only one value of the spin -- as opposed to two, each differing in sign from the other -- signals reflection non-invariance. Of course, the lack of this symmetry in our theory is already evident from the Lagrangian, which contains reflection non-invariant structures: $\varepsilon^{\alpha\beta\gamma}$, $\bar{\psi}\psi$. One may regain a P and T conserving system by working with a doublet of models, one with masses (μ,m), the other with $(-\mu,-m)$, and defining parity and time-inversion to include a field interchange.

Feynman-Dyson perturbation theory is straight-forwardly carried out. It is both infrared and ultraviolet finite in the Landau gauge, where the free propagators read

$$D_{\mu\nu}(p) = \frac{-i}{p^2 - \mu^2 + i\varepsilon}\left[P_{\mu\nu}(p) - i\mu\,\varepsilon_{\mu\nu\alpha}\,p^\alpha/p^2\right] \tag{3.12a}$$

$$S(p) = \frac{i}{\not p - m} \tag{3.12b}$$

Consequently, this is the only non-trivial field theory which is known to possess a perturbation expansion entirely free of divergences -- not even normal ordering need be performed, provided Lorentz and gauge invariance are maintained.

B. Non-Abelian Theory

The 3-dimensional mass term can be generalized to a non-Abelian gauge theory. The gauge field Lagrange density is

$$\mathcal{L} = \frac{1}{2g^2}\,\mathrm{tr}\,F^{\mu\nu}F_{\mu\nu} - \frac{\mu}{2g^2}\,\varepsilon^{\mu\nu\alpha}\,\mathrm{tr}\left(F_{\mu\nu}A_\alpha - \tfrac{2}{3}A_\mu A_\nu A_\alpha\right) \tag{3.13}$$

We use a matrix notation

$$A_\mu = g\,T^a A^a_\mu$$

$$F_{\mu\nu} = g\,T^a F^a_{\mu\nu} = \partial_\mu A_\nu - \partial_\nu A_\mu + [A_\mu, A_\nu] \tag{3.14}$$

which employs the representation matrices of the group.

$$[T^a, T^b] = f^{abc}\,T^c \tag{3.15}$$

The coupling constant is g, while μ/g^2 is dimensionless. The field equations which follow from (3.13) are gauge covariant,

$$\mathcal{D}_\mu F^{\mu\nu} + \frac{\mu}{2} {}^*F^\nu = 0 \tag{3.16a}$$

$$\mathcal{D}_\mu = \partial_\mu + [A_\mu, \ ] \tag{3.16b}$$

and from our previous consideration of the non-interacting limit ($g = 0$), we know that μ indeed provides a mass for the field. The dual field

$$\begin{aligned} {}^*F^\mu &= \tfrac{1}{2} \varepsilon^{\mu\alpha\beta} F_{\alpha\beta} \\ F^{\alpha\beta} &= \varepsilon^{\alpha\beta\mu} \, {}^*F_\mu \end{aligned} \tag{3.17}$$

satisfies the Bianchi identity

$$\mathcal{D}_\mu {}^*F^\mu = 0 \tag{3.18}$$

as a consequence of the definitions (3.14b) and (3.17), or alternatively as a consequence of the field equation (3.16a). The dual of (3.16a) is

$$\mathcal{D}_\alpha {}^*F_\beta - \mathcal{D}_\beta {}^*F_\alpha - \mu F_{\alpha\beta} = 0 \tag{3.19a}$$

and another covariant divergence converts this, with the help of (3.16 a) and the Ricci identity $[\mathcal{D}_\alpha, \mathcal{D}_\beta] = [F_{\alpha\beta}, \ ]$ into

$$(\mathcal{D}_\alpha \mathcal{D}^\alpha + \mu^2) {}^*F^\mu = \varepsilon^{\mu\alpha\beta} [{}^*F_\alpha, {}^*F_\beta] \tag{3.19b}$$

which is the non-Abelian analogue of (3.6b).

$\mathcal{L}$ is not invariant against gauge transformations; rather it changes by a total derivative. Consider a finite transformation

$$A_\mu \rightarrow U^{-1} A_\mu U + U^{-1} \partial_\mu U \tag{3.20a}$$

The response of the action to the gauge transformation (3.20a) is

$$\begin{aligned} \int d^3x\, \mathcal{L} \longrightarrow \int d^3x\, \mathcal{L} &+ \frac{\mu}{g^2} \int d^3x\, \varepsilon^{\alpha\beta\mu} \partial_\alpha \operatorname{tr} [\partial_\beta U U^{-1} A_\mu] \\ &+ \frac{\mu}{3g^2} \int d^3x\, \varepsilon^{\alpha\beta\gamma} \operatorname{tr} (\partial_\alpha U U^{-1} \partial_\beta U U^{-1} \partial_\gamma U U^{-1}) \end{aligned} \tag{3.20b}$$

The second term on the right-hand side, which is manifestly a total divergence, is the analogue of the Abelian term (3.5b). We shall only consider gauge transformations which tend to the identity at temporal and spatial infinity.

$$U(x) \xrightarrow[x\to\infty]{} \pm I \tag{3.21}$$

This restriction is made to avoid convergence problems in (3.20). Also, it reflects our assumption of asymptotic space-time uniformity. With Eq. (3.21), we may conclude that the A-dependent surface integral in (3.20) vanishes. The last term in (3.20), which has no Abelian analog, can also be converted to a surface integral once the integrand is rewritten as a total derivative. This can be made manifest by introducing an explicit parametrization for U. We choose the gauge group to be SU(2) [more generally, we consider a SU(2) subgroup of the gauge group] and make use of the exponential parametrization.

$$\begin{aligned} U(x) &= \exp T^a \Theta^a(x) \\ T^a &= \sigma^a/2i \\ \Theta^a &= \hat{\Theta}^a |\Theta| \end{aligned} \tag{3.22}$$

It follows that

$$\int d^3x\, \mathcal{L} \longrightarrow \int d^3x\, \mathcal{L} + \mu \frac{8\pi^2}{g^2} w(U) \tag{3.23}$$

where we have introduced the "winding number" of the gauge transformation U.

$$\begin{aligned} w(U) &= \frac{1}{24\pi^2} \int d^3x\, \varepsilon^{\alpha\beta\gamma}\, \mathrm{tr}\left[\partial_\alpha U U^{-1} \partial_\beta U U^{-1} \partial_\gamma U U^{-1}\right] \\ &= \frac{-1}{16\pi^2} \int d^3x\, \varepsilon^{\alpha\beta\gamma} \varepsilon^{abc} \partial_\alpha \left[\hat{\Theta}^a \partial_\beta \hat{\Theta}^b \partial_\gamma \hat{\Theta}^c \right. \\ &\qquad \left. \times \left(|\Theta| - \sin|\Theta|\right)\right] \end{aligned} \tag{3.24}$$

It will be shown in the next Lecture that w(U), though given by a surface integral, is not zero, but takes an integer value which characterizes the homotopic equivalence class to which U belongs. Only for homotopically trivial U's -- those continuously deformable to I -- does w(U) vanish. These considerations are, of course, familiar from the analysis of topological structure in 4-dimensional Yang-Mills theory.[21] That they should reappear in the 3-dimensional theory is not surprising, in view of the further mathematical/topological connections which we shall draw in the fourth Lecture.

We conclude that the action is not gauge invariant, but changes by $\mu(8\pi^2/g^2)\, w(U) = \mu(8\pi^2/g^2) \times$ integer. However in quantum mechanics, it is the exponential of the action, $\exp i \int d^3x \mathcal{L}$, that should be gauge invariant. Hence a change in the action, can be tolerated only if it is an integral multiple of 2π. Consequently the requirement of gauge invariance gives a quantization condition on the dimensionless ratio $4\pi\mu/g^2$.

$$4\pi \frac{\mu}{g^2} = n \qquad\qquad n = 0, \pm 1, \ldots \tag{3.25}$$

A Euclidean formulation leads to the same conclusion. The functional integral requires $\exp - \int d^3x \mathcal{L}$ to be gauge invariant, but the mass term's contribution to the action is purely imaginary; a factor of i appears when the continuation to imaginary time [Euclidean space] is performed. The winding number is a world scalar; hence it takes the same integer value regardless of the space's signature. The quantization condition (3.25) follows as before; it is entirely due to the internal group.

In the next Lecture we shall explore the topological setting for this remarkable result, which is somewhat analogous to Dirac's monopole quantization condition. Also we show there how the quantum theory becomes inconsistent in the absence of quantization, and how similar considerations enforce an unexpected quantization condition in a 4-dimensional example.

Let me speculate concerning the physical significance of the mass term. As I remarked in the first, introductory Lecture, finite temperature perturbation theory, for simple models, suggests that in the high temperature limit a 3-dimensional version of that same model should come into play. Of course such a dimensional reduction makes no reference to non-perturbative phenomena. Moreover, for a non-Abelian gauge theory, which we know to be rich in non-perturbative effects, neither the 4-dimensional finite temperature perturbation theory, nor zero temperature perturbation theory in three dimensions make sense. We offered one possible resolution of the difficulty: a resummation of the perturbative series produces terms which are not analytic in the coupling constant. However, with the discovery of the 3-dimensional mass term, another idea presents itself: we conjecture that the high temperature limit of a non-Abelian, 4-dimensional gauge theory is governed by a 3-dimensional, massive yet gauge invariant Yang-Mills theory, of the type that I am here describing.

There is no derivation of this fact; but neither can it be falsified, since naive perturbation theory does not exist and we do not have sufficient control over the formalism to extract non-perturbative behavior. Confronting such a hiatus, we invoke the principle of "naturalness" to aid in constructing the effective Lagrangian. The 3-dimensional effective Lagrangian should possess all terms with quantum numbers of the 4-dimensional theory, whose high temperature asymptote is being described. According to theis criterion, the gauge invariant mass should be present, since its reflection non-invariance mirrors the reflection non-invariance of the 4-dimensional θ vacua. Indeed, when discussing the topological setting of our mass term in the next Lecture, we shall see an intimate mathematical connection between the mass term and the quantity responsible for the θ vacua. However, it is not known at present, whether this mathematical relationship can be the basis for a physical derivation.

If we accept the gauge invariant mass as a proper term in the effective Lagrangian which summarizes high temperature behavior of physical non-Abelian gauge theories, we get another bonus, beyond infrared regularity. Owing to the quantization condition (3.25), the mass becomes evaluated in terms of the coupling constant and if we recall the connection between the 3-and 4-dimensional coupling constants, we find

$$\mu = \frac{g^2}{4\pi} n = \frac{e^2 T}{4\pi} n = \alpha T n \tag{3.26}$$

The integer structure to μ is most fascinating. A non-vanishing mass presumably arises from a non-vanishing θ, and discontinuities in the former for the 3-dimensional model are suggestive of different phases in the latter for the 4-dimensional theory. That different values of θ correspond to different phases has been occasionally suggested. Clearly it would be most satisfying if more understanding of these speculative ideas could be obtained.

Finally let me conclude by remarking that perturbation theory for gauge invariant massive Yang-Mills theory has been performed. Though much more complicated than in the Abelian case, the one loop results have been evaluated.[7] It appears that infrared divergences are absent as in the Abelian theory; ultraviolet finiteness should also be guaranteed by the super-renormalizable interaction. So probably this too is a finite theory.

IV. TOPOLOGICAL CONNECTION

I have repeatedly stated that the 3-dimensional gauge invariant mass term carries a topological significance, and that its quantization may be understood as arising from non-trivial topological properties. In this Lecture I explain this.

To appreciate the full topological setting, it is best to review the topological origins of θ-vacua in 4-dimensional Yang-Mills theory; this will also let us recognize the intimate mathematical connection between the 4-dimensional structures responsible for θ-vacua and those leading to the mass in three dimensions, thus lending support to the conjecture that the latter arises from the former in a high temperature reduction. Moreover, as will be seen, understanding fully the topological relationships in 4-space will also provide us all the information we need in lower dimensions. After retelling the 4-dimensional story, we give the 3-dimensional one, and we conclude by returning to four dimensions, where a quantization condition may again be established.

A. θ-Vacua in Four Dimensions

One may recognize the occurrence of a hidden angle in 4-dimensional gauge theories from a functional integral approach[22], or from a Hamiltonian formulation.[23] The relationship between the two is instructive and helps us understand the 3-dimensional phenomenon.

(i) Functional integral formulation

In a [Euclidean space] functional integral formulation, we begin with the formula

$$Z = \int \mathcal{D}A^\mu \exp - \int d^4x \left(-\frac{1}{2} \mathrm{tr}\, F^{\mu\nu} F_{\mu\nu} + \frac{i\Theta}{16\pi^2} \mathrm{tr}\, {}^*F^{\mu\nu} F_{\mu\nu} \right) \tag{4.1}$$

The measure $\mathcal{D}A^\mu$ includes all the requisite Faddeev-Popov factors and signifies integration over all gauge potential configurations. The term proportional to θ, involving the dual field

$$\begin{aligned} {}^*F^{\mu\nu} &= \frac{1}{2}\varepsilon^{\mu\nu\alpha\beta}F_{\alpha\beta} \\ F^{\alpha\beta} &= \frac{1}{2}\varepsilon^{\alpha\beta\mu\nu}\,{}^*F_{\mu\nu} \end{aligned} \tag{4.2}$$

is inserted because it is a possible addition which does not affect equations of motion, since it is a total divergence.

$$-\frac{1}{16\pi^2}\,\mathrm{tr}\,{}^*F^{\mu\nu}F_{\mu\nu} = \partial_\mu X^\mu \tag{4.3a}$$

$$X^\mu = -\frac{1}{16\pi^2}\varepsilon^{\mu\alpha\beta\gamma}\,\mathrm{tr}\left(F_{\alpha\beta}A_\gamma - \frac{2}{3}A_\alpha A_\beta A_\gamma\right) \tag{4.3b}$$

The factor i in (4.1) insures that in Minkowski space the action is real. [Here θ must not be confused with the infinitesimal gauge function $\theta^a = \theta^a|\theta|$.]

Next it is recognized that the integration over all A's may be divided into homotopically inequivalent classes, labelled by the integers. This labelling is established in the following manner. We impose the condition that potentials be everywhere regular in the finite plane, and that at large distances they approach a pure gauge.

$$A_\mu \xrightarrow[x\to\infty]{} U^{-1}\partial_\mu U \tag{4.4}$$

Here U is a matrix, which depends only on the variables describing the boundary of our space. We take the space to be S_4, because we compactify the physically occurring R_4, and work with the SU(2) gauge group [or with an SU(2) subgroup of the gauge group]. Hence U is a unitary, 2 × 2 matrix of unit determinant; it is defined on S_3, the boundary of S_4. [Although the space is compactified for purposes of the topological analysis, the integrals which we shall explicitly perform remain on the infinite space.]

It is now apparent that the matrices U provide a mapping from S_3 to SU(2). Since the three parameters that specify the SU(2) group also form an S_3 space, we are speaking of the mapping

$$S_3 \rightarrow S_3 \tag{4.5}$$

But it is known that such mappings fall into distinct homotopy classes, labelled by the integers. Mathematically this is expressed by the statement

$$\Pi_3(SU(2)) = \Pi_3(S_3) = \begin{pmatrix} \text{group of} \\ \text{all integers} \end{pmatrix} \tag{4.6a}$$

More generally it is true that

$$\Pi_i(S_i) = \begin{pmatrix} \text{group of} \\ \text{all integers} \end{pmatrix} \tag{4.6b}$$

[$\Pi_i(X)$ stands for the group of mappings of S_i into X.] Matrices U, belonging to the n'th homotopy class can be continuously deformed into each other, but not into matrices lying in a different class. Those in the trivial, n = 0, class are deformable to the identity and are called "small" gauge functions; the others, not deformable to the identity, are called "large" gauge functions.

Consequently, by virtue of (4.4), the gauge potentials are also classified by the integers, which describe the gauge functions that control A's asymptote. When A_μ is a member of the trivial class, it is any [non-singular] potential that vanishes faster than x^{-1} at large distances, while the n-instanton configuration is a member of the n'th class.[24]

Finally one can show that the integral of $*F^{\mu\nu}F_{\mu\nu}$ gives an analytic expression for the integer n which labels the topological class to which the potentials belong -- it is the "Pontryagin index."[25]

$$n = -\frac{1}{16\pi^2}\int d^4x \, \mathrm{tr} \, {*F^{\mu\nu}} F_{\mu\nu} \tag{4.7}$$

A way of establishing (4.7) is to use (4.3) and cast (4.7) into integrals over boundaries.

$$-\frac{1}{16\pi^2}\int d^4x\, tr\, {}^*F^{\mu\nu}F_{\mu\nu} = \int d^4x\, \partial_\mu X^\mu = \int d^3\vec{x}\, X^0 \Big|_{\tau=-\infty}^{\tau=\infty} + \int_{-\infty}^{\infty} d\tau \int dS^i X^i \tag{4.8}$$

[Although we are in Euclidean space, we arbitrarily call one of the coordinates, "time", τ and number it 0; the remaining coordinates define "space".] According to (4.4), at the boundaries the potentials are pure gauges. Also we shall always assume that at spatial infinity the gauge functions U become trivial.

$$U \xrightarrow[\tau\to\infty]{} \pm I \qquad \tau = |\vec{x}| \tag{4.9}$$

Hence we can rewrite (4.8) as

$$-\frac{1}{16\pi^2}\int d^4x\, tr\, {}^*F^{\mu\nu}F_{\mu\nu} = N(\infty) - N(-\infty) \tag{4.10}$$

where N(τ) is an integral over the U's, at fixed τ.

$$N(\tau) = \frac{1}{24\pi^2}\int d^3\vec{x}\, \varepsilon^{ijk}\, tr\left[\partial_i U U^{-1} \partial_j U U^{-1} \partial_k U U^{-1}\right] \tag{4.11}$$

This is exactly the winding number w(U), defined in (3.24). [There the formula is in Minkowski 3-space; here in Euclidean 3-space. But the difference is of no consequence since the expression is a world scalar.] We shall show that N(τ) is an integer; this validates (4.7) and proves that the winding number is an integer.

The gauge functions at fixed τ depend on the remaining three spatial variables. Also the condition (4.9) means that points at infinity are all identified with each other, since U is a constant there. Hence the 3-space is S_3, and the matrices U provide mappings of S_3 onto SU(2) which, as we have stated above, are labelled by the integers. Indeed an analytic expression for that integer is precisely (4.11), as we now demonstrate.

The integrand in (4.11) is a total divergence, as is seen from an exponential parametrization for U. The formulas were already given in (3.22) and (3.24). We record these equations here again.

$$U(\vec{x}) = e^{T^a \Theta^a(\vec{x})}$$
$$T^a = \sigma^a/2i, \quad \Theta^a = \hat{\Theta}^a |\Theta| \tag{4.12}$$

$$\mathcal{N} = w(U) = -\frac{1}{16\pi^2}\int d^3\vec{x}\, \varepsilon^{ijk} \varepsilon^{abc} \partial_i \Big[\hat{\Theta}^a \partial_j \hat{\Theta}^b \partial_k \hat{\Theta}^c \times \big(|\Theta| - \sin|\Theta|\big)\Big] \tag{4.13}$$

Moreover, as we are interested in gauge functions U which satisfy (4.9), $|\theta|$ -- the magnitude of θ^a -- approaches a multiple of 2π.

$$|\Theta| \xrightarrow[T\to\infty]{} 2\pi n_1 \tag{4.14}$$

Consequently (4.13) can be rewritten as a surface integral of unit vectors $\hat{\theta}^a$.

$$w(U) = -\frac{n_1}{8\pi}\int dS^i \varepsilon^{ijk} \varepsilon^{abc} \left[\hat{\Theta}^a \partial_j \hat{\Theta}^b \partial_k \hat{\Theta}^c\right] \tag{4.15}$$

The unit vectors $\hat{\theta}^a$ depend on the two angular variables (α,β) which parametrize the 2-dimensional surface over which the integral (4.15) is performed; i.e. they are defined on S_2. Also since the $\hat{\theta}^a$'s themselves define a 2-sphere, they give a mapping of S_2 to S_2. So once again we appeal to (4.6) and conclude

$$\Pi_2(S_2) = \left(\begin{array}{c}\text{group of}\\ \text{all integers}\end{array}\right) \tag{4.16}$$

The integer classifying the homotopy of this mapping is explicitly given by the integral in the right-hand side of (4.15). To see this, we parametrize the $\hat{\theta}^a$'s.

$$\hat{\Theta}^1 = \sin\omega \cos\varphi$$
$$\hat{\Theta}^2 = \sin\omega \sin\varphi$$
$$\hat{\Theta}^3 = \cos\omega \qquad \pi \lesssim \omega \leq 0, \; 2\pi \lesssim \varphi \leq 0 \tag{4.17}$$

The angles ω and ϕ are functions of α and β. It is easy to check that

$$\frac{1}{8\pi}\int dS^i \varepsilon^{ijk} \varepsilon^{abc} \hat{\Theta}^a \partial_j \hat{\Theta}^b \partial_k \hat{\Theta}^c =$$
$$\frac{1}{4\pi}\int_0^{2\pi} d\beta \int_0^{\pi} d\alpha \sin\omega \left(\frac{\partial\omega}{\partial\alpha}\frac{\partial\varphi}{\partial\beta} - \frac{\partial\omega}{\partial\beta}\frac{\partial\varphi}{\partial\alpha} \right) \tag{4.18}$$

Apart from sign, the quantity in parentheses is the Jacobian of the transformation from (α,β) to (ω,ϕ). Hence the above integral is just the integer n_2 which counts the [signed] number of times (ω,ϕ) range over their S_2 sphere as (α,β) range once over theirs. Thus

$$\frac{1}{8\pi}\int dS^i \varepsilon^{ijk} \varepsilon^{abc} \hat{\Theta}^a \partial_j \hat{\Theta}^b \partial_k \hat{\Theta}^c = n_2 \tag{4.19}$$

This then shows that N and w(U) are indeed integers.

After the topological detour by which we proved (4.7), we return to the functional integral (4.1) and separate it into individual integrals over the topological classes to which A_μ belongs.

$$Z = \sum_{n=-\infty}^{\infty} Z_n$$
$$Z_n = \int \mathcal{D}A_n^\mu \exp - \int d^4x \left(-\tfrac{1}{2}\,\mathrm{tr}\, F^{\mu\nu}F_{\mu\nu} + \frac{i\theta}{16\pi^2}\,\mathrm{tr}\, {}^*F^{\mu\nu}F_{\mu\nu} \right)$$
$$= e^{in\theta} \int \mathcal{D}A_n^\mu \exp - \int d^4x \left(-\tfrac{1}{2}\,\mathrm{tr}\, F^{\mu\nu}F_{\mu\nu} \right)$$
$$Z = \sum_{n=-\infty}^{\infty} e^{in\theta} \int \mathcal{D}A_n^\mu \exp - \int d^4x \left(-\tfrac{1}{2}\,\mathrm{tr}\, F^{\mu\nu}F_{\mu\nu} \right) \tag{4.20}$$

The above functional integral demonstration that in Yang-Mills theory an angle θ is present, which measures the relative phases of contributions from different topological sectors, leaves the skeptical student with some questions. Is it clear that one must add a surface term to the action in (4.1)? What is the relevance of all these topological properties of classical, well-behaved fields to the quantum field theory, which involves either operators in the operator formulation, or c-numbers without regularity requirements in a functional integral formulation? To answer these questions and to provide an alternative derivation, we now turn to a Hamiltonian description of the Yang-Mills theory, where again the angle will emerge as a hidden parameter.

(ii) Hamiltonian formulation

A canonical Hamiltonian formulation for a gauge theory is always given in the Weyl gauge [$A^0 = 0$], so named because Weyl was the first to use it for electrodynamics.[26] In the Yang-Mills application, the energy, <u>viz</u>. the Hamiltonian, is

$$H = \frac{1}{2}\int d^3\vec{x}\left(\vec{E}_a^2 + \vec{B}_a^2\right)$$

$$E_a^i = F_a^{io}\,, \quad B_a^i = -\frac{1}{2}\varepsilon^{ijk}F_a^{jk} \tag{4.21}$$

The canonical variables are the spatial components of the vector potentials $\vec{A}_a$, and their canonical momenta $\vec{\Pi}_a$, which coincide with the [negative] electric field, since we take the Lagrange density to be conventional.

$$\mathcal{L} = -\frac{1}{4}F_a^{\mu\nu}F_{a\,\mu\nu} \tag{4.22}$$

$$\Pi_a^i = \frac{\delta\mathcal{L}}{\delta\dot{A}_a^i} = F_a^{oi} = -E_a^i \tag{4.23}$$

Consequently these quantities satisfy canonical commutation relations, the only non-vanishing one being

$$\left[E_a^i(\vec{x}),\ A_b^j(\vec{y})\right] = i\delta_{ab}\,\delta^{ij}\,\delta(\vec{x}-\vec{y}) \tag{4.24}$$

[These are equal-time commutation relations, but we shall always suppress a common time argument.]

When the time development of $\vec{A}_a$ and $\vec{E}_a$ is studied with the help of the Hamiltonian equations of motion,

$$\begin{aligned} i\left[H, \vec{A}_a\right] &= \dot{\vec{A}}_a \\ i\left[H, \vec{E}_a\right] &= \dot{\vec{E}}_a \end{aligned} \tag{4.25}$$

and the commutators (4.25) are evaluated with the help of (4.24), it is found that the spatial component of the field equation

$$\mathcal{D}_\mu F^{\mu\nu} = 0 \tag{4.26}$$

is regained; <u>viz</u>. Ampere's law is a Hamiltonian equation of motion.

$$\dot{\vec{E}}_a = \vec{\nabla}\times\vec{B}_a - g\,\varepsilon_{abc}\,\vec{A}_b\times\vec{B}_c \tag{4.27}$$

But the temporal component, <u>viz</u>. Gauss' law

$$\vec{\nabla}\cdot\vec{E}_a - g\,\varepsilon_{abc}\,\vec{A}_b\cdot\vec{E}_c = 0 \tag{4.28}$$

does not follow, because it is a fixed-time constraint on the canonical variables.

To proceed, one ignores for the moment the absence of Gauss' law, and remains with the Hamiltonian system given by Eqs. (4.21), (4.24) and (4.25). This quantum mechanical theory is completely consistent, but does not yet define a Yang-Mills model because Gauss' law is missing. It is observed that the dynamical system possesses a Noether symmetry: equations of motion are invarant against transforming the canonical variables in the following way.

$$\vec{A} \rightarrow \vec{A}' = U^{-1} \vec{A} U - U^{-1} \vec{\nabla} U$$
$$\vec{E} \rightarrow \vec{E}' = U^{-1} \vec{E} U \qquad (4.29)$$

Here we have used matrix notation and U is a time-independent SU(2) matrix.Of course, this is just the gauge freedom which remains once A^0 has been set to zero, but we are viewing (4.29) as an ordinary Hamiltonian symmetry. When U is given by (4.12) and θ^a is infinitesimal, the generator of infinitesimal transformations is F.

$$F = -\frac{1}{g} \int d^3\vec{x}\ \theta^a(\vec{x})\, G_a(\vec{x}) \qquad (4.30a)$$

$$G_a = \vec{\nabla} \cdot \vec{E}_a - g\, \varepsilon_{abc} \vec{A}_b \cdot \vec{E}_c \qquad (4.30b)$$

G_a is not zero, since we do not have Gauss' law, but it does commute with the the Hamiltonian (4.21), as can be easily verified from (4.24).

Finally, Gauss' law (4.28) is implemented, and the Yang-Mills theory is regained, by requiring that physical states $|\Psi\rangle$ be annihilated by the generator F, or equivalently, since θ^a is arbitrary, by G_a.

$$G_a |\Psi\rangle = 0 \qquad (4.31a)$$

It follows that physical states are invariant under those finite transformations which are formed by iterating the infinitesimal one.

$$e^{iF} |\Psi\rangle = |\Psi\rangle \qquad (4.31b)$$

The meaning of Eqs. (4.31) is most easily understood in a Schrödinger, fixed time representation. The state $|\Psi\rangle$ is taken to be a wave functional depending on A, and the canonical momentum is realized by functional differentiation.

$$E^i_a(\vec{x}) |\Psi\rangle = i \frac{\delta}{\delta A^i_a(\vec{x})} \Psi(\vec{A}) \qquad (4.32)$$

Then (4.31a) is recognized as a functional differential equation that must be satisfied by $\Psi(\vec{A})$. The equation is integrable, since the commutators of G_a's close.

$$\left[\frac{1}{g} G_a(\vec{x}), \frac{1}{g} G_b(\vec{y})\right] = i\varepsilon_{abc} \frac{1}{g} G_c(\vec{x})\delta(\vec{x}-\vec{y}) \tag{4.33}$$

Also, since

$$e^{iF}\Psi(\vec{A}) = \Psi(\vec{A}') \tag{4.34a}$$

where $\vec{A}'$ is the gauge transformation of $\vec{A}$ as in (4.29) with U given by (4.12), it is clear that the content of (4.31b) is that the wave functional is gauge invariant.

$$\Psi(\vec{A}') = \Psi(\vec{A}) \tag{4.34b}$$

However, it is important to appreciate that the above discussion applies only to those finite gauge transformations which can be obtained by iterating [i.e. exponentiating] the infinitesimal one. Clearly for these, the gauge function U is deformable to the identity. [The deformation is given by the iteration.] Therefore the above is restricted to small gauge transformations. But gauge transformations involving large gauge functions, cannot be deformed by iterating an infinitesimal transformation and we cannot assert that the wave functionals remain invariant when these "large" gauge transformations are performed.

According to the previous discussion, the U's fall into homotopy classes labelled by the intergers n. We may call the operator which implements a gauge transformation in the n'th homotopy class $\mathcal{G}_n$, but only $\mathcal{G}_0$ has the representation e^{iF}. For all n, we know that $\mathcal{G}_n$ is unitary and commutes with the Hamiltonian, since it generates a symmetry transformation. From general quantum mechanical principles, we conclude that physical states are eigenvectors of $\mathcal{G}_n$, with an eigenvalue which is a pure phase.

$$\mathcal{G}_n \Psi(\vec{A}) = \Psi(\vec{A}') = e^{-i\Theta_n} \Psi(\vec{A}) \tag{4.35a}$$

Owing to the additive nature of the gauge function's homotopic characterization, it is clear that $\theta_n = n\theta$ and we have the final result.

$$\mathcal{G}_n \Psi(\vec{A}) = \Psi(\vec{A}') = e^{-in\theta} \Psi(\vec{A}) \tag{4.35b}$$

This is the origin of the θ-angle in a Hamiltonian formulation of the theory.

One may make contact between the above Hamiltonian description, based on the conventional Lagrange density (4.22), and the previous functional integral formulation, based on the modified Lagrangian density in (4.1), by the following reasoning.

It is possible to remove the phase from the transformation law (4.35), provided we can find a functional of $\vec{A}$, call it $W(\vec{A})$, which changes by n when a gauge transformation in the n'th homotopy class is performed.

$$W(\vec{A}') = W(\vec{A}) + n \tag{4.36}$$

For then we can write

$$\Psi(\vec{A}) = e^{-i\theta W(\vec{A})} \Phi(\vec{A}) \tag{4.37}$$

and $\Phi(\vec{A})$ is gauge invariant against all gauge transformations.

$$\mathcal{G}_n \Phi(\vec{A}) = \Phi(\vec{A}') = \Phi(\vec{A}) \tag{4.38}$$

Such a functional may indeed be constructed. It is

$$W(\vec{A}) = -\frac{1}{16\pi^2}\int d^3\vec{x}\, \varepsilon^{ijk}\, tr\left[F_{ij} A_k - \tfrac{2}{3} A_i A_j A_k\right] \tag{4.39}$$

Note that $W(\vec{A})$ is just the spatial integral of X^0, with X^0 defined in (4.3b). Gauge transformations change W(A) to [compare (3.20)]

$$W(\vec{A}') = W(\vec{A}) + \frac{1}{8\pi^2}\int d^3\vec{x}\, \varepsilon^{ijk} \partial_i tr\left[A_j \partial_k U U^{-1}\right] + \frac{1}{24\pi^2}\int d^3\vec{x}\, \varepsilon^{ijk}\, tr\left(\partial_i U U^{-1} \partial_j U U^{-1} \partial_k U U^{-1}\right) \tag{4.40}$$

The second term, involving the A-dependent integral, vanishes owing to sufficiently rapid fall-off at large distances, which is guaranteed by (4.9) [and by appropriate restrictions on the large-distance asymptote of the vector potentials]. The last term is the integer which labels U's homotopy classes, as we demonstrated in (4.11)-(4.19). Thus (4.36) is true.

A universal phase factor can be removed from all wave functionals at the expense of adding a total time derivative to the Lagrangian. [A simple quantum mechanical argument shows that changing the phase of all wave functions is achieved by adding to the Lagrangian a total time derivative.] Thus if the quantum theory, based on wave functionals Ψ, is obtained from a Lagrangian $L_\Psi = \int d^3\vec{x}\,\mathcal{L}$, the one based on wave functionals $\Phi = e^{-i\theta W(\vec{A})}\Psi$ is obtained from the Lagrangian

$$L_\Phi = \int d^3\vec{x}\;\mathcal{L} + \theta \frac{d}{dt} W(\vec{A}) \tag{4.41a}$$

But we also have

$$W(\vec{A}) = \int d^3\vec{x}\; X^0$$

$$\frac{d}{dt} W(\vec{A}) = \int d^3\vec{x}\,\frac{\partial}{\partial t} X^0 = \int d^3\vec{x}\;\partial_\mu X^\mu = -\frac{1}{16\pi^2}\int d^3\vec{x}\,\mathrm{tr}\,{}^*F^{\mu\nu}F_{\mu\nu} \tag{4.41b}$$

In the derivation, we have added a spatial integral of a total divergence; this is allowed, since we assume sufficiently rapid fall-off at spatial infinity. Eqs. (4.41) show that we can have gauge invariant states, provided the Lagrange density is modified as in (4.1).

(iii) Mathematical terminology

Finally, let me record some mathematical nomenclature. As we have seen in the above 4-dimensional discussion -- but it is also generally true in any even number of dimensions -- the gauge invariant Pontryagin density $\mathcal{P}_{2n}$, integrated over the even dimensional space, is an invariant that characterizes the topological class to which the gauge fields belong. The 4-dimensional example may be supplemented by a 2-dimensional one.

$$\mathcal{P}_2 = \frac{1}{2\pi} {}^*F = \frac{1}{4\pi} \varepsilon^{\mu\nu} F_{\mu\nu} \tag{4.42a}$$

$$\mathcal{P}_4 = -\frac{1}{16\pi^2} \operatorname{tr} {}^*F^{\mu\nu} F_{\mu\nu} \tag{4.42b}$$

[The 2-dimensional Pontryagin density arises in 2-dimensional QED and is responsible for mass generation in that model.][27] Since their integral is a topological invariant, these gauge invariant objects can also be written as total derivatives of gauge variant quantities.

$$\mathcal{P}_{2n} = \partial_\mu X^\mu_{2n} \tag{4.43}$$

We have seen the 4-dimensional formula before in (4.3); the two dimensional one is trivial.

On odd-dimensional spaces Pontryagin classes do not exist. But one may construct another topological quantity, called the "Chern-Simons secondary characteristic class."[28] This is gotten by integrating one component of $X^\mu{}_{2n}$ over the 2n - 1 dimensional space which does not include that component. As we have seen, the integral is gauge invariant against homotopically trivial gauge transformations; otherwise, it changes by the winding number of the transformation.

Thus we recognize that the functional integral analysis in 4-space makes use of the Pongryagin class. The Hamiltonian, fixed-time analysis uses the Chern-Simons secondary characteristic class, which is appropriate to three dimensions. W(A) of (4.39) is precisely the Chern-Simons quantity. Moreover, the 2-dimensional Pontryagin class is responsible for mass generation in a 2-dimensional example, while the 3-dimensional Chern-Simons characteristic gives our 3-dimensional mass, which we now examine further.

B. Mass Term of 3-Dimensional Gauge Theories

The Chern-Simons secondary characteristic class, which we met in the fixed-time analysis of the 4-dimensional gauge theory, is also proportional to the mass term in the action of the 3-dimensional gauge theory.

$$-\frac{\mu}{2g^2}\int d^3x\, \varepsilon^{\mu\nu\alpha}\, \mathrm{tr}\left(F_{\mu\nu}A_\alpha - \frac{2}{3}A_\mu A_\nu A_\alpha\right) = \mu\,\frac{8\pi^2}{g^2}\,W(A) \tag{4.44}$$

Consequently we understand the mass term's gauge transformation properties (3.21) to be precisely those of a Chern-Simons quantity, as we demonstrated above. [The fact that there we are in Euclidean space and here in Minkowski space is immaterial, since (4.44) is a world scalar, insensitive to the metric of the space.]

[This observation about the mass term in Yang-Mills theory has an immediate parallel in the construction of a topological mass for 3-dimensional gravity from the 4-dimensional *RR Hirzebruch-Pontryagin density. But this subject is outside the scope of my lectures, hence those interested are referred to the literature.][29]

We now present a functional integral and a Hamiltonian analysis[30] of the mass quantization condition (3.25).

(i) Functional integral formulation

The functional integral for a 3-dimensional, massive gauge theory [in Euclidean space] is given by

$$Z = \int \mathcal{D}A^\mu \exp -\left\{\int d^3x \left(-\frac{1}{2}\mathrm{tr}\, F^{\mu\nu}F_{\mu\nu}\right) - i\mu\,\frac{8\pi^2}{g^2}\,W(A)\right\} \tag{4.45}$$

As the functional integration ranges over all gauge potentials, for any given A^μ it also encounters its gauge copies A'^μ. Remembering that we are in 3-space, we recall that the gauge functions fall into homotopically distinct classes labelled by the integers. Now the usual gauge fixing prescriptions, which are [implicitly] contained in $\mathcal{D}A^\mu$, remove gauge copies arising from the homotopically trivial, small gauge transformations. [Recall that Faddeev-Popov procedures are formulated in infinitesimal terms.] However, there seems to be no way of removing gauge copies in (4.45) arising from non-trivial, large gauge transformations. Thus we may write Z as

$$Z = \sum_{n=-\infty}^{\infty} Z_n \tag{4.46}$$

Here Z_0 results by performing an integration over A^μ, with no gauge copies; Z_1 results from integrating over A^μ related by a large gauge transformation in the first homotopy class to those occurring in the integral for Z_0; <u>etc</u>. [We emphasize that (4.46) is a different decomposition than in the similar appearing 4-dimensional formula (4.20). There the integrations for the different Z_n's range over different topological, Pontryagin classes of gauge potentials, which are not gauge copies of each other. Here the integrations for the different Z_n's range over large gauge copies of the vector potentials, that determine Z_0.] But it is clear that once we have determined Z_0, Z_n for $n \neq 0$, may be evaluated by changing variables in the functional integral from A^μ to A'^μ which is defined to be the gauge transform of A^μ with a large gauge function of the nth homotopy class. Such a change of variables does not affect the measure nor the usual action, since both are gauge invariant. In the mass term, W(A) changes according to (3.21), so we get

$$Z = Z_0 \sum_{n=-\infty}^{\infty} e^{in\mu \frac{8\pi^2}{g^2}} \tag{4.47}$$

Now we see that if the mass term is not quantized, the infinite sum vanishes, by destructive interference. On the other hand, when the quantization condition holds, the sum becomes $\sum_{n=-\infty}^{\infty} 1$, which, though infinite, may be harmlessly cancelled by a normalizing denominator in the definition of Z.

Thus the result: the massive gauge theory vanishes in the absence of mass quantization. Admittedly, the argument is exceedingly formal, relying on the functional integral, which is not well defined. Also one wonders whether there may be some way of restricting the integration so that only Z_0 is kept and mass quantization is avoided. In order to remove such doubts, we now turn to a Hamiltonian derivation.

(ii) Hamiltonian formulation

The crucial equation in the Hamiltonian formulation is Gauss' law, just as it was for the analysis of θ-vacua in 4-dimensional gauge theories. In terms of Weyl gauge [$A^0 = 0$] canonical variables $\vec{A}_a$ and $\vec{\Pi}_a$, [the latter no longer coincides with $-\vec{E}_a$, owing to the time-derivative couplings in the mass term's contribution to the Lagrange density] the generator of infinitesimal spatial gauge transformations is

$$F = -\frac{1}{g}\int d^2\vec{x}\; \theta^a(\vec{x})\, G_a(\vec{x}) \tag{4.48a}$$

$$G_a = -\vec{\nabla}\cdot\vec{\Pi}_a + g\,\varepsilon_{abc}\,\vec{A}_b\cdot\vec{\Pi}_c - \frac{\mu}{2}\,\vec{\nabla}\times\vec{A}_a \tag{4.48b}$$

Gauss's law, which is unconventional due to the mass term, is imposed in the quantum theory by requiring that G_a annihilate physical states.

$$G_a |\Psi\rangle = 0 \tag{4.49a}$$

This is a functional differential equation, which the physical wave functionals $\Psi(\vec{A})$ must satisfy. In the Schrödinger representation we may write the finite version of (4.49a) as

$$e^{iF}\Psi(\vec{A}) = \Psi(\vec{A}) \tag{4.49b}$$

The purely spatial, fixed-time Hamiltonian formulation involves gauge functions that depend on 2-dimensional spatial variables. Since $\Pi_2\ (SU(2)) = \Pi_2\ (S_3)$ is trivial, there is only one homotopy class for the 2-dimensional gauge transformations -- they are all small. Hence (4.49b) describes the response of physical states to all gauge transformations, since all are deformable to the identity.

Eq. (4.49a) is locally integrable, as even the modified G_a continues to satisfy the commutation relations (4.33). But global integrability needs to be examined further, because there remains a topological subtlety. The 3-dimensional statement that $\Pi_3(SU(2))$ is non-trivial, implies that the space of the 2-dimensional gauge functions is not simply connected. To see this, let us consider any 3-dimensional U which is not deformable to the identity. We now view U as defining a family of 2-dimensional matrices depending on the spatial 2-vector x and on the parameter τ. As $\tau \to \pm\infty$, $U \to I$; as a function of τ, our family describes a loop which starts and ends at the identity. Yet this loop cannot be shrunk to a point, since by hypothesis U is not deformable to the identity!

As a consequence, even though the functional differential equation (4.49a) is locally integrable, by virtue of (4.33), there is no asurance that it will be globally integrable, owing to the multiple connectivity of the function space. To be explicit, let us evaluate the left-hand side of (4.49b). Use of the canonical commutation relations gives

$$e^{iF}\,\Psi(\vec{A}) = e^{i\Omega(\theta)}\,\Psi(\vec{A}') \tag{4.50a}$$

where

$$\Omega(\theta) = -\frac{\mu}{g^2}\int d^2\vec{x}\,\varepsilon^{ij}\,\mathrm{tr}\left(A^i\partial_j U U^{-1}\right) + \frac{\mu}{2g^2}\int d^2\vec{x}\,\varepsilon^{ij}\,\varepsilon^{abc}\,\hat{\theta}^a\,\partial_i\hat{\theta}^b\,\partial_j\hat{\theta}^c\left(|\theta| - \sin|\theta|\right) \tag{4.50b}$$

and $\vec{A}'$ is the gauge transform $\vec{A}$ with $U = \exp(\sigma^a/2i)\theta^a$. Formulas (4.50) state that the transformation law for wave functionals under the Hamiltonian symmetry which changes $\vec{A}$ to $\vec{A}'$ requires not only gauge transforming the argument of the wave functional, but also there is phase change Ω.

This complicated transformation law reflects a general quantum mechanical phenomenon. Whenever a symmetry transformation on canonical variables changes the Lagrangian by a total time derivative of a function of coordinates only, that function appears as a phase in the transformation law for the quantum mechanical wave function.

In our case the transformation law for the Lagrange density (3.20) implies that under a spatial gauge transformation the Lagrangian changes according to

$$L = \int d^2\vec{x}\, \mathcal{L} \to L + \frac{\mu}{g^2} \int d^2\vec{x}\, \varepsilon^{\alpha\beta\mu}\, \partial_\alpha \operatorname{tr}\left(\partial_\beta U U^{-1} A_\mu\right)$$
$$+ \frac{\mu}{3g^2} \int d^2\vec{x}\, \varepsilon^{\alpha\beta\gamma} \operatorname{tr}\left(\partial_\alpha U U^{-1} \partial_\beta U U^{-1} \partial_\gamma U U^{-1}\right)$$

(4.51a)

Using the identity (3.24) and dropping spatial surface terms leaves

$$L \to L - \frac{d}{dt}\Omega \tag{4.51b}$$

This provides an alternative derivation of (4.50).

According to (4.44b), the right-hand side of (4.50a) must be equal to $\Psi(\vec{A})$.

$$e^{iF}\Psi(\vec{A}) = e^{i\Omega(\theta)}\Psi(\vec{A}') = \Psi(\vec{A}) \tag{4.52}$$

This shows that in the massive gauge theory wave functionals are not gauge invariant, rather they acquire a phase when a gauge transformation is performed. Let us now examine this equation for that family of gauge transformations, which we discussed above as evidence for non-simple connectivity of the gauge function space. The family is parametrized by τ and at $\tau = \pm\infty$, the gauge function is the identity, hence $\vec{A}' = \vec{A}$. It follows from (4.52) that

$$e^{i\Omega(\theta)}\Big|_{\tau=-\infty} = e^{i\Omega(\theta)}\Big|_{\tau=\infty} = 1$$

(4.53a)

which also implies that

$$\Omega(\theta)\Big|_{\tau=-\infty}^{\tau} = 2\pi \times \text{integer}$$

(4.53b)

since we may assume that at $\tau = -\infty$, $U = I$ because θ^a, and therefore also $\Omega(\theta)$ vanish, but at $\tau = \infty$, $U = I$, yet θ^a and $\Omega(\theta)$ do not vanish. Examining (4.50b), we see that the first contribution to Ω vanishes whenever $U = I$; the second depends on θ^a, and cannot be written in terms of U. However, (4.53b) may be rewritten, with the addition of an innocuous surface term, as

$$\Omega(\theta)\Big|_{\tau=-\infty}^{\tau=\infty} = \frac{\mu}{2g^2}\int_{-\infty}^{\infty} d\tau \frac{\partial}{\partial \tau}\int d^2\vec{x}\, \varepsilon^{ij}\varepsilon^{abc}\, \hat{\theta}^a \partial_i \hat{\theta}^b \partial_j \hat{\theta}^c \times \left(|\theta| - \sin|\theta|\right)$$

$$= -\mu \frac{8\pi^2}{g^2} w(U)$$

(4.53c)

Since w(U) is itself an integer, (4.53b) implies the mass quantization condition, which is now recognized as a consequence of Gauss' law: without mass quantization, the functional differential equation which implements Gauss' law is not globally integrable.

(iii) Discussion

With an eye towards applying these ideas in four dimensions, let us summarize the key reasons for quantizing the mass term. First, we have the non-trivial homotopic structure of 3-dimensional gauge transformations: $\Pi_3(SU(2))$ = group of all integers. Second, there is a term in the action which is invariant against small gauge transformations but not against large ones. Consequently, unless a quantization condition is enforced, the functional integral vanishes, owing to cancellations between contributions coming from non-trivial gauge copies of the vector potential. Alternatively, in a Hamiltonian formulation, the non-trivial structure of $\Pi_3(SU(2))$, implies that the gauge function space is not simply connected. Unless the mass term is quantized the differential equation which implements the modified Gauss' law, [the modification arises from the mass term] cannot be globally interacted.

C. Parameter Quantization in a 4-Dimensional Gauge Theory.

The mechanism which is responsible for mass quantization has an analogue in a 4-dimensional, SU(2) gauge theory.[3] Let us state first that $\Pi_4(SU(2)) = \Pi_4(S_3)$ = cyclic group of two integers. Hence 4-dimensional gauge functions fall into two homotopically distinct classes, and correspondingly the space of fixed-time, 3-dimensional gauge functions is not simply connected.

The functional integral for the pure gauge theory is completely gauge invariant; therefore the doubling due to gauge copies is innocuous. However, it has been observed that coupling massless Weyl fermions with only left-handed [or only right-handed] interactions to the gauge field changes the situation. Consider N species of massless fermion doublets in left-handed interaction with the gauge field, and perform the functional integration over the fermions.

$$Z = \int \mathcal{D}A^\mu \mathcal{D}\bar{\Psi}_i \mathcal{D}\Psi_i \exp - \int d^4x \left(-\frac{1}{2}\operatorname{tr} F^{\mu\nu}F_{\mu\nu} + \sum_{i=1}^{N} \bar{\Psi}_i \gamma^\mu D_\mu \Psi_i\right)$$
$$= \int \mathcal{D}A^\mu \det{}^{N/2}(\gamma^\mu D_\mu) \exp - \int d^4x \left(-\frac{1}{2}\operatorname{tr} F^{\mu\nu}F_{\mu\nu}\right) \tag{4.54}$$

Here D_μ is the covariant derivative in the fundamental [doublet] representation. The fermion determinant comes with the power 1/2 for each species because the spinors satisfy the Weyl condition: $\gamma_5\psi = \pm\psi$ -- there are half as many components as in a Dirac spinor. The functional integral can now be separated into two pieces; the first represents a functional integration which does not range over any gauge copies, in the second the gauge potentials are related to those in the first integral by a large gauge transformation.

$$Z = Z_+ + Z_- \tag{4.55}$$

The non-trivial effect now comes from the unexpected circumstance that the square root of the fermion determinant is not invariant against large gauge transformations; rather a minus sign occurs. Put in another way, evaluating the square root requires chosing one of its two branches. But it can be shown that whatever choice is made in one class, the opposite choice must be made in the other class.

The proof of this property of the fermion determinant will not be given here; those interested should consult the research literature.[3] However the consequence is clear.

$$Z = \left(1 + (-1)^N\right) Z_+ \tag{4.56}$$

So unless the number of fermion doublets is even, the functional integral vanishes. This then is the quantization condition: N/2 must be an integer.

A Hamiltonian version of the argument has not been given. Presumably the Gauss law constraint, which in the presence of the fermion charge density ρ_i^a, $i = 1,\ldots,N$ reads

$$\left(\vec{\mathcal{D}}^{ab}\cdot\vec{E}^b - \sum_{i=1}^{N} \rho_i^a\right)|\Psi\rangle = 0 \tag{4.57}$$

cannot be globally integrated for odd N, since the space is multiply connected. However a detailed calculation, analogous to the 3-dimensional one, is lacking, owing to the difficulty of defining a Hilbert space on which the fermion operators $\rho_i{}^a$ should act.

V. CONCLUSION

We have seen that 3-dimensional gauge theories offer much interesting insight into the workings of gauge-invariant quantum field theory. Many further investigations can be envisioned: one wants techniques for studying possible non-analytic terms in gauge invariant amplitudes. Since these can occur only in high orders, our direct resummation method becomes prohibitively lengthy. The approach through dimensional regularization[16] seems promising. It would be worthwhile to establish in greater detail the relation of this method to condensed matter procedures.[17]

The topological mass terms and the associated Chern-Simons structures give evidence that gauge theories in odd-dimensional spaces offer unexpected possibilities. Since odd-dimensional gauge theories are of interest in supersymmetry, in supergravity, in Kaluza-Klein theory and in other attempts at dimensional reduction, there is obvious interest in understanding the gauge theoretic options in this context.

Quantization of parameters for consistency of the quantum theory is familiar from the Dirac monopole example. The 3-dimensional massive gauge theory[7] and the 4-dimensional chiral SU(2) theory[3] are novel instances of this phenomenon. One would like to settle the question whether the former is indeed relevant in a high temperature reduction, as we have conjectured, and one wants to ascertain whether there are any physically interesting applications of the latter. [The usual constraint that axial vector anomalies be absent already seems to eliminate 4-dimensional theories which are globally inconsistent.]

Yet another interesting thread connects the present work with the general question of massive, yet gauge invariant vector fields. Since the advent of unification schemes based on Yang-Mills models, this question has become phenomenologically important. Most frequently it is addressed in the context of the Higgs mechanism, which however is frequently viewed to be aesthetically and phenomenologically unattractive. Attempts to replace the scalar Higgs fields by fermion, anti-fermion bound states[31] have not produced thus far convincing and acceptable models.[32] It is therefore interesting to observe that in the two examples of gauge invariant, massive vector fields without Higgs scalars -- *viz.* the original Schwinger model of massless, 2-dimensional spinor electrodynamics and the present 3-dimensional, topologically massive gauge fields -- the vector meson mass arises from a topological mechanism and not directly from fermion condensates.[33] Is there a topological reason, as yet undiscovered, for massive, gauge invariant vector particles in 4-dimensions?

ACKNOWLEDGMENT

This research was supported in part by funds provided by the U.S. Department of Energy, under contracts DE-AC02-76ER03069 and DE-AC02-76CH00016. I am grateful for the hospitality extended to me at the Brookhaven National Laboratory, where these lectures were prepared.

References

1. The work summarized here was performed in collaboration with S. Deser and S. Templeton.
2. Other recent investigations of 3-dimensional gauge theories are by R. Feynman, Nucl. Phys. B188, 479 (1981); I. Singer, Physica Scripta 24, 817 (1981).
3. E. Witten, Princeton University preprint.
4. S. Weinberg, in "Understanding the Fundamental Constituents of Matter" (A. Zichichi, Ed.), Plenum, New York, NY, 1978; A. Linde, Rep. Progr. Phys. 42, 389 (1979); D. Gross, R. Pisarski and L. Yaffe, Rev. Mod. Phys. 53, 43 (1981).
5. See e.g. L. Dolan and R. Jackiw, Phys. Rev. D9, 3320 (1974).
6. R. Jackiw and S. Templeton, Phys. Rev. D23, 2291 (1981); S. Templeton, Phys. Lett. 103B, 134 (1981) and Phys. Rev. D24, 3134 (1981).
7. S. Deser, R. Jackiw and S. Templeton, Phys. Rev. Lett. 48, 975 (1982) and Ann. Phys. (NY) 140, 372 (1982).
8. R. Jackiw, Ph.D. Thesis, Cornell University 1966 (unpublished); Super-renormalizable, massless scalar theories have also been studied by K. Symanzik, Lett. Nuovo Cimento 8, 772 (1973); M. Bergere and F. David, Saclay preprint DPh-T/82-16 (unpublished). In both investigations non-analyticity in the coupling constant is established.
9. Super-renormalizable gauge theories were examined by G. 't Hooft, in "Field Theory and Strong Interactions", Acta Physica Austriaca, Suppl. XXII, (P. Urban, Ed.), Springer-Verlag, Wien, 1980; my interest in the subject was reawakened by this investigation. Subsequent research is by Jackiw and Templeton, Ref. (6); T. Appelquist and R. Pisarski, Phys. Rev. D23, 2305 (1981).

10. Because we include fermions, this model cannot be viewed as the infinite temperature reduction of a 4-dimensional theory. However, because the charged fermions provide interactions, the model is analogous to, but simpler than, 3-dimensional Yang-Mills theory, which could arise from a finite temperature 4-dimensional theory. The Yang-Mills model is discussed below.

11. J. Cornwall, R. Jackiw and E. Tomboulis, Phys. Rev. D10, 2428 (1974).

12. Appelquist and Pisarski, Ref. 9.

13. Templeton, Ref. 6.

14. H. Bethe and E. Salpeter, "Quantum Mechanics of One- and Two-Electron Atoms", Plenum, New York, NY, 1977.

15. H. Pagels, Phys. Rep. 16C, 219 (1975).

16. E. Guendelman and Z. Radulovic, MIT preprint (in preparation).

17. This is somewhat analogous to Bogoliubov's method of "compensating for dangerous [infrared divergent] graphs" by a canonical transformation, which isolates the zero momentum mode of a scalar field theory. For a review see A. Novaco, J. Low Temp. Phys. 2, 465 (1970). I thank V. Emery for calling my attention to this analogy. Note also that a constant gauge field plays a role in summing the dominant infrared logarithms, see Eqs. (2.19) and Ref. 13.

18. These results have been verified in an independent calculation by Appelquist and Pisarski, Ref. 9. Also O. Kalashnikov and V. Klimov, Yad. Fiz. 33, 848 (1981), [Sov. J. Nucl. Phys. 33, 443 (1981)] have studied, in lowest order, the finite temperature, 4-dimensional vacuum polarization tensor and ghost propagator. They have determined the high temperature asymptote. The result is precisely (2.27), which provides a non-trivial example of the high temperature dimensional reduction.

19. Deser, Jackiw and Templeton, Ref. 7; W. Siegel, Nucl. Phys. B156, 135 (1979); R. Jackiw and S. Templeton, Phys. Rev. D23, 2291 (1981); J. Schonfeld, Nucl. Phys. B185, 157 (1981); R. Jackiw in "Asymptotic Realms of Physics", MIT Press, Cambridge, MA, (A. Guth, K. Huang, R. Jaffe, Eds.), 1983; H. Nielsen and H. Woo, (unpublished).

20. H. Nielsen and P. Olesen, Nucl. Phys. B61, 45 (1973).

21. R. Jackiw, Rev. Mod. Phys. 52, 661 (1980).

22. G. 't Hooft, Phys. Rev. Lett. 37, 8 (1976); C. Callan, R. Dashen and D. Gross, Phys. Lett. B63, 334 (1976).

23. R. Jackiw and C. Rebbi, Phys. Rev. Lett. 37, 172 (1976); for a review see Ref. 21.

24. A. Belavin, A. Polyakov, A. Schwartz and Y. Tyupkin, Phys. Lett. B59, 85 (1975); R. Jackiw, C. Nohl and C. Rebbi, Phys. Rev. D15, 1642 (1977); M. Atiyah, V. Drinfeld, N. Hitchin and Y. Manin, Phys. Lett. A65, 185 (1978).

25. The Pontryagin index entered physics in the seminal paper by Belavin, Polyakov, Schwartz and Tyupkin, Ref. 24.

26. H. Weyl, "The Theory of Groups and Quantum Mechanics", Dover, New York, NY, 1950.

27. Jackiw, Ref. 19.

28. S. Chern, "Complex Manifolds without Potential Theory", 2nd. ed. Springer-Verlag, Berlin, 1979.

29. Deser, Jackiw and Templeton, Ref. 7; S. Deser, DeWitt Festschrift, to appear.

30. The Hamiltonian analysis was provided by J. Goldstone and E. Witten; I thank them for discussion.

31. R. Jackiw and K. Johnson, Phys. Rev. D8, 2386 (1973); J. Cornwall and R. Norton, Phys. Rev. D8, 3338 (1973).

32. For a summary see E. Farhi and R. Jackiw, "Dynamical Gauge Symmetry Breaking", World Scientific, Singapore, 1982.

33. These topological mechanisms for gauge vector meson mass generation are reviewed by Jackiw, Ref. 19.

THE THERMODYNAMICS OF STRONGLY INTERACTING MATTER

Helmut Satz
Fakultät für Physik
Universität Bielefeld
Germany

ABSTRACT

We survey the thermodynamics of strongly interacting matter, as predicted by quantum chromodynamics. At low values of temperature and density, we find hadronic matter, while at high values the system becomes a primordial plasma of non-interacting massless quarks and gluons. Separating the two regimes is a transition region in which colour deconfinement and chiral symmetry restoration take place. We obtain preliminary values for the transition parameters and discuss the possibility of observing plasma formation in high energy heavy ion collisions.

I. Introduction

Changes of state are among the most familiar and yet most striking instances of collective behaviour. Do they also occur in strongly interacting matter? Will a sufficient increase in density lead us out of the realm of normal nuclear matter, made up of nucleon constituents, into a new world, in which quarks and gluons form a plasma of primordial matter?

Such a transition has been discussed ever since the advent of the quark model of hadrons. Today the pursuit of these ideas, the physics of strongly interacting matter at very high density, is emerging more and more as an autonomous field of research[1)]. It brings together problems and features from nuclear physics, particle physics, and statistical physics, In cosmology, it is essential for an understanding of the early stages of our universe.

The basis of this field is quantum chromodynamics; with QCD, we have today a serious and so far "uncontradicted" candidate for the theory of strong interactions. This provides us with the possibility of deriving strong interaction thermodynamics. Phenomenological models of various types have been employed for many years to discuss a possible transition from nuclear to quark matter; they always have two phases as input and obtain the transition by construction. The lattice formulation of QCD[2] now for the first time allows us to deduce the phases of strongly interacting matter and the transition behaviour from one basic theory. QCD thermodynamics predicts the existence of a quark-gluon plasma at high temperature as well as that of hadronic matter at low temperature; between these regimes lies a transition region, in which - coming from low T - colour is deconfined and chiral symmetry restored.

II. Phenomenology

Quantum chromodynamics describes the interaction of quarks and gluons - the strong interaction. Quarks and gluons are not observed as free objects in the physical vacuum: in non-interacting form, as free particle states, we only see hadrons. The low density limit of QCD thermodynamics must therefore lead to a system of non-interacting hadrons. On the other hand, sufficiently energetic "hard" hadron-hadron or hadron-lepton scattering processes indicate independent interactions of quarks and gluons with each other or with leptons: strong interactions become weak at high energies. This novel feature, "asymptotic freedom", is in fact provided by QCD. To obtain such a behaviour, quarks and gluons must have an additional intrinsic degree of freedom, "colour", which makes gluon-gluon interactions possible in a gauge-invariant framework. In potential language, the confining force between quarks rises linearly with increasing separation; at short distances, however, the potential becomes Coulomb-like. For the high-density limit of QCD, we therefore expect a plasma of non-interacting, "Debye-screened" quarks and gluons; here, in addition, an overall shift in energy between vacuum and plasma ground state has to be taken into account.

Let us use these limiting forms to construct a simple two-phase picture of strongly interacting matter. Our thermodynamic variables are the temperature T, the baryonic chemical potential μ, and the volume V; μ specifies the baryon number density. Here and in the following we shall for simplicity always restrict ourselves to non-strange hadrons (pions and nucleons) and quarks (u and d).

For zero baryon number $(\mu = 0)$, we have at low temperatures basically a gas of pions. For a non-interacting system of massless pions, the pressure is given by

$$P_H(T, \mu = 0) = \frac{\pi^2}{90} \times 3 \times T^4 \quad , \tag{2.1}$$

taking into account the three charge states of the pion. The corresponding energy density is

$$\varepsilon_H(T, \mu = 0) = \frac{\pi^2}{30} \times 3 \times T^4 \quad . \tag{2.2}$$

At sufficiently high temperatures, we expect a plasma of non-interacting quarks and gluons, with the pressure

$$P_Q(T, \mu = 0) = \frac{\pi^2}{90} [2 \times 8 + \frac{7}{8} \times 2 \times 2 \times 2 \times 3] T^4 - B \quad . \tag{2.3}$$

Here both gluons (first term in square brackets) and quarks (second term) have two spin degrees of freedom; in accord with the SU(3) gauge structure of QCD, gluons have eight and quarks three colour degrees of freedom. There are, for $\mu = 0$, both quarks and antiquarks present, and we include, as mentioned, two flavours. Finally, the (positive) bag constant B accounts for the ground state shift in the plasma; to provide confinement at low density, the physical vacuum excerts a pressure on the system. - The corresponding energy density becomes

$$\varepsilon_Q(T, \mu = 0) = \frac{\pi^2}{30} [2 \times 8 + \frac{7}{8} \times 2 \times 2 \times 2 \times 3] T^4 - B \quad . \tag{2.4}$$

To satisfy the thermodynamic requirement of minimal free energy, the system must be in the state of higher pressure. For low T , this is the pion gas, for high T , the plasma; see fig. 1. From $P_H(T_c, 0) = P_Q(T_c, 0)$ we find

$$T_c = [(45/17\pi^2) B]^{1/4} \simeq 0.72 \, B^{1/4} \tag{2.5}$$

for the transition temperature at $\mu = 0$. From the description of hadronic spectra, we have $145 \text{ MeV} \leq B^{1/4} \leq 235 \text{ MeV}$; using $B^{1/4} \simeq 190$ MeV , we obtain $T_c \simeq 140$ MeV . The transition arrived at in this way is by construction of first order, with

$$\varepsilon_Q(T_c, 0) - \varepsilon_H(T_c, 0) = 2B \tag{2.6}$$

as the latent heat per unit volume. The resulting energy density is also shown in fig. 1.

Let us now apply the same simple model to the case of cold strongly interacting matter $(T = 0)$. In the hadronic phase, we now have a completely degenerate Fermi gas of nucleons, whose pressure and energy density are given by

$$P_H(T = 0, \mu) = \frac{1}{24\pi^2} \times 2 \times 2 \times \mu^4 \quad , \tag{2.7}$$

$$\varepsilon_H(T=0,\ \mu) = \frac{1}{8\pi^2} \times 2 \times 2 \times \mu^4 \quad , \tag{2.8}$$

with protons and neutrons of two possible spin orientations. The baryon number density is given by

$$n_B = 2\mu^3/3\pi^2 \quad . \tag{2.9}$$

At high baryon number density, we expect a cold plasma, whose pressure is given by

$$P_Q(T=0,\ \mu) = \frac{1}{24\pi^2} \times 2 \times 2 \times 3 \times \mu_Q^4 - B \tag{2.10}$$

including two flavour and three colour degrees of freedom for the quarks; B again denotes the vacuum pressure on the system. The quark density is given by

$$n_Q = 2\mu_Q^3 / \pi^2 \tag{2.11}$$

in terms of the quark chemical potential μ_Q ; from $n_B = n_Q / 3$, we obtain

$$\mu_Q = \mu \tag{2.12}$$

so that the subscript Q on the chemical potential in eq. (2.10) can be dropped. The energy density of the plasma is

$$\varepsilon_Q(0,\ \mu) = \frac{1}{8\pi^2} \times 2 \times 2 \times 3 \times \mu^4 - B \quad . \tag{2.13}$$

We obtain the critical point by requiring minimal thermodynamic potential, which again means highest pressure. From $P_H(0,\ \mu_c) = P_Q(0,\ \mu_c)$ we get

$$\mu_c = (3\pi^2\ B)^{1/4} \simeq 2.33\ B^{1/4} \tag{2.14}$$

which implies

$$n_c = 2(3\pi^2)^{-1/4}\ B^{3/4} \simeq .85\ B^{3/4} \tag{2.15}$$

for the critical density. Using as above $B^{1/4} \simeq 200$ MeV , this yields

$$n_c \simeq 5n_o \tag{2.16}$$

with $n_o = 0.17\ \text{fm}^{-3}$ denoting standard nuclear density. In comparison, the baryon number density of a inside a nucleon is about $3n_o$, so that the value (2.16) seems not unreasonable. - Finally, we obtain, as in eq. (2.6),

$$\varepsilon_Q(0,\ \mu_c) - \varepsilon_H(0,\ \mu_c) = 2B \tag{2.17}$$

for the latent heat of the transition at $T = 0$.

Extrapolating these results to intermediate temperatures and chemical

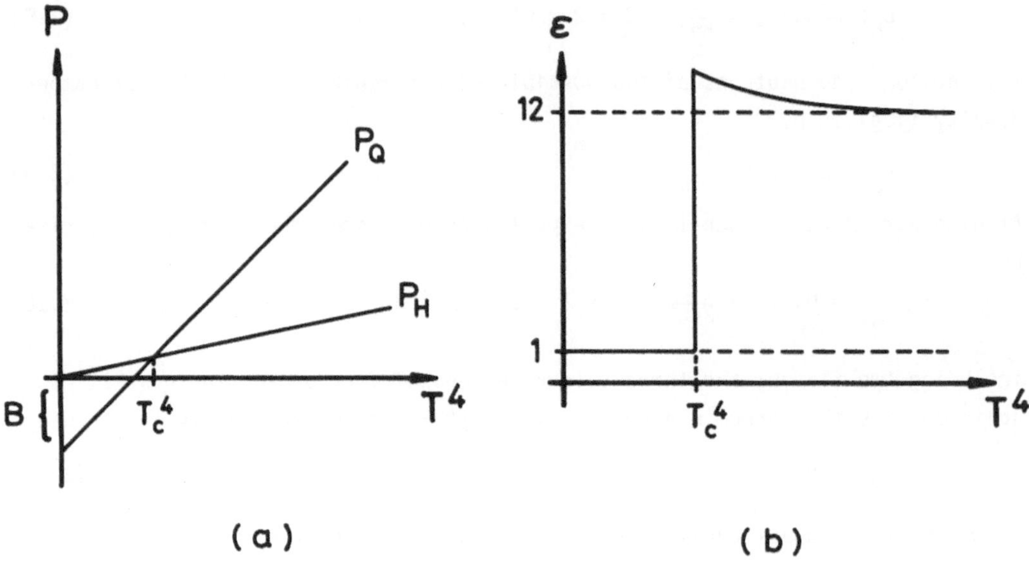

Figure 1 : Pressure (a) and energy density (b) in a two-phase ideal gas model for strongly interacting matter.

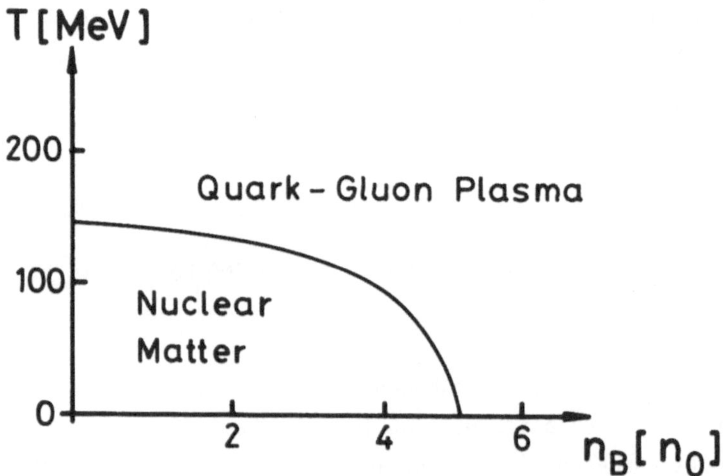

Figure 2 : Phase diagram in a two-phase ideal gas model for strongly interacting matter; n_B is a baryon number density in units of standard nuclear density n_o .

potentials, we obtain the phase diagram shown in fig. 2. It clearly presents only the rudiments of the picture: the inclusion of internuclear forces and/or resonances in the hadronic phase, and the inclusion of perturbative terms in the plasma would certainly modify the values of the transition parameters. Nevertheless, the basic reason for the transition would be expected to remain: we have more degrees of freedom in the plasma phase, causing it to dominate at high temperatures and pressures. Because of B, we have a higher pressure "normalization" in the hadronic phase, leading it dominate at low temperatures and pressures.

In any phenomenological model, such as the one considered here, the transition is of course obtained by construction. The basic question for QCD therefore is: can we derive both critical behaviour and limiting phases from one fundamental description? In the following section, we shall see that this is indeed the case.

III. The Phase Structure of QCD

The Lagrangian density of quantum chromodynamics is given by

$$\mathcal{L}(A, \psi, \bar{\psi}) = - \frac{1}{4} [\partial_\mu A^a_\nu - \partial_\nu A^a_\mu - g f^a_{bc} A^b_\mu A^c_\nu]^2 + + \bar{\psi}^k_f (i\not\partial - g \not{A}_a \lambda^a) \psi^f_k \tag{3.1}$$

in terms of the gluon fields A^a_μ and the quark spinors ψ^k_f. Here the f_{abc} are the structure constants of the colour gauge group, whose generators λ_a satisfy $[\lambda_a, \lambda_b] = i f^c_{ab} \lambda_c$. The gluonic colour indices a,b,c run from one to eight for colour SU(3), those for the quarks (k) from one to three ("red, white and blue"). As we shall here consider only u and d quarks, the flavour index f only takes on two values.

If we would set the structure constants f_{abc} equal to zero, we would recover quantum electrodynamics; it is the non-abelian nature, with the gluon-gluon interaction in eq. (3.1), which distinguishes QCD. In contrast to electrodynamics, we thus have in chromodynamics an interacting theory even if we leave out quarks altogether. The resulting Yang-Mills theory in fact already exhibits many of the essential features of the full theory and can therefore be taken as a model to introduce both formalism and evaluation techniques of QCD thermodynamics. We shall make use of this and first consider purely gluonic matter; subsequently, we shall go on to full QCD with quarks.

The theory introduced with the Lagrangian (3.1) is an interacting relativistic quantum field theory; so far, the only general non-perturbative way to solve such a theory, is provided by the lattice regularization approach of K. Wilson[2]. Together with the Monte Carlo evaluation technique pioneered by M. Creutz[3], it will form the basis for our treatment of QCD thermodynamics.

The partition function for a quantum system described in terms of fields $A(x)$ by a Hamiltonian $H(A)$ is defined as

$$Z = \mathrm{Tr}\,\{\exp - \beta H\} \quad , \tag{3.2}$$

where $\beta^{-1} = T$ is the physical temperature. The conventional lattice formulation is obtained from this in three steps, which we shall now briefly sketch for the Yang-Mills system

$$\mathcal{L}(A) = -\frac{1}{4}\,[\partial_\mu A_\nu^a - \partial_\nu A_\mu^a - g\, f_{bc}^a\, A_\mu^b A_\nu^c]^2 \quad . \tag{3.3}$$

First, the partition function Z is rewritten in form of a path integral[4]

$$Z(\beta, V) = \int [dA]\, \exp\, \{\int_0^\beta d\tau \int_V d^3x\, \mathcal{L}[A(x,\tau)]\} \quad , \tag{3.4}$$

where $\mathcal{L}[A(x,\tau)]$ is the Euclidean density, with $\tau = it$, periodic in τ . The three-dimensional integral of the Hamiltonian formulation $(H \sim \int d^3x\, \mathcal{H}(x))$ thus becomes an asymmetric four-dimensional integral, with the "special" dimension measuring the temperature.

In the next step, we replace the Euclidean $x - \tau$ continuum by a finite lattice, with N_σ sites and spacing a_σ in the spatial part, N_β sites and spacing a_β in the temperature direction. The integrals in the exponent of eq. (3.4) now become sums, and we have $V = (N_\sigma a_\sigma)^3$, $\beta = N_\beta a_\beta$. The thermodynamic limit requires $N_\sigma \to \infty$ at fixed a_σ ; the continuum limit is obtained by a_σ , $a_\beta \to 0$ with fixed $N_\beta a_\beta$, which forces also $N_\beta \to \infty$. The success of the approach rests on the (lucky) facts that already rather small lattices $(N_\sigma \sim 5 - 10$, $N_\beta \sim 3 - 5)$ seem to be asymptotic, and such that scale changes (changes in lattice spacings) can be connected to changes in the coupling strength g by the renormalization group relation, indicating continuum behaviour.

In the last step, we replace the gauge field "variable" $A_\mu((x_i + x_j)/2)$ associated to the link between two adjacent sites i and j by the gauge group element

$$U_{ij} = \exp\,\{\, -i(x_i - x_j)^\mu A_\mu\, (\frac{x_i + x_j}{2})\,\} \quad , \tag{3.5}$$

where $A_\mu(x) = \lambda_a A^a_\mu(x)$. With this transformation, the partition function becomes

$$Z(\beta, V) = \int \prod_{\{links\}} dU_{ij} \exp\{-S(U)\} \quad , \tag{3.6}$$

where the lattice action is, for colour SU(N), given by

$$S(U) = \frac{2N}{g^2} \{ \frac{a_\beta}{a_\sigma} \sum_{\{P_\sigma\}} [1 - \frac{1}{N} \mathrm{Re\,Tr}\, U_{ij} U_{jk} U_{kl} U_{li}]$$

$$+ \frac{a_\sigma}{a_\beta} \sum_{\{P_\beta\}} [1 - \frac{1}{N} \mathrm{Re\,Tr}\, U_{ij} U_{jk} U_{kl} U_{li}] \} \quad . \tag{3.7}$$

Here the sum $\{P_\sigma\}$ runs over all purely spacelike lattice plaquettes (ijkl) , while $\{P_\beta\}$ runs over all those with two spacelike and two "temperature-like" links. - If we insert eq. (3.5) in eq. (3.6/3.7) and expand for small lattice spacings $(|x_i - x_j| \to 0)$, then we recover in leading order the starting form (3.4). - In eq. (3.7), we have kept the colour gauge group general, since the behaviour of the SU(2) Yang-Mills system[5] is presently known with greater precision than that for the SU(3) system[6]. We shall therefore generally consider both; they appear to provide basically the same thermodynamics.

From eq. (3.6/3.7), the energy density

$$\varepsilon \equiv (-1/V)\, (\partial \ln Z/\partial\beta)_V = -(N_\sigma^3 N_\beta a_\sigma^3)^{-1}\, (\partial \ln Z/\partial a_\beta)_{a_\sigma} \tag{3.8}$$

is found to be[5])

$$\varepsilon \simeq 2N\, (N_\sigma^3 N_\beta a_\sigma^3 a_\beta g^2)^{-1} \{ \langle \frac{a_\beta}{a_\sigma} \sum_{\{P_\sigma\}} [1 - \frac{1}{N} \mathrm{Re\,Tr}\, UUUU] \rangle$$

$$- \langle \frac{a_\sigma}{a_\beta} \sum_{\{P_\beta\}} [1 - \frac{1}{N} \mathrm{Re\,Tr}\, UUUU] \rangle \} \tag{3.9}$$

with $\langle\rangle$ denoting the usual thermodynamic average

$$\langle X \rangle \equiv \{\int \prod dU\, e^{-S(U)}\, X(U)\} \,/\, \{\int \prod dU\, e^{-S(U)}\} \quad . \tag{3.10}$$

Eq. (3.9) is our starting point for the Monte Carlo evaluation of gluon thermodynamics.

The evaluation is now carried out as follows. The computer simulates an $N_\sigma^3 \times N_\beta$ lattice; for convenience we choose $a_\sigma = a_\beta = a$. Starting from a given ordered (all $U = 1$, "cold start") or disordered (all U random, "hot start") initial configuration, successively each link is assigned a new element

U' , chosen randomly with the weight exp $\{-S(U)\}$. One traverse of this procedure through the entire lattice is called one iteration. In general, it is found that five hundred or so iterations provide reasonable first indications about the behaviour of the energy density (3.9), but for some precision one should have more. The results shown here for colour SU(2) are obtained with typically around three thousand iterations, after which we observe quite stable behaviour; the SU(3) results are generally based on a few hundred iterations. The work was done with $N_\sigma = 7,9,10$ for $N_\beta = 2,3,4,5$; apart from expected finite lattice size effects[7] there was no striking N_σ dependence of ε , suggesting that in general the thermodynamic limit is reached. To give at least some intuitive grounds for this, note that a $10^3 \times 3$ lattice has about 12,000 link degrees of freedom.

As result of the Monte Carlo evaluation, we obtain for a lattice of given size (N_σ, N_β) the energy density ε as function of g . In the continuum limit, g and the lattice spacing a are for colour SU(N) related through

$$a\,\Lambda_L = (11Ng^2 / 48\pi^2)^{-51/121} \exp\{-24\pi^2 / 11Ng^2\} \quad ; \tag{3.11}$$

this relation is found by requiring a dimensional parameter Λ_L to remain constant under scale changes accompanied by corresponding changes in coupling strength. Hence once we are in the region of validity of the continuum limit, eq. (3.11) gives us the connection between g and a . Since $(N_\beta a)^{-1}$ is the temperature in units of Λ_L , we then have the desired continuum form of $\varepsilon(\beta)$.

In fig. 3, we show the resulting energy density ε as function of the temperature T , for both SU(2)[5] and SU(3)[6]. We first note that at high temperatures, the results of the Monte Carlo evaluation agree quite well with the anticipated Stefan-Boltzmann form

$$\varepsilon/T^4 = \begin{cases} \pi^2/5 & SU(2) \\ 8\pi^2/15 & SU(3) \end{cases} \tag{3.12}$$

Let us now got to lower T , concentrating on the SU(2) case. At about $T = 50\,\Lambda_L$, ε drops sharply. The derivative of ε gives us the specific heat, shown in fig. 4a. At $T \simeq 43\,\Lambda_L$, it has a singularity-like peak, which signals the transition from bound to free gluons. With Λ_L taken in physical units, this gives us $T_c \simeq 180 - 200$ MeV ; for SU(3), we find similarly $T_c \simeq 160 - 180$ MeV . How do we know that it is deconfinement which occurs here? One can study the behaviour of a static $q\bar{q}$ pair immersed in a gluon system of temperature T [8,9]; the free energy F of an isolated quark then serves to

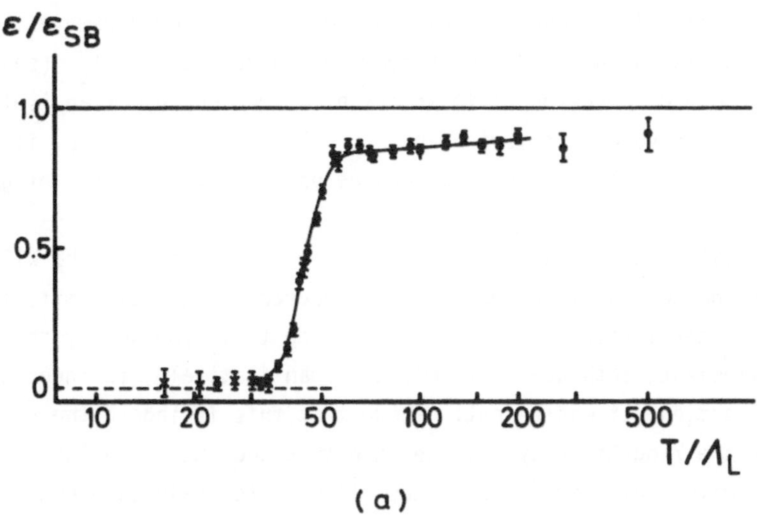

(a)

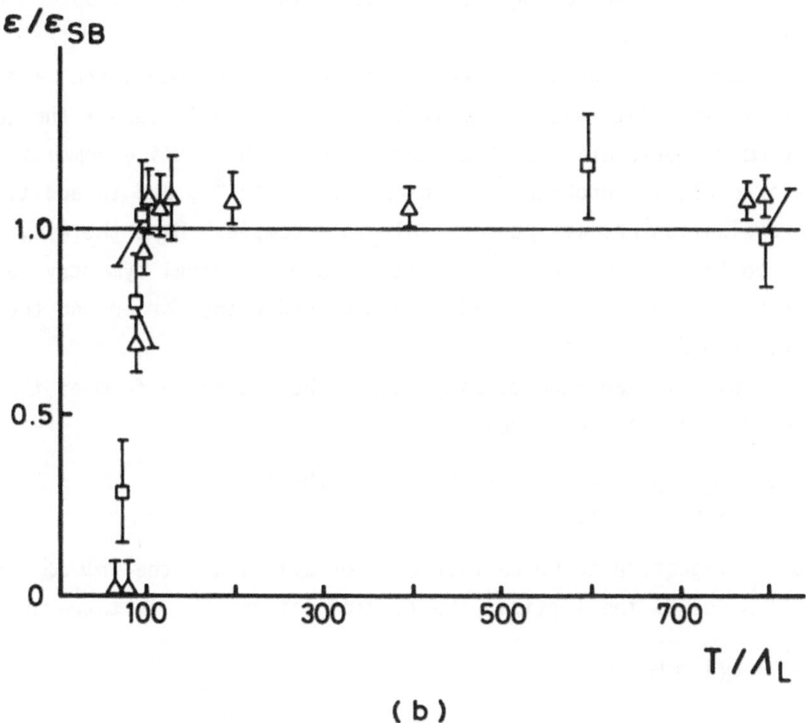

(b)

Figure 3 : Energy density of the Yang-Mills system, normalized to the ideal gas value ε_{SB}, (a) for SU(2) colour group, from ref. 5), and (b) for SU(3) colour group, from ref. 6).

define the thermal Wilson loop $\langle L \rangle = \exp\{-\beta F\}$ as order parameter. It is found that $\langle L \rangle$ is essentially zero below and non-zero above T_c (see fig. 4b). Since $\langle L \rangle = 0$ corresponds to an infinite free energy of an isolated colour source, we have confinement below T_c . In accord with this, it can also be shown that for $T < T_c$ the system behaves essentially as a gas of gluonium states[10].

All lattice results presented here were obtained with the Wilson form (3.7) of the action, which provides the correct continuum limit. There are, however, other lattice actions which also do this, and we may therefore ask if deconfinement, both qualitatively and quantitatively, is independent of the choice of action. It was recently shown that this is indeed the case[11].

For the Yang-Mills system, we have thus seen that the lattice formulation together with Monte Carlo techniques allow us to evaluate gluon thermodynamics over the whole temperature range. The resulting behaviour shows the expected two-phase nature: at low temperatures, we have a hadronic resonance gas of gluonium states; heating brings us to a deconfinement transition and beyond that to an ideal gluon gas.

We now want to extend our considerations to include quarks and antiquarks. We shall see that this brings in a basically new feature - the question of chiral symmetry restoration at high temperature. The lattice fomulation encounters as a result the problem of species doubling[2,12], and in addition the Monte Carlo evaluation becomes considerably more complex. Nevertheless, first results both on the full QCD energy density[13] and on chiral symmetry restoration[13,14] have now appeared; we shall first consider the former and then return to chiral symmetry questions.

For the full Lagrange density (3.1), the Euclidean form of the partition function on the lattice is now given by

$$Z = \int \prod_{\text{links}} dU \prod_{\text{sites}} d\psi\, d\bar{\psi}\; e^{-S^G(U) - S^F(U,\,\psi,\,\bar{\psi})} \tag{3.13}$$

with the dU integration to be carried out for all links, the $d\psi\, d\bar{\psi}$ integrations for all sites of the lattice. The fermion action S^F is taken in the form

$$S^F = \bar{\psi}(1 - KM)\,\psi \quad , \tag{3.14}$$

$$M_\mu = (1-\gamma_\mu)\, U_{nm}\, \delta_{n,m-\hat{\mu}} + (1+\gamma_\mu)\, U^+_{mn}\, \delta_{n,m+\hat{\mu}} \quad , \tag{3.15}$$

while the gluon part S^G is given by eq. (3.7); the coupling between quarks and gluons is given by the "hopping parameter" $K(g^2)$. The integration over

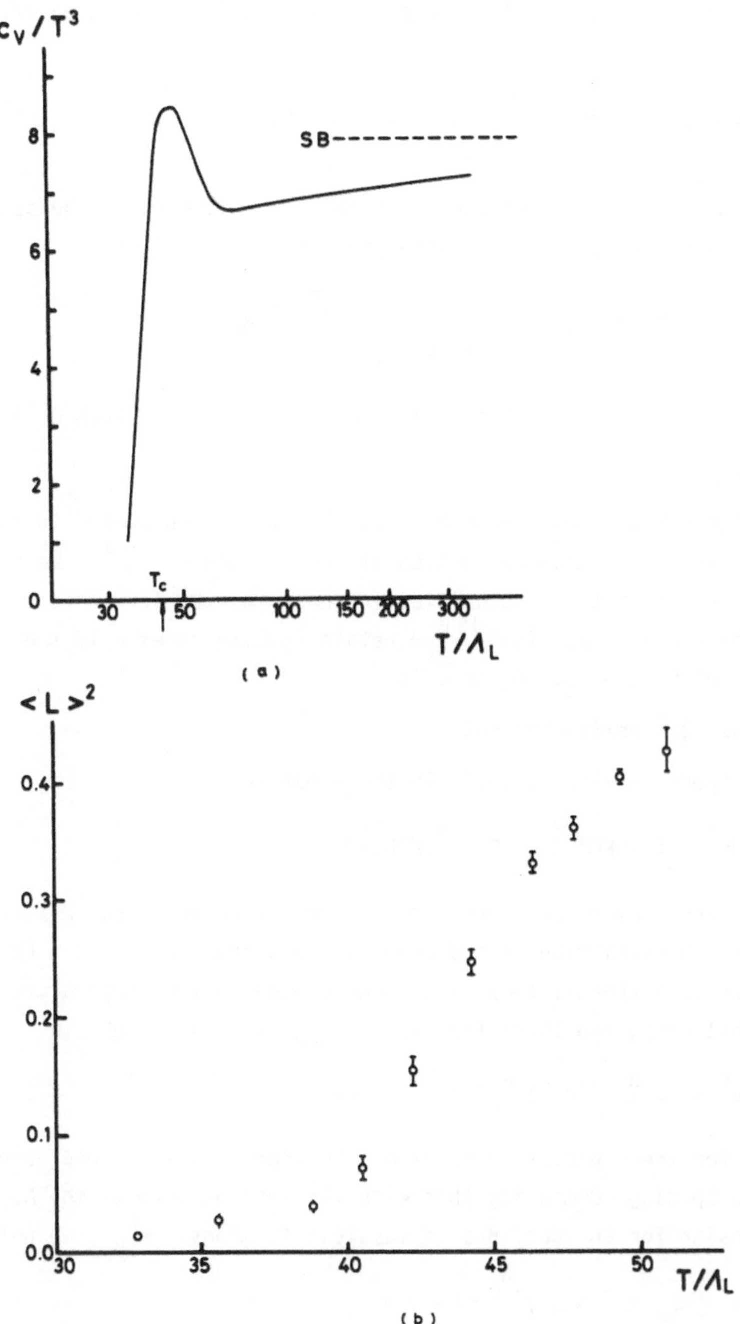

Figure 4 : Specific heat (a) and squared order parameter (b) for the SU(2) Yang-Mills system, from ref. 5).

the anticommuting spinor fields can be carried out[15] to give an effective boson form

$$Z = \int \prod_{\text{links}} dU \, d^{-S^G(U)} \det(1 - KM) \quad . \tag{3.16}$$

The energy density ε is obtained from this Z ; it becomes the sum $\varepsilon = \varepsilon^G + \varepsilon^F$ of a pure gluon part and a quark-gluon part[13]

$$\varepsilon^F \equiv -\xi^2 (N_\sigma^3 N_\beta a_\sigma^4 Z)^{-1} \int \prod_{\text{links}} dU \, e^{-S^G(U)} \det Q \times$$

$$\times \left\{ \frac{3K(g^2)}{4} \mathrm{Tr}(M_0 Q^{-1}) - \frac{K(g^2)}{4} \sum_{\mu=1}^{3} \mathrm{Tr}(M_\mu Q^{-1}) \right\} \tag{3.17}$$

with $Q \equiv 1 - KM(U)$.

The computational problem beyond what is encountered in the pure Yang-Mills case lies in the evaluation of $\det Q$ and of Q^{-1} . We shall here use the expansion of these quantities in powers of the fermionic coupling K ("hopping parameter expansion")[16], and retain in both cases only the leading term. For $\det Q$ the leading term is

$$\det Q = \det(1 - KM) \simeq 1 \tag{3.18}$$

("quenched approximation"), while in the expansion

$$Q^{-1} = [1 - KM]^{-1} = \sum_{\ell=0}^{\infty} K^\ell [M(U)]^\ell \quad , \tag{3.19}$$

because of gauge invariance, the first contribution to $\mathrm{Tr}(Q^{-1}M)$ arises for the shortest non-vanishing closed loop obtained from $M(U)$ U . For $N_\beta = 2$ and 3 , this is a thermal loop, i.e., one closed in the temperature direction; hence in that case, the first term is $\ell = N_\beta - 1$, and we obtain

$$\varepsilon^F a^4 \simeq \frac{3}{4} [K(g^2)]^{N_\beta} 2^{N_\beta + 2} \langle L \rangle \tag{3.20}$$

with $\langle L \rangle$ for the expectation value of the thermal Wilson loop, and a for the lattice spacing. Comparing this with the leading term of the hopping parameter expansion for an ideal gas of massless fermions, ε^F_{SB} , we get

$$\varepsilon^F / \varepsilon^F_{SB} = [8K(g^2)]^{N_\beta} \langle L \rangle / N \tag{3.21}$$

since for the ideal gas $K = 1/8$, $\langle L \rangle = N$.

Taking $K(g^2)$ from a numerical evaluation[17] and using the Monte Carlo data[6] for $\langle L \rangle$, we obtain for the SU(3) case the ratio $\varepsilon^F/\varepsilon^F_{SB}$ shown in fig. 5. We note that the energy density takes on its asymptotic value for $T \gtrsim 100\ \Lambda_L$; around $T \sim 80\ \Lambda_L$ (~160 MeV) , there is a sharp drop, corresponding to the onset of confinement.

For the SU(2) case, the restriction to the leading term of the hopping parameter expansion (3.19) has been removed[18]; including all terms up to order 50 results in the energy density shown in fig. 5b. We note that the qualitative features of fig. 5a persist.

In fig. 6, we show finally the overall energy density ε/T^4 for full QCD with colour SU(3), obtained by combining the results for ε^F with those for the pure Yang-Mills system. We conclude that full quantum chromodynamics with fermions indeed appears to lead to the deconfinement behaviour observed in the study of Yang-Mills systems alone. In particular, we note that at temperatures $T \gtrsim 2T_c$ essentially all constituent degrees of freedom have been "thawed".

Quantum chromodynamics, for massless quarks a priori free of dimensional scales, contains the intrinsic potential for the spontaneous generation of two scales: one for the confinement force coupling quarks to form hadrons, and one for the chiral force binding the collective excitations to Goldstone bosons. These two lead in thermodynamics to two possible phase transitions, characterized by two critical temperatures, T_c and T_{ch} . Above T_c , the density is high enough to render confinement unimportant: hadrons dissolve into quarks and gluons. Above T_{ch} , chiral symmetry is restored, so that quarks must be massless. For T below both T_c and T_{ch} , we have a gas of massive hadrons; for T above both T_c and T_{ch} , we have a plasma of massless quarks and gluons. Conceptually simplest would be $T_c = T_{ch}$; the possibility $T_c > T_{ch}$ appears rather unlikely[19]. On the other hand, $T_c < T_{ch}$ would correspond to a regime of unbound massive "constituent" quarks, as they appear in the additive quark model for hadron-hadron and hadron-lepton interactions[20]. The question of deconfinement vs. chiral symmtery restoration thus confronts us with one of the most intriguing aspects of quark-gluon thermodynamics.

The fermionic action of Wilson used in the last section avoids species doubling at the cost of chiral invariance. Even an ideal gas of massless quarks in this formulation is not chirally invariant, since the expectation value $<\psi\bar{\psi}>$ is always different from zero. It has therefore been suggested[21] to use the difference between this "Stefan-Boltzmann" value and the corresponding QCD value for Wilson fermions as the physically meaningful order parameter: it would vanish when the behaviour of a non-interacting system of massless fermions is reached.

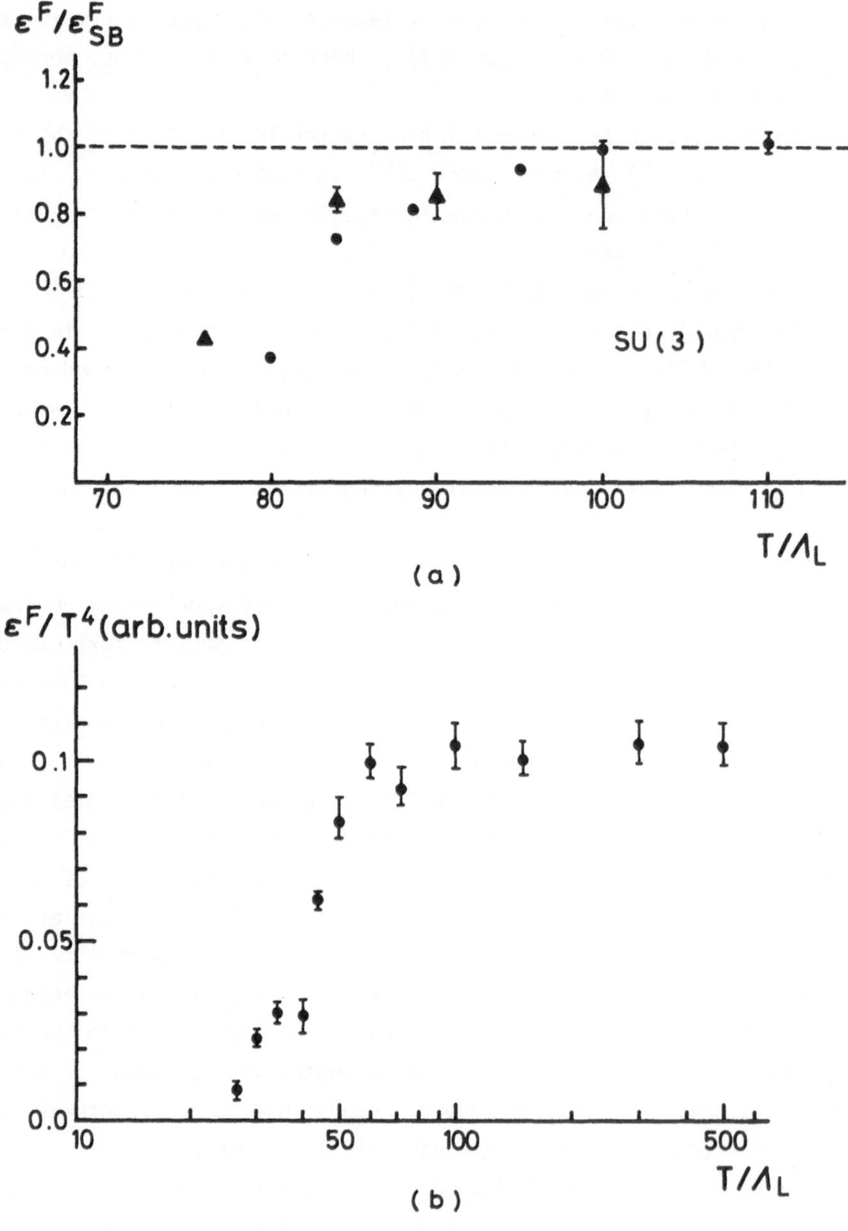

Figure 5a : Energy density of the fermion sector, normalized to the ideal gas value ε^F_{SB} , for SU(3) Wilson fermions, leading term hopping parameter expansion, from ref. 13).

Figure 5b : Energy density for SU(2) Wilson fermions, hopping parameter expansion up to order 50 , from ref. 18).

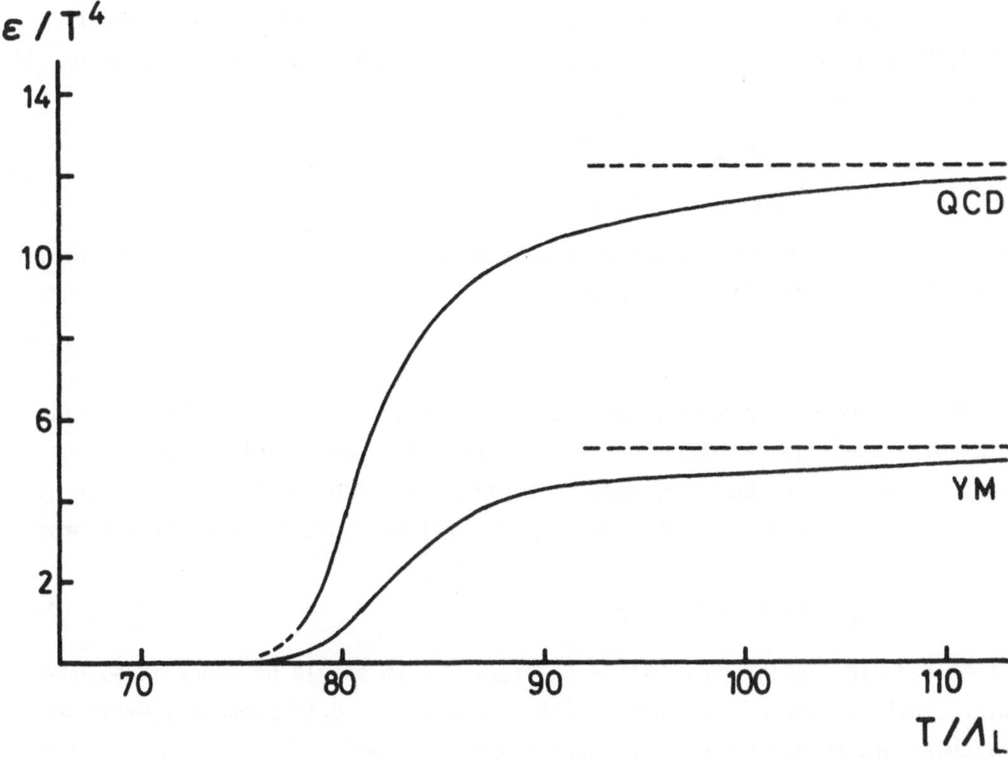

Figure 6 : Energy density of full QCD, compared to that of the SU(3) Yang-Mills system, from ref. 13).

In fig. 7 we show this order parameter as calculated for colour SU(2)[22] and SU(3)[13], in leading power of the hopping parameter expansion. It is non-zero up to

$$T_{ch} \simeq \begin{cases} 60\ \Lambda_L & SU(2) \\ 100\ \Lambda_L & SU(3) \end{cases} , \tag{3.22}$$

and vanishes for higher temperatures. This suggests chiral symmetry restoration slightly above deconfinement, with

$$T_{ch} / T_c \simeq 1.3 \quad . \tag{3.23}$$

It remains open at present to what extent this will be modified by the inclusion of virtual quark loops, or if there are any significant finite lattice effects. Using for the SU(2) case a chirally invariant action with the resulting species doubling, it was found in ref. 14) that chiral symmetry restoration occurs at

$$T_{ch} = (0.55 \pm 0.07)\ \sqrt{\sigma} \quad ; \tag{3.24}$$

this leads to similar conclusions on T_{ch}/T_c. It should be emphasized, however, that in view of possible finite size effects, both presently available calculations do not exclude the possibility $T_{ch} = T_c$.

In the lattice evaluation of QCD thermodynamics, we have calculated all physical quantities in terms of the dimensional lattice scale Λ_L. To convert Λ_L into physical units, we just have to measure one of these physical observables. String tension considerations give for Yang-Mills systems

$$\Lambda_L = \begin{cases} (1.1 \pm 0.2) \times 10\ \ \sqrt{\sigma} = (4.4 \pm 0.8)\ \text{MeV}\ ^{22)} \\ (1.3 \pm 0.2) \times 10^{-2} \sqrt{\sigma} = (5.2 \pm 0.8)\ \text{MeV}\ ^{23)} \end{cases} \tag{3.25}$$

in case of colour SU(2) and

$$\Lambda_L = (5.0 \pm 1.5) \times 10^{-3} \sqrt{\sigma} = (2.0 \pm 0.6)\ \text{MeV}\ ^{24)} \tag{3.26}$$

for colour SU(3). The deconfinement temperature is found to be

$$T_c = (38\ ^{8)} - 43\ ^{5)})\ \Lambda_L \tag{3.27}$$

for SU(2) and

$$T_c = (75\ ^{25)} - 83\ ^{6)})\ \Lambda_L \tag{3.28}$$

for SU(3). Taking the average of eq. (3.25), we have

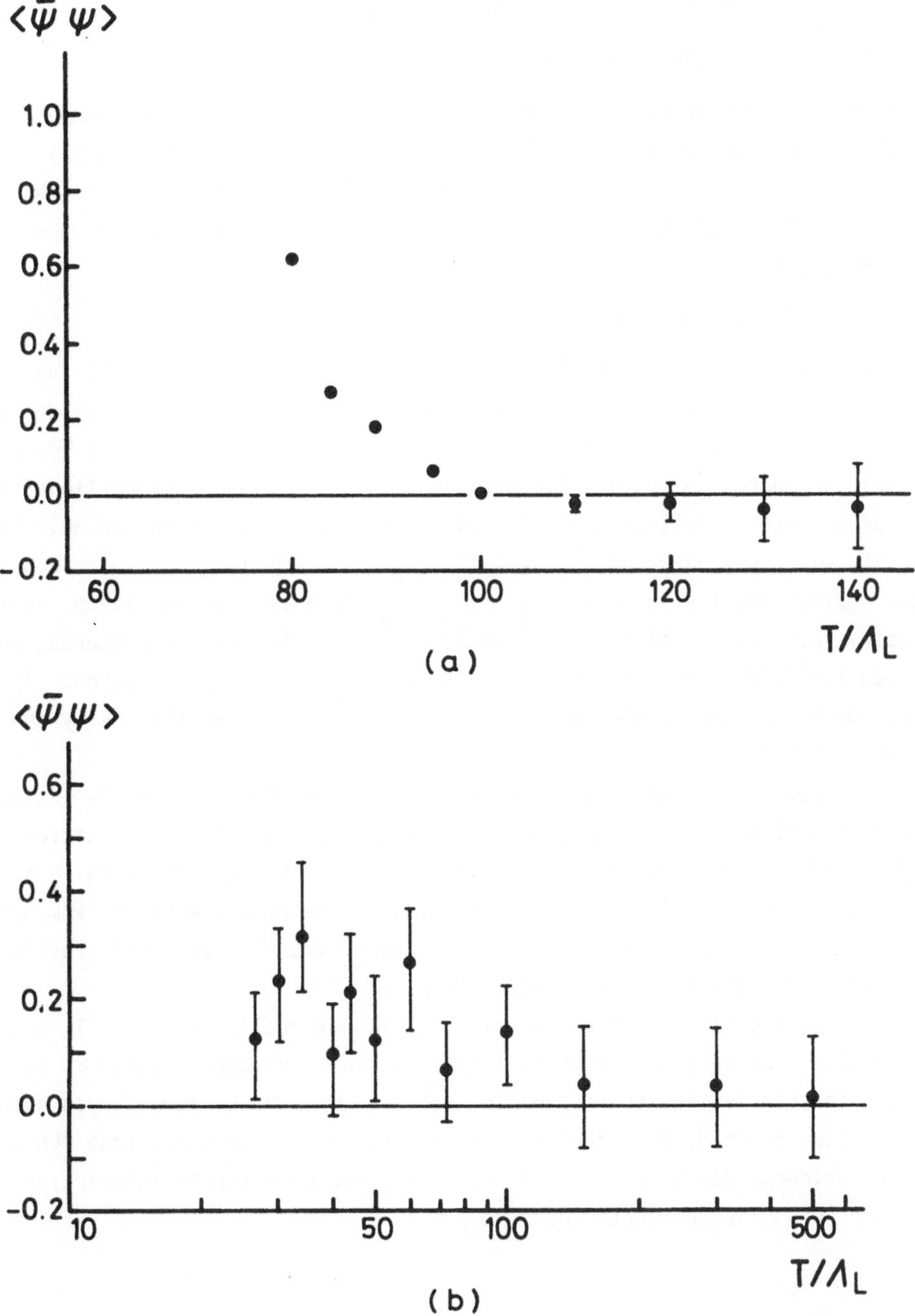

Figure 7 : Chiral symmetry order parameter, (a) for SU(3) Wilson fermions, from ref. 13), and (b) for SU(2) Wilson fermions, from ref. 18).

$$T_c = \begin{cases} [(170-210) \pm 30] \text{ MeV} & SU(2) \\ [(150-170) \pm 50] \text{ MeV} & SU(3) \end{cases} \tag{3.29}$$

and thus little or no dependence of T_c on the colour group. The temperature for chiral symmetry restoration is accordingly given by relation (3.23).

From eq. (3.29) and the form of fig. 7, we can now estimate the energy density values at the two transition points. For the SU(3) Yang-Mills case, we obtain

$$\varepsilon(T_c) \simeq 200 - 300 \text{ MeV/fm}^3 \quad , \tag{3.30}$$

where we have assumed that the turn-over in ε occurs at about half the Stefan-Boltzmann value. This range, corresponding roughly to hadronic energy density, seems physically quite reasonable. It is not known at present if and how much it would be increased by the introduction of quarks; a shift proportional to that of the Stefan-Boltzmann limit would double the value of eq. (3.30). This suggests twice standard nuclear density (n_0 = 150 MeV/fm^3) as lower and four times nuclear density as upper bound for the deconfinement transition. Chiral symmetry restoration, if it occurs at only slightly higher temperatures, requires considerably higher energy densities. Just a small increase beyond T_c brings up to the top of the Stefan-Boltzmann "shelf", where the energy density is above 2 GeV/fm^3 .

Our basic conclusion in this chapter is certainly that the lattice formulation of quantum chromodynamics appears to be an extremely fruitful approach to the thermodynamics of strongly interacting matter. It is so far the only way to describe within one theory the whole temperature range from hadronic matter to the quark-gluon plasma. It leads to deconfinement and provides first hints on chiral symmetry restoration.

We are still at the beginning. It is not really clear if $T_c \neq T_{ch}$, finite size scaling near the phase transitions has not been studied at all for $T \neq 0$, and the lattice thermodynamics of systems with non-zero baryon number has not been touched. Nevertheless, there seems to emerge today from QCD ever growing evidence for a two or three state picture of strongly interacting matter such as we have presented here.

Acknowledgement

It is a pleasure to thank J. Engels, R.V. Gavai, M. Gyulassy, F. Karsch and L. McLerran for stimulating discussions on various aspects of the topic.

References

1) See e.g., Quark Matter Formation and Heavy Ion Collisions, M. Jacob and H. Satz (Eds.), World Scietific Publ. Co., Singapore, Nov. 1982

2) K. Wilson, Phys. Rev. D10 (1974) 2445; in New Phenomena in Subnuclear Physics, A. Zichichi (Ed.), Plenum Press, New York 1977 (Erice 1975)

3) M. Creutz, Phys. Rev. D21 (1980) 2308

4) C. Bernard, Phys. Rev. D9 (1974) 3312

5) J. Engels, F. Karsch, I. Montvay and H. Satz, Phys. Lett. 101B (1981) 89; J. Engels, F. Karsch, I. Montvay and H. Satz, Nucl. Phys. B205 [FS5] (1982) 545

6) I. Montvay and E. Pietarinen, Phys. Lett. 110B (1982) 148; I. Montvay and E. Pietarinen, Phys. Lett. 115B (1982) 151

7) J. Engels, F. Karsch and H. Satz, Nucl. Phys. B205 [FS5] (1982) 239

8) L.D. McLerran and B. Svetitsky, Phys. Lett. 98B (1981) 195; L.D. McLerran and B. Svetitsky, Phys. Rev. D24 (1981) 450

9) J. Kuti, J. Polônyi and K. Szlachânyi, Phys. Lett. 98B (1981) 199

10) J. Engels, F. Karsch, I. Montvay and H. Satz, Phys. Lett. 102B (1981) 332

11) R.V. Gavai, Nucl. Phys. B [FS], in press (BI-TP 82/11, 1982); R.V. Gavai, F. Karsch and H. Satz, "Universality in Finite Temperature Lattice QCD", Bielefeld Preprint BI-TP 82/26, 1982

12) L. Susskind, Phys. Rev. D16 (1977) 3031

13) J. Engels, F. Karsch and H. Satz, Phys. Lett. 113B (1982) 398

14) J. Kogut, M. Stone, H. Wyld, J. Shigemitsu, S. Shenker and D. Sinclair, Phys. Rev. Lett. 48 (1982) 1140

15) T. Matthews and A. Salam, Nuovo Cimento 12 (1954) 563; 2 (1955) 120

16) C.B. Lang and H. Nicolai, Nucl. Phys. B200 [FS4] (1982) 135; A. Hasenfratz and P. Hasenfratz, Phys. Lett. 104B (1981) 489

17) A. Hasenfratz, P. Hasenfratz, Z. Kunszt and C.B. Lang, Phys. Lett. 110B (1982) 289

18) J. Engels and F. Karsch, to be published

19) E.V. Shuryak, Phys. Lett. 107B (1981) 103; R.D. Pisarski, Phys. Lett, 110B (1982) 155

20) H. Satz, Phys. Lett. 25B (1967) 27 and Phys. Lett. 25B (1967) 220; H. Satz, Nuovo Cim. 37A (1977) 141

21) C.B. Lang and H. Nicolai, Nucl. Phys. B200 [FS4] (1982) 135

22) G. Bhanot and C. Rebbi, Nucl. Phys. B180 [FS2] (1981) 469

23) M. Creutz, Phys. Rev. Lett. 45 (1980) 313

24) M. Creutz, Phys. Rev. Lett. 45 (1980) 313; E. Pietarinen, Nucl. Phys. B190 [FS3] (1981) 239

25) K. Kajantie, C. Montonen and E. Pietarinen, Z. Phys. C9 (1981) 253

THE QUARK-GLUON PLASMA

Larry McLerran
Physics FM-15
University of Washington
Seattle, Washington

and

Research Institute for Theoretical Physics
University of Helsinki
Siltavuorenpenger 20C
Helsinki 17 Finland

SECTION 1

1. INTRODUCTION

These lectures present features of high temperature and baryon number density hadronic matter as it might be described by QCD, and describe the possible formation of such matter in ultra-relativistic nucleus-nucleus collisions. The confinement-deconfinement transition, and the chiral symmetry breaking transition are discussed using order parameters. The current understanding of the dynamics of these phase changes is reviewed.

The first lecture begins by presenting naive arguments for expecting phase transitions at several times the energy density of hadronic matter. These phase transitions are then discussed in terms of order parameters. For Yang-Mills theories in the absence of dynamical quarks the exponential of the negative of the free energy of an isolated static test quark is employed to measure the confinement-deconfinement phase transition. When dynamical quarks are present, N-ality order parameters are introduced which measure the probability of the difference of quarks and anti-quarks being non-zero mod N for an SU(n) gauge theory. The chiral symmetry restoration transition is extensively discussed in the remainder of the lecture. I consider a massless doublet of quarks and review the U(2)×U(2) symmetry of such a theory. Order parameters which measure the breaking of chiral U(1) and SU(2) symmetries are derived.

The dynamics of these phase transitions, and the relationship between confinement-deconfinement and the breakdown or realization of a global dynamical Z_N symmetry, is the subject of the second lecture. An analogy between high temperature and density Yang-Mills theory in 3+1 dimensions and a 3-dimensional Z_N spin theory is used to infer the order of the deconfinement-confinement transition. What little is known about the dynamics of chiral symmetry breaking is discussed in the remainder of the lecture. Some warnings are posted for using the quenched approximation to measure chiral symmetry breaking in lattice Monte-Carlo computations.

The last lecture reviews how hadronic matter might be formed in ultra-relativistic nuclear collisions. The space time dynamics of the Landau hydrodynamical model and the inside-outside cascade model are reviewed. The hydrodynamic equations for an inside-outside cascade model in 1+1 dimensions is derived, and the straightforward generalizations of these equations to 3+1 dimensions is discussed.

SECTION 2

LECTURE I: SOME GENERAL FEATURES OF PHASE TRANSITIONS IN FINITE TEMPERATURE AND DENSITY YANG-MILLS THEORY

The conjecture that a finite temperature and baryon number density hadronic system undergoes a confinement-deconfinement phase transition has its origins in extremely simple pictures of hadrons as composites of quarks and gluons. At some energy density of hadronic matter, hadrons will on the average overlap with nearby hadrons. The hadrons are then no longer confined to individual hadrons, and become free to wander unconfined throughout the hadronic matter. This transition might occur at an energy density about that of a proton

$$\mathcal{E} \sim \frac{M_p}{\frac{4}{3}\pi r_p^3} \sim 500 \text{ MeV/fm}^3, \tag{2.1}$$

where M_p is a proton mass and r_p is the proton radius which I shall take as the r.m.s. charge radius. This energy density is only a few times that of nuclear matter,

$$\mathcal{E}_{N.M.} \sim 150 \text{ MeV/fm}^3. \tag{2.2}$$

A confinement-deconfinement transition therefore might occur in the cores of neutron stars, $\mathcal{E} \sim 10\text{-}20\,\mathcal{E}_{N.M.}$, in ultra-relativistic heavy ion collisions, $\mathcal{E} \gtrsim 10\text{-}30\,\mathcal{E}_{N.M.}$, and almost certainly occurred in the early universe.

The drastic change in the properties of hadronic matter which results from the overlap of hadrons may induce other phase changes in matter. The symmetry which when broken allows the nucleon to be massive and the pion essentially massless, that is chiral symmetry, may become restored in hadronic matter at some energy density of the order of that of confinement-deconfinement. In zero temperature quark-gluon matter, liberated quarks might bind into Cooper pairs to produce a superconductor or a superfluid.

At energy densities far above these phase transition energy densities, hadronic matter very probably becomes an almost ideal gas of quarks and gluons. This conjecture follows from the asymptotic freedom of QCD. At short distances the effective strength of quark-gluon interactions is weak. At high energy densities, quarks and gluons interact most often at short distances, and the effect of these "weak" interactions should be small compared to free particle effects.(1) This should be true for the free energy which measures bulk properties of the quark-gluon matter. There may be small non-perturbative contributions arising from long distance interactions, but the dominant contribution should be that of an ideal gas. Some quantities may, of course, be sensitive to non-perturbative effects, for example, the long distance behavior of correlation functions.

The behavior of an ideal gas of particles is, of course, boring and dull. As a QCD test, a measurement which would verify this ideal gas behavior would be interesting. The most interesting phenomena in quark-gluon matter should nevertheless occur at densities near those of phase transitions. The remainder of this lecture is a discussion of these phase transitions.

The confinement-deconfinement phase transition is clearly illustrated in an SU(N) Yang-Mills theory without dynamical quarks.[(2-3)] A static test quark placed in a vacuum of this theory has infinite energy and is confined. In very high energy density gluonic matter, this quark is not confined and has finite free energy. The confinement-deconfinement phase transition occurs when this free energy becomes finite.

The free energy difference of system with a static test quark at position $\vec{r}$ and the system with no quark is determined by the thermal expectation value of the local operator $L(\vec{r})$,

$$e^{-\beta F} = \langle L \rangle \ . \tag{2.3}$$

The operator L is an order parameter for the phase transition since

$$\begin{aligned} \langle L \rangle &= 0 \iff \text{Confinement} \\ \langle L \rangle &\neq 0 \iff \text{Deconfinement.} \end{aligned} \tag{2.4}$$

At low energy densities, $\langle L \rangle$ should be zero for a finite temperature range $0 \leq T \leq T_c$ and is finite for $T \geq T_c$.

When dynamical fermions are included in the grand canonical ensemble of a Yang-Mills theory, the free energy of an isolated static test quark is no longer a useful parameter for characterizing confinement. Light dynamical fermions contribute to this ensemble and may form a bound state with the test quark.[(4)] Since this bound state has finite energy, $\langle L \rangle$ is finite in both the confined and deconfined phases of the theory.

To see how an order parameter arises in an SU(N) theory with quarks, consider a one flavor theory.[(5)] The color singlet bound states of this theory always passes an integer multiple of N of quarks minus antiquarks. With the baryon charge

$$Q_B \equiv \frac{1}{N} \int d^3x\, \bar{\psi}(x)\gamma^0\psi(x) \ , \tag{2.5}$$

the N-ality operators are defined as

$$\nu_k \equiv \frac{1}{N} \sum_{j=0}^{N-1} e^{2\pi i j(Q-k/N)} , \tag{2.6}$$

with $k = 0,\ldots,N-1$. These operators are useful because they project on to states where the difference of quarks minus antiquarks is k mod N. The operators ν_k are order parameters since for $k \neq 0$

$$\langle\nu_k\rangle = 0 \qquad \Leftrightarrow \qquad \text{Confinement} \tag{2.7}$$

$$\langle\nu_k\rangle \neq 0 \qquad \Leftrightarrow \qquad \text{Deconfinement.} \tag{2.8}$$

In the confined phase N-ality non-singlet configurations of quarks have infinite energy, hence Eq. (2.7), but no such restriction applies to the deconfined phase. Notice that at very high energy densities $\nu_j = \nu_k \neq 0$ for all j and k since there should be equal probability of having configurations with a finite number of excess quarks.

An interesting consequence of this N-ality scheme for classifying phase transitions is that for SU(N) theories, $N \geq 4$, there may be a variety of confinement-deconfinement transitions. In an SU(4) theory, one might imagine first liberating N-ality quartets to form a system of confined N-ality doublets, and the doublets might become liberated at an even higher energy density.

Hadronic matter in a finite size spatial volume illustrates the application of the order parameters ν_j. At low energy density, the constituents of this matter are N-ality singlet hadrons. The N-ality, $\langle\nu_j\rangle$, $j \neq 0$, is therefore zero for all configurations of the matter except those where a hadron is close to the surface of the volume. In this situation, part of the hadron is inside and part is outside of the volume and a non-zero contribution to $\langle\nu_j\rangle$ may result. With a typical hadron radius R, hadron density η, and surface area S, this contribution is

$$\langle\nu_j\rangle \sim \eta R S. \tag{2.9}$$

As the energy density of the hadronic matter increases, an energy density is approached where hadrons begin coalescing into large hadrons. At the confinement-deconfinement transition, the average radius of these giant hadrons $R \to \infty$, and a contribution to $\langle\nu_j\rangle$, $j\neq 0$ may be obtained as a volume effect. To isolate volume from surface effects, spatial boundary conditions which conserve the N-ality flux at the surface must be imposed so that the surface contribution vanishes. The $\langle\nu_j\rangle$ changes from zero to a non-zero value at the confinement-deconfinement transition.

The vanishing or non-vanishing of $\langle\nu_j\rangle$ may be understood by analogy to the vanishing or non-vanishing of $\langle L\rangle$ for a static test quark. The quantity $\langle\nu_j\rangle$ gives the exponential of the negative of the free energy of a N-ality singlet hadronic system in the presence of j excess dynamical quarks. As was the case for $\langle L\rangle$, $\langle\nu_j\rangle$ should vanish for the confined phase. In the next lecture, we shall see that this behavior of $\langle\nu_j\rangle$ is the consequence of a global, dynamical Z_N symmetry.[(3,6-8)]

The chiral symmetry breaking-restoration phase transition is another phase transition of hadronic matter which might, or might not, be directly related to a confinement-deconfinement transition.[(9-10)] Almost nothing is understood about the dynamics of this transition. To initiate a discussion of this phase transition, consider QCD in the chiral (massless) SU(2) flavor symmetric limit. We shall ignore heavy quark flavors since quarks such as charm, bottom, and top are too heavy to influence the dynamics of hadronic matter at the temperatures where a phase transition is expected, and since strange quarks influence on the chiral dynamics is also weak but estimable by techniques which may be extended from the considerations here. For example, chiral SU(3) might be used as an approximate symmetry of hadronic matter, and corrections to this symmetry might be computed as an expansion in the strange quark mass.

Chiral SU(2) flavor symmetry is an invariance of the massless QCD Lagrangian under combined phase, flavor γ^5-phase and γ^5-flavor rotations. With τ_I the generator of isospin rotations, these transformations are

$$\psi \to e^{i\theta}\psi \tag{2.10}$$

$$\psi \to e^{i\vec{\alpha}\cdot\vec{\tau}_I}\psi \tag{2.11}$$

$$\psi \to e^{i\bar{\theta}\cdot\gamma_5}\psi \tag{2.12}$$

$$\psi \to e^{i\vec{\alpha}\cdot\vec{\tau}_I\gamma_5}\psi, \tag{2.13}$$

where ψ is a quark field. The electromagnetic, quark flavor, axial U(1), and axial quark flavor currents,

$$J^\mu = \bar{\psi}\gamma^\mu\psi \tag{2.14}$$

$$J_5^\mu = \bar{\psi}\gamma^\mu\gamma^5\psi \tag{2.15}$$

$$J_a^\mu = \bar{\psi}\gamma^\mu\tau_I^a\psi \tag{2.16}$$

$$J_{5a}^\mu = \bar{\psi}\gamma^\mu\gamma^5\tau_I^a\psi \tag{2.17}$$

are formally conserved as a consequence of Eqs. (2.10-2.13),

$$\partial_\mu J^\mu = 0 \tag{2.18}$$

$$\partial_\mu J_5^\mu = 0 \tag{2.19}$$

$$\partial_\mu J_a^\mu = 0 \tag{2.20}$$

and

$$\partial_\mu J_{5a}^\mu = 0. \tag{2.21}$$

The axial U(1) current, Eq. (2.19), is anomalous, however, and quantum corrections to the QCD Lagrangian give [11]

$$\partial_\mu J_5^\mu = \frac{1}{8\pi^2} F^{\mu\nu}F^d_{\mu\nu}, \tag{2.22}$$

where

$$F^d_{\mu\nu} = \frac{1}{2} \varepsilon_{\mu\nu\lambda 6} F^{\lambda 6} \tag{2.23}$$

The chiral U(1) symmetry appears to be broken, but the situation is, however, not quite so simple. The contribution to the right hand side of Eq. (2.22) is the divergence of a current. This current might be added to the left hand side of Eq. (2.22) to define a conserved axial U(1) current. The problem with this procedure is that this current is gauge variant. We might think this would provide no difficulty when we consider the charge corresponding to this current since gauge transformations which approach the identity at spatial infinity might not affect the charge. The spectrum of states would possess a U(1) symmetry. This does not however happen since gauge transformations which wind as they approach spatial infinity may change the charge.[12] The symmetry of the Lagrangian under these big gauge transformations nevertheless might break, and the chiral U(1) symmetry might again be a symmetry of the spectrum of states. The fact that the chiral U(1) symmetry is broken follows from the explicit computation of 't Hooft who showed that the effects of instantons do indeed break this symmetry.[13] Instantons and their ilk provide the only known mechanism for breaking chiral U(1). Since the effect of instantons is exponentially small in weak coupling, e^{-1/g^2}, we might expect chiral U(1) to become a good approximate symmetry of QCD at high temperatures and densities since g^2 is effectively small.

To illustrate the effects of the symmetries of Eqs. (2.10-2.13), imagine an approximation to QCD for which these symmetries are exact. This approximation might be QCD to all orders in perturbation theory ignoring the effects of instantons. Such an approximation would be very good for weak coupling since instantons and their ilk have strength e^{-1/g^2}.

In this approximation, the mutually commuting charges and Casimirs associated with Q, Q_5, $\vec{Q}$, and $\vec{Q}_5$ label states. To find these Casimirs and generators, consider the charge algebra

$$[Q,Q_5] = [Q,\vec{Q}] = [Q,\vec{Q}_5] = [Q_5,\vec{Q}] = [Q_5,\vec{Q}_5] = 0, \tag{2.24}$$

$$[Q_i,Q_j] = i\varepsilon_{ijk}\, Q_k, \tag{2.25}$$

$$[Q_{5i},Q_{5j}] = i\varepsilon_{ijk}\, Q_k, \tag{2.26}$$

$$[Q_i,Q_{5j}] = i\varepsilon_{ijk}\, Q_{5k}. \tag{2.27}$$

This algebra decomposes into U(1)×U(1)×SU(2)×SU(2) under the change of variables

$$\vec{Q}^{\pm} = \frac{1}{2}(\vec{Q} \pm \vec{Q}_5), \tag{2.28}$$

and states are labeled by $q, q_5, q^{\pm 2}$ and $q_z^{\pm}$.

The quantities $q, q_5, q^{\pm 2}$ and $q_z^{\pm}$ have simple physical interpretations. Let $N_u^{\pm}$, $N_{\bar{u}}^{\pm}$, $N_d^{\pm}$ and $N_{\bar{d}}^{\pm}$ be the number of positive (+) or negative (-) chirality (helicity) quarks of flavor up, anti-up, down, and anti-down. These variables solve

$$q = \frac{1}{2}\{q \pm q_5\} \tag{2.29}$$

$$q^{\pm} = \frac{1}{2}\{N_u^{\pm} - N_{\bar{u}}^{\mp} + N_d^{\pm} - N_{\bar{d}}^{\mp}\}\ , \tag{2.30}$$

and

$$q_z^{\pm} = \frac{1}{2}\{N_u^{\pm} - N_{\bar{u}}^{\pm} - N_d^{\pm} + N_{\bar{d}}^{\mp}\}\ . \tag{2.31}$$

The casimirs $q^{\pm 2}$ which label representations of $\vec{Q}_{\pm}$ may be inferred from the maximum and minimum values of $q_{\pm}^3$ in any multiplet.

The simplest state labeled by these charge operators is the

vacuum which has vanishing $q, q_5, q_\pm^2$ and $q_{z\pm}$. The next simplest states are those states whose spatial wave function is independent of the interchange of valence quarks. Our basic assumption in characterizing simple states is that if one valence quark has a definite helicity, the remaining valence quarks have the same helicity. In order to have particle states of definite parity, we shall have to take simple linear combinations of these states since parity takes $Q_5 \leftrightarrow -Q_5$ and $\vec{Q}^\pm \leftrightarrow \vec{Q}^\mp$.

A nucleon is a state of 2 up and 1 down or 1 up and 2 down quarks. Labeling states by $|q, q_5, q^+, q_z^+, q^-, q_z^-\rangle$, the proton and neutron are

$$|p\rangle = \frac{1}{\sqrt{2}} \{|3,3;\tfrac{1}{2},\tfrac{1}{2};0,0\rangle + |3,-3;0,0;\tfrac{1}{2},\tfrac{1}{2}\rangle\} , \tag{2.32}$$

and

$$|n\rangle = \frac{1}{\sqrt{2}} \{|3,3;\tfrac{1}{2},-\tfrac{1}{2};0,0\rangle + |3,-3;0,0;\tfrac{1}{2},-\tfrac{1}{2}\rangle\}. \tag{2.33}$$

This state is obviously an isodoublet by the rules for addition of angular momentum

$$\vec{Q} = \vec{Q}_+ + \vec{Q}_-. \tag{2.34}$$

As a consequence of Eqs. (2.32-2.33), massive nucleons are parity doubled. This follows by multiplying Eqs. (2.32-2.33) by either Q_5 or Q_{5z}, and is a consequence of either U(1) or chiral SU(2) symmetry. This doubling theorem is, however, evaded for massless nucleons. For massless nucleons, states may never be labeled in a rest frame and nucleons always have non-zero spatial momentum. Parity eigenstates always involve linear combinations of helicity eigenstates and plane waves of opposite momenta. For an undoubled nucleon, there are two such linear combinations. Multiplication by Q_5 or Q_{5z} interchanges these linear combinations but does not imply an intrinsically doubled spectrum.

The next simplest state is a pion. The π^+ is composed of a u quark and a d anti-quark, so that π^+ is in the $(\frac{1}{2},\frac{1}{2})$ representation of SU(2)×SU(2). The odd parity pion wave functions are

$$|\pi^+\rangle = \frac{1}{\sqrt{2}} \{|0,2;\tfrac{1}{2},\tfrac{1}{2};\tfrac{1}{2},\tfrac{1}{2}\rangle - |0,-2;\tfrac{1}{2},\tfrac{1}{2};\tfrac{1}{2},\tfrac{1}{2}\rangle\} \tag{2.35}$$

$$|\pi^-\rangle = \frac{1}{\sqrt{2}} \{|0,2;\tfrac{1}{2},-\tfrac{1}{2};\tfrac{1}{2},-\tfrac{1}{2}\rangle - |0,-2;\tfrac{1}{2},-\tfrac{1}{2};\tfrac{1}{2},-\tfrac{1}{2}\rangle\} \tag{2.36}$$

and

$$|\pi^0\rangle = \frac{1}{2}\{|0,2;\tfrac{1}{2},\tfrac{1}{2};\tfrac{1}{2},-\tfrac{1}{2}\rangle + |0,2;\tfrac{1}{2},-\tfrac{1}{2};\tfrac{1}{2},\tfrac{1}{2}\rangle$$

$$- |0,-2;\tfrac{1}{2},-\tfrac{1}{2};\tfrac{1}{2},\tfrac{1}{2}\rangle - |0,-2;\tfrac{1}{2},\tfrac{1}{2};\tfrac{1}{2},-\tfrac{1}{2}\rangle\}. \qquad (2.37)$$

The complicated structure for $|\pi^0\rangle$ arises since the π^0 is an I=1 linear combination of a $u\bar{u}$ and a $d\bar{d}$ pair.

If we perform a chiral SU(2) transformation on $|\pi^{\pm}\rangle$ by multiplying by any component of $\vec{Q}_5$, we either produce zero or some multiple of a 0^+ isoscalar state, G,

$$|G\rangle = \frac{1}{2}\{|0,2;\tfrac{1}{2},\tfrac{1}{2};\tfrac{1}{2},-\tfrac{1}{2}\rangle - |0,2;\tfrac{1}{2},-\tfrac{1}{2};\tfrac{1}{2},\tfrac{1}{2}\rangle$$

$$- |0,-2;\tfrac{1}{2},\tfrac{1}{2};\tfrac{1}{2},-\tfrac{1}{2}\rangle + |0,-2;\tfrac{1}{2},-\tfrac{1}{2};\tfrac{1}{2},\tfrac{1}{2}\rangle . \qquad (2.38)$$

A paity doublet, G',π', of this G,π system is produced by multiplying $|G\rangle$ and $|\pi\rangle$ by Q_5. The set of parity doubled states and the symmetries which connect them are shown in Table I.

Needless to say, this spectrum of particles bears little resemblance to reality. Some modification of this unrealistic spectrum arises as a consequence of a breakdown of chiral U(1) invariance. This breakdown is signaled by the anomaly in the U(1) axial vector current. This breaking will split π',G' quartet from the π,G quartet.

At finite energy density in hadronic matter, the effects of instantons, which are presumably responsible for driving this symmetry breaking, decrease as the energy density increases. If the ($\vec{\pi}$,G) and ($\vec{\pi}$',G') multiplet existed in matter up to relatively weak coupling, their mass splitting would rapidly disappear as the temperature increased. In any case, the spectrum of states and their parity doubled partners should approach one another as the energy density increases. This might be tested by computing

$$\Delta_{\pm} \equiv \frac{1}{2} \ (1 \pm P)Q_5{}^2 \ . \qquad (2.39)$$

The parity operator is P and Q_5 is the chiral charge operator.

Even with a breakdown of the chiral U(1) symmetry, the spectrum of Table I is still unrealistic. This spectrum becomes realistic after a dynamical breakdown of chiral SU(2).

If we take

$$G_{op} = \sum_{\text{flavors}} \bar{\psi}\psi \tag{2.40}$$

as a quantum operator which creates a G meson, the breakdown of chiral SU(2) is signalled by a condensation of these mesons

$$G = \langle G_{op} \rangle \neq 0. \tag{2.41}$$

The pion field does not condense the vacuum so that

$$\vec{\pi} \equiv \bar{\psi}\gamma^5 \tau_I \psi, \tag{2.42}$$

satisfies

$$\langle \vec{\pi} \rangle = 0. \tag{2.43}$$

In infinite temperature and density hadronic matter, pion condensation might occur and $\langle \vec{\pi} \rangle$ might also be non-zero.

The generation of expectation values for G_{op} and $\vec{\pi}$ is analogous to the spontaneous breakdown of rotational invariance and magnetization in spin systems. If chiral symmetry was unbroken, G_{op} and $\vec{\pi}$ could not have expectation values since these operators rotate among themselves under chiral SU(2) rotations. In a magnet, rotational invariance is broken by the spontaneous generation of an expectation value for a spin operator

$$\langle S_z \rangle \neq 0. \tag{2.44}$$

The spectrum of states is no longer rotationally invariant since the Hilbert space of states generated by the magnet when magnetized along any fixed axis of magnetization is orthogonal to the Hilbert space corresponding to any other axis. The analogous situation for chiral SU(2) breaking is that the states which are chiral rotations of the states in a Hilbert space of fixed chiral properties

$$|S'\rangle = e^{i\vec{\alpha}\cdot\vec{Q}_5} |S\rangle \tag{2.45}$$

are orthogonal. The expectation value

$$\mathcal{N}(\vec{\alpha}) \equiv \langle e^{i\vec{\alpha}\cdot\vec{Q}_5} \rangle \tag{2.46}$$

is therefore dual to G in the sense that $\mathcal{N}(\vec{\alpha}) \neq 0$ in the chirally symmetric phase and is zero in the broken phase for $\vec{\alpha} \neq 0$.

The pions become the Goldstone bosons of chiral SU(2) symmetry breaking. There is no Goldstone boson of chiral U(1) symmetry breaking since there is an anomaly in the U(1) current. At finite hadronic matter density, the concept of mass loses its meaning, and the concept of a Goldstone boson requires clarification. What we shall mean by a Goldstone boson is a collective excitation of the hadronic matter which, in the limit that its spatial momentum goes to zero, has zero energy.

To see how these Goldstone bosons arise, consider the partially Fourier transformed real time response function

$$\begin{aligned} \mathcal{J}^{\mu}_{ij}(q) &\equiv \langle J^{\mu}_{5i}(q)\pi_j(0)\rangle \\ &= \delta_{ij}\{q^0 N^0(q^0,|\vec{q}|)\delta^{\mu 0} + \vec{q}^{\,k} N(q^0,|\vec{q}|)\delta^{\mu k}\} \; . \end{aligned} \tag{2.47}$$

The pion field, $\pi(0)$, is evaluated at zero spatial coordinate. This quantity vanishes in the chiral SU(2) symmetric phase since it is generated by a chiral rotation from $\langle J^{\mu}_{i}(q)G(0)\rangle$ which vanishes by parity and isospin invariance. Taking the divergence of Eq. (2.47) gives an equation which relates energy and spatial momentum of pion excitations,

$$q^0 = |q|\sqrt{N/N^0} \tag{2.48}$$

and barring pathological kinematic singularities of N and N^0, we conclude that there is a Goldstone pion. The nature of the kinematic singularities of $\sqrt{N/N^0}$ should, of course, be studied to make this conclusion firmer.

SECTION 3

LECTURE II: DYNAMICS OF PHASE TRANSITIONS IN HADRONIC MATTER

In this lecture, I shall discuss the dynamics of the phase transitions in hadronic matter as described by QCD, emphasizing what is known about confinement-deconfinement transitions. These transitions are first discussed for finite temperature SU(N) Yang-Mills theory in the absence of dynamical quarks. The confinement-deconfinement transition will be restated as a transition between a globally Z_N symmetric and asymmetric phase. The mechanism of generating a phase transition as first suggested by N. Weiss(14) and later used by Svetitsky and Yaffe(15) to infer the order of finite temperature transitions is discussed. These considerations are generalized for the case where dynamical fermions are included in the QCD Lagrangian. This lecture ends with some brief comments on what little is known about chiral symmetry breaking, and some warnings about employing the quenched approximation in lattice Monte-Carlo computations to study this problem.

The confinement-deconfinement phase transition in quarkless SU(N) Yang-Mills theory is signaled by a finite value for the free energy of a static test quark. This free energy is computed from the thermal expectation value of a spatially local order parameter $L(\vec{r})$ as

$$e^{-\beta F_q(\vec{r})} = \langle L(\vec{r})\rangle. \tag{3.1}$$

The parameter $\vec{r}$ labels the static test quark position, and F is the difference in free energy for the system with from that without a test quark. As shown in Appendix A, $L(\vec{r})$ is the Polyakov string operator

$$L(\vec{r}) = \frac{1}{N}\,\mathrm{Tr}\,\exp\{i\int_0^\beta dt\ \tau\cdot A^o(\vec{r},t)\}. \tag{3.2}$$

The corresponding free energy of an assembly of quarks at positions $\vec{r}_1,\cdots,\vec{r}_{N_q}$ and anti-quarks at $\vec{r}_1',\cdots,\vec{r}_{N_{\bar{q}}}'$ is

$$e^{-\beta F(\vec{r}_1,\cdots,\vec{r}_{N_q};\vec{r}_1',\cdots,\vec{r}_{N_{\bar{q}}}')} = \langle L(\vec{r}_1)\cdots L(\vec{r}_{N_q})L^\dagger(\vec{r}_1')\cdots L^\dagger(\vec{r}_{N_{\bar{q}}}')\rangle \tag{3.3}$$

The singular nature of the force between a quark and anti-quark in the confined phase of QCD follows from Eqs. (3.1) and (3.3). If a quark and anti-quark are at positions $\vec{r}_1$ and $\vec{r}_2$,

$$e^{-\beta F(\vec{r}_1,\vec{r}_2)} = \langle L(\vec{r}_1)L^\dagger(\vec{r}_2)\rangle. \tag{3.4}$$

If $|\vec{r}_1 - \vec{r}_2|$ is large, this expectation value of product of operators clusters into

$$\lim_{|\vec{r}_1-\vec{r}_2|\to\infty} e^{-\beta F(\vec{r}_1,\vec{r}_2)} = \langle L(0)\rangle^2. \tag{3.5}$$

In the confined phase, $\langle L\rangle = 0$, and the long distance force is singular.

The issue of the existence of a confinement-deconfinement phase transition may be reformulated as the issue of whether a global, dynamical Z_N symmetry of gluon interactions is broken or realized. This global symmetry is an invariance of the finite temperature QCD action under gauge transformations which have periodic logarithmic time derivative at $t = 0$ and $t = \beta$ and are periodic up to an element of the center of the gauge group. Here β is the inverse temperature. For a review of finite temperature field theory, see H. Satz's lectures. With such a gauge transformation, these requirements are

$$V(\vec{r},\beta) = C_N^j\, V(\vec{r},0), \tag{3.6}$$

$$V^{-1}(\vec{r},\beta)\, \frac{d}{dt} V(r,t)\Big|_{t=\beta} = V^{-1}(\vec{r},0)\, \frac{d}{dt} V(r,t)\Big|_{t=0}, \tag{3.7}$$

and the center element is

$$C_N^j = e^{2\pi ij/N}. \tag{3.8}$$

Under such a transformation

$$L(\vec{r}) \to e^{2\pi ij/N}\, L(\vec{r}). \tag{3.9}$$

The periodic boundary conditions on the gluon fields for $t=0$ and $t=\beta$ are maintained by this transformation.

If this symmetry is dynamically realized, the free energy of any N-ality non-singlet configuration of quarks and gluons is divergent since

$$e^{-\beta F_{N_q N_{\bar{q}}}} \to e^{-2\pi ij/N}\, e^{-\beta F_{N_q N_{\bar{q}}}} \tag{3.10}$$

implies either

$$e^{-\beta F_{N_q N_{\bar{q}}}} = 0, \tag{3.11}$$

or

$$N_q - N_{\bar{q}} = IN \tag{3.12}$$

where I is some integer. If this dynamical symmetry is spontaneously broken, N-ality non-singlet configurations of quarks may have finite face energy.

The Polyakov string will itself tend to an element of the center of the gauge group at high enough temperatures. In the deconfined phase, the free energy of an isolated quark becomes small, so long as an ultimate cutoff is imposed or self-energies are removed, so that $L \to 1$. $L \to 1$ is the physical limiting value for L, but Z_N symmetry allows $L \to e^{2\pi ij/N}$. In the deconfined phase L freezes into one of these center elements.

Nathan Weiss proposed that the confinement-deconfinement phase transition could be understood as a condensation of Z_N domains as measured by L.(14) In different domains, L is close to different elements of the center. The confined phase of QCD corresponds to a condensate of domains, and $\langle L \rangle = 0$. In the deconfined phase, the system is in one domain and $\langle L \rangle \neq 0$. This simple picture is very nice, although there is no Monte-Carlo data which support such a picture.(16) This picture, if valid, is presumably true only in the neighborhood of a confinement-deconfinement transition and might be difficult to probe in computer simulations.

The scenario of Weiss has recently been elaborated and developed by Svetitsky and Yaffe,(15) who provide plausible arguments for the order of confinement-deconfinement transitions in non-Abelian gauge theories. Their argument proceeds by postulating that all the degrees of freedom of an SU(N) gauge theory may be integrated out of the functional integral save degrees of freedom described by Polyakov strings. The effect of this integration supposedly only makes the effective coupling constant temperature dependent, $g \to g(T)$. The Polyakov strings depend only on space, and the remaining theory is a 3-dimensional theory with a global Z_N symmetry. This theory is in the same universality class as Z_N spin models. The effective temperature of the 3-dimensional spin system may be shown to be

$$T_{eff} = g^2(T). \tag{3.13}$$

As T increases, T_{eff} decreases and the 3-dimensional Z_N spin system orders leading to deconfinement. For SU(3) theories, the Z_3 theory has a first order phase transition,(17) and a first order transition is expected, but suggests a second order transition in SU(2) gauge theories. Monte-Carlo data suggest a second order transition for SU(2) theories, but the data for SU(3) is inconclusive.(18)

The confinement-deconfinement phase transition for QCD in the presence of dynamical fermions can also be understood as a breakdown or realization of a global Z_N symmetry. Recall that the N-ality operator ν_k is a projection operator to states of definite N-ality. This N-ality operation also projects the partition function onto a sum of terms which have simple behaviors under gauge transformations which are periodic up to an element of the center. This projection is

$$\begin{aligned} Z &= \mathrm{Tr}\ e^{-\beta H} \\ &= \sum_{k=0}^{\infty} \mathrm{Tr}\ \nu_k\ e^{-\beta H} \\ &= \sum_{k=0}^{\infty} Z_k . \end{aligned} \tag{3.14}$$

To derive the transformation properties of the Z_k's, consider the functional integral representation for Z after the fermions have been integrated out. The effect of fermions is embodied in the determinant of the fermion Green's function in an external gluon field. This determinant may be expanded in sums of products of loops such as

$$U_L = \mathrm{Tr}\ P\ e^{i\oint_L d\ell^\mu \tau\cdot A_\mu} . \tag{3.15}$$

These loops either close within the four volume (β,V), or may close by wrapping around the four volume several times in the Euclidian time direction. The number of loops wrapping in a positive sense minus those wrapping in a negative sense mod N gives the N-ality. A contribution of N-ality j therefore transforms as $e^{2\pi ijk/N}$ under a gauge transformation which is periodic up to $e^{2\pi ik/N}$. The transformation of Z_j is therefore

$$Z_j \to e^{2\pi ijk/N}\ Z_j . \tag{3.16}$$

The Polyakov string L always transforms as an N-ality one operator. The free energy of a static quark is

$$\langle L\rangle = \sum_{k=0}^{N-1} \mathrm{Tr}\ L\ \nu_k\ e^{-\beta H}, \tag{3.17}$$

and is finite both in the confined and the deconfined phase since there is always a singlet, k=N-1, contribution of Eq. (3.17). This is a consequence of the fact that in the confined phase L measures the free energy of bound states of a heavy quark and a light dynamical quark.

The Z_N transformations of Eq. (3.16) guarantee that $Z_k=0, k\neq0$, if these transformations may be dynamically realized. These transformations however might not be dynamically realized. The various contributions to Z_k generated by non-periodic gauge transformations come from field configurations with different Z_N properties. In the Z_N symmetric phase, the functional integral for Z goes over fields A^μ/Z_N and their Z_N copies. The Z_N symmetry broken phase arises if the functional integral becomes "frozen in" to a fixed A^μ/Z_N sector.

The underlying symmetries which characterize the confinement-deconfinement transition in the absence or presence of dynamical fermions are very similar. The bulk properties of the system, such as the specific heat, may very well have similar properties for QCD with, or without, dynamical fermions.

The dynamics of the chiral symmetry phase transition is more difficult to understand than that of the confinement-deconfinement transition. An argument of Casher and Kogut suggests that chiral symmetry is always broken in the confined phase of QCD.[19] The argument proceeds by first assuming chiral symmetry is unbroken. A pion consists of a bound state of a quark and anti-quark of fixed chirality. The quark and anti-quark have opposite momentum in the pion rest frame. The confining force makes the trajectories of the quark and anti-quark turn around after the quark and anti-quark separate a distance of more than a fermi. The confining force is spin zero, and the quark and anti-quark must flip helicity. The chirality of the configuration changes. Chiral invariance and confinement therefore seem incompatible.

The simplest possible scenario for chiral symmetry restoration is if it is restored as a consequence of deconfinement. If chiral symmetry breaking is a consequence of the condensations of σ-mesons, then when confinement disappears, the σ-mesons might ionize. Since quarks are fermions, they cannot condense. This argument is not too convincing, however, since even if confinement disappears, the σ-mesons might remain as bound states until some higher energy density. To be honest, the mechanism of chiral symmetry breaking is not well understood, and the relationship, if any, between the energy density of confinement and that of chiral symmetry restoration is not understood.

There has been a recent claim based on a quenched Monte-Carlo calculation that claims the temperature of chiral symmetry restoration and that of deconfinement are quite different.[20] This conclusion is based on the comparison of data between their calculation and that of Kuti et al.[3] The conclusion is much less dramatic if their data are compared with that of Engels et al.[3]

There are also fundamental problems with applying the quenched approximation to finite temperature systems with fermions.[(21)] For example, the Z_2 transformation in an SU(2) gauge theory which flips $L \to -L$, changes the statistics of fermions from Fermi-Dirac to Bose-Einstein. This follows because such a transformation shifts the discrete frequency spectrum of the fermion propagator by a half integer. To see this, consider the propagator with $A^o = 0$. Under the gauge transformation,

$$V = e^{2\pi i t \tau_3/\beta}, \quad A^o_a \to \frac{2\pi}{\beta}\,\delta_{a3},$$

and the frequency spectrum shifts. If the fermion determinant is present in the functional integral, these Bose-Einstein modes are suppressed since the determinant vanishes. The fermion determinant therefore is essential to give fermions a distribution typical of Fermi-Dirac statistics! Far above the confinement-deconfinement transition, these gauge transformations are not dynamically allowed, and by measuring $\langle L \rangle$, one may determine that one is in the correct Z_N configuration so that fermions have Fermi-Dirac distributions. Near and below the confinement-deconfinement transition, the situation is hazy.

If a study of chiral symmetry breaking is attempted in the quenched approximation, even more problems arise. Instantons give a contribution to the functional integral which generates a singularity in $\bar{\psi}\psi$. The fermion determinant cancels this singularity, and if the number of quark flavors is greater than 1, $\bar{\psi}\psi \to 0$ since there are more zeros of the determinant than can be compensated for by $\bar{\psi}\psi$. If the determinant is ignored, $\bar{\psi}\psi$ will be non-zero and, in fact, divergent in the continuum limit at all temperatures. On a lattice in a fixed spatial volume, $\bar{\psi}\psi$ will be finite, but non-zero for all temperatures. At and below the confinement-deconfinement temperature, topological excitations may play an important role in the quark-gluon dynamics. It seems very difficult to disentangle the dynamics of chiral symmetry breaking from artifacts of the quenched approximation.

In principle, these problems of the quenched approximation may be resolved by systematically including the effects of the fermion determinant. A first tentative step in this direction has been taken by Engles et al., but much work remains before the water will be crystal clear.[(22)]

SECTION 4

LECTURE III: SPACE-TIME PICTURES OF ULTRA-RELATIVISTIC NUCLEUS-NUCLEUS COLLISIONS

The phase transitions described in the previous lectures occur at energy densities several times that of nuclear matter. There are several situations where this matter might occur in nature. Hadronic matter existed at energy densities many orders of magnitude greater than that of nuclear matter in the early universe. The confinement-deconfinement transition and the chiral transition certainly played a role in the evolution of the early universe, but the consequences of this role are difficult to abstract from the state of the present universe. This issue has not yet been fully addressed, and as our knowledge of hadronic matter increases, consequences of these phase changes might be found.

In the cores of neutron stars, cold, dense hadronic matter exists at energy densities $\mathcal{E} \sim 10\text{-}20\ \mathcal{E}_{N.M.}$. Models of neutron stars with quark cores have been made, and, using reasonable assumptions about the energy density of the confinement-deconfinement transition, quark cores as large as $r \sim 8$ km are found for neutron stars with radii $r \sim 10$ km.[23] The bulk properties of these quark stars are depressingly similar to those of neutron stars constructed from conventional nuclear matter equations of state. There is some hope of detecting hypothetical quark cores of neutron stars by measuring cooling rates,[24] but the data and computations on this issue are inconclusive.[25]

Ultra-relativistic nucleus-nucleus collisions provide a promising environment where high density hadronic matter might be produced and studies.[26] This lecture provides a review of recent theoretical developments in our understanding of the space-time dynamics of these collisions.[27-30] There are several key questions which are addressed by space-time pictures of nucleus-nucleus collisions. What energy densities are achieved in these collisions? How long does matter produced in these collisions exist at high energy density? What are the experimental signals for the production of new states of matter? The first two of these questions will be discussed in this lecture. The present wisdom concerning the third very important question is reviewed by Van Hove in Ref. (31).[31]

Much of the remainder of this lecture is a retelling of relevant sections of the last paper in Ref.(28) and in Ref.(30).

These space-time pictures may be applied to head-on collisions between large nuclei of equal baryon number A at asymptotically large center-of-mass energies. We shall typically take uranium as an example, and consider center-of-mass per nucleon energies $E_{cm} > 30$ GeV. Head-on collisions will be taken as those collisions with an impact parameter less than the range of the nuclear force, $b < 1$ fm.

Head-on collisions are not extremely rare, since geometrical considerations show that 1/2% of all uranium-uranium collisions are head-on. These collisions are not easily confused with peripheral collisions. Assuming the multiplicities in nuclear collisions grow as the nuclear baryon number, A, the multiplicity in a head-on collision between heavy nuclei at $E_{cm} > 30$ GeV is $n \sim 10^3$-10^4. Statistical fluctuations in peripheral collisions might only rarely simulate an event with so large a multiplicity. In addition, head-on collisions will be signaled by a violent, complete disintegration of projectile and target nuclei.

An elegent, simple space-time model of nuclear collisions is provided by Landau's hydrodynamical model. In the simplest version of this model, the collision of two nuclei is studied in the center-of-mass frame.[32] The two nuclei appear in this frame as two Lorentz contracted pancakes flying toward one another at near the velocity of light (Fig. 1). The thickness of these two nuclei is

$$\Delta X = \frac{R_N}{\gamma} \tag{4.1}$$

where $\gamma = E_{cm}/2M$ is the energy per nucleon of each nucleus and R_N is the rest frame nuclear radius. When these two nuclei collide, they stick together and produce a distribution of hot hadronic matter with thickness $\Delta X_m = 2\Delta X$ (Fig. 2). This matter then undergoes hydrodynamic expansion according to Landau's equations (Fig. 3). The outward flow of matter is primarily along the axis of the beam of nuclei, and most of the particle production takes place during the initial collision when two nuclei stick together. This collision and subsequent expansion may be represented by the light cone diagram of Fig. 4.

The energy density achieved in such a model is

$$\varepsilon \simeq \gamma^2 \; \varepsilon_{NM}, \tag{4.2}$$

where γ is the energy density of nuclear matter. The factor of γ^2 arises from compression of the nuclei, and a factor of γ for the total energy carried by each nucleus. The achieved energy density grows quadratically with center-of-mass energy. For $E_{cm} > 30$ GeV, $\varepsilon >$ $200\ \varepsilon_{NM}$. Since the density of matter inside a proton,

$$\varepsilon_p \simeq \frac{1}{\frac{4}{3}\,\pi r_p^3} \sim 450\ \mathrm{MeV/fm}^3 \tag{4.3}$$

where r_p is the r.m.s. proton radius, is only a few times the energy density of nuclear matter, a quark-gluon plasma might form. Since ε rises linearly with center-of-mass energy, tremendously high energy densities arise at Isabelle energies, and in this simple version of the Landau model the formation of a quark-gluon plasma would seem certain.

This conceptually simple and attractive version of the Landau hydrodynamical model has difficulty, however, explaining many features of conventional hadronic interactions. The leading particle effect in hadron-hadron interactions suggests that two hadrons would very rarely stick together and then bounce off one another in high energy interactions. In most collisions, the hadrons apparently pass through one another. This transparency is most dramatic in high energy hadron-nucleus interactions. The projectile hadron must pass through many mean free paths of the target nucleus. The distribution of the scattered projectile and inelastically produced particles with momenta close to that of the projectile is nevertheless very similar to that of hadron-hadron interactions. The hadron projectile behaves almost as if it passed through the target nucleus and only scattered once. Another problem for simple hydrodynamical models of hadronic interactions is approximate scaling. Approximate scaling gives the total multiplicity as some power of the logarithm of the center-of-mass energy. Simple hydrodynamical models typically have the multiplicity proportional to a power of the center-of-mass energy. Such a proportionality appears to be at odds with SPS data on $\bar{p}p$ collisions.(33) Finally, the observed jet structure in e^+e^- collisions is not simply explained by a hydrodynamical model.

The quark-parton model provides an alternative description of hadronic interactions which incorporates the leading particle effect, transparency in hadron nucleus interactions, approximate scaling, and jets in e^+e^- collisions. The successful application of this model to hadron-nucleus interactions provides a guide for a corresponding application to nucleus-nucleus collisions.(27,34-35)

The qualitative features of the quark-parton model of nucleus-nucleus collisions which distinguish it from the Landau hydrodynamical model are simply understood. The dynamics of the central region is best understood when a nucleus-nucleus collision is analyzed in the center-of-mass frame. In this frame, the target and projectile nucleus are Lorentz contracted pancakes with a limiting thickness $\Delta X \sim 1$ fm which approach one another at near the velocity of light (Fig. 5).[29] This limiting thickness is a consequence of the Heisenberg uncertainty principle. The low longitudinal momentum component of the nuclear wave function must have a spatial extent $X \sim 1/p^{\parallel}$. The low momentum, wee parton component of the nuclear wave function is composed of gluons and quark-antiquark pairs, or, in a different base of states, virtual pions. These degrees of freedom have momentum $p \sim 200$ MeV corresponding to $\Delta X \sim 1$ fm. The higher momentum components of the nuclear wave function have smaller partial extent. The valence quark, or alternatively, nucleon component of the nuclear wave function has a spatial extent of $\Delta X \sim R/\gamma$, as was the case for the Landau hydrodynamical model.

When these two nuclei pass through one another, the low momentum component of the nuclear wave function interacts strongly and comes to rest in the center-of-mass frame. The higher momentum components interact less strongly and pass through one another (Fig. 6). This low momentum component which has been scraped off from the two nuclei may now begin a hydrodynamic expansion. Since the low momentum component of the nuclear wave function (or, for that matter, of a nucleons wave function) is approximately independent of the center-of-mass energy, and since the low energy interactions which are primarily responsible for stopping this component are also independent of E_{cm}, the energy density of matter scraped from the nuclei into this space-time region is approximately energy independent.

After the low momentum component of the nuclear wave functions have interacted, the nuclei continue to inelastically produce matter. The low momentum matter which has been scraped away from the nuclei interacts with higher longitudinal momentum components of the nuclear wave functions. This interaction is strong so long as the relative momenta of the scraped away matter and the components of the nuclear wave function are small. These higher momentum components are formed in a time $\tau \sim R_o$, where $R_o \sim 1$ fm, in their own rest frame. This formation time is dilated in the center-of-mass frame, $\tau \sim p^{\parallel} R_o$, since these components have longitudinal momenta $p^{\parallel}$. These components arise from a region of spatial extent $\Delta X \sim R_N/p^{\parallel}$ in the Lorentz contracted nuclei. The matter forms in an inside-outside cascade (Fig. 7).

The inside-outside cascade may be represented by a light cone diagram (Fig. 8). Matter is produced at the edges of the forward light come and propagates forward in time. Matter was formed primarily at the apex of the light cone in the Landau hydrodynamical model. After formation the matter may undergo hydrodynamical expansion.[29]

The energy density of the matter forming at the edge of the light cone may be measured in a frame co-moving with this matter. In the simplest parton model, this energy density is taken to be independent of its position on the light cone. If the total multiplicity of centrally produced particles in nucleus-nucleus collisions is proportional to $A^{2/3+\alpha}$, the energy density at the edge of the light cone will be proportional to A^{α}.

After a time $\tau \sim \frac{E_{cm} R_0}{2}$, the valence quark, or nucleon, component of the nuclear wave function begins to materialize. The formation of this matter is most simply described in the rest frame of one of the nuclei. The target nucleus sees a Lorentz contracted projectile nucleus with a limiting thickness $\Delta X \sim R_0$ in this frame (Fig. 9). The target is, of course, not moving and not Lorentz contracted.

As the projectile nucleus passes through the target, the low momentum components of its wave function interact strongly with the target. This component heats the target. The amount of heat should be approximately energy independent at asymptotic projectile energies. The projectile nucleus also compresses the target. This target probably forms a "fireball" which moves off in the direction of the projectile. The nucleons in this "fireball" may acquire longitudinal momenta typical of hadron-nucleus interactions. The Lorentz factor of the "fireball" is, therefore, $\gamma \sim 1.5$-2.0 (Figs. 10-11). After the projectile passes through the target, heat begins to materialize in the central region. This materialization was described in the previous paragraphs.

In the next few paragraphs, I shall present the results of a semi-quantitative analysis of ultra-relativistic nuclear collisions.[28-30] I shall later describe the derivation of these results. Possible techniques for refining these computations are the subject of the closing paragraphs.

The analysis of nucleus-nucleus collisions simplifies in rapidity variables,

$$y = \ln \frac{E + p^{\parallel}}{m} , \tag{4.4}$$

where E is a particle's energy, $p^{\parallel}$ is its longitudinal momentum, and m is its transverse mass. The fragmentation regions of the nuclei are identified by particles with rapidities close to that of the target and projectile nuclei. For heavy nuclei, such as uranium, the width of the nucleus fragmentation region is $\Delta y \sim 3$-4. For nuclei with baryon number A,

$$\Delta y \simeq \ln A^{1/3} + \text{constant}. \tag{4.5}$$

The region of rapidity not included in the fragmentation region is the central region. Heavy nuclei, such as uranium, must have $E_{cm} > 30$ GeV before a central region is kinematically allowed.

If the multiplicity in the central region grows as the nuclear baryon number, A, and if the multiplicity continues to rise with energy as it does at ISR energies, the energy density of hot, dense matter in the central region may be as high as 2 GeV/fm^3 at ISR energies and 4 GeV/fm^3 at Isabelle energies. If the multiplicity grows as $A^{2/3}$, the energy density is probably too small, $\varepsilon \sim 300\text{-}600$ MeV/fm^3 to produce a plasma, unless the coefficient of the $A^{2/3}$ term is anomalously large. Some data from hadron-nucleus collision indicate that if the multiplicity grows as $A^{2/3}$, the coefficient is indeed anomalously large, and even for uranium-uranium collisions the difference between an A and $A^{2/3}$ multiplicity growth is only a factor of two.(36) The energy density in the central region might be $\varepsilon \sim 1\text{-}2$ GeV/fm^3 in this case.

Hot, dense baryon-rich "fireballs" of hadronic matter may form in the fragmentation regions. The energy density of this matter may be $\varepsilon \sim 2$ GeV/fm^3, assuming the multiplicity grows as the baryon number A in the fragmentation region. This growth is consistent with hadron-nucleus collision data.

The dependence of the energy densities on baryon number of the nuclei is $A^{1/3}$ if multiplicities grow as A and constant if they grow as $A^{2/3}$. There is recent cosmic ray data on nucleus-nucleus collisions which are consistent with a multiplicity growth propotional to A.(37) In Figure 12, a pseudo-rapidity plot for the interaction of a 10 TeV/nucleon calcium nucleus with, presumably, a carbon nucleus is shown. About 600 charged particles are produced. The density of particles in the central region is consistent with a $\log^2 s$ multiplicity growth and growth proportional to A, but is inconsistent with $A^{2/3}$ and $\log^2 s$ (if the coefficient of $A^{2/3}$ is one). There is also preliminary data on a high energy silicon-emulsion nucleus interaction.(37) This event appears to have even higher multiplicity, producing a situation in which a formation of a quark-gluon plasma is even more likely.

A distinction between matter produced in the fragmentation regions and the central region is found in the baryon number density. This density is the difference between the number of baryons and antibaryons per unit volume. There should be only a small baryon number density in the central region. If a quark-gluon plasma forms in both the central and fragmentation region, the study of the transition region between the fragmentation region and central region would allow a study of the dependence of characteristics of the quark-gluon plasma on baryon number density.

A hypothetical rapidity distribution for baryon number (baryons minus antibaryons) for a head-on collision between nuclei of baryon number A is shown in Fig. 13a. The baryon number is concentrated in a nucleus fragmentation region of width $\Delta y \sim 2\text{-}3$. The heights of these fragmentation regions are proportional to A.

A hypothetical rapidity distribution for mesons in this collision is shown in Fig. 13b. If the height of the central region was proportional to $A^{2/3}$, it would be clearly separated from the fragmentation region at Isabelle energies. The width of the fragmentation region for pions is $\Delta y \sim 3\text{-}4$.

A feature of nucleus-nucleus collisions which distinguishes them from hadron-hadron and hadron-nucleus collisions is the extremely large number of particles which are produced in the collision. If enough particles are produced in the primeval distribution of hot, hadronic matter, and if the matter stays hot and dense long enough, the constituents of the matter will come into thermal equilibrium. The characteristic time scales for matter in the central region is $\tau \sim 2$ fm and is independent of baryon number. This is the characteristic time for the matter to decrease its energy density by a factor of two. The characteristic time scale in the fragmentation region is 2 fm $< \tau < A^{1/3}$ fm. This time depends on whether the dominant cooling mechanism is expansion generated from momentum gradients present in the initial formation of matter as the nuclei collide, or by radiation from an expanding "fireball" present in the fragmentation region.

Estimates which employ the parton model indicate that kinetic equilibrium is established in the fragmentation region.[(28)] The establishment of kinetic equilibrium requires that the momentum space distribution of particles is thermal in the region of momentum space most occupied by particles. The conclusion that kinetic equilibrium is attained is also verified by perturbative QCD computations.[(38)] These perturbative calculations also suggest that chemical equilibrium may also be achieved.[(38-39)] Chemical equilibrium requires that the relative population densities of different particle flavors is thermal. The matter in the central region also probably achieves kinetic equilibrium if the energy density is $\varepsilon > 1\text{-}2$ GeV/fm^3.

The parton model picture of hadronic interactions leads to the results of the preceding paragraphs. An important feature of this picture is the inside-outside development of cascades in hadronic interactions. This cascade development explains trahsparency in hadronucleus interactions, and is a consequence of time dilation and an intrinsic formation time for hadronic matter.

The inside-outside cascade may be understood by studying the ronic interaction shown in Fig. 14. A massless fragment is inelastically produced in the interaction between a projectile and target hadron. After a time t, the projectile and fragment have separated distances

$$\Delta r_\perp \sim \frac{p_\perp}{p_\parallel} \tau \tag{4.6}$$

and

$$\Delta r_\parallel \sim \left(\frac{p_\perp}{p_\parallel}\right)^2 \tau \ . \tag{4.7}$$

In a frame co-moving with the fragment, that is, a local zero longitudinal momentum frame, the distances are

$$\Delta r_\perp \sim \Delta r_\parallel \sim \frac{p_\perp}{p_\parallel} \tau \ . \tag{4.8}$$

A Lorentz invariant generalization of this result for massive particles follows from the replacement $p_\perp \to \sqrt{p_\perp^2 + m^2}$.

The fragment may be properly included as part of the projectile hadron's wave function if the separation between fragment and projectile is $\Delta r_\perp$, $\Delta r_\parallel < 1$ fm in a frame co-moving with the fragment. This is the intrinsic formation time for hadronic matter. This time is dilated in the laboratory frame. The time for a fragment to materialize in a cascade in this frame is $\tau \sim p_\parallel$, so that the slow fragments materialize before the fast fragments.

This cascade development may be elegantly re-cast in terms of rapidity variables.[28] The Lorentz γ factor and velocity of a particle with rapidity y is

$$\gamma = \frac{\sqrt{p_\perp^2 + m^2}}{m} \quad \cosh y \tag{4.9}$$

$$v = \tanh y \ . \tag{4.10}$$

The time it takes for a particle to form in its rest frame is $R_o \sim 1$ fm. The time in an arbitrary frame is obtained by Lorentz boosting and using Eqs. (4.6-4.7). The density of particles when they first materialize is easily found from Eqs. (4.9)-(4.10). Consider a pion which has just been inelastically produced in a hadron-hadron collision (Fig.15). We shall consider the rest frame of this pion. A nearest neighbor pion materializes at the space-time coordinates $X \sim t \sim (\gamma + \sqrt{\gamma^2 - 1}) R_o = R_o e^y$.

The spatial separation between these pions is

$$\Delta X \sim R_o e^y. \tag{4.11}$$

Since the rapidity density of pions in the central region is $dN/dy \sim 2$ for pp interactions at ISR energies,

$$\Delta X \sim \frac{1}{2} \text{ fm}. \tag{4.12}$$

At Isabelle energies

$$\Delta X \sim \frac{1}{4} \text{ fm}. \tag{4.13}$$

There is a large momentum difference $\Delta p_{\parallel}$, between these pions since

$$p_{\parallel} = m_{\pi} \sinh y \tag{4.14}$$

which is numerically

$$\frac{dp_{\parallel}}{dX} = 200 \text{ MeV/fm-pion}. \tag{4.15}$$

These simple pictures demonstrate the difficulty of forming a thermalized distribution of hadronic matter in ordinary hadronic interactions. The pions materialize at a distance of $\Delta X \sim 1/2$ fm from one another. As they materialize, they are flying away from one another at near the velocity of light. It would be very difficult for these pions to interact sufficiently to produce a thermal distribution. In high multiplicity events, this situation is somewhat less difficult.

The production of matter in a nucleus-nucleus collision is similar to that in hadron-hadron collisions. An essential difference is that the multiplicities are much higher. Many "strings" of matter are produced by the large number of nucleon-nucleon interactions.(40) Each "string" is surrounded by many other "strings", and the matter from different "strings" may thermalize (Fig. 16).

In the central region of nucleus-nucleus collisions, matter is composed of inelastically produced particles. The cross-sectional area of the matter in the central region is $S \sim \pi R^2 \sim 4\ A^{2/3}$. Assuming the height of the central region in nucleus-nucleus collision is proportional to A, the number density of pions is

$$n \sim \frac{1}{2} A^{1/3} \text{ pions/fm}^3 \tag{4.16}$$

at ISR energies and

$$n \sim A^{1/3} \text{ pions/fm}^3 \tag{4.17}$$

at Isabelle energies. If each pion has an average energy $E \sim 600$ MeV in a local $p_{\parallel} = 0$ frame, the energy densities are $\varepsilon \sim 2$ GeV/fm^3 at ISR energies and $\varepsilon \sim 4$ GeV/fm^3 at Isabelle energies. These energies are smaller, and A independent if the total multiplicity in the central region $n \sim A^{2/3}$.

The characteristic time scale for the expansion of matter in the central region follows from the momentum gradient of Eq. (4.15) or Eq. (4.10) for the velocity. The time it takes for pions to increase their separation by a factor of two is $\tau \sim 2$ fm.

The energy densities achieved in the fragmentation region in nucleus-nucleus collisions arise from inelastic production of matter which becomes trapped with the nucleons, and by compression of the nucleons. The low momentum particles, which will be thought of as pions, may be produced and trapped in the nuclei. These low momentum pions have longitudinal momentum $p_{\parallel} < R_N/R_o p_{\perp}$. Estimates of the number of pions trapped in the nucleus fragmentation region give about 3-4 pions/nucleon for uranium-uranium collisions at ISR energies. At Isabelle energies, this number would be increased by perhaps a factor of two arising from scaling violations. The number of pions should be proportional ln $A^{1/3}$ since linearly increasing $A^{1/3}$ linearly increases the rapidity interval in which pions may become trapped.

The energy trapped per nucleon is $E/N \sim 3$-4 GeV at ISR energies. This energy might be a factor of two higher at Isabelle energies. Uncertainties in the evaluation of this energy are about a factor of two. This number follows from assuming the energy per trapped pion is $E \sim 600$ MeV and per nucleon $E \sim 1$ GeV.

After the target nucleons are struck, they acquire longitudinal momentum. This longitudinal momentum is approximately that of a pp collision, corresponding to a $\gamma \sim 2$. The nucleons and trapped pions may form a baryon-rich fireball which moves down the beampipe with a $\gamma \sim 2$.

As the projectile nucleus traverses the target, imparting longitudinal momentum to the target nucleons, the target is compressed. As shown in Fig. 16, this compression results from the sequential encounters of the projectile nucleus with the target nucleons. After the left-most target nucleon is encountered, this nucleon acquires a velocity v. It travels a distance vR before the second nucleon is struck. In the lab frame, the apparent compression is $1/(1-v)$. In the rest frame of the struck nucleus, the compression is

$$C = \frac{1}{\gamma(1-v)} \simeq 2\gamma. \tag{4.18}$$

The energy density in the fragmentation region is estimated from the compression C, the trapped energy per baryon, E/N, and the energy density of nuclear matter as

$$\varepsilon \sim \rho_o \; C \; E/N \sim 2 \; \text{GeV/fm}^3. \tag{4.19}$$

This energy density might be as much as a factor of two higher at Isabelle energies.

The characteristic time scale for expansion in the fragmentation region may be larger than that of the central region. In the central region, the matter expands with characteristic time scale $\tau \sim 2$ fm. In the fragmentation region, expanding heat runs into nucleonic matter which is almost locally at rest in the rest frame of the "fireball" formed in the fragmentation region. If the nucleons could absorb the momenta gradients in the heat, the "fireball" would not be initially expanding. In this case, the fireball would cool by thermal emission from its surface. The characteristic time scale for this expansion is the time it takes a sound wave to cross a compressed nuclear diameter. For $v_s^2 \sim 1/3$, this time is $\tau \sim A^{1/3}$ fm. The characteristic time scale for expansion may therefore be in the range 2 fm $< \tau < A^{1/3}$ fm.

The description of nucleus-nucleus collisions advocated in the previous paragraphs is phenomenological. A proper theoretical treatment of the collision should address at least two issues: the approach to equilibrium from a non-equilibrium distribution of matter in the colliding nuclei, and the subsequent hydrodynamic expansion of the matter after it has achieved equilibrium. The approach to equilibrium might be studied using transport theory.[41] Perhaps the collisions which are most responsible for the approach to equilibrium are energetic enough that perturbative QCD, or at least weak coupling QCD, might be appropriate. Applying QCD to transport theory is, however, extremely difficult since infrared divergences arise from the masslessness of the gluons.[41] The proper application of QCD to the approach to equilibrium will be a major project, and will take considerable hard work to carry through.

The primary effect of the pre-equilibrium processes is to generate a distribution of matter which is locally in thermal equilibrium with energy density ε_0 after some time τ_0. This distribution provides initial conditions for hydrodynamic equations,

$$\partial_\mu T^{\mu\nu} = 0, \tag{4.20}$$

where $T^{\mu\nu}$ is the stress-energy tensor.[29,32] The relationship between energy density and pressure are needed to compute $T^{\mu\nu}$. This relationship may be evaluated using Monte-Carlo techniques.[3] The hadronic matter distribution might therefore be computable, with currently available theoretical methods, from the time equilibrium is achieved until the time hadronic matter decouples! This time corresponds to an energy density of nuclear matter. This situation may be unique in the study of hadronic collisions.

The inside-outside cascade model of hadron-hadron interactions determines the initial distribution of matter in a nucleus-nucleus collision. The inside-outside cascade is simply understood in a 1 space- 1 time dimensional hadron-hadron collision, where a projectile with velocity $v \approx 1$ strikes a target at rest with $x=t=0$. The fragments of the collision follow classical trajectories,

$$u^\mu = \frac{x^\mu}{\tau} , \tag{4.21}$$

where x^μ is the fragment coordinate, u^μ is its two velocity, and τ is its proper time

$$\tau^2 = x_\mu x^\mu = (t+x)(t-x). \tag{4.22}$$

These fragments materialize at a proper time

$$\tau = \tau_0 \approx 1 \text{ fm}, \tag{4.23}$$

and do not interact before this time. As described above, this formation time explains transparency in hadron-nucleus collisions since the fast hadron projectile fragments materialize outside the target nucleus due to Lorentz time dilation of the formation time.

Since the fragments follow classical trajectories, the momentum and coordinate of a fragment are in one-to-one correspondence. The rapidity y is

$$y = \frac{1}{2} \ln \frac{E+p_L}{E-p_L} = \frac{1}{2} \ln \frac{t+x}{t-x} . \tag{4.24}$$

This classical relationship violates the Heisenberg uncertainty principle, and ignores quantum fluctuations in cascade development. This classical treatment however may be adequate when averaged over events, and may even be appropriate for nucleus-nucleus collisions at fixed impact parameter. The large multiplicities in nucleus-nucleus collisions may suppress the effects of fluctuations.

The absence of fragment interaction for $\tau < \tau_0$ simplifies the description of nucleus-nucleus collisions. If the central multiplicity in a head-on collision of two nuclei of equal baryon number A scales as A, each nucleon in the target generates an independent cascade. The coordinate and momentum of each fragment are specified at materialization. These fragments act as sources for the energy-momentum tensor and baryon number current in the hydrodynamic equations.

The energy-momentum tensor and baryon number current generated by a single nucleon-nucleon collision in an inside-outside cascade are

$$T^{\mu\nu} = [(\varepsilon(y,\tau) + \rho(y,\tau) u^{\mu}(y,\tau) u^{\nu}(y,\tau) - g^{\mu\nu}\ \rho(y,\tau)]\theta(\tau-1) \qquad (4.25)$$

and

$$J_B^{\mu} = u^{\mu}(y,\tau) n_B(y,\tau)\theta(\tau-1). \qquad (4.26)$$

Here the step function, $\theta(\tau-1)$, reflects the formation time in an inside-outside cascade, and we have taken $\tau_0 = 1$ fm. The quantities ε, ρ, and n_B are the energy density, pressure, and baryon number density measured in a frame co-moving with the matter distribution. The two velocity of the distribution is u^{μ}.

The Eqs. (4.25)-(4.26) correspond to the $T^{\mu\nu}$ and J_B^{μ} generated in a single nucleon-nucleon collision, and are used to illustrate how an inside-outside cascade generates sources for a matter distribution. We do not advocate a hydrodynamical treatment of hadron-hadron or hadron-nucleus collisions, only for nucleus-nucleus collisions.

The source terms for the hydrodynamic equations are inferred by taking the two-divergence of Eqs. (4.25)-(4.26). We use the conservation of $T^{\mu\nu}$ and J_B^{μ} for $\tau>1$, and Eq. (4.21) to obtain

$$\partial_{\mu} T^{\mu\nu} = x^{\nu}\ \varepsilon(y,\tau)\delta(\tau-1) \qquad (4.27)$$

and

$$\partial_{\mu} J_B^{\mu} = n(y,\tau)\delta(\tau-1). \qquad (4.28)$$

The energy density and baryon number density at formation time are related to single particle inclusive distributions for particles of species i as

$$\varepsilon(y,1) = \sum_j \bar{m}_i \frac{dNi(y)}{dy}, \qquad (4.29)$$

and

$$n(y,1) = \frac{dN_B(y)}{dy}. \qquad (4.30)$$

When applied to three space dimension collisions, these inclusive distributions may approximately be those observed in nucleon-nucleon interactions since particle interactions may only slightly modify these primeval distributions. Similarly, $\bar{m}_i$ is estimated as the transverse mass averaged over the p_T distribution:

$$\bar{m}_i \approx (m_i^2 + \langle p_T^2 \rangle_i)^{1/2} \quad . \tag{4.31}$$

In reality, however, these primeval distributions may have to be slightly reparameterized and fit to experimental data when applied to nucleus-nucleus collisions.

In order to extend Eqs. (4.27)-(4.28) to equal A nucleus-nucleus collisions, let us first consider the target fragmentation region. The Lorentz contracted beam nucleus passes through the target nucleus (Fig. 18) with velocity $v_B \approx 1$ interacting with its nucleons at positions $x' = v_B t' \approx t'$. Each target nucleon generates independently its source of $T^{\mu\nu}$ and J^μ_B after unit proper time has elapsed after the interaction at $t' = x'$, i.e., along the curve $(t-x')^2-(x-x')^2 = (t+x-2x')(t-x) = 1$. The source terms in the hydrodynamic equations

$$\partial_\mu T^{\mu\nu} = \Sigma^\nu \tag{4.32}$$

$$\partial_\mu J^\mu_B = \sigma \tag{4.33}$$

for nucleus-nucleus collisions in the target fragmentation region are then simply obtained by integrating Eqs. (4.27)-(4.30) over x' within the nucleus. The result is

$$\Sigma^\nu = n_0\, \chi^\nu \sum_i \bar{m}_i \left.\frac{dNi}{dy}\right|_{y=-\log(t-x)} \theta(t,x) \tag{4.34}$$

$$\sigma = n_0 \frac{1}{t-x} \left.\frac{dN_B}{dy}\right|_{y=-\log(t-x)} \theta(t,x) \tag{4.35}$$

where $\theta(t,x)$ is one in the region (Fig. 18) bounded by the three curves $\tau^2 = (t+x)(t-x)=1$, $t-x=1$, $t+x=1/(t-x)+ 4R$, and zero otherwise. The nuclear baryon number density is $n_0 = A/2R$ and the quantity χ^ν is given by

$$\chi^0 = \frac{1}{2}[\frac{1}{(t-x)^2} + 1] \tag{4.36}$$

$$\chi^1 = \frac{1}{2}[\frac{1}{(t-x)^2} - 1]. \tag{4.37}$$

Figure 19 shows the source region on the τ,y plane. The magnitude of the source depends (Eqs. (4.36)-(4.37)) only on t-x and is thus constant along the curves $\tau = (t-x)e^y$ and increases when y increases. One sees explicitly how Σ^ν effectively approaches $A\delta(\tau-1)$ (Eq. (4.27)) far from the target fragmentation region.

The above result is also valid in the central region if the central rapidity density of produced particles scales as A. Then the sources generated by each target nucleon add independently, as assumed in the above derivations. If the central rapidity density does not sacle as A, the considerations described above must be modified. The simplest modification is to multiply the source term for $T^{\mu\nu}$ by $A^{\alpha(y)-1}$ which parameterizes the ratio of multiplicities of a nucleus-nucleus collision to that of a nucleon-nucleon collision. Some modification of the fragmentation region will also be required arising from the constraint of energy conservation. The central region and fragmentation regions should still be simple to understand, however, since $\alpha(y)$ should be constant in these regions, even though the transition region between these regions is complicated.

The properties of the beam fragmentation region clearly have to follow from the symmetry of the collision and Lorentz invariance. To derive the sources Σ^ν and σ in the beam fragmentation region, we first Lorentz transform from the target rest frame to the center of mass frame. The derivation of Σ^ν and σ is valid for $x<0$ which is on the target side in the center of mass frame. To derive Σ^ν and σ for $x>0$, we simply replace $\Sigma^\pm \equiv \Sigma^0 \pm \Sigma^1 \leftrightarrow \Sigma^\mp$ and $x^\pm \leftrightarrow x^\mp$, since the source should be symmetric under this reflection. For this to be a correct procedure, the overlap of the target and beam source regions has to be negligible. The criterion for this is from Fig. 19 seen to be $2R/\gamma_B \ll 1$. Otherwise, for instance, the total baryon number does not integrate to 2A but to something significantly less.

Our final result for the hydrodynamic equations describing globally A+A collisions is given by the two equations (4.32) and the single equation (4.33). The source terms in these equations are given by Σ^0 and Σ^1 in Eq. (4.34) and σ in Eq. (4.35), symmetrized to contain the beam source region. These source terms correspond to entropy production by inelastic particle production. The hydrodynamic equations in the absence of these sources generate isentropic expansion, and all entropy production arises from these sources. The quantities to be solved are $\varepsilon(t,x)$, $v(t,x)$, $n_B(t,x)$, and $p(t,x)$ from the equation of state. From these predictions for various experimentally measurable quantities like dilepton production rates can be calculated.

In general, the above equations have to be solved numerically. Two physically interesting cases can be obtained analytically. Firstly, assume y_B is very large and neglects the beam and target source regions entirely. Then the source terms are proportional to $\delta(\tau-1)$ and if the source intensity $\delta(y,1) = \varepsilon(1)$ is further assumed to be independent of y, the solution is

$$\varepsilon(t,x) = c_s^2\, p(t,x) = \frac{\varepsilon(1)}{(\tau/\mathrm{fm})^{1+c_s^2}}$$

$$v(t,x) = \frac{x}{t}\ ,\ n_B(t,x) = 0. \tag{4.38}$$

This coincides (for $c_s^2 = 1/3$) with the central region solution in Ref. 27. As is clear from Fig. 19, the beam and target source regions should, however, be included for accelerator energies realistic at present.

The second case is relevant for the hydrodynamic compression of the target volume. Assume that the Eqs. (4.32) give a very relativistic constant velocity field $u^0 = \gamma_{FB}$, $u^1 = \gamma_{FB}v_{FB}$, $u^0 + u^1 = e^{y_{FB}} >> 1$. Then the solution for $n_B(t,x)$ is found to be proportional to $e^{y_{FB}}$, the compression factor discussed in Ref. 28.

SECTION 5

APPENDIX

This Appendix is a transcription of a derivation in Ref. (3).

To calculate the free energy of a configuration of static quarks and anti-quarks, we introduce the operators $\psi_a^\dagger(\vec{r}_i,t)$ and $\psi_a(\vec{r}_i,t)$ which create and annihilate static quarks of color a at position $\vec{r}_i$ and time t, along with their charge conjugate fields $\psi_a^{c\dagger}$ and ψ_a^c for anti-quarks. These fields satisfy the equal time anti-commutation relations

$$\{\psi_a(\vec{r}_i,t),\ \psi_b^\dagger(\vec{r}_j,t)\} = \{\psi_a^c(\vec{r}_i,t),\ \psi_b^{c\dagger}(\vec{r}_j,t)\} = \delta_{ij}\delta_{ab}, \tag{5.1}$$

with all other equal time anti-commutators set to zero.

The quark fields obey the static time evolution equation

$$\{\frac{1}{i}\frac{\partial}{\partial t} - \tau\cdot A^o(\vec{r}_i,t)\}\psi(\vec{r}_i,t) = 0, \tag{5.2}$$

so that

$$\psi(\vec{r}_i,t) = T \exp i \int_0^t dt'\ \tau\cdot A^o(\vec{r}_i,t'))\psi(\vec{r}_i,0). \tag{5.3}$$

The symbol T in this equation denotes a time-ordered exponential.

The operators ψ and ψ^c may be employed to obtain an expression for the free energy of a configuration of N_q quarks and $N_{\bar{q}}$ anti-quarks. This is given by

$$\exp -\beta F(\vec{r}_1,\cdots\vec{r}_{N_q},\vec{r}_1'\cdots\vec{r}_{N_{\bar{q}}}') = \frac{1}{N^{N_q+N_{\bar{q}}}} \sum_s \langle s|e^{-\beta H}|s\rangle, \tag{5.4}$$

with the summation indicated over all states $|s\rangle$ with heavy quarks at $\vec{r}_1\cdots\vec{r}_{N_q}$ and heavy anti-quarks at $\vec{r}_1'\cdots\vec{r}_{N_{\bar{q}}}'$. The factors of N are to cancel the degeneracy factors in this summation introduced by the color labels of the static quarks. Introducing the quark fields gives this free energy as

$$e^{-\beta F_{N_q N_{\bar{q}}}} = \frac{1}{N^{N_q+N_{\bar{q}}}} \sum_{s'} \langle s'| \sum_{a,b} \psi_{a_1}(\vec{r}_1,0)\cdots\psi_{a_{N_q}}(\vec{r}_{N_q},0)\psi_{b_1}^c(\vec{r}_1,0)$$
$$\cdots\psi_{b_{N_{\bar{q}}}}^c(\vec{r}_{N_{\bar{q}}}',0)e^{-\beta H}\ \psi_{a_1}^\dagger(\vec{r}_1,0)\cdots\psi_{a_{N_q}}^\dagger(\vec{r}_{N_q},0)\psi_{b_1}^{\dagger c}(\vec{r}_1,0)$$
$$\cdots\psi_{b_{N_q}}^{\dagger c}(\vec{r}_{N_q}',0)|s'\rangle, \tag{5.5}$$

where now the sum is over all states $|s'\rangle$ with no heavy quarks. Since $e^{-\beta H}$ generates Euclidian time translations, i.e.,

$$e^{\beta H}\, 0(t) e^{-\beta H} = 0(t+\beta), \tag{5.6}$$

for any operator $0(t)$, Eq. (5.6) becomes

$$e^{-\beta F_{N_q N_{\bar q}}} = \frac{1}{N^{N_q + N_{\bar q}}} \sum_{s'} |s'\rangle \sum_{a,b} e^{-\beta H}\, \psi_{a_1}(\vec r,\beta)\psi^{\dagger}_{a_1}(\vec r_1,0)$$

$$\cdots \psi_{a_{N_q}}(\vec r_{N_q},\beta)\psi^{\dagger}_{a_{N_q}}(\vec r_{N_q},0)\psi^{c}_{b_1}(\vec r_1,\beta)\psi^{\dagger}_{b_1}(\vec r_1,0)$$

$$\cdots \psi^{c}_{b_{N_{\bar q}}}(\vec r\,'_{N_{\bar q}},\beta)\psi^{\dagger c}_{b_{N_{\bar q}}}(\vec r\,'_{N_{\bar q}},0)|s'\rangle\ . \tag{5.7}$$

Using Eq. (5.3) and its charge conjugate, together with Eq. (5.1), and introducing the Wilson line as

$$L(\vec r) = \frac{1}{N}\,\mathrm{tr}\, T \exp\left(i \int_0^{\beta} dt\ \tau\cdot A^{o}(\vec r,t)\right), \tag{5.8}$$

the free energy $F_{N_q,N_{\bar q}}$ is

$$e^{-\beta F_{N_q N_{\bar q}}} = \langle \mathrm{Tr}[e^{-\beta H}\, L(\vec r_1)\cdots L(\vec r_N) L^{\dagger}(\vec r\,'_1)\cdots L^{+}(\vec r\,'_{N_q})]\rangle . \tag{5.9}$$

ACKNOWLEDGMENTS

I have benefited from conversations with many people during the development of the ideas presented in these lectures. I especially benefited from conversations with R. Anishetty, C. De Tar, J. Bjorken, L.S. Brown, J. Gunion, M. Gyulassy, L. van Hove, M. Jacob, K. Kajantie, J. Kapusta, P. Koehler, J. Kuti, H. Miettinen, R. Pisarski, H. Satz, B. Svetitsky, W. Willis, and L. Yaffe. I particularly thank K. Kajantie, with whom I collaborated on the work presented in the last section, C. De Tar, with whom I collaborated on the derivation N-ality order parameters, and who critically read a first version of these lectures. I also thank the hospitality of the Research Institute for Theoretical Physics where these lectures were written.

This work was supported in part by the U.S. Department of Energy.

REFERENCES

1. J.C. Collins and M.J. Perry, Phys. Rev. Lett. 34, 1353 (1975).
2. A.M. Polyakov, Phys. Lett. 72B, 477 (1978); L. Susskind, Phys. Rev. D20, 2610 (1979).
3. L. McLerran and B. Svetitsky, Phys. Lett. 98B, 195 (1981), Phys. Rev. D24, 450 (1981); J. Kuti, J. Polonyi, and K. Szlachanyi, Phys. Lett. 98B, 199 (1981); J. Engels, F. Karsch, H. Satz, and I. Montvay, Phys. Lett. 101B, 89 (1981).
4. E. Fradkin and S.H. Shenker, Phys. Rev. D19, 3682 (1979).
5. C. DeTar and L. McLerran, Helsinki University Preprint HU-TFT-T-82-27.
6. G. 't Hooft, Nucl. Phys. B153, 141 (1979).
7. N. Weiss, Phys. Rev. D25, 2667 (1982); Phys. Rev. D24, 475 (1981).
8. L.G. Yaffe and B. Svetitsky, Cornell University Preprint CLNS-82/530.
9. T.D. Lee and G.C. Wick, Phys. Rev. D9, 2291 (1974).
10. R.D. Pisarski, Phys. Lett. 110B, 155 (1982); E.V. Shuryak, Phys. Lett. 107B, 103 (1981).
11. S.L. Adler, Phys. Rev. 177, 2426 (1969); J.S. Bell and R. Jackiw, Nuovo Cimento 60A, 47 (1969).
12. C.G. Callan, R.F. Dashen, and D.J. Gross, Phys. Lett. 63B, 334 (1976); R. Jackiw and C. Rebbi, Phys. Rev. Lett. 37, 172 (1976).
13. G. 't Hooft, Phys. Rev. D14, 3432 (1976).
14. N. Weiss, Can. J. Phys. 59, 1686 (1981).
15. L.G. Yaffe and B. Svetitsky, Phys. Rev. D26, 963 (1982).
16. J. Kripfganz, Proceedings of Workshop on Quark Matter Production and Heavy Ion Collisions, Bielefeld, West Germany, May 10-14 (1982).
17. H.W.J. Blöte and R.H. Swendsen, Phys. Rev. Lett. 43, 799 (1979); B. Nienhuis, E.K. Riedel, and M. Schick, Phys. Rev. B23, 6055 (1981); F.Y. Wu, Rev. Mod. Phys. 54, 235 (1982).
18. I. Montvay and E. Pietarinin, Phys. Lett. 110B, 148 (1982); 115B, 151 (1982); K. Kajantie, C. Montonen, and E. Pietarinen, Zeit. Phys. C9, 253 (1981).
19. K. Johnson, private communication.
20. J. Kogut, M. Stone, H.W. Wyld, J. Shigemitsu, S.H. Shenker, and D.K. Sinclair, Phys. Rev. Lett. 48, 1140 (1982).
21. L. Yaffe, private communication.
22. J. Engels, F. Karsch, and H. Satz, Phys. Lett. 113B, 398 (1982).
23. W.B. Fechner and P.C. Joss, Nature 274, 347 (1978).
24. A. Burrows, Phys. Rev. Lett. 44, 1640 (1980); N. Iwamoto, Phys. Rev. Lett. 44, 1637 (1980).
25. G. Baym, Proceedings of Statistical Mechanics of Quarks and Hadrons, September 17-29, Bielefeld, Germany (1980).
26. Proceedings of Workshop on Quark Matter Production and Heavy Ion Collisions, May 10-14, Bielefeld, West Germany (1982).
27. J.D. Bjorken in "Current-Induced Reactions", Proceedings of the International Summer Institute on Theoretical Physics, Hamburg, West Germany (1975), ed. by J.G. Körner, G. Kramer, and D. Schildknecht (Springer, New York, 1976) p. 93.
28. R. Anishetty, P. Koehler, and L. McLerran, Phys. Rev. D22, 2793 (1980); L. McLerran, Proceedings of 5th High Energy Heavy Ion Study, May 18-22 (1981), Lawrence Berkeley Laboratory, Berkeley, California.
29. J.D. Bjorken, FERMILAB-PUB-82/44-THY (1982).
30. K. Kajantie, Proceedings of Workshop on Quark Matter Production and Heavy Ion Collisions, May 10-14, Bielefeld, West Germany (1982); K. Kajantie and L. McLerran, Helsinki University Preprint HU-TFT-82-24.
31. L. Van Hove, Proceedings, Workshop on Quark Matter Production

and Heavy Ion Collisions, May 10-14, Bielefeld, West Germany (1982).
32. L. Landau, Proceedings of the Academy of Sciences, USSR, Physical Series, Vol. 17, p. 51 (1953); L.V. Shuryak, Phys. Lett. 78B, 150 (1978).
33. UAS-Collaboration CERN Phys. Lett. 107B, 310, 315 (1981); UA-1 Collaboration CERN Phys. Lett. 107B, 320 (1981).
34. K. Gottfried, Proceedings of Fifth International Conference on High Energy Physics and Nuclear Structure, Uppsala (1975).
35. A. Bialas, Proceedings of the First Workshop on Ultra-relativistic Nuclear Collisions (1979).
36. A. Mueller, Proceedings of the 1981 Summer Workshop, High Energy Heavy Ion Physics 1BNL 51443, Vol. 2, pp. 618-653 (1981).
37. T. Burnett et al., Proceedings of AIP Conference Proceedings "$\bar{p}p$ Collider Physics" Madison, Wisconsin (1981).
38. K. Kajantie and H. Miettinen, Zeitshrift für Physik C, 341 (1981).
39. T. Biro and J. Zimanyi, Phys. Lett. 113B, 6 (1982); B. Muller and J. Rafelski, Phys. Rev. Lett. 48, 1066 (1982).
40. A. Bialas, Proceedings of 17th Recontre de Moriond, Les Ares, France, March 14-26, 1982.
41. Sai-ping Li and L. McLerran, University of Washington Preprint 40048-82-PT9 (1982).
42. A. Casher, J. Kogut, and L. Susskind, Phys. Rev. 10, 732 (1974).

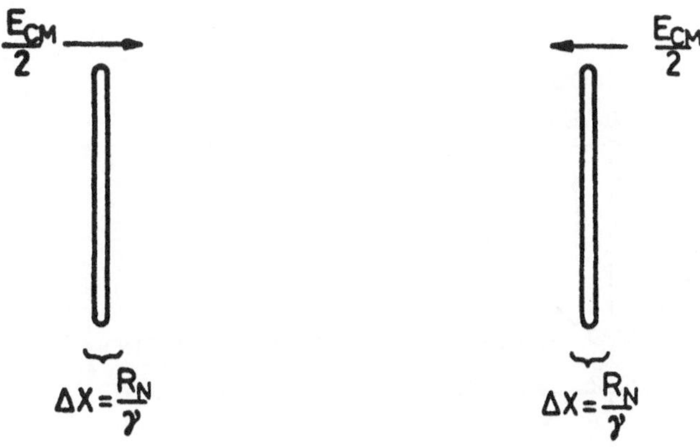

Fig. 1. Two nuclei of thickness $\Delta X \simeq RN/\gamma$ approaching one another in the center-of-mass frame. This thickness is appropriate for the simplest version of Landau's hydrodynamical model of the collision.

Fig. 2. The two nuclei sticking together immediately after a collision in the simplest version of Landau's hydrodynamical model.

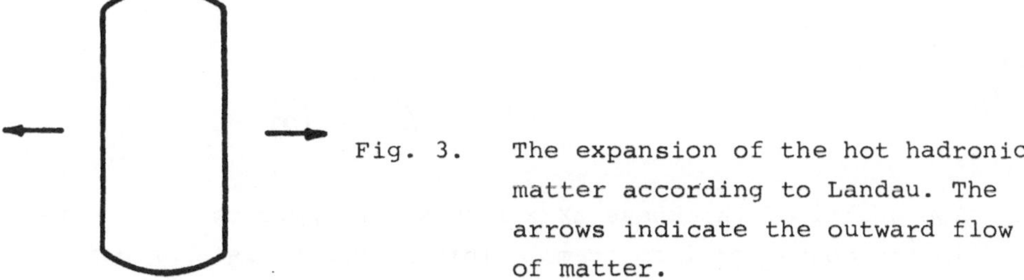

Fig. 3. The expansion of the hot hadronic matter according to Landau. The arrows indicate the outward flow of matter.

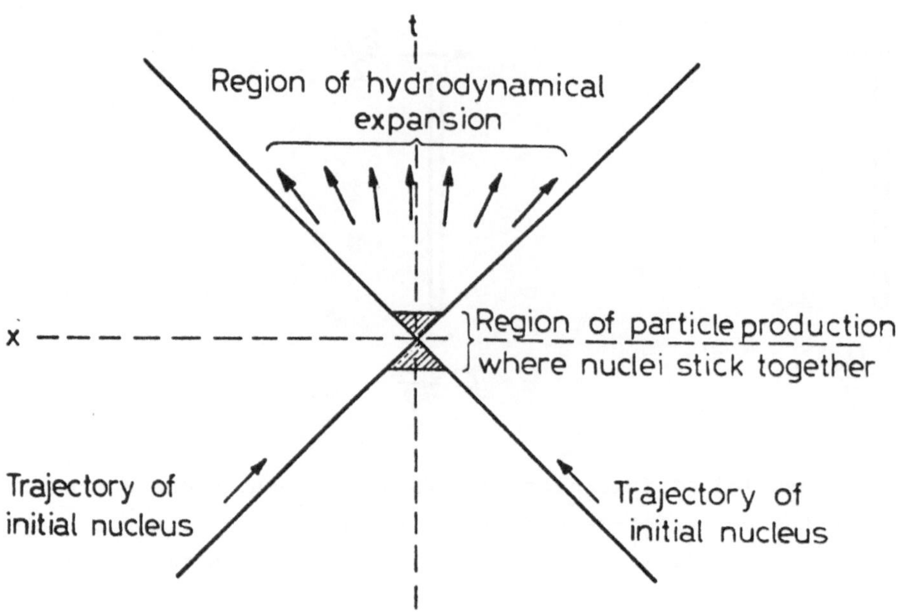

Fig. 4. A light cone diagram of a nuclear collision in the simplest version of Landau's model.

Fig. 5. Two nuclei of thickness $\Delta X \simeq 1$ fm which approach one another in the center-of-mass frame. This thickness is appropriate for the quark-parton model of nucleus-nucleus collisions.

Fig. 6. The two nuclei after passing through one another. The shaded area represents heat formed in the central region.

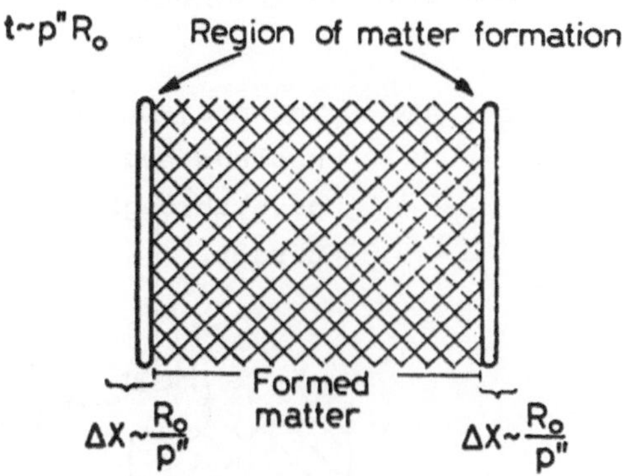

Fig. 7. The two nuclei forming an inside-outside cascade at time $\tau \sim p\ R_o$.

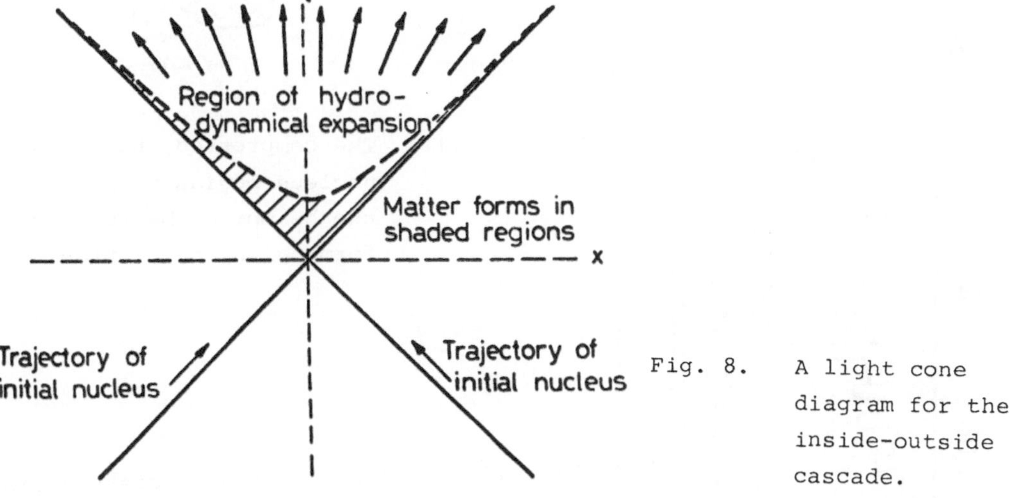

Fig. 8. A light cone diagram for the inside-outside cascade.

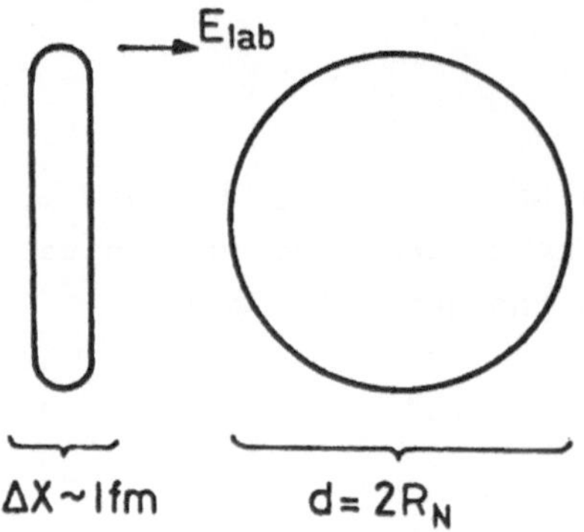

Fig. 9. The projectile nucleus with limiting thickness $\Delta X \sim 1$ fm approaches the target nucleus.

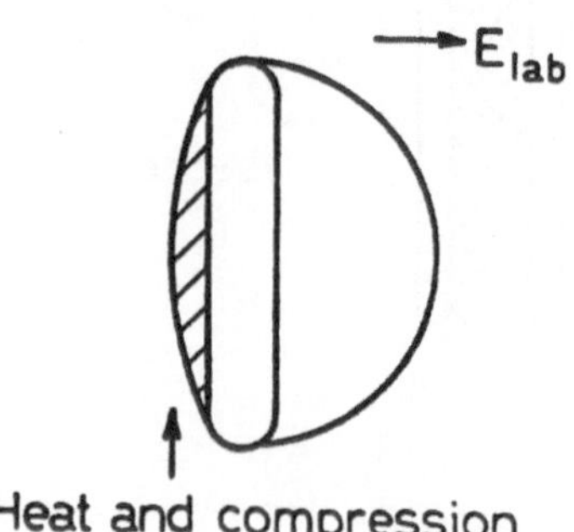

Fig. 10. The projectile nucleus begins to pass through the target.

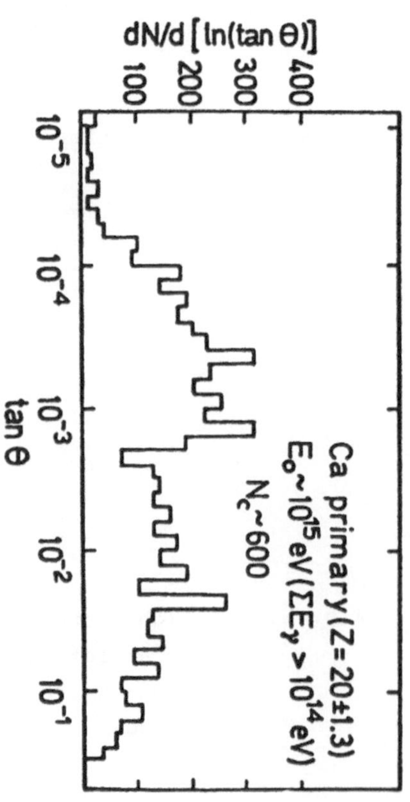

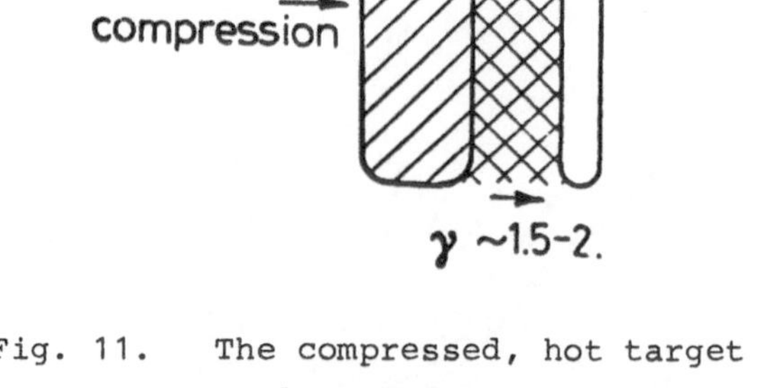

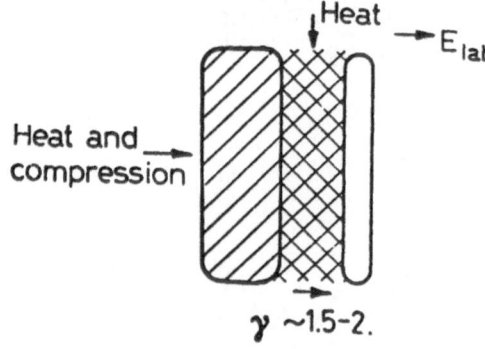

Fig. 11. The compressed, hot target nucleus begins to move down the beampipe. Matter begins forming in the central region.

Fig. 12. A pseudo-rapidity distribution for a 10^3 TeV calcium nucleus-emulsion event.

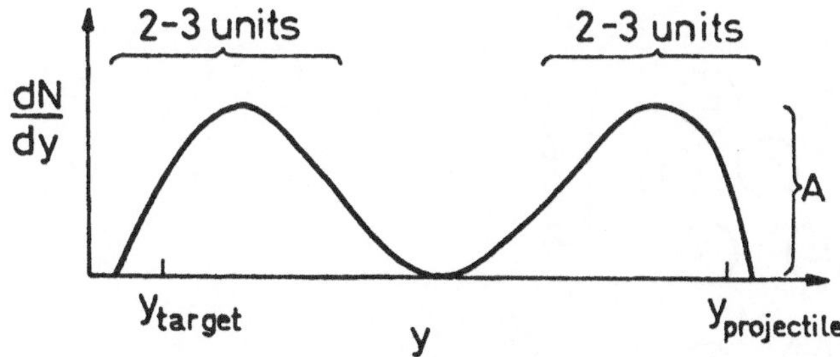

Fig. 13. Rapidity distributions for head-on nucleus-nucleus collisions:

(a) The baryon number (nucleons minus antinucleons)

(b) The meson distribution assuming heights in the central region proportional to A (——) and $A^{2/3}$ (----).

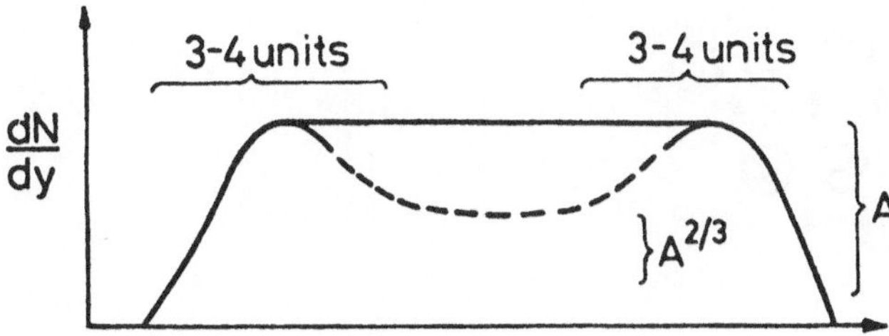

Fig. 14. Inelastic particle production.

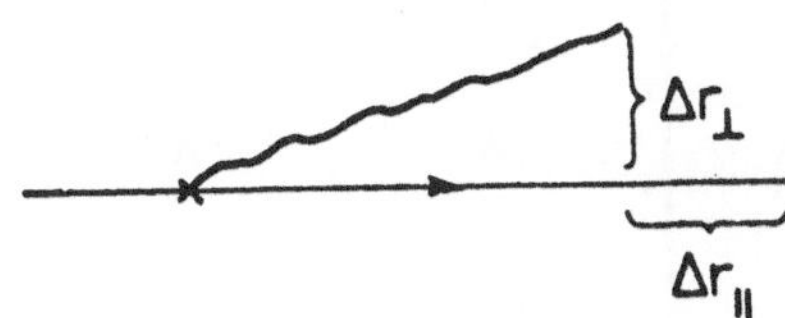

Fig. 15. Materialization of pions in hadron-hadron collisions. The arrows represent the direction of particle momenta.

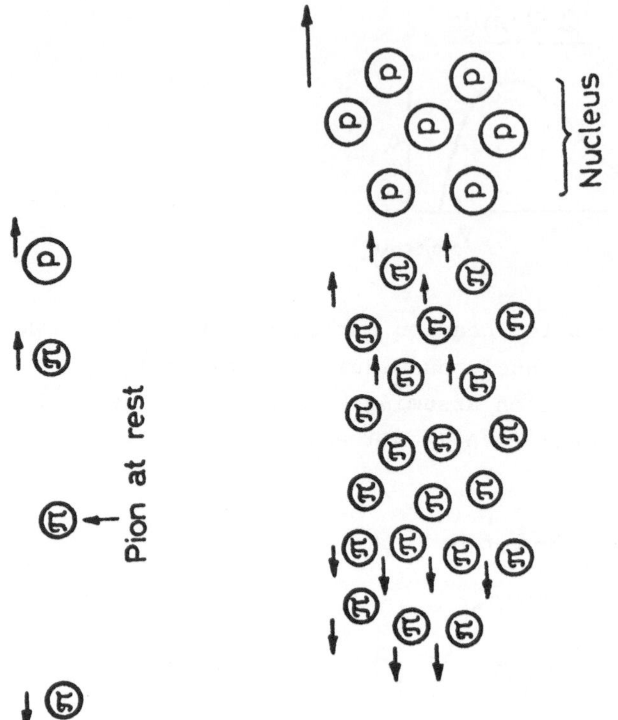

Fig. 16. The matter produced from many nucleon-nucleon interactions. Thermalization arises from the interaction of matter produced by many different nucleons.

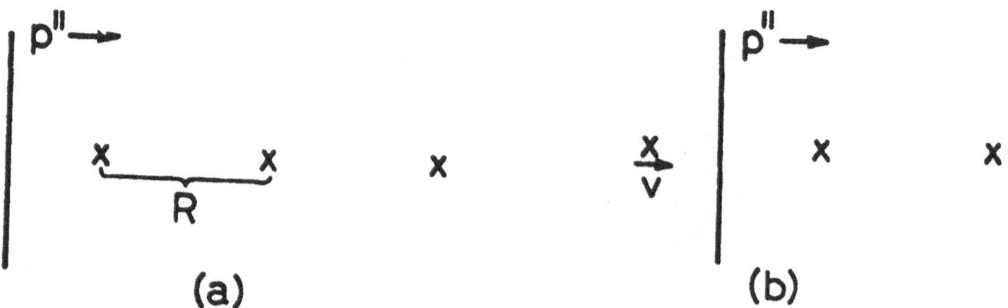

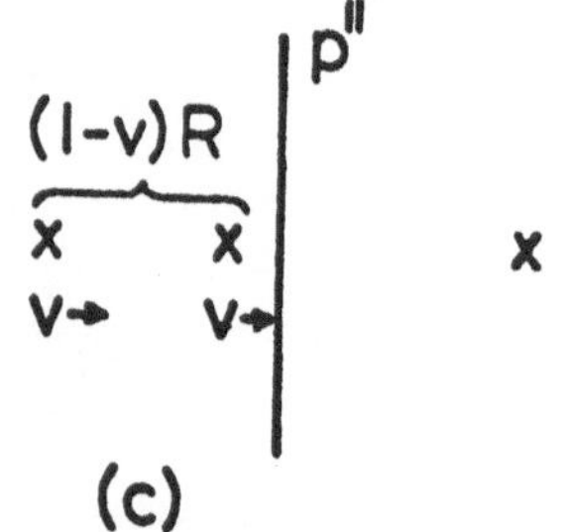

Fig. 17. Compression of the target nucleus
(a) Before encountering the first nucleon
(b) After encountering the first nucleon
(c) After encountering the second nucleon.

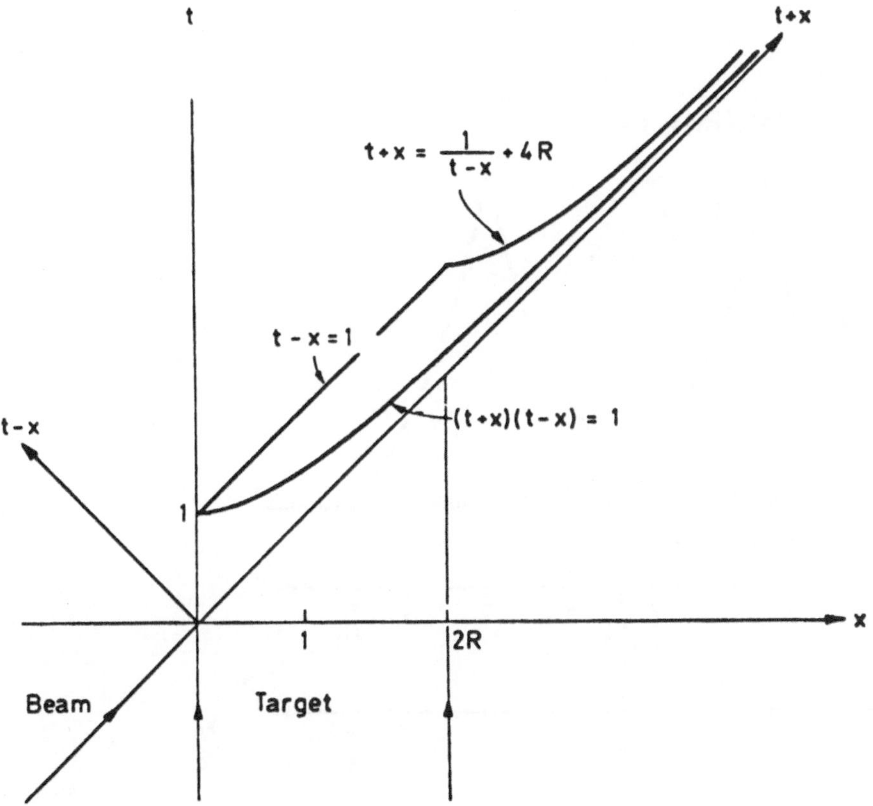

Fig. 18. A nucleus-nucleus collision in the space-time diagram in units of fm and fm/c. The target of thickness 2R is at rest and the beam, moving with velocity $v_B \gtrsim 1$, is Lorentz contracted to a negligible thickness $2R/\gamma_B$. The curves and the straight line t-x=1 bound the source region in Eqs. (4.34)-(4.35).

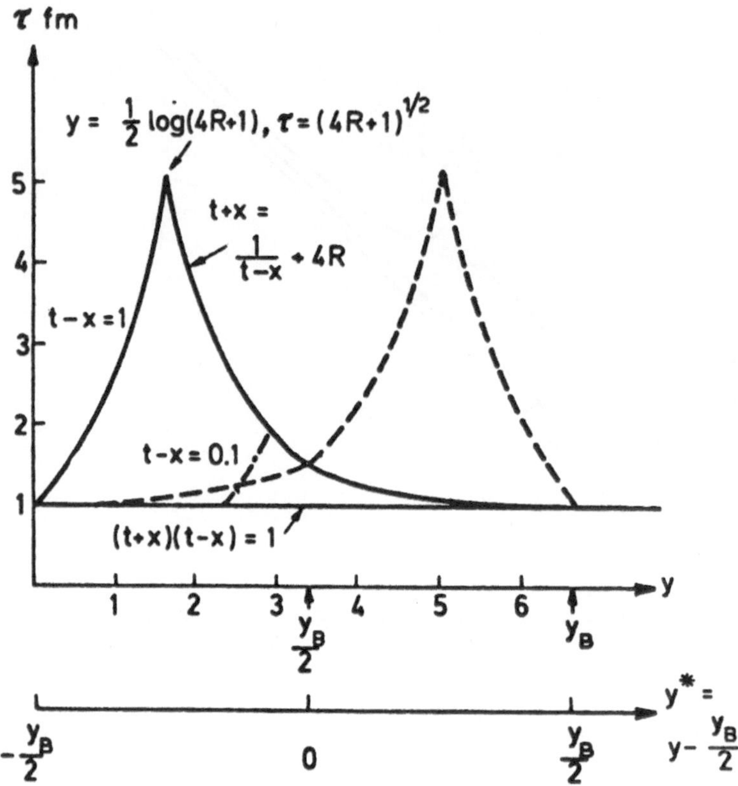

Fig. 19. The source region of Fig. 1 on the proper time-rapidity plane (solid line). The numbers refer to a U+U collision of 400 GeV/A energy on a fixed target. The dashed line shows the beam source region obtained by symmetrising around $y^* = 0$, y^* = CMS rapidity.

TABLE I. Spectrum of states in a U(2) × U(2) chirally symmetric world.

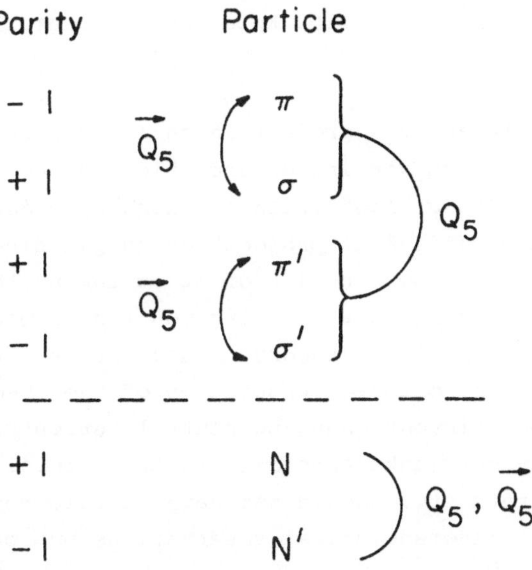

RANDOM DYNAMICS

H.B. Nielsen
Niels Bohr Institutet and NORDITA
Blegdamsvej 17
DK-2100 København Ø, Danmark

Now I shall talk about a project which I and a series of collaborators and others have been working on and which may be called "random dynamics" or "laws without law" (with an expression due to J.A. Wheeler[1]) coined for his not so different project) or "explaining various symmetries (etc.)". One may also conceive of it as taking the principle of naturalness[2]), i.e. saying that coupling constants, masses or any other (bare) parameters of the field theory should not have to be finetuned in order to explain the observed properties of the theory (such as the nonexistance of weak flavour changing neutral currents), to its extreme. In fact one usually only asks according to the naturalness principle that the coupling constants etc. should not have to take any special value to explain remarkable phenomenological observations but only be restricted by the symmetries. We want to go a bit further and not even assume the symmetries. Rather we shall attempt to derive the latter.

The list of my collaborators on this project is: N. Brene[3]), S. Chadha[4]),[5]),[6]), (A. Conkey), C.D. Förster[5,7]), (C.D. Froggatt[8,9,10])), M. Lehto[11]) (C. Litwin[5,6])), M. Ninomiya[5,6,7,11,12]), J. Picek[13,14]), S. Shenker[7]). The names put in parentheses here are the ones whose work lies somewhat on the borderline of the project of "explaining symmetries". For instance C. Litwin, S. Chadha and I[5,6]) have "derived" quantum mechanics or rather the linearity of the time-development equation - which is then interpreted as the Schrödinger equation - for a random first order differential time development equation after long time as the most likely. J. Picek and I[13,14]) have investigated somewhat phenomenologically possibilities of Lorentz-invariance breaking for weak interactions. What A. Conkey, C.D. Froggatt and I[8,9,10]) have been doing could really be called "random dynamics" but not "deriving symmetries" since we compute lepton and quark masses and generalized Cabibbo angles in a model in which there are ascribed random charges to left- and right-handed fermion fields.

Let us begin by asking how can we expect a final result of searching the laws of nature to look like. We may expect to find them either simple or complicated. That is to say we may expect one of the following two possibilities:

I A very simple and beautiful model or scheme will be found to be true. An example of such an end to the search for fundamental laws would be SO(8) supergravity. Another example is the by now old Heisenberg-Ivanenko model which also has the intention of being a model for all of physics. Nowadays it is mainly Dürr[15] and his collaborators that develop this model. However one might be worried if any theory would be beautiful and simple enough to deserve being "the" model for all of physics.

II The fundamental laws of nature are extremely complicated, so complicated that it might be better to say that there are no laws at all. Then it might be better to assume that the true law of nature is one randomly chosen among a large class of possible laws of nature. Then one must hope that the laws as we know them now will come out in some limits such as the long wave length and low energy limit.

It is this latter possibility that I want to talk about in this part of my talk and which I want to make propaganda for.

At first one might remark that even if there was a fundamental law of all physics that was simple from one point of view, it is conceivable that it could be so complicated from the way we are thinking that we might indeed do best by considering the law a randomly chosen one. That situation would be analogous to that the number π, say, can be considered succesfully as random from some point of view. In fact the decimals in $\pi = 3,14159...$ are for most practical purposes good random numbers; one tenth of them being 1, one tenth 2 etc. when you take a large sector of decimals in π. So even though π is from one point of view simple it being the ratio of the circumferance of a circle to its diameter it can from another point of view be considered as random. Analogously one might have a chance of getting good results by assuming the fundamental laws of nature to be random even if they were in fact not - but had some simplicity in a way very different from the way we look at them. This possibility means that even if the "project of random dynamics" turns out explaining a lot of phenomenologically correct physics we cannot be sure that the fundamental laws of nature are random from all points of view. It also means, however, that even if the fundamental laws of nature are not fully random (from all points of view) it might not be total waste of time to work on "random dynamics". It might in fact anyway explain several phenomena.

Let me now give some further arguments for studying the random dynamics idea. In fact let me give arguments that it might be a true idea:

I In R. Peccei's talks[16] we have heard how one is driven toward a quite complicated picture with preons and technicolor when one wants to make a model that naturally explains the phenomena observed up till now. In Peccei's scheme we meet more and more degrees of freedom as we go up in energy scale. After the well-established particles - to which you may also count Z_0 and $W^{\pm}$ - we shall meet the technicolor Yang-Mills fields and techniquarks etc., next the metacolor Yang-Mills binding preons to form bound states some of which are the quarks and leptons. Also there comes at much higher energy grand unification. Indeed it is becoming quite complicated. That indicates that the fundamental physics - taken as what goes on at some fundamental scale assumed to be say the Planck mass scale - is rather complicated and thus might indeed be simplest described if it could be considered random.

II Analogy with "old" laws of nature such as Hooke's law, Newtonian mechanics, law of thermal expansion of a body and Ohm's law indicates that many such phenomenogical laws have an explanation rather independent of what the fundamental physics is since they are just Taylor expansion approximations.

Let us as the final example take Hooke's law: Let us think of a piece of material, a solid, in the form of a stick. Its length is denoted by l and it is supposed that we pull it with a force F attempting to extend it. Now it is assumed that the length l is an analytical function of F, i.e. l = l(F). It is of course the length at which the situation is static we talk about. That l(F) is an entire function is of course an assumption - that might be justified by appeal to a more detailed knowledge of the physics of the stick - but it is one that is expected to be true under a very large class of models for the atomic structure of the stick. It does not matter much in practice whether the stick is crystalline or of glass for example - in both cases l(F) remains analytic. Now it can easily be seen that for a small enough F we can expand l(F) around the value F = 0 and obtain a linear expansion

$$l(F) = l_0 + \text{const.} \cdot F \tag{1}$$

where l_0 = l(0) is the length at no force. The constant

$$\text{const.} = \frac{dl}{dF}(F=0) \tag{2}$$

is the constant in Hooke's law, which we have now obtained.

It is obvious how this type of argumentation can also be used to derive e.g. the rule that the length of a stick increases with an amount proportional to a temperature increase (provided the change is small enough.

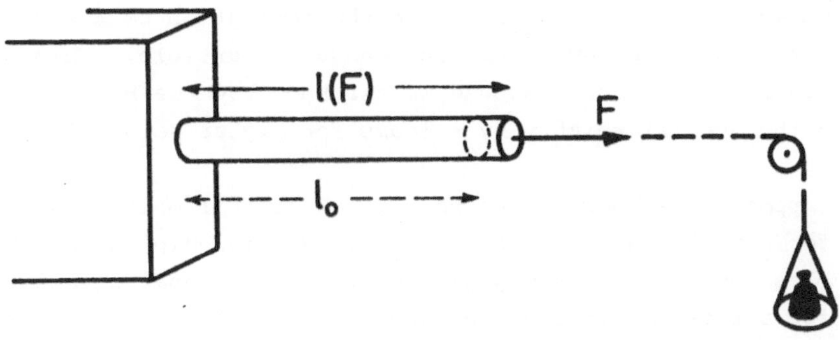

Fig. 1a. A stick being pulled in by the force F provided by a weight.

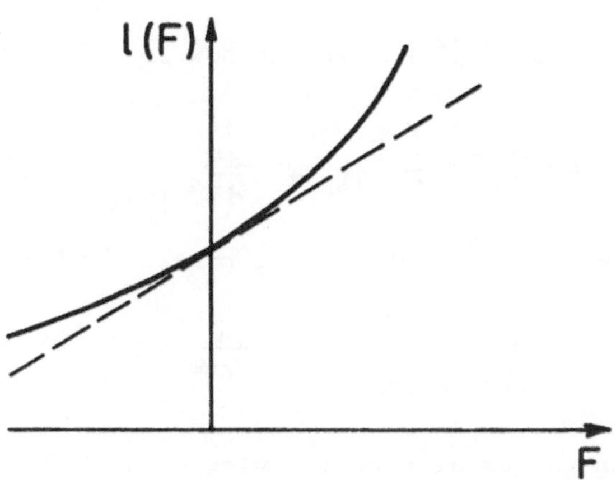

Fig. 1b. Curve illustrating the relation (-) between length $\ell(F)$ of the stick and force F together with the linear approximation to this relation (---).

Let us as a second example take the derivation of the Hamiltonian for a free low momentum (i.e. slow enough) particle. This example may be worth thinking about because we all know how the Hamiltonian - according to the special relativity - looks for higher velocities or momenta, too.

Supposing we know the principles of general mechanics the Hamiltonian for a particle is assumed to be a function $H(\vec{p}, \vec{x})$ of the dynamical variables "momentum" $\vec{p}$ and position $\vec{x}$. Translational invariance tells us that for a free particle - i.e. one not interacting with other things - the Hamiltonian cannot depend on the position $\vec{x}$. So we must have

$$H(\vec{p}, \vec{x}) = H(\vec{p}) \tag{3}$$

Next rotational invariance tells us that it can also not depend on the direction of the momentum $\vec{p}$ but only on its size. Assuming that $H(\vec{p}, \vec{x})$ is analytic we then find

$$H(\vec{p}, \vec{x}) = H(\vec{p}^{\,2}) \tag{4}$$

is only a function of $\vec{p}^{\,2}$ and we can now for "small" momenta $\vec{p}$ make use of a Taylor-expansion

$$H(\vec{p}, \vec{x}) = H_0 + \frac{dH}{d\vec{p}^{\,2}}\bigg|_{\vec{p}^{\,2}=0} \cdot \vec{p}^{\,2} + \ldots \tag{5}$$

The constant $H_0 = H(\vec{p}^{\,2} = 0)$ is not important and we may ignore it. Calling

$$\frac{dH}{d\vec{p}^{\,2}}\bigg|_{\vec{p}^{\,2}=0} = \frac{1}{2m} \tag{6}$$

we thus obtain the usual nonreletivistic Hamiltonian

$$H = \frac{\vec{p}^{\,2}}{2m} \tag{7}$$

for a free particle.

This "derivation" of the Hamiltonian of a free particle is of course not the historic way of finding it since Newton's study of the motion of particles predates the very concept of a Hamiltonian. However we may see from our knowledge of the energy - or Hamiltonian - of a free relativistic particle

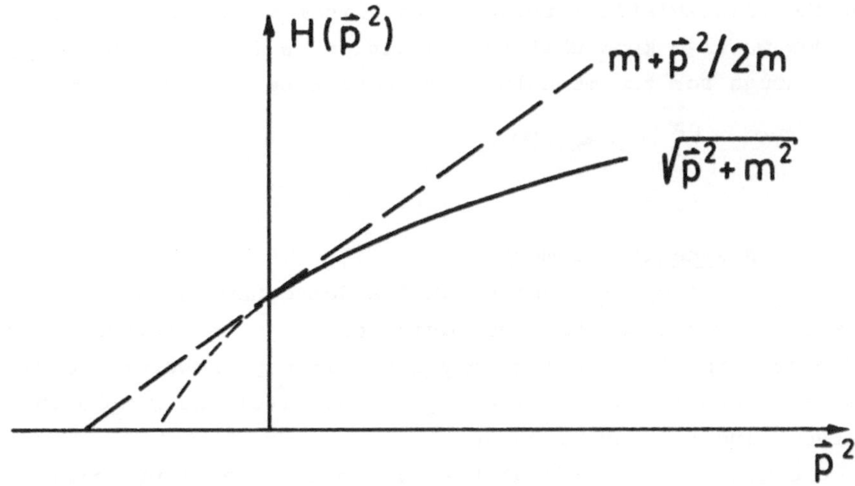

Fig. 2. The relation between the Hamiltonian $H(\vec{p})$ and the momentum square $\vec{p}^2$ for a free particle together with the linear approximation to this relation.

$$H = \sqrt{\vec{p}^2 + m^2} \tag{8}$$

that the simple form $\frac{\vec{p}^2}{2m}$ of the nonrelativistic (i.e. low momentum) free particle Hamiltonian is in fact understandable only as a Taylor expansion the relativistic formula being rather more complicated.

Now we also know what the condition is for the momentum $\vec{p}$ to be small enough for the momentum Hamiltonian being a good approximation:

$$|\vec{p}| << m \tag{9}$$

III Aesthetic argument.

As a third type of argument for "random dynamics" we may say that the alternative might be that one would once - in the future - find the true fundamental model for all of physics. It may, however, be difficult to imagine a fundamental model being so beautiful and simple that it would deserve being the law of nature.

One might perhaps dream that assuming several principles there would only be one consistent model left. An example of such a dream would be the bootstrap as developped by G. Chew. If one does not even impose symmetries such that a fundamental cut-off is excluded it seems however that there are, many self-consistent quantum field theories possible. Thus you need at least some assumptions before the theory will be unique. Will these principles have a chance of being beautiful enough?

IV Attempts to derive modern laws

As a fourth argument for that all laws may be derived from some random ones we may give the list of what we have already achieved:

1) Quantum mechanics (rather linearity of Schrödinger equation) from the limit of late times. (C. Litwin, S. Chadha & H.B. N.)

2) Lorentz (and rotational) invariance in the infrared limit. (S. Chadha, M. Ninomiya & H.B. N.)

3) Zero-mass photons (or Yang-Mills) in some phase. (D. Forster, M. Ninomiya & H.B.N.[7]) and independently S. Shenker[7])
[Kalb-Ramond in a phase is not so much an agreement with experiment since Kalb-Ramond fields have not been observed (M. Lehto, M. Ninomiya & H.B.N.)]

4) Gauge invariance in the infrared limit. (J. Iliopoulos, Nano-

poulos & Tamaros)

5) Gravity in a phase of a ("lego") lattice model. (Work in progress by M. Lehto, (D. Förster), M. Ninomiya & H.B. N.)

6) Explaining very crudely

$$\frac{\log m_\tau/m_\mu}{\log m_\mu/m_e} \simeq 0.6 \tag{10}$$

(and analogously for quarks) from random quantum number assignments. (C.D. Froggatt and H.B. Nielsen[8,9]).

7) Connected center of the gauge-group observable at "low" energy compared to the fundamental scale. (N. Brene & H.B. N.[3])

Even more speculatively we hope to explain why there are 1+3 dimensions[4] and why the gauge group should be of the type

R x SU(2) x SU(3) x ...x SU(P)/H

where the series of SU(n) groups have n running through the prime numbers up to some prime P. This group is the factor group consisting of cosets of a certain discrete subgroup H of the center of the group R x SU(2) x ... x SU(P) to be explained below.

8) There are several symmetries of strong and electromagnetic interactions which are understandable from the field theory model at least if you assume that the number of right-handed and left-handed representations under SU(3) and the electromagnetic are equal for each combination of representations.

That this should be the case follows from the need to avoid Adler anomaly troubles with the unitarity and the renormalizability of the gauge theories provided one assumes

a) that the electric charge of the fermions is always less than or equal to 1

b) that the color SU(3) representation is at most a triplet (i.e. it is triplet, antitriplet or singlet), and

c) the charge quantization rule will follow from the group U(3)). (This remark is part of an unpublished work with S. Chadha.)

The symmetries gotten out of the field theory model this way are parity P, charge conjugation C and timereversal T.

The CPT -theorem means that CPT is gotten out of any local field theory obeying Lorentz invariance. So also weak interactions should show CPT -invariance.

Symmetries such as Gell-Mann's SU(3), chiral symmetry SU(3) x SU(3) or SU(2) x SU(2) or just isospin SU(2) come about because the masses of the u, d and even s -quarks are small relative to the strong interaction scale Λ_{QCD}.

Various ways of obtaining symmetries

Most of the "laws of nature" (but not all) that we can "derive" have the character of being symmetry laws, i.e. they really tell that the action giving the dynamics of the physics we observe is invariant under a symmetry transformation. Now how can a symmetry appear without being there at the outset? It might sound a bit surprising that one can get a symmetry if one is not assuming it to be there in the fundamental dynamics i.e. in the fundamental action, but that is what we claim, and we would even like to speculate that all the symmetries discovered phenomenologically have indeed appeared by some mechanism or another.

We have in mind three mechanisms for generating symmetries:

I A notationally introduced (purely formal) symmetry may become relevant in some phase (for the vacuum).

An example of this method is the appearance of gauge invariance according to D. Förster, M. Ninomiya, H.B. Nielsen[7], and S. Shenker[7].

II Renormalization group β-function can imply the suppression of symmetry breaking terms towards the infrared.

E.g. Gauge invariance the J. Iliopoulos - Nanopoulos - Tamaros - way or Lorentz invariance the way of S. Chadha, M. Ninomiya and myself.

III The symmetry may be valid for all renormalizable field theories with given other symmetries.

E.g. flavour conservation in Q.C.D.

Let me now go to a somewhat more detailed discussion of these methods of explaining symmetries.

I Definitionally introduced symmetries

The typical example of how a symmetry can appear in this definitional way is a theory with the same degrees of freedom as a lattice gauge theory but with gauge noninvariant terms in the fundamental action.

The fundamental field variables of such a non-gauge invariant model with the degrees of freedom of a lattice gauge theory is a field

$$U(\bullet\!\!-\!\!\bullet) \tag{11}$$

taking values in a group G, the "gauge group" and being defined on the links (linepieces) of a lattice which we for simplicity think of as a cubic lattice. That is to say we consider a lattice the points of which correspond to elements of Z^d where d is the dimension of space-time and Z is the set of integers. We think of Euclideanized space-time. That is to say we consider an imaginary time so that we really consider a d = 4 dimensional statistical mechanical system with the action $\frac{1}{\hbar}S$ taking the place of $\frac{H}{T}$ where H is the energy and T the temperature (times Boltzmann constant). The Z^d -lattice is imagined imbedded in the d dimensional space-time (d = 4 in experimentally relevant case.) The links are given by pairs of neighboring points x, y $\in$ Z^d and on each of the links is defined a fundamental variable $U(\overset{x}{\bullet}\!\!-\!\!\overset{y}{\bullet})$ taking values in the group G, i.e. $U(\overset{x}{\bullet}\!\!-\!\!\overset{y}{\bullet}) \in G$.

When one makes gauge theories one requires from the action S (or the energy H if one thinks of a statistical mechanical system) that it shall be invariant under the gauge transformation

$$U(\overset{x}{\bullet}\!\!-\!\!\overset{y}{\bullet}) \rightarrow \Lambda(\overset{x}{\bullet})\, U(\overset{x}{\bullet}\!\!-\!\!\overset{y}{\bullet})\, \Lambda(\overset{y}{\bullet})^{-1} \tag{12}$$

for any pair (x,y) of neighboring points on the lattice Z^d. Here the gauge transformation function Λ is defined on the lattice points and takes values $\Lambda(x)$ in the group G. Requiring this gauge invariance one gets typical actions of the form

$$S = \sum_{\square} \frac{\beta}{2}\, \mathrm{Tr}\,(U_{\square}) \tag{13}$$

where

$$U_{\square} = U(\overset{x}{\bullet}\!\!-\!\!\overset{y}{\bullet})\, U(\overset{y}{\bullet}\!\!-\!\!\overset{z}{\bullet})\, U(\overset{z}{\bullet}\!\!-\!\!\overset{w}{\bullet})\, U(\overset{w}{\bullet}\!\!-\!\!\overset{x}{\bullet}) \tag{14}$$

is the product (of the group G) of the link variable $U(\bullet\!\!-\!\!\bullet)$ corresponding to the four links surrounding a plaquette $\square$ being a square with unit sides spanned by four points of the lattice Z^d. The coefficient

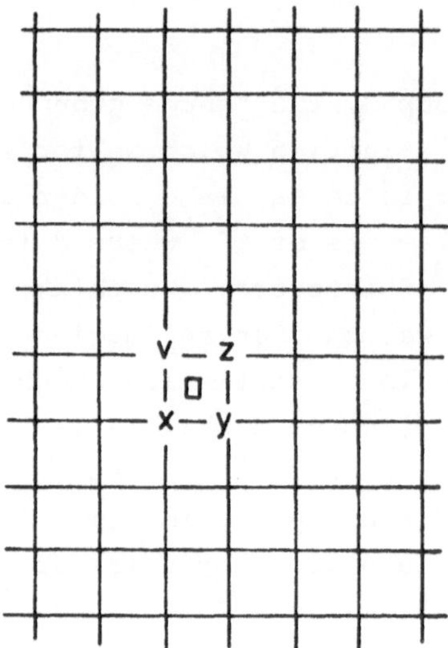

Fig. 3 . Two dimensional lattice with the names of sites used in formula (14) for the plaquette denoted □ .

β is the inverse coupling constant squared. The summation $\sum_{\square}$ runs over all the plaquettes i.e. all the small squares. The trace Tr denotes the trace when $U_{\square}$ is identified with the matrix of some representation of G of the group element $U_{\square}$. In general one may take the action to be given by a sum of terms like eq. (13) with the trace taken for various representations of G

$$S = \sum_{\square} \sum_{r} \frac{\beta_r}{2} \mathrm{Tr}_r (U_{\square}) \tag{15}$$

and it would also be possible to include terms corresponding to conglomerates of plaquettes. The latter are traces or linear combinations of traces of products of link variables around a conglomerate of plaquettes. To ensure reality of the action we should take β_r's for conjugate representations to be complex conjugate or we can take the real part of the expression given for S.

Now, however, it is our point not to assume gauge invariance for the fundamental action a priori. Therefore we add further terms which are not gauge invariant such as for example some functions of the link variable $U(\bullet\!-\!\bullet)$ summed over the links

$$\sum_{\bullet-\bullet} h\left(U(\bullet\!-\!\bullet) \right) \tag{16}$$

As the simplest example we may take the group G to be the abelian group U(1) (the set of complex numbers with norm unity), and then the action to be

$$S = \beta \sum_{\square} \mathrm{Re}\, U_{\square} + K \sum_{\bullet-\bullet} \mathrm{Re}\, U(\bullet\!-\!\bullet) \tag{17}$$

The first term is invariant under the gauge transformation while the second one - the one with coefficient K - is not. So this model is <u>not</u> gauge invariant and we must interprete all the degrees of freedom as having physical significance. For a gauge theory in contrast only combinations of field variables invariant under gauge transformations have physical signifigance.

Now the surprise is that even such a non-gauge invariant model has a phase - a whole region in the space of values for the parameters β and K - in which there are long range correlations corresponding to a massless photon-like particle. This is a surprise because one may classically follow E.A. Ivanov and V.I. Ogievietsky[17)] in considering the photon a Goldstone boson for the spontaneous breakdown of local gauge invariance. At least it is well-known that the lack of a mass term for the photon is a consequence of gauge invariance. In fact one sees immediately that a term $\frac{1}{2} m^2 A_\mu^2$ added to the continuum theory Lagrangean $- \frac{1}{4} F_{\mu\nu}^2$ of usual electrodynamics will spoil the gauge invariance under

$$A_\mu \to A_\mu + \partial_\mu \lambda \tag{18}$$

Therefore we would expect that since there is no gauge invariance in the lattice model with the action eq. (17) there would also be no gauge invariance in the continuum limit and thus no reason for there being no photon mass term $\frac{1}{2} m^2 A_\mu^2$. So why should there not be one?

How comes this surprise of a zero-mass photon in non-gauge invariant model about?

The point is that one can reformulate the model so that there is a gauge invariance, although a rather formal one. In fact we introduce a superfluous number of new variables $U_h(\bullet\!-\!\bullet)$ defined on the links just as $U(\bullet\!-\!\bullet)$ and $\Omega_h(\cdot)$ defined on the sites i.e. the points of Z^d. Both fields $U_h(\quad)$ and $\Omega_h(\cdot)$ take values in the group G or U(1) in the simplified case. They are introduced by writing

$$U(\overset{x}{\bullet}\!-\!\overset{y}{\bullet}) = \Omega_h(\overset{x}{\bullet})\, U_h(\overset{x}{\bullet}\!-\!\overset{y}{\bullet})\, \Omega_h(\overset{y}{\bullet})^{-1} \tag{19}$$

where $U(\overset{x}{\bullet}\!-\!\overset{y}{\bullet})$ is the fundamental variable attached to the link $\overset{x}{\bullet}\!-\!\overset{y}{\bullet}$. Since we have introduced more new variables $\Omega_h(\bullet)$ and $U_h(\bullet\!-\!\bullet)$ (with an index h denoting "human" in contrast to God's variables" U) than the original ones $U(\bullet\!-\!\bullet)$ there is the possibility of changing around the new variables without changing $U(\bullet\!-\!\bullet)$. It is such a change in the variables $\Omega_h(\cdot)$ and $U_h(\bullet\!-\!\bullet)$ given by

$$U_h(\overset{x}{\bullet}\!-\!\overset{y}{\bullet}) \to \Lambda(\overset{x}{\bullet})\, U_h(\overset{x}{\bullet}\!-\!\overset{y}{\bullet})\, \Lambda(\overset{y}{\bullet})^{-1} \tag{20}$$

and

$$\Omega_h(\overset{x}{\bullet}) \to \Omega_h(\overset{x}{\bullet})\, \Lambda(\overset{x}{\bullet})^{-1} \tag{21}$$

and leaving the original field $U(\bullet\!-\!\bullet)$ invariant

$$\begin{aligned} U(\overset{x}{\bullet}\!-\!\overset{y}{\bullet}) \to & \left[\Omega_h(\overset{x}{\bullet})\Lambda(x)^{-1}\right]\left[\Lambda(x)\, U_h(\overset{x}{\bullet}\!-\!\overset{y}{\bullet})\, \Lambda(y)^{-1}\right]\cdot \\ & \left[\Omega_h(y)\Lambda(y)^{-1}\right]^{-1} \\ = & \ \Omega_h(\overset{x}{\bullet})\, U_h(\overset{x}{\bullet}\!-\!\overset{y}{\bullet})\, \Omega_h(\overset{y}{\bullet})^{-1} = U(\overset{x}{\bullet}\!-\!\overset{y}{\bullet}) \end{aligned} \tag{22}$$

which we shall take as the gauge tranformation. The gauge function is $\Lambda(x) \in G$ (or U(1) in the simple case).

Let us stress that this gauge transformation eqs. (20-21) is a priori only a transformation of the formalism $\Omega_h(\cdot)$ and $U_h(\cdot\!-\!\cdot)$ which we have introduced ourselves and thus it is a very formal symmetry operation. Since the action originally depended only on the field variable $U(\cdot\!-\!\cdot)$ it will of course be invariant under the gauge symmetry (eqs. 20-21) when (eq. 19) is substituted into it.

In this way we have reformulated the original model in terms of $U(\cdot\!-\!\cdot)$ not having gauge invariance, as a gauge invariant one in terms of $\Omega_h(\cdot)$ and $U_h(\cdot\!-\!\cdot)$ containing a kind of lattice charged field $\Omega(\cdot)$ extra. A priori the field $\Omega_h(\cdot)$ has only one real number degree of freedom per site while there is for a genuine charged field a complex one equivalent to two real degrees of freedom per site. By what we may call block spinning we can, however, form some blockspin field

$$\varphi(\text{block}) = \sum_{x \in \text{block}} \Omega_h(x) \tag{23}$$

which is defined on blocks of lattice points and which takes on complex values. We are thinking of $\Omega_h(\cdot) \in U(1)$ as a complex number with norm unity. We have divided the set of sites Z^d up into blocks each consisting of say λ^d sites forming a d-dimensional cube with sidelength λ-1. The idea of such a blockspinning is that for considerations of wavelengths long compared to λ lattice constants one may use the field variable φ(block) instead of $\Omega_h(\cdot)$ thereby deriving an effective and approximate model replacing the model considered at first. In order to make the definition of the block spin field variable φ(block) well-defined we shall choose what may be called Lorentz gauge on the lattice requiring

$$\prod_{\substack{\cdot\!-\!\cdot \text{ links}\\ \text{going out}\\ \text{from a}\\ \text{given}\\ \text{site.}}} U(\cdot\!-\!\cdot) = 1 \tag{24}$$

inside the blocks.

The field φ(block) is very much like a (scalar) charged field. It is complex and it will under a gauge transformation given by a gauge function $\Lambda(\cdot)$ depending only on the block to which the site belongs transform just like a charged field i.e.

$$\varphi(\text{block}) \rightarrow \varphi(\text{block}) \cdot \Lambda(\text{block}) \tag{25}$$

Now this charged field φ(block) may or may not take a nonzero expectation value in a Lorentz like gauge where (eq. 24) is valid all over. That is to say φ(block) may be a Higgs field or just an "ordinary" charged (scalar) field with "ordinary" positive mass squared ($m^2 > 0$). The important point for us is the possibility

$$\langle \varphi(\text{block}) \rangle = 0 \tag{26}$$

so that there is no Higgsing and thus a massless photon is obtainable for a whole phase of value combinations (β,K) of the coefficients in the action (eq. 17). That there are indeed both phases, the Higgs one and zero photon mass one, has been argued by Fradkin and Shenker[18] and by Rabinovici et al.[19]. It is argued that a lattice field theory given by an action

$$S = \beta \sum_{\square} \mathrm{Re}\, U_{h\square} + K \sum_{\underset{x\ y}{\cdot\!-\!\cdot}} \mathrm{Re}\left[\Omega_h(\overset{x}{\cdot})\, U_h(\overset{x\ y}{\cdot\!-\!\cdot})\, \Omega_h(\overset{y}{\cdot})^{-1}\right] \tag{27}$$

(derived from (eq. 17) by insertion of (eq.19)) has the two phases by considering the model along the boundaries of the phase diagram i.e. for extreme values 0 or ∞ of the parameters β and K. For example, for K = 0 one gets (essentially) usual gauge invariant lattice U(1) electrodynamics and it is known in $d > 3$ dimensions that for β larger than a certain critical value β_c one finds a spin wave phase (= a phase with a zero mass photon) while for $\beta < \beta_c$ one has a "confining" electrodynamics with no infinite range correlations i.e. with no zero mass particles. On another boundary line $\beta = \infty$ the $U_h(\cdot\!-\!\cdot)$ -field freezes in for a gauge like (eq. 24) used all over the lattice. That is to say the field $U_h(\cdot\!-\!\cdot)$ will only be able to oscillate very little in the large β limit and we shall have in fact

$$U_h(\cdot\!-\!\cdot) = 1 \tag{28}$$

for $\beta = \infty$ in the Lorentz like gauge (eq. 24). Thus along this boundary the $\Omega_h(\cdot)$ -field effectively decouples and forms an x-y -model itself. An x-y model is given by a field on a lattice with values in U(1) just as $\Omega_h(\cdot)$ and an action

$$K \sum_{\overset{\bullet\!\!-\!\!\bullet}{x\ y}} \mathrm{Re}\left[\Omega_h(\overset{x}{\cdot})\,\Omega_h(\overset{y}{\cdot})^{-1}\right] \tag{29}$$

The x-y -model has a phase transition separating a "cold" phase with $K > K_c$ and a nonzero expectation value for Ω_h and a "hot" one with $K < K_c$ and $\langle\Omega_h\rangle = 0$ for $d > 2$. Here K_c is the critical K-value.

Somehow there is a phase transition curve in the two-dimensional phase diagram of (β,K) -values connecting the two just mentioned phase transitions on the boundaries of this diagram $(\beta,K) = (\beta_c,0)$ and $(\beta,K) = (\infty,K_c)$. The corner region of (β,K) pairs delinated by this curve and having rather large (but not necessarily ∞) β and rather small (but not necessarily 0) K is the zero mass photon phase.

The existence of a whole region - a whole phase - of this type might be made plausible by remarking that even if β is large but not infinite the Ω_h -field will behave much like the x-y -model and thus for K small enough will have only short range correlations. Then it will be no nonzero expectation value of Ω_h nor of φ(block) and we will have the unhiggsed phase. Thus there is for large β and small but even nonzero K a zero photon mass phase.

The suggested phase diagram is drawn in a compactified form:

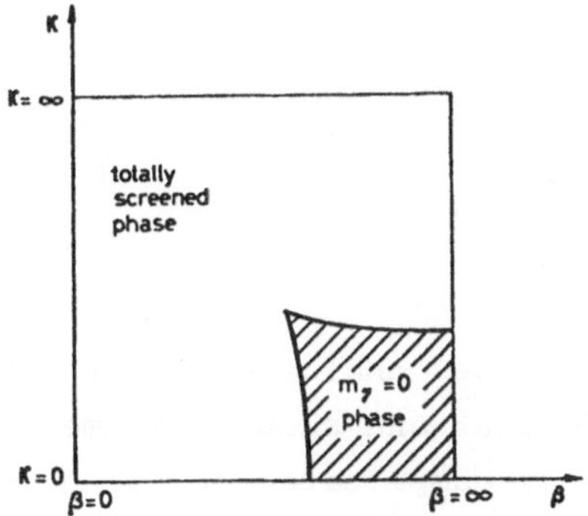

The zero mass photon is massless because of the gauge invariance under the transformations (eqs. 20-21). In spite of this symmetry being formal it seems to be able to cause a photon to be massless.

This is the idea of our "definitionally introduced symmetry" method of obtaining symmetries.

One first introduces it completely artificially by defining some new variables (in our example Ω_h and U_h) in such a way that there exists in a trivial way a gauge symmetry (in our example (eqs. 20-21)). Next one shows that there exists a phase (in our case the one with rather large β and rather small K) in which this trivial gauge symmetry nevertheless has an important consequence a massless photon-like particle, say.

The connection with the above mentioned philosophy of random dynamics hangs on the appearance of the massless photon in a whole phase, a whole region in the (β,K) -space. The point is that we imagine the fundamental parameters (= coupling constants) such as β and K in our simplified model take random values. Now if there is a whole region of (β,K) -values giving a certain property (e.g. the massless photon) then there is finite nonzero probability for this property occurring in nature. When we here say that β and K are taken to be random it means that we take some "reasonable" probability measure in the space of pairs (β,K) and assume that the world was created with (β,K) given as a couple of random numbers taken from a distribution with this "reasonable" probability measure. We assume that a whole region (a whole phase) under this measure has a finite and non-zero probability. In fact we would by a "reasonable measure" think of one of the form

$$P(\beta, K) \; dK \, d\beta \tag{30}$$

with

$$P(\beta, K) > 0 \tag{31}$$

for all (β,K).

We have not immediately argued that there must exist a massless photon from this model. To do that one should assume that the fundamental lattice theory is a very complicated one with many parameters containing several approximately decoupled copies of the model we just described. That is to say we imagine a model having n fields $U_i(\cdot\!-\!\cdot)$ with i = 1,2,...,n asigned to each link and the interaction has the form

$$S = \sum_{i=1}^{n} \left[\beta_i \sum_{\square} \mathrm{Re}\, U_{i\square} + K_i \sum_{\cdot-\cdot} \mathrm{Re}\, U_i(\cdot\!-\!\cdot) \right] \tag{32}$$
$$+ \left[\text{terms giving interactions between fields with different } i\text{-indices} \right]$$

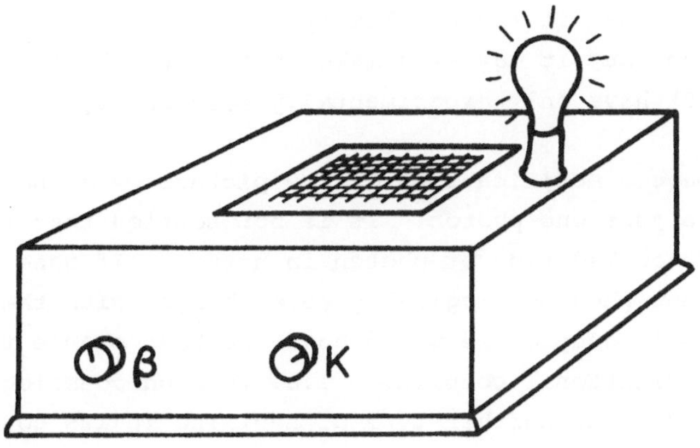

Fig. 5. The main point of the work by D. Förster, M. Ninomiya and myself is that it is not necessary to finetune the parameters β and K in order to make the lattice theory have massless light.

So there are at least 2n parameters $(\beta_1,\ldots,\beta_n,K_1,\ldots,K_n)$ but one may think of further ones in the "terms giving interactions between fields with different i-indices". If one speculates that the latter terms are not so important it seems likely that a (nonzero) number of the (β_i,K_i) pairs give rise to zero mass photons. In such a picture we might thus claim that it is very likely that there exists indeed a zero-mass photon, but it may - to make up for it - be a problem why we (until now?) have only experimentally seen one type of massless photon.

Without severe modifications of the picture we do not see why there should be just one photon. It is not settled that there might not exist a decoupled type of photon in nature. If none of the particles we know phenomenologically were charged with the type of charge to which it coupled we would have the best chance to see it through its gravitational coupling - i.e. through cosmological effects. It would act much like another type of neutrino in its cosmological implications. But that we would hardly be able to detect.

It has been critized by M. Veltman[20] that a model in which gauge invariance is broken at high energy - at the ultraviolet cutoff scale - has essentially no chance of agreeing with experiment. The point of M. Veltman is that in a diagram contributing to photon-photon scattering with a loop of $W^\pm$ -bosons a huge suppression is taking place due to gauge invariance. So if there is a gauge invariance breaking at a cutoff scale identified with say the Planck energy scale $E_{pl} = \sqrt{G^{-1}h\, c^5} = 1.9562\cdot 10^{16}$ erg $= 1.22098\cdot 10^{28}$ eV $= 1.22098\cdot 10^{19}$ GeV there will be an appreciable and easily observable photon-photon cross section due to intermediate states with W-bosons. Since the photon photon scattering cross section is safely known to be very small the possibility of severe gauge invariance breaking at the Planck scale must be excluded. Let us stress that Veltman to perform his arguments used a W -propagator corresponding to a Hilbert-space metric that is positive definite. We can say that he assumed a W-propagator as one might expect in a physically healthy theory having no negative probabilities.

One might, and Veltman indeed did, expect that his arguments were a challenge to the above described work by D. Förster, M. Ninomiya and myself[7] (and independently by S. Shenker[7]). However, the formal gauge invariance (eqs. 20-21) is exact even at the cutoff scale because it is purely due to our definition (eq. 19) of the new (= "human") variables Ω_h and U_h. Therefore there is in our model no breaking of the gauge invariance at all and therefore no reason for any abnormal photon-photon scattering cross section. The usual gauge invariance implied suppression will take place. In order to properly discuss Veltman's argument in our model we should consider the non-abelian Yang-Mills version of it, i.e. we should take G = U(2) (or U(1) x SU(2)) corresponding to the Weinberg-Salam -model let us say. But it is not difficult to extend our model to the non-abelian case. It will still be true that the gauge invariance is by definition and thus exact at all scales.

Thus our picture for understanding gauge invariance is after all not in conflict with Veltman's argumentation. This argumentation is rather an attack on the way of explaining gauge invariance put forward by I. Iliopoulos, T. Tamaros and D.V. Nanopoulos[21]. Their explanation rather belongs to the next method (number II) of obtaining symmetries - in fact to the one using the renormalization group β-function. That means that they present a model in which there is no gauge invariance at high energy, but that the renormalization group β-functions for various coupling constants allowed when gauge invariance is not assumed are such that they make gauge invariance appear approximately in the infrared (i.e. low energy) limits. So this model is indeed one in which gauge invariance is broken at the high energy (ultraviolet) scale but approximately kept at low energy and thus it is of the type critized by M. Veltman. There is, however, no logical contradiction because the Iliopoulos-Nanopoulos-Tamaros -model has photons with all four components of the A_μ -field being a priori present but some components leading to negative Hilbert-space metric states. It is in fact like Gupta-Bleuler- type formalism with respect to the metric of the Hilbert space. So a photon polarized with the A_μ -field in time like direction (i.e. A_0 numerically larger than $|\vec{A}|$) leads to a negative metric state. Since Veltman assumed no negative metric states there is no contradiction with the Iliopoulos-Nanopoulos-Tamaros understanding of why we have gauge invariance, except that it is of course a somewhat weak point from a physical point of view to have a nonpositive definite Hilbert-space.

Gravitation

A further example of obtaining a symmetry by the definitional way is a project on which M. Lehto, M. Ninomiya and I (and at a very early stage D. Förster) are working. As a continuation of the above described work by D. Förster, M. Ninomiya and myself[7)] (and independently S. Shenker[7)]) it is natural to think that if one can obtain electrodynamics and Yang-Mills theories without finetuning there should also be the possibility of obtaining another gauge like theory, the Einstein theory of gravitation. One expects that the masslessness of the graviton is connected with the invariance of the general relativity formalism under reparametrization of space-time coordinates. It is this reparametrization invariance that should take the place of the gauge invariance. So we shall make reparametrization invariance come true by definition by introduction of some new ("human") variables.

In fact we take the starting point from a fundamental theory that is a kind of lattice gravity theory in which there is a priori no manifold at all. Thus there is at first also no reparametrizations of course. But then in the next step we introduce some integration variables among which there is a variable set $\xi^{\mu}(P)$ for each site P of the lattice asigning to it a point on a manifold, that comes at the end to make up the space time manifold.

Let us only sketch what we attempt to do rather shortly since it is only work in progress:

1) First we think of the fundamental lego-lattice model by writing its partition function

$$Z = \sum_{n=0}^{\infty} \sum_{\mathcal{Y} \subseteq \mathbb{P}\times\mathbb{P}} \prod_{Y \in \mathcal{Y}} d\ell(Y) \quad e^{-S_{fund}(n,\mathcal{Y},\ell)} \tag{33}$$

where

$$\mathbb{P} = \{1, 2, \ldots, n\} \tag{34}$$

Here n is the number of lattice sites (or points) and $\mathbb{P}$ thus the set of names of them. The idea of the set of pairs (P_1, P_2) of points in $\mathbb{P}$ called $\mathcal{Y}$ is that it is the set of those pairs connected by a link. In this way $\mathcal{Y}$ becomes identified with the set of links in the lattice. The summation over the variable $\mathcal{Y}$ means that the way the lattice points $P \in \mathbb{P}$ are linked together is a dynamical variable $\mathcal{Y}$.

It is because of this linking being dynamical we call the lattice a "lego" one. It refers to the legobuilding blocks, the lattice is one that can be build together in many ways. For simplicity we have left out some extra degrees of freedom that should interact with $\mathcal{Y}$ in such a way as to arrange that the large scale topology of the lattice becomes rather simple when conceived as approximating a four dimensional manifold. The variables $l(Y)$ asigned to the links $Y \in \mathcal{Y}$ are meant to approximate the lengths of the links in question. There might be imagined other variables but we are not writing them for simplicity. In fact the model put forward is inspired by the Regge calculus idea[22] of approximating a general relativity space-time manifold by a piecewise flat space-time consisting of simplicies.

The lego lattice shall be essentially a lattice gravity theory much like the one considered by R.M. Williams and M. Rocek[23].

2) Next we introduce a series of new ("human") variables by some trivially correct operations consisting typically in multiplying by unity written in form like

$$1 = \int \mathcal{D}g_{\mu\nu} \; \delta(g_{\mu\nu} - g_{\mu\nu}^{RC}) \tag{35}$$

Here the functional integration over the dummy (= integration) variable is supposed to be given according to the idea of A. Polyakov[24] by using his metric in the space of all metric tensorfields. The δ-function $\delta(g_{\mu\nu} - g_{\mu\nu}^{R.C.})$ is the corresponding functional δ-function. Actually we shall like to take it as a somewhat smeared (not sharp) functional δ-function. The metric tensor $g_{\mu\nu}^{R.C.}$ is taken to be a rather comlicated construction from the variables $\mathcal{Y}$ and l of the lattice theory and from some further new variables $\xi^\mu(P)$. These $\xi^\mu(P)$ for $P \in \mathbb{P}$ asign to every lattice point P a coordinate set $\xi^\mu(P)$. The $g_{\mu\nu}^{R.C.}$ is the metric tensor field in a Regge calculus piecewiese flat metric space time using the $\xi^\mu(P)$'s as the coordinates of the corner points of the simplexes with side lengths given by the variables $l(Y)$.

Also over the variables $\xi^\mu(p)$ we introduce an integration $\prod_{P=1}^{n} \int d\xi^\mu(P) \sqrt{g(\xi^\mu(P))}$. We must make an effort to keep all the time reparametrization invariance but that should be possible although we may have to take $g_{\mu\nu}^{R.C.}$ to depend also on $g_{\mu\nu}$ and thus we are in danger of getting corrections to (eq. 35).

3) After that we imagine that all other variables of integration than $g_{\mu\nu}$ are integrated out. That is to say we imagine that the summation over n and $\mathcal{Y}$ and the integration over the l's and the ξ^μ are performed at first for fixed values of the field $g_{\mu\nu}$. The result may

be written as

$$Z = \int \mathcal{D} g_{\mu\nu}\, e^{-S_{eff}(g_{\mu\nu})} \tag{36}$$

where the effective action $S_{eff}(g_{\mu\nu})$ is obtained from integrating out the remaining variables

$$e^{-S_{eff}(g_{\mu\nu})} = \sum_n \prod_{p=1}^{n} \int d\xi^\mu(p) \sqrt{g(\xi)}$$

$$\cdot \sum_{\mathcal{Y} \subseteq \mathbb{P}\times\mathbb{P}} \prod_{\gamma \in \mathcal{Y}} \int d\ell(\gamma) \cdot e^{-S_{fund}(\mathcal{Y}, n, \ell)} \tag{37}$$

$$\cdot\ \delta\left(g_{\mu\nu} - g_{\mu\nu}^{Rc}(\mathcal{Y}, n, \ell, g_{\mu\nu})\right) \Big/ \prod_{p \in \mathbb{P}} \int d\xi^\mu(p) \sqrt{g(\xi)}$$

4) It is not a priori obvious at all that the effective action for the theory written only in terms of $g_{\mu\nu}$ (i.e. $S_{eff}(g_{\mu\nu})$ is local in the sense of being of the form

$$S = \int \mathcal{L} \sqrt{g}\, d^4x \tag{38}$$

On the contrary it is almost certainly not. There is, however, some chance that it can be of a form like (eq.38) provided we allow the Lagrangian L(x) not only to depend on $g_{\mu\nu}$ and its derivatives in the point x^μ (over which one integrates) but also on $g_{\mu\nu}(y)$ for y in some finite radius neighborhood and even allow for some very weak dependence falling off exponentially as the distance of x^μ to y^μ increases. It is in fact our main point to show that such an extremely much weakened form of locality may hold, and that it can be "dynamically stable". By it being "dynamically stable" is meant that it remains true even if minor but finite changes are performed in the parameters (= coupling constants) if the original action $S_{fund.}$

If indeed one can show that there is - hopefully with nonzero probability - an $S_{fund.}$ so that the corresponding $S_{eff}(g_{\mu\nu})$ is of the form (eq.38) with a finite range L, then in the long wave length limit there must exist a local approximation to this L. Then we are very close to have produced Einsteinian theory of gravitation in this limit. In the infrared limit one expects the terms having coefficients of dimension mass to as high power as possible to dominate. Reparametrization invariance strongly restricts what terms there can be in a local Lagrangian density and after the cosmological term - a constant contribution to L - the next term in the order of dominance according to dimension is a term proportional to the scalar curvature kR. It has a coefficient k of dimension mass to the second power (or length to minus second power, remember $\hbar = c = 1$). But kR is precisely the Lagrangian for Einstein gravitation.

In order to claim that one really got Einstein gravity theory one should presumably also exclude the possibility of a phase with $g_{\mu\nu} = 0$ for example.

But it is crucial that we can at least argue for the weak form of locality that the Lagrangian density for S_{eff} should be of finite range only. It turns out that such a weak form of locality will follow from the cluster property for the field theory for which the partition funetion is the expression (eq. 37). By the cluster property is here meant that the expectation value for a product of fields

$$\langle \varphi(x_1^\mu)\varphi(x_2^\mu)\cdots\varphi(x_k^\mu)\rangle \tag{39}$$

goes to the product of the expectation values

$$\langle \varphi(x_1^\mu)\varphi(x_2^\mu)\cdots\varphi(x_\ell^\mu)\rangle\langle\varphi(x_{\ell+1}^\mu)\cdots\varphi(x_m^\mu)\rangle \cdots\langle\varphi(x_{h+1}^\mu)\cdots\varphi(x_{k-1}^\mu)\varphi(x_k^\mu)\rangle \tag{40}$$

the points $x_1, x_2, \ldots, x_k$ approach a configuration in which they form a set of widely separated clusters $x_1^\mu, x_2^\mu, \ldots, x_\ell^\mu$, $x_{\ell+1}^\mu, \ldots, x_m^\mu$, ... $x_{h+1}^\mu, \ldots, x_{k-1}^\mu, x_k^\mu$.

We suppose that we can apply a theorem by L. Gross[25)] that tells that under some conditions one can prove the cluster property for a rather general type of lattice field theory.

So we hope to show that it is indeed not so difficult to obtain the Einstein theory of gravitation from lego lattice type theory without finetuning.

Supposing this project succeeds we would obtain not only a gravitational theory but also - one might say - an explanation for why one have in flat space-time the symmetries that can be considered special cases of reparametrization invariance. These symmetries are translational invariance and Lorentz invariance.

In case the explanation of Lorentz invariance and translational invariance is that they are special cases - for flat space time - of a by definition exact reparametrization invariance these symmetries must be exact. Thus this is an alternative to the explanation of why there is Lorentz invariance to be given below using the renormalization group method. A difference in prediction between the two methods is that the renormalization group method only explains an approximate Lorentz invariance while the "definitional method" (I) leads to exact Lorentz invariance.

Kalb-Ramond field

A third example of obtaining a symmetry the "definitional way" is a model for which M. Lehto, M. Ninomiya and I[11)] made a mean field calculation, a lattice theory which could be called a non-gauge invariant Kalb-Ramond model. The Kalb-Ramond field theory may be described as a modification of electrodynamics. While electrodynamics is described by a four-potential $A_\mu(x)$ the Kalb-Ramond theory is described by an antisymmetric tensor-potential $A_{\mu\nu}(x) = -A_{\nu\mu}(x)$. Corresponding to the electromagnetic field $F_{\mu\nu}(x) = \partial_\mu A_\nu(x) - \partial_\nu A(x)$ of electrodynamics the Kalb-Ramond theory has a three-indices antisymmetric tensor $F_{\mu\nu\rho} = \partial_{[\mu}A_{\nu\rho]} = \partial_\mu A_{\nu\ \rho} + \partial_\nu A_{\rho\mu} + \partial_\rho A_{\mu\nu}$. Corresponding to gauge invariance for electrodynamics the Kalb-Ramond theory has invariance under $A_{\mu\nu}(x) \to A_{\mu\nu}(x) + \partial_\mu \Lambda_\nu(x) - \partial_\nu \Lambda_\mu(x)$ where $\Lambda_\mu(x)$ is a set of four arbitrary functions defined on space-time.

When we go over to a lattice model the Kalb-Ramond $A_{\mu\nu}(X)$ - field gets replaced, by a $U(\square)$ -field defined on the plaquettes $\square$ of the lattice and taking values in the abelian group $U(1) = \{z \in C \,|\, |z| = 1\}$, and the gauge functions $\Lambda_\mu(x)$ are replaced by a gauge function $\Lambda(\bullet\!\!-\!\!\bullet)$ defined on the links $\bullet\!\!\rightarrow$ of the lattice and also taking values in the group $U(1)$. The Kalb-Ramond gauge transformation on the lattice then is

$$U({}_1\overset{2}{\underset{4}{\square}}{}_3) \to U({}_4\overset{1}{\underset{3}{\square}}{}_2)\, \Lambda(\cdot\overset{1}{-}\cdot)\, \Lambda(\vdots{}^{2})\, \Lambda(\cdot\overset{3}{-}\cdot)\, \Lambda(\vdots{}^{4}) \tag{43}$$

Invariance under this gauge transformation is achieved if the action is only depending on cube-variables $U_{\boxdot}$, derived variables, defined as products of the fundamental plaguette variables $U(\square)$ for the six plaguettes surrounding the cube $\boxdot$ in question.

Like Förster, Ninomiya, I and Shenker considered a lattice electron dynamics in which the gauge invariance was broken (eq. 17) by a term $K \Sigma \mathrm{Re} U$ (•—•) Lehto, Ninomiya and I considered a lattice Kalb-Ramond model in which the action also has a term breaking gauge invariance. That is to say we considered an a action which written with a notation suggesting the analogy to (eq. 17) is

$$S = \beta \sum_{\boxdot} \mathrm{Re}\, U_{\boxdot} + K \sum_{\square} \mathrm{Re}\, U(\square) \qquad (44)$$

A gauge invariant Kalb-Ramond theory has a massless particle (analogous to the photon but with only one physical component in 3+1 dimensions). By use of the mean field approximation technique we found that there exists a phase i.e. a range for the parameters (β, K) such that there is a zero mass particle. This parameter range includes values with $K \neq 0$ so that there is no gauge invariance a priori of the type (eq. 43). In spite of that we can introduce a new set of variables such that there is a symmetry of this type for trivial reasons and we even show that it in a phase leads to a zero mass particle.

The mean field technique is essentially one in which the fields such as $U_h(\square)$ are approximated by their average (their mean) in the thermodynamic (or vacuum) state. We shall not go in detail with it, but only remark that it gets more reliable the higher the dimension. Therefore the existence of the zero mass particle phase is only reliable for high enough dimension of space time. In fact Robert B. Pearson found that the Kalb-Ramond model is in a confined phase in four or less than four dimensions. A confined phase means one analogous to the hoped for confinement in Q.C.D. In the confined phase there is no massless particle and thus our mean field calculation was in fact not true in four or less dimensions.

The renormalization group method II

Let us now consider the next method of obtaining symmetries. That is the one consisting in that effective scale depending coupling constants may send towards more symmetric values in the infrared. The philosophy of this method is that the "fundamental" scale of energy is much higher than the today experimentally accessible one. If the "fundamental" energy scale is the Planck-energy scale this will indeed be the case since even LEP-energies are small compared to the Planck energy.

An example of a derivation of an approximate symmetry by this renormalization group method is that of Lorentz invariance by S. Chadha and myself using non-Lorentz invariant quantum electrodynamics. We set up the most general quantum electrodynamics-like theory obeying a series of assumptions among which we do not include the principle of relativity (Lorentz invariance). What we do assume is: Quantum fields are A_μ boson field and Ψ Weyl fermion fields with 8 real components (or 4 complex ones).

1) translational invariance
2) gauge invariance under

$$A_\mu(x) \rightarrow A_\mu(x) + \partial_\mu \Lambda(x) \tag{45}$$

3) chiral invariance in the sense that we consider Weyl particles each having its own type of "flavour". (This assumption we may avoid by a more detailed model[6].) That means we consider a "massless" fermion model.

4) renormalizability in the sense that we include only terms in the Lagrangian density L having a coefficient of dimension mass to a non-negative power [this is meant counting the dimension of A_μ as that of mass and ψ as mass to the $\frac{3}{2}$- power].

5) we assume quantum mechanics or rather that we have quantum field theory.

Under these assumptions the most general form of the Lagrangian density is

$$\begin{aligned}\mathcal{L} = & -\frac{1}{4}\eta^{\mu\nu\rho\sigma} F_{\mu\nu}(x) F_{\rho\sigma}(x) \\ & -\frac{1}{2}\psi\gamma^0\gamma^a\left[e^\mu_{+a}\frac{1+iq\gamma_5}{2} + e^\mu_{-a}\frac{1-iq\gamma_5}{2}\right]\cdot \\ & \cdot\frac{1}{i}D_\mu\psi + \eta^{\mu\nu\lambda}_{woo} F_{\mu\nu} A_\lambda\end{aligned} \tag{46}$$

Here one has not allowed terms linear in A_μ. We have used here a notation in which ψ is a Majorana field $\psi^+ = \psi$ but with an extra charge degree of freedom so that ψ has 8 rather than four components and so that a charge matrix q can be constructed, $q = \begin{pmatrix} 0 & 1\!\!1 \\ -1\!\!1 & 0 \end{pmatrix}$ The two different η's, $\eta^{\mu\nu\rho\sigma}$ and $\eta^{\mu\nu\lambda}$ and the vierbeins e^{μ}_{+a} and e^{μ}_{-a} are sets of coupling constants (or parameters) as are the electric charge and the mass in usual electrodynamics. Although this Lagrangian (eq. 46) may formally look Lorentz invariant because of all the indices being properly contracted it is indeed not so because of attempted Lorentz transformation shall be a transformation of the fields ψ and A_μ only but not of the "coupling constants" $\eta^{\mu\nu\lambda}$, $\eta^{\mu\nu\rho\sigma}$, e^{μ}_{+a}, and e^{μ}_{-a}. The gauge covariant derivative is denoted

$$D_\mu = \partial_\mu - ieqA_\mu \tag{47}$$

where e is the electric charge. [One has a choice of notation either to have e as a coupling constant and require some normalization of $\eta^{\mu\nu\rho\sigma}$, or to have no e.] The coupling constants $\eta^{\mu\nu\rho\sigma}$ can be taken to obey the symmetry properties

$$\begin{aligned} \eta^{\mu\nu\rho\sigma} &= -\eta^{\nu\mu\rho\sigma} = -\eta^{\mu\nu\sigma\rho} \\ &= \eta^{\nu\mu\sigma\rho} = \eta^{\rho\sigma\mu\nu} \end{aligned} \tag{48}$$

under permutations of the indices.

The "coupling constants" $\eta^{\mu\nu\lambda}_{Woo}$ must be antisymmetric under permutation of the three indices

$$\begin{aligned} \eta^{\mu\nu\lambda}_{Woo} &= -\eta^{\nu\mu\lambda}_{Woo} = -\eta^{\mu\lambda\nu}_{Woo} = -\eta^{\lambda\nu\mu}_{Woo} \\ &= \eta^{\nu\lambda\mu}_{Woo} = \eta^{\lambda\mu\nu}_{Woo} \end{aligned} \tag{49}$$

in order that the part of the action arising from $\eta^{\mu\nu\lambda}_{Woo} F_{\mu\nu} A_\lambda$ be gauge invariant. Strictly speaking there is also the possibility of inserting a "coupling constant" $\xi^{\mu}_{\pm}$ into the covariant derivative, replacing it

$$D_\mu \rightarrow D_\mu + i\,\xi_{\pm\mu} \tag{50}$$

These $\xi_{\pm\mu}$ are different for the two handednesses, it being $\xi_{+\mu}$ for the term with the $\frac{1+iq\gamma_5}{2}$ projection matrix and $\xi_{-\mu}$ for the $\frac{1-iq\gamma_5}{2}$ projector. However, we can transform away such $\xi_{\pm\mu}$'s by redefining the field $\psi(x)$ by the replacement

$$\psi(x) \rightarrow e^{-i\left(\xi_{+\mu}\frac{1+iq\gamma_5}{2}+\xi_{-\mu}\frac{1-iq\gamma_5}{2}\right)\cdot x^\mu} \cdot \psi(x) \tag{51}$$

and thus we have for simplicity left out $\xi_{+\mu}$ and $\xi_{-\mu}$.

Such $\xi_{\pm\mu}$'s and the coupling constant $\eta^{\mu\nu\lambda}_{Woo}$ have dimension of mass while the other "coupling constants" $e^{\mu}_{\pm a}$ and $\eta^{\mu\nu\rho\sigma}$ are dimensionless. Supposing that the "coupling constants" (= parameters of theory) are of order of magnitude unity when the fundamental scale units (presumable Planck-units) are used we will at experimental scales find $\xi_{\pm\mu}$ and $\eta^{\mu\nu\lambda}_{Woo}$ to be huge while the dimensionless $\eta^{\mu\nu\rho\sigma}$ and $e^{\mu}_{\pm a}$ are of the order unity seen at all scales at first (renormalization group effects may modify this). The $\xi_{\pm\mu}$ could be transformed away so that we do not have to bother about them. If, however, a term

$$\eta^{\mu\nu\lambda}_{Woo} F_{\mu\nu} A_\lambda \tag{52}$$

is present in the Lagrange-density it is expected to dominate at low energies compared to the fundamental scale. If space time dimension had been 3 = 2+1 this term (eq.52) would be Lorentz invariant since it would be just the topological mass term discussed by R. Jackiw[26)] at this school. In four space-time dimensions, however, this term (eq.52) breaks Lorentz invariance. Since if present it would presumably be dominant one needs an argument why such a non-covariant topological mass term is not allowed. Of course we should not argue it away by reference to Lorentz invariance since that would be using what we want to show. One might argue that requiring both symmetry under parity P and under time reversal T would forbid such a term. However, I do not like to assume P and T either since they are not good symmetries for weak interactions (with $K^0_L \to \pi\pi$ included) and I would like to predict at the end that also weak interactions should have at least approximate Lorentz invariance. So I prefer that we argue that a non-covariant topological mass term (or Woo term as I called it in ref. 5 because C.H. Woo from Maryland pointed it out to me) is not gauge invariant if there are magnetic monopoles in the model. That the gauge invariance may break down if monopoles are included is connected with that the Bianchi identities (i.e. the Maxwell equations following from the mere existence of the A_μ)

$$\varepsilon^{\mu\nu\rho\sigma} \partial_\nu F_{\rho\sigma} = 0 \tag{53}$$

have to be used to show the gauge invariance of the Woo-term action (the topological mass term action)

$$S_{Woo} = \int \eta^{\mu\nu\lambda}_{Woo} F_{\mu\nu} A_\lambda \, d^4x \tag{54}$$

In fact the gauge transformation of this term is

$$\begin{aligned} \eta^{\mu\nu\lambda}_{Woo} F_{\mu\nu} A_\lambda &\rightarrow \eta^{\mu\nu\lambda}_{Woo} F_{\mu\nu} (A_\lambda + \partial_\lambda \Lambda) \\ &= \eta^{\mu\nu\lambda}_{Woo} F_{\mu\nu} A_\lambda - \Lambda \eta^{\mu\nu\lambda}_{Woo} \partial_\lambda F_{\mu\nu} \\ &\quad + \partial_\lambda [\eta^{\mu\nu\lambda}_{Woo} F_{\mu\nu} \Lambda] \end{aligned} \tag{55}$$

for the gauge transformation (eq. 45). The term $\partial_\lambda[\eta^{\mu\nu\lambda}_{Woo} F_{\mu\nu} \Lambda]$ is a total divergence and thus contributes to the action (eq.54) as only a boundary (or surface) term coming from the infinitely far away boundary of integration over d 4. We neglect such boundary terms and thus the gauge invariance of S_{Woo} depends on whether the term $\Lambda \eta^{\mu\nu\lambda}_{Woo} \partial_\lambda F_{\mu\nu}$ in (eq. 55) is zero (or at least a total divergence). Because of the antisymmetry condition (eq. 49) for $\eta^{\mu\nu\lambda}_{Woo}$ one may write

$$\eta^{\mu\nu\lambda}_{Woo} = \eta_{Woo\,\chi} \, \varepsilon^{\chi\mu\nu\lambda} \tag{56}$$

in terms of four coupling constants called $\eta_{Woo\,\chi}$. But then

$$\eta^{\mu\nu\lambda}_{Woo} \partial_\lambda F_{\mu\nu} = \eta_{Woo\,\chi} \, \varepsilon^{\chi\mu\nu\lambda} \partial_\lambda F_{\mu\nu} = 0 \tag{57}$$

because of the Bianchi densities (eq. 53). Thus the non-covariant topological mass term (eq. 52) gives a gauge invariant construction S_{Woo} to the action provided the Bianchi identities (eq. 53) hold.

However, if there exist monopoles in the model the Bianchi identities (eq. 53) are replaced by

$$\varepsilon^{\mu\nu\rho\sigma} \quad \partial_\nu F_{\rho\sigma} = J^\mu_{mon.} \tag{58}$$

where $J^\mu_{mon.}$ is the monopole magnetic charge density. So the gauge invariance of (eq. 54) is spoiled by the existence of magnetic monopoles and one thus seems to be forced to take

$$\eta^{\mu\nu\lambda}_{Woo} = 0 \tag{59}$$

For instance if there is fundamentally a lattice theory - as suggested above in arguing for gauge invariance the "definitional way" - there is effectively also monopoles in the model because the lattice Bianchi identify

$$\prod_{\substack{\square \text{ in a} \\ \text{cube}}} U_\square = 1 \tag{60}$$

for an abelian gauge group U(1) only implies

$$\sum_{\substack{\square \text{ in a} \\ \text{cube}}} F_\square = 0 \pmod{2\pi} \tag{61}$$

if we define the magnetic flux $F_\square$ through the plaquette $\square$ by

$$U_\square = e^{iF_\square} \tag{62}$$

That is to say one obtains on the lattice only the Bianchi identity corresponding to the continuum theory condition

$$\frac{1}{2} \int_\Sigma d\sigma^{\mu\nu} F_{\mu\nu} = 0 \pmod{2\pi} \tag{63}$$

Here Σ is a closed two-dimensional surface and $d\sigma^{\mu\nu}$ is the infinitesimal surface element along it. But the condition (eq. 63) will precisely allow magnetic monopoles with a quantized magnetic charge, and thus the lattice in fact has monopoles on it. Thus if there is a fundamental lattice there are effectively monopoles and the non-covariant topological mass term looses its gauge invariance. Let us suppose that we have in this way talked away this term and obtained (eq. 59). This kind of argument used to get rid of $\eta^{\mu\nu\lambda}_{Woo}F_{\mu\nu}A_\lambda$ is really rather reminiscent of the third method of obtaining symmetries, the one consisting in deriving from other symmetries and renormalizability.

But now we are in the position to use the method II - the renormalization group method - to obtain Lorentz invariance in the low energy limit. In fact now the only remaining parameters (= coupling constants) in the Lagrangian (eq. 46), $e^\mu_{\pm a}$ and $\eta^{\mu\nu\rho\sigma}$, are dimensionless.

So the model - massless non-covariant electrodynamics - is now formally (or classically) invariant under scale transformations. If you have a classical solution for the time development of the fields $A_\mu(x)$ and $\psi(x)$ you can find another one in which the field configuration is magnified corresponding to a homoteti of space-time. In fact the magnification symmetry may be written

$$\begin{aligned} A_\mu(x^\nu) &\rightarrow A_\mu(h\cdot x^\nu)\cdot h \\ \psi(x^\nu) &\rightarrow \psi(h\cdot x^\nu)\cdot h^{3/2} \end{aligned} \tag{64}$$

where the scaling factor h can be a positive number.

Like Q.C.D. with massless quarks (that is also a scale invariant theory formally) the non-covariant Q.E.D. without $\eta^{\mu\nu\lambda}_{Woo}F_{\mu\nu}A_\lambda$ -term is quantum mechanically not truly scale invariant. In fact it is well-known that an effective or renormalized coupling constant is defined in Q.C.D. in a scale dependent manner $\alpha_s(\mu)$ and turns out (when quantum corrections are taken into account) indeed to depend upon renormalization point μ and thus indeed to be scale dependent. In quite an analogous way the renormalized "coupling constants" $\eta^{\mu\nu\rho\sigma}_{ren.}(\mu)$ and $e^\mu_{ren.\pm a}(\mu)$ of the non-covariant electrodynamics (also without the $\eta^{\mu\nu\lambda}_{Woo}F_{\mu\nu}A_\lambda$ -term) are indeed scale dependent.

In fact S. Chadha and I[6a)] computed (NBI-HE-82-42) their scale dependence to lowest order perturbation theory and found a dependence indeed to be there.

The renormalized vierbein for say the $\frac{1+iq\gamma_5}{2}$ - projected part $e^{\mu}_{ren.+a}(\mu)$ taken at the renormalization point μ might be defined in terms of the inverse propagator $\bar{G}_{+}^{-1}(p)$ for this way projected electron field. By the propagator $\bar{G}_{+}(p)$ we mean the full $\frac{1+iq\gamma_5}{2}$-projected propagator, i.e. it includes quantum corrections. It is considered a function of the four momentum p^{μ} and thus is strictly speaking the Fourier-transform of the propagator in x-representation.

We define an average for any function $\Theta(p)$ which, like $\bar{G}^{-1}(p)$ say, depends on a four momentum p_{μ} and can be extended into Euklideanized p_{μ} -value (i.e. with p^0 made imaginary) by

$$\langle \Theta(p) \rangle_{\mu} = \frac{\int (d^4p)_{Eucl.}\ \delta(p^2_{Eucl.} - \mu^2)\ \Theta(p)}{\int (d^4p)_{Eucl.}\ \delta(p^2_{Eucl.} - \mu^2)} \tag{65}$$

I.e. $\langle \Theta(p) \rangle_{\mu}$ is an average of $\Theta(p)$ over momentum values of magnitude equal to (the renormalization point) μ. We can then define the renormalized or running vierbein by

$$\left\langle \frac{\partial \bar{G}_{+}^{-1}(p)}{\partial p_{\nu}} \right\rangle = \gamma^{a}\, e^{\nu}_{ren.+a}(\mu) \tag{66}$$

and analogously for the $\frac{1-iq\gamma_5}{2}$ -projection

$$\left\langle \frac{\partial \bar{G}_{-}^{-1}(p)}{\partial p_{\nu}} \right\rangle = \gamma^{a}\, e^{\nu}_{ren-a}(\mu) \tag{67}$$

Similarly if $\Delta_{\mu\nu}(p)$ denotes the full propagator for the photon i.e. for the field $A_{\mu}(x)$ we may define

$$\left\langle \frac{\partial^2 \left(\Delta_{..}(p) \right)^{-1\ \mu\nu}}{\partial p_{\rho}\, dp_{\sigma}} \right\rangle = \eta^{\rho\mu\sigma\nu}_{ren}(\mu) \tag{68}$$

Now for any "coupling constant" g say one defines

$$\beta_g = \frac{\partial g(\mu)}{\partial \log \mu} \tag{69}$$

so that for example

$$\beta_\eta^{\mu\nu\rho\sigma}\,(\eta^{\alpha\beta\gamma\delta},\,e^{\alpha}_{+a},\,e^{\tau}_{-b}) \overset{\text{def.}}{=} \frac{d\eta^{\mu\nu\rho\sigma}_{ren}(\mu)}{d\log\mu} \tag{70}$$

We (Chadha and I)[6a] calculated to lowest order perturbation (in the strength of interaction of the photon with the electron) and to lowest order in the breaking of Lorentz invariance the β-functions and found

$$\beta_\eta^{\mu\nu\rho\sigma} = -\frac{\alpha}{6\pi}\left(g^{\mu\rho}_{ren+}\,g^{\nu\sigma}_{ren+} - g^{\mu\sigma}_{ren+}\,g^{\nu\rho}_{ren+}\right) - \frac{\alpha}{6\pi}\left(g^{\mu\rho}_{ren-}\,g^{\nu\sigma}_{ren-} - g^{\mu\sigma}_{ren-}\,g^{\nu\rho}_{ren-}\right) \tag{71}$$

for the $\beta_\eta^{\mu\nu\rho\sigma}$, where we have defined the metric "tensors" correspon-ding to the vierbeins $e^{\mu}_{\pm a}$ by say

$$g^{\mu\nu}_{ren+}(\mu) = e^{\mu}_{ren+f}(\mu)\,g^{ff'}\,e^{\nu}_{ren+f'}(\mu) \tag{72}$$

and

$$g^{\mu\nu}_{ren-}(\mu) = e^{\mu}_{ren-f}(\mu)\,g^{ff'}\,e^{\nu}_{ren-f'}(\mu) \tag{73}$$

Here $g^{ff'}$ is just the pure number metric matrix

$$g^{ff'} = \begin{pmatrix} -1 & & & \\ & 1 & & \\ & & 1 & \\ & & & 1 \end{pmatrix} \tag{74}$$

In (eq. 71) the finestructure constant is denoted by

$$\alpha = \frac{e^2}{4\pi} \tag{75}$$

but actually there is no need for having a coupling constant e (and thus an α) because we can choose to normalize the $A_\mu(x)$ -field in such a way that the covariant derivative (eq. 47) is written instead

$$D_\mu = \partial_\mu - iqA_\mu \tag{76}$$

That is to say we have called the previous eA_μ by just A_μ. This then requires $\eta^{\mu\nu\rho\sigma}$ to be scaled by a factor $\frac{1}{e^2}$ but that is all right. Actually a notation with an explicit e is ackward because it would suggest that there were more "coupling constants" than is actually the case. Normalizing the A_μ -field this way means that we put $e = 1$ and thus $\alpha = \frac{1}{4\pi}$ in (eq. 71).

The for our main purpose of explaining Lorentz invariance most important feature of the calculational result (eq. 71) is the sign of the β-function relative to the sign of what $\eta^{\mu\nu\rho\sigma}$ would be in the Lorentz invariant case

$$\eta^{\mu\nu\rho\sigma}\Big|_{\substack{\text{in Lorentz}\\ \text{invariant case}}} = \frac{1}{2}\left[g^{\mu\rho}_{(\gamma)}\, g^{\nu\sigma}_{(\gamma)} \right. \\ \left. g^{\mu\sigma}_{(\gamma)}\, g^{\nu\rho}_{(\gamma)} \right] \tag{77}$$

This was for the covariant derivative having an e, i.e. (eq. 47). In case we use the notation with the simpler covariant derivative (eq. 76) we rather have

$$\eta^{\mu\nu\rho\sigma}\Big|_{\text{in Lorentz invariant case}} = \frac{1}{2e^2}\left[g^{\mu\rho}_{(\gamma)}\, g^{\nu\sigma}_{(\gamma)} - g^{\mu\sigma}_{(\gamma)}\, g^{\nu\rho}_{(\gamma)} \right] \tag{78}$$

We have put an index (γ) on the metric tensor $g^{\mu\nu}_{(\gamma)}$ to suggest the possibility that a free photon might behave Lorentz invariant by itself, namely if $\eta^{\mu\nu\rho\sigma}$ is of the form (eq. 77 or eq. 78), without the full model being Lorentz invariant. In fact the metric $g^{\mu\nu}_{(\gamma)}$ could be a special one for the photon only while the metrics $g^{\mu\nu}_{+}$ and $g^{\mu\nu}_{-}$ for the two Weyl fields (γ^5 -projections) for the electron could be different.

We see that the terms on the right hand side of equation (eq.71) are just some negative number multiplied by expressions of the form (eq. 77) or (eq. 78) but with the electron metrics $g^{\mu\nu}_{\pm}$ replacing the metric $g^{\mu\nu}_{(\gamma)}$ in (eq. 77) and (eq. 78). Now the philosophy was that at high energy i.e. for large values of μ, say μ equal to the fundamental scale the "coupling constants" are given "at random" and therefore most likely not in conformity with Lorentz invariance. So we imagine $\eta^{\mu\nu\rho\sigma}_{\text{ren.}}(\mu_{\text{fund.}})$ for a large value of μ to be given and ask how will $\eta^{\mu\nu\rho\sigma}_{\text{ren.}}(\mu)$ develop as we take the renormalization point μ smaller and smaller? We see from (eq. 71) that

$$\begin{aligned} \eta^{\mu\nu\rho\sigma}_{\text{ren}}(\mu) \simeq{}& \eta^{\mu\nu\rho\sigma}_{\text{ren}}(\mu_{\text{fund.}}) \\ &+ \log\frac{\mu_{\text{fund.}}}{\mu}\cdot\frac{\alpha}{6\pi}\Big[g^{\mu\rho}_{\text{ren}+}(\mu)\, g^{\nu\sigma}_{\text{ren}+}(\mu) \\ &- g^{\mu\sigma}_{\text{ren}+}(\mu)\, g^{\nu\rho}_{\text{ren}+}(\mu) + g^{\mu\rho}_{\text{ren}-}(\mu)\, g^{\nu\sigma}_{\text{ren}-}(\mu) \\ &- g^{\mu\sigma}_{\text{ren}-}(\mu)\, g^{\nu\rho}_{\text{ren}-}(\mu) \Big] \end{aligned} \tag{79}$$

and thus that $\eta^{\mu\nu\rho\sigma}_{\text{ren.}}(\mu)$ (at the lower μ-value) contains two extra terms which are a positive coefficient $\log\frac{\mu_{\text{fund.}}}{\mu}\cdot\frac{\alpha}{3\pi}$ times the covariant forms of $\eta^{\mu\nu\rho\sigma}$ corresponding to the metrics $g^{\mu\nu}_{\text{ren.}\pm}(\mu)$. It is not important which argument μ is used in these metrices defined by (eq. 72) and (eq. 73) to the accuracy of this crude relation (eq. 79), it should be something between $\mu_{\text{fund.}}$ and μ. It follows from (eq. 79) that the lower we go in energy i.e. the smaller μ, the more $\eta^{\mu\nu\rho\sigma}_{\text{ren.}}(\mu)$ comes to be influenced by the terms proportional to $\log \frac{\mu_{\text{fund.}}}{\mu}$ and being of form covariant. Provided the two metrics $g^{\mu\nu}_{\text{ren.}\pm}(\mu)$ are close to each other dominance of these terms would mean that $\eta^{\mu\nu\rho\sigma}_{\text{ren.}}(\mu)$ is Lorentz invariant even with an average metric between the two electron metrics so that the theory would be covariant then.

We also calculated how the vierbeins $e^{\mu}_{\text{ren.}\pm a}(\mu)$ or rather the associated metric tensors $g^{\mu\nu}_{\text{ren.}\pm a}(\mu)$ behave as function of the renormalization point μ and found

$$\beta_g{}^{\mu\nu} = \frac{\partial g^{\mu\nu}_{\text{ren}-}(\mu)}{\partial \log \mu}$$

$$= -\frac{\alpha}{3\pi}\left[4\,\delta\eta_{\text{ren}-}{}^{\cdot\mu}{}_{\cdot}{}^{\nu}(\mu) - g^{\mu\nu}_{\text{ren}-}(\mu)\,\delta\eta_{\text{ren}-}{}^{\cdot x}{}_{\cdot x}(\mu)\right] \tag{80}$$

Here we have defined $\delta\eta^{\mu\nu\rho\sigma}_{\text{ren.}}(\mu)$ by

$$\eta^{\mu\nu\rho\sigma}_{\text{ren}}(\mu) = \frac{1}{2}\left[g^{\mu\rho}_{\text{ren}-}(\mu)\,g^{\nu\sigma}_{\text{ren}-}(\mu) - g^{\mu\sigma}_{\text{ren}-}(\mu)\,g^{\nu\rho}_{\text{ren}-}(\mu)\right] + \delta\eta^{\mu\nu\rho\sigma}_{\text{ren}-}(\mu) \tag{81}$$

so that it measures how much $\eta^{\mu\nu\rho\sigma}_{\text{ren.}}(\mu)$ deviates from what it should have been if it should be covariant using as the metric tensor the one $g^{\mu\nu\rho\sigma}_{\text{ren.}}(\mu)$ for the $\frac{1-q\gamma_5}{2}$ -projected electron field.

The significance of the β-function (eq. 80) is that as we go towards smaller values of μ i.e. towards the infrared the effective (or renormalized) metric $g^{\mu\nu}_{\text{ren}-}(\mu)$ approaches a metric that can be derived from $\eta^{\mu\nu\rho\sigma}_{\text{ren.}}(\mu)$ and which we may call $g^{\mu\nu}_{\text{ren.}(\gamma)}(\mu)$. In fact we define $g^{\mu\nu}_{\text{ren.}(\gamma)}(\mu)$ by

$$\eta^{\mu\nu\rho\sigma}_{\text{ren}}(\mu) = \frac{1}{2}\left[g^{\mu\rho}_{\text{ren}(\gamma)}(\mu)\,g^{\nu\sigma}_{\text{ren}(\gamma)}(\mu) - g^{\mu\sigma}_{\text{ren}(\gamma)}(\mu)\,g^{\nu\rho}_{\text{ren}(\gamma)}(\mu)\right] + \delta\eta^{\mu\nu\rho\sigma}_{\text{ren}(\gamma)}(\mu) \tag{82}$$

together with the orthogonality condition

$$\delta\eta^{\mu\nu\rho\sigma}_{ren(\gamma)}(\mu)\, g_{ren(\gamma)\mu\rho}(\mu) = 0 \tag{83}$$

where $g_{ren(\gamma)\mu\rho}(\mu)$ is by definition the inverse matrix of $g^{\mu\nu}_{ren(\gamma)}(\mu)$. The idea of this orthogonality condition is in fact that one may derive it approximately (in the limit in which Lorentz invariance is only weakly broken) from the requirement that $g^{\mu\nu}_{ren(\gamma)}(\mu)$ be constructed so that for the given $\eta^{\mu\nu\rho\sigma}_{ren}(\mu)$ the rest $\delta\eta^{\mu\nu\rho\sigma}_{ren(\gamma)}(\mu)$ be as small as possible. In that sense $g^{\mu\nu}_{ren(\gamma)}(\mu)$ is defined to be that metric tensor which can to the best approximation reproduce the true $\eta^{\mu\nu\rho\sigma}_{ren}(\mu)$ when inserted in the expression for a covariant $\eta^{\mu\nu\rho\sigma}$ i.e. in (eq. 77), say. To the approximation that Lorentz invariance is only weakly broken so that we only have to include up to first order terms in say $g^{\mu\nu}_{ren-}(\mu) - g^{\mu\nu}_{ren(\gamma)}(\mu)$ one can calculate from (eq. 80) that

$$\beta_{g-}^{\mu\nu} = \frac{4\alpha(\mu)}{3\pi} \cdot \left[g^{\mu\nu}_{ren-}(\mu) - g^{\mu\nu}_{ren(\gamma)}(\mu) \right] \tag{84}$$

It is easily seen that the sign in this equation (eq. 84) is such that when we decrease the renormalization point or energy scale the metric $g^{\mu\nu}_{ren-}(\mu)$ tends towards the "metric for the photon" $g^{\mu\nu}_{ren(\gamma)}(\mu)$. Thus we see that the various metrices $g^{\mu\nu}_{ren-}(\mu)$, $g^{\mu\nu}_{ren(\gamma)}(\mu)$ and analogously also $g^{\mu\nu}_{ren+}(\mu)$ tend towards each other as μ goes towards the infrared.

So although Lorentz invariance was not a good symmetry for high energies, i.e. for large μ, it would be so approximately at low energy.

When we here have talked about Lorentz invariance, rotational invariance was included in it. Especially rotational invariance but also Lorentz invariance has been checked experimentally with high accuracy and it is actually not so easy to obtain sufficient accuracy by this method of deriving rotational invariance especially if we assume an appreciable breaking at the Planck scale say. In fact a Michelson-Morley type experiment with laser light has for example given a bound of the order of 10^{-15} for directional anisotropy of space. (A. Brillet and J.L. Hall, Improved Laser Test of the Isotropy of Space. Phys. Rev. Lett. 42, number 9, p. 549 (1979)). The ratio of the fundamental to the experimental may be taken to be

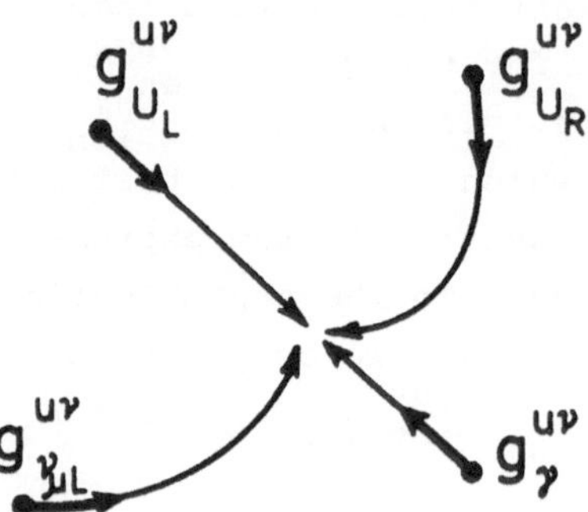

Fig. 6. Illustration of the direction of motion due to renormalization group effects of the renormalized metrics as a "time" $t = -\log \mu$ passes. They will end up in the infrared together.

$E_{planck}/10$ GeV say, it is 10^{18}. So as μ decreases by 18 orders of magnitude we would need the Lorentz invariance breaking to decrease by 15 orders of magnitude. That would require β-functions of the order of 1 (in fact of the order $\frac{15}{18}$ for the Lorentz invariance improving part), but that would require coupling constants to correspond to rather strong interactions i.e. $\alpha(\mu)$ the fine structure constant should be of order unity. If we only had quantum electrodynamics this would not work, but even if we allow for non-covariant Yang-Mills theory and use the calculation of the β-function for such a theory by M. Ninomiya and myself[12] it still is not very easy to get accurate Lorentz invariance (including rotational invariance).

Ninomiya and I[12] considered a non-covariant non-abelian gauge theory making assumptions essentially the same as the ones of the non-covariant electrodynamics by S. Chadha and myself[6a]. But Ninomiya and I required gauge invariance under the non-abelian gauge transformation

$$g\mathbb{A}_\mu(x) \longrightarrow \Lambda(x)\, g\mathbb{A}_\mu(x)\, \Lambda^{-1}(x) + \Lambda^{-1}(x)\, \partial_\mu \Lambda(x) \qquad (85)$$

instead of the abelian one (eq. 45). Here $\mathbb{A}_\mu(x)$ means a matrix in a representation of the gauge group,

$$\mathbb{A}_\mu(x) = \lambda_a A_\mu^a(x) \qquad (86)$$

where the λ_a's make up a basis for the Lie algebra in this representation and $A_\mu^a(x)$ are the Yang-Mills potentials. The gauge transformation function $\Lambda(x)$ is an arbitrary function with matrix values representing elements of the group G and defined on the Minkowski space.

The result again turned out to be such that Lorentz invariance is more accurate the lower the energy scale. That is to say the renormalized $\eta_{r\,ren}^{\mu\nu\rho\sigma}(\mu)$ approaches a covariant form towards the infrared. This is derived from the non-abelian non-covariant β-function

$$\beta_\eta^{\mu\nu\rho\sigma} = \frac{d\eta_{ren}^{\mu\nu\rho\sigma}(\mu)}{d\log\mu} =$$

$$= \frac{C_2(G)}{4\pi^2} \cdot 4\pi\alpha \left[\frac{11}{6} \eta^{\mu\nu\rho\sigma}_{ren}(\mu) + \frac{7}{6} \delta\eta^{\mu\nu\rho\sigma}_{ren}(\mu) \right.$$

$$\left. - \frac{7}{3} \delta\eta^{\mu\gamma\rho}_{ren\ \gamma}\, g^{\nu\sigma} + \frac{1}{2} \delta\eta^{\alpha\beta}_{ren\ \alpha\beta}\, g^{\mu\rho} g^{\nu\sigma} \right]$$

$$- \sum \frac{T(R)}{12\pi^2} \cdot 4\pi\alpha \left[g^{\mu\rho}_{ren\,R}(\mu)\, g^{\nu\sigma}_{ren\,R}(\mu) \right.$$

$$\left. - g^{\mu\sigma}_{ren\,R}(\mu)\, g^{\nu\rho}_{ren\,R}(\mu) \right] \qquad (87\ \text{contd.})$$

Here $g_{\mu\nu}$ denotes an arbitrarily chosen metric not too far from the other metrics ,especially not so far from the photon metric. Defining $\delta\eta^{\mu\nu\rho\sigma}_{ren}$ by

$$\eta^{\mu\nu\rho\sigma}_{ren} = \frac{1}{2}\left(g^{\mu\rho} g^{\nu\sigma} - g^{\mu\sigma} g^{\nu\rho}\right) + \delta\eta^{\mu\nu\rho\sigma}_{ren} \qquad (88)$$

the dependence on $g^{\mu\nu}$ in formula (eq. 87) should actually be absent to the accuracy to which we have calculated i.e. only including first order terms in deviate from covariance with $g^{\mu\nu}$ as a common metric tensor.

The quadratic Casimir operator for the gauge group G was denoted by $C_2(G)$ and defined by

$$C_2(G)\, \delta^{ab} = f^{acd} f^{bcd}$$

where f^{abc} are the structure constants for the Lie algebra of G.

The symbol T(R) is a quantity characteristic to the representation R of a (Weyl) fermion and is defined by

$$T(R)\, \delta^{ab} = \mathrm{Tr}\,(\sigma^a \sigma^b)$$

where σ^a denotes under the representation R the representative of the a'th Lie algebra basis vector. For example G = SU(N) and R = fundamental N-dimensional representation we have

$$C_2(SU(N)) = N$$

$$T(R) = \frac{1}{2}$$

The metrics for the various types of (Weyl) fermions in the last term in (eq. 87) are normalized to unit determinant

$$\det\left(g^{\mu\nu}_{ren}(\mu) \right) = -1 \tag{89}$$

The statement (symmetrized according to (eq. 48)) means that the formula (eq. 87) is only true after antisymmetrization in μ and ν and in ρ and σ and symmetrization in the pair (μ,ν) and the pair (ρ,σ) of the right-hand side.

Formula (eq. 87) is made with $\alpha = \frac{g^2}{4\pi}$ just being considered a constant, even independent of the renormalization point μ. We may, however, use a notation so that $\eta^{\mu\nu\rho\sigma}_{ren}(\mu)$ is normalized by say

$$\det\left(g^{\mu\nu}_{ren(\gamma)} \right) = 1 \tag{90}$$

and an overall scaling of $\eta^{\mu\nu\rho\sigma}_{ren}(\mu)$ is absorbed into a coupling constant $g_{ren}(\mu)$ or $\alpha_{ren}(\mu) = \frac{g_{ren}(\mu)^2}{4\pi}$. Then a variation with scale μ of $\eta^{\mu\nu\rho\sigma}_{ren}(\mu)$, that is of the form of increasing or decreasing all the (a priori 4^4, but in reality) 20 or 21 components of $\eta^{\mu\nu\rho\sigma}_{ren}(\mu)$ by the same factor, is reformulated as a variation of $\alpha_{ren}(\mu)$ instead. The first term in the bracket of the Yang-Mills part (first term) of the β-function (eq. 87) $\frac{11}{6}\,\eta^{\mu\nu\rho\sigma}_{ren}(\mu)$ represents precisely such an overall scaling of $\eta^{\mu\nu\rho\sigma}_{ren}(\mu)$ as μ varies. Going to the notation with absorbing such dependence into the variation of $\alpha(\mu)$ therefore this term disappears, and furher the part arising from the fermion interaction is replaced by one proportional to the deviation of the metrics. In fact one obtains in this normalized notation

$$\begin{aligned}
\beta_\eta^{\mu\nu\rho\sigma} &\equiv \frac{d\eta^{\mu\nu\rho\sigma}_{ren.,norm.}(\mu)}{d\log\mu} \\
&= \frac{C_2(G)\,\alpha_{ren}(\mu)}{\pi}\cdot\Big[\frac{7}{6}\,\delta\eta^{\mu\nu\rho\sigma}_{ren.,norm.}(\mu) \\
&\quad -\frac{7}{3}\,\delta\eta^{\mu\gamma\rho}_{ren.,norm.\,\gamma}(\mu)\,g^{\nu\sigma} + \frac{1}{2}\,\delta\eta^{\alpha\beta}_{ren.,norm.\,\alpha\beta}\,g^{\mu\rho}g^{\nu\sigma}\Big] \\
&\quad -\sum\frac{T(R)}{3\pi}\,\alpha_{ren}(\mu)\Big[\delta g^{\mu\rho}_{ren.R}(\mu)\,g^{\nu\sigma} + g^{\mu\rho}\,\delta g^{\nu\sigma}_{ren.R}(\mu) \\
&\quad -\delta g^{\mu\sigma}_{ren.R}(\mu)\,g^{\nu\rho} - g^{\mu\sigma}\,\delta g^{\nu\rho}_{ren.R}(\mu)\Big]
\end{aligned} \tag{91}$$

Here we have defined

$$\delta g^{\mu\nu}_{ren,R}(\mu) = g^{\mu\nu}_{ren,R}(\mu) - g^{\mu\nu} \tag{92}$$

and we note that in the Lorentz invariant case the expression (eq. 91) will be zero.

It is essentially the positive sign on the term $\frac{7}{6}\,\delta\eta^{\mu\nu\rho\sigma}_{ren\ norm}(\mu)$ in (eq. 91) that tells that also the quantum corrections from the Yang-Mills interaction makes Lorentz invariance become more accurate the lower the energy scale (μ).

The technique used to perform β-function calculations like these is straightforward perturbation theory. M. Ninomiya and I used the background gauge method following rather tightly a covariant calculation performed by N.K. Nielsen[27].

Gauge invariance the β-function way

A third example of the application of the renormalization group method to obtain a symmetry is the achievement of gauge invariance for electrodynamics with spin $\frac{1}{2}$ particles by J. Iliopoulos, D.V. Nanopoulos and T.N. Tamaros[21].

Not assumming gauge invariance, but conserving parity, charge conjugation, Lorentz invariance and translational invariance the Lagrangian for interaction of a vector field $A_\mu(x)$ with the Dirac field $\psi(x)$ to be considered becomes

$$\begin{aligned}\mathcal{L} = {} & \bar{\psi}(i\not{\partial} - m)\psi - \tfrac{1}{2}(\partial_\mu A_\nu)^2 \\ & + \tfrac{1}{2}\mu^2 A_\mu^2 - e\bar{\psi}\gamma_\mu\psi A^\mu \\ & - \tfrac{1}{4} g A_\mu^4 + \xi(\partial_\mu A^\mu)^2\end{aligned} \tag{93}$$

Here the coupling parameters (coupling constants) are μ^2, e, g, and ξ. The term $\xi(\partial_\mu A^\mu)^2$ determines in what gauge one obtains the model at the end and is not considered important and in fact left out by Iliopoulos et al.[21].

The photon mass μ has dimension mass and we do not expect that to disappear in the infrared limit unless we put it to zero. But what can be hoped for is that the ratio $\frac{g}{e^2}$ goes to zero in the infrared limit so that the gauge non-invariant term gA_μ becomes negligible to a low energy observer. This is in fact what is true since they find

$$\pi^2 \beta_{e^2} = \frac{e^4}{6}$$
$$\pi^2 \beta_g = \frac{e^2 g}{3} + 2g^2 \tag{94}$$

and thus putting $\eta = \frac{g}{e^2}$ that

$$\beta_\eta \sim \frac{1}{6}\eta + 2\eta^2 \tag{95}$$

It should be remarked that the non-gauge invariant spinor Q.E.D. has to be quantized with an indefinite Hilbert-space. It was this model which I mentioned as being of a type critized by M. Veltman above[20].

It is interesting that for a scalar field interacting with a vector field it is not true that gauge invariance appears in the infrared limit. But one may say that only the fermion (spinor) Q.E.D. is fundamental so that it is enough that such a theory gives gauge invariance in the low energy limit. T.N. Tamaros informed me that a non-gauge invariant Yang-Mills theory similar to Q.E.D. becomes gauge invariant towards the infrared.

Symmetries appearing in renormalizable Lagrangian

The third method of obtaining symmetries which I mentioned was that a renormalizable theory with some symmetries may automatically have other ones.

As an example we think of baryon number conservation in the standard model i.e. the combined model of SU(3) - color Yang-Mills and the Weinberg-Salam uniting weak and electromagnetic interactions with quarks and leptons. The gauge group of the standard model has the Lie algebra of U(1) x SU(2) x SU(3) (we shall see below that it is reasonable to consider S(U(2) x U(3)) the gauge group). By construction of a renormalizable Lagrangian obeying SU(3) color gauge symmetry from quark fields in the triplet representation one can only produce terms with both a $\bar{\psi}$ and a ψ factor. In fact the allowed terms are

$$\bar{\psi} \gamma^\mu D_\mu \psi \tag{96}$$

and

$$\varphi^{(+)} \bar{\psi} \frac{1 \pm \gamma_5}{2} \psi \tag{97}$$

where D_μ is the covariant derivative and φ is the Higgs field. If it was not for the U(1) and SU(2) gauge symmetries there could also be a mass term

$$m \bar{\psi} \psi \tag{98}$$

With only terms of the type $\bar{\psi} \ldots \psi$ the number of quarks minus the number of antiquarks will be conserved. But that is three times the baryon number and thus the baryon number must be conserved.

We also mentioned as an example of this method chiral invariance SU(2) x SU(2) or SU(3) x SU(3) including isospin SU(2) and Gell-Mann's SU(3) symmetry respectively. This should be gotten out of Q.C.D. but is not a perfect case because it is necessary to add as an assumption that the quark masses are small. Preferably one should add some symmetries enough to suppress the quark masses.

We refer to S. Weinberg[28)] for these symmetry understandings.

Also there is no way of breaking say parity in Q.C.D. once the representations of both right and left handed quarks as triplets are given. S. Chadha and I (unpublished) once argued that requiring representations smallest possible and that Adler anomalies should not spoil renormalizability only this parity conserving type of model gets allowed.

What gauge group should we see?

We have above argued for that symmetries may appear out of fundamental laws of nature that are somewhat random.
It would be nice to be able to predict something about what symmetries should come out. For instance the "derivations" of gauge invariance seem a priori not to be so sensitive to which gauge group one wants to "derive". I would therefore like at the end to mention a recent work by N. Brene, and myself[3)] attempting to argue why the standard model gauge group S(U(2) x U(3)) should be favoured.

First let us follow L. Michel's idea of asigning a meaning to the group and not only to the Lie algebra[29]. It is a general belief that the standard model described by the Lie algebra of

$$R \times SU(2) \times SU(3) \tag{99}$$

gives a good description of the physics at experimental energy scales at least. All the fields phenomenologically observed-quarks, leptons and even the Higgs field-belong not only to representations of the Lie algebra of (eq. 99) but also of a certain group called

$$S\left(U(2)\times U(3)\right) = R\times SU(2)\times SU(3)/\text{"}Z\text{"} \tag{100}$$

where "Z" is to the group of integers Z isomorphic discrete subgroup of the center of R x SU(2) x SU(3) generated by

$$(2\pi, -\mathbb{1}^{2\times 2}, e^{i2\pi/3}\,\mathbb{1}^{3\times 3}) \in R\times SU(2)\times SU(3) \tag{101}$$

The point is that for all representations with the electric charge obeying the usual quantization rule

$$Q = -\tfrac{1}{3}\cdot \text{"triality"} \pmod 1 \tag{102}$$

the element (eq. 101) is represented by the unit matrix

$$\begin{aligned} &\rho(2\pi, -\mathbb{1}, e^{i2\pi/3}\mathbb{1}) \\ &= e^{i2\pi y}\, e^{i2\pi A}\, e^{i\frac{2\pi}{3}\cdot\text{"triality"}}\,\mathbb{1} = \mathbb{1} \end{aligned} \tag{103}$$

so that all elements in a coset of the group "Z" are represented by the same matrix for the representations obeying (eq. 102). Here y is the weak hypercharge, A the weak isospin (denoting the representation of the SU(2) Lie algebra) and y is normalized so that the electric charge is

$$Q = y + A_3 \tag{104}$$

where A_3 is the third component of the weak isospin. The "triality" is a characteristic property of an SU(3) representation; for a triplet it is 1 and for a representation occurring in the composition of m triplets it is counted modulo 3 equal to m.

The table shows the quantum numbers of the left-handed Weyl particles of the Standard Model.

Left-handed Weyl fermions	R Y	SU(2) A	SU(3) $\underline{r}$	Q = Y+A_3	triality
$\begin{pmatrix}\nu\\e^-\end{pmatrix}_L \begin{pmatrix}\nu_\mu\\\mu^-\end{pmatrix}_L \begin{pmatrix}\nu_\tau\\\tau^-\end{pmatrix}_L$	-1/2	-1/2	$\underline{1}$	$\begin{pmatrix}0\\-1\end{pmatrix}$	0
e^+_L μ^+_L τ^+_L	1	0	$\underline{1}$	1	0
$\begin{pmatrix}u\\d^c\end{pmatrix}_L \begin{pmatrix}c\\s^c\end{pmatrix}_L \begin{pmatrix}t\\b^c\end{pmatrix}_L$	1/6	$\frac{1}{2}$	$\underline{3}$	$\begin{pmatrix}2/3\\-1/3\end{pmatrix}$	1
$\bar{u}_L$ $\bar{c}_L$ $\bar{t}_L$	-2/3	0	$\underline{\bar{3}}$	-2/3	-1
$\bar{d}_L$ $\bar{s}_L$ $\bar{b}_L$	1/3	0	$\underline{\bar{3}}$	1/3	-1
Higgs field $\begin{pmatrix}\varphi_0\\\varphi_{-1}\end{pmatrix}$	-1/2	1/2	$\underline{1}$	$\begin{pmatrix}0\\-1\end{pmatrix}$	0

One can say that the charge quantization rule (eq. 102) can be explained if one assumes that the gauge group is S(U(2) x U(3)) = R x SU(2) x SU(3)/"Z" rather than just R x SU(2) x SU(3) or U(1) x SU(2) x SU(3).

In a sense if is via this group S(U(2) x U(3) charge quantization is explained in the grand unified Georgi-Glashow[30)] model SU(5) because the subgroup of the group SU(5) with the Lie algebra of R x SU(2) x SU(3) is just S(U(2) x U(3)) = R x SU(2) x SU(3)/"Z" . If one just has a continuum Yang-Mills theory (i.e. not lattice theory) it is only the Lie algebra that matters except for Michel's idea that the representations of fermions or other fields are restricted in a way depending on which of the Lie group with a given Lie algebra is used. However, one may give the Lie group a more direct meaning by postulating a lattice theory. In fact the fundamental link variables U(•—•) of lattice gauge theory take on values in the Lie group. If we have the group S(U(2) x U(3)) = R x SU(2) x SU(3)/"Z" and postulate a lattice - as suggested by our "derivation" of gauge symmetry the definitional way - one can only have representations obeying charge quantization (eq. 102). So if we ignore grand unified theories we may say that the experimental observationa of the charge quantization (eq. 102) is suggestive of the existence of a lattice theory, so that the group rather than only the Lie algebra has a physical significance from the outset and thus can explain (eq. 102).

So let us consider the gauge group S(U(2) x U(3)) = R x SU(2) x SU(3)/"Z" phenomenologically suggested and speculate on what properties of just this group have made it the one we should observe.

Some of the properties of this group SU(U(2) x U(3)) which N. Brene and I[3] may hope to have a chance to "explain" are:

1) Center (S(U(2) x U(3)) is nontrivial.

By the center of a group G is understood the subgroup Center (G) of G consisting of those elements $c \in G$ commuting with all elements $g \in G$. i.e.

$$\text{Center}(G) = \{ c \in G \mid cg = gc \quad \forall g \in G \} \tag{105}$$

That the center is nontrivial means that it is not consisting of the unit element i.e.

$$\text{Center}\left(S\left(U(2) \times U(3) \right) \right) \neq \{1\} \tag{106}$$

In fact

$$\text{Center}\left(S\left(U(2) \times U(3) \right) \right) \cong U(1) \tag{107}$$

2) The center is topologically connected.

Connected may mean here curve-connected i.e. that there exist a curve running inside the center Center(S(U(2) x U(3))) connecting any two points (= elements) in this center.

Since the center is isomorphic to U(1) and the (topological) group U(1) is (topologically) connected it is seen that Center (S(U(2) x U(3))) is connected.

3) The gauge group is not fully a cross product of other groups. I.e. it is not of the form

$$A \times B \tag{108}$$

where A and B are other (Lie) groups.

If it was not for the division by the discrete subgroup "Z" of the center in the expression

$$R \times SU(2) \times SU(3) \ / \ "Z" \qquad (109)$$

the group would have been of the cross product form A x B. But the factor group R x SU(2) x SU(3)/"Z" cannot be factorized as a cross product.

4) The β-functions for the Lie algebras of R, SU(2) and SU(3) are rather small.

What we mean by this is that quadratic Casimirs $c_2(R)$, $c_2(SU(2))$, $c_2(SU(3))$ are the smallest you have in some sense: $c_2(R) = 0$ and equation (eq. 89) tells us that for SU(N) groups the quadratic Casimir is N and thus smallest for the smallest N-values.

In fact we can show that under the restrictions of the other principles the values N = 2 and 3 are the smallest (remember that of course SU(1) = 1 is trivial.

5) No top of hat point singularity of the space of conjugacy "classes".

The meaning of this statement is that the space the points of which are the conjugacy classes of the phenomenological gauge group S(U(2) x U(3)) lacks a certain type of singularity when considered a metric space (or Riemannian manifold). One can on Lie groups define a metric - for simple groups in an apart from normalization unique way - such that it is invariant under both right and left multiplication. This metric - making the group a metric space - induces also a distance between pairs of classes of conjugate elements of the group, thus making the set of conjugacy classes into a metric space too. The conjugacy classes are the classes corresponding to the equivalence relation ~ defined by two elements of the group f, g being conjugate

$$f \sim g \iff \exists\, h \text{ in the group } [f = h g h^{-1}] \qquad (110)$$

The conjugacy class metric space corresponding to the group S(U(2) x U(3)) has in a neighborhood of every point (i.e. every conjugacy class) which is isometric with a - by flat hyperplanes delimited - - piece of a four dimensional euclidean space. So the only singularities are corners, edges etc. However, the metric space of conjugacy classes for the group SU(4) of rotations in a four dimensional euclidean space form rather a two dimensional space with the shape of a paper hat glued together from a square cut to half along a diagonal.

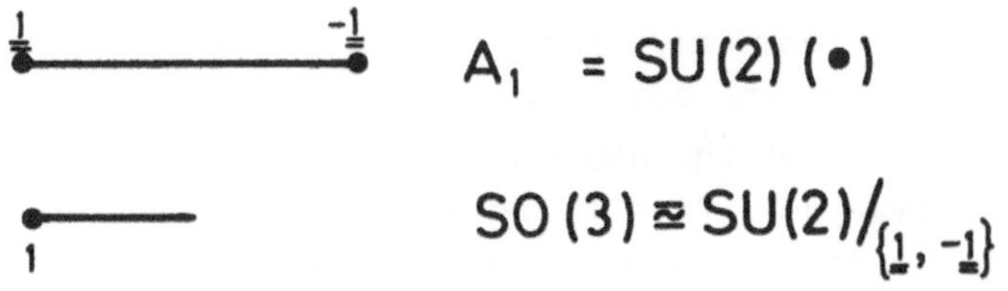

1 —— -1 $A_1 = SU(2)\,(\bullet)$

1 —— $SO(3) \cong SU(2)/_{\{1,\,-1\}}$

Fig. 7. Conjugacy class spaces for rank one groups. • center class (•) Dynkin diagram.

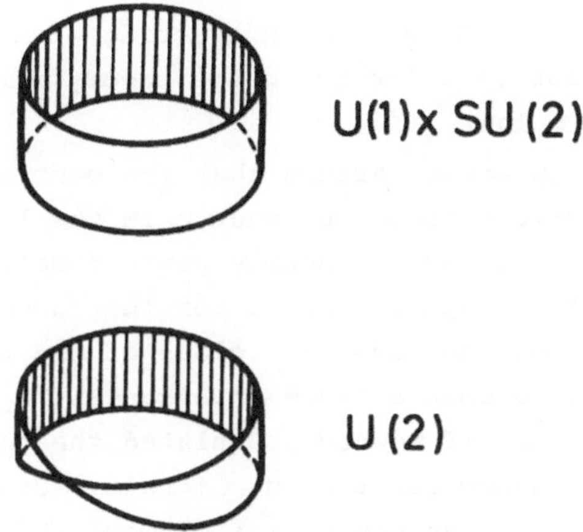

Fig. 9. Conjugacy class spaces for a couple of nonsemisimple groups both with the Lie algebra of R̀ x SU(2). The conjugacy classes consisting of a single center element each are represented by the points on the edges on this figure. For U(2) the figure is the Möbius band and has a connected edge (= center here). For U(1) x SU(2) the center falls into two connectedness components.

This SU(4) -conjugacy class space has a singularity in its interior i.e. away from the boundary (curve). This interior singularity is the top point of the paper hat and we therefore refer to a singularity of this type as a "top of hat" singularity.

The main purpose of mentioning this property 5) of no top of hat singularity of S(U(2) x U(3)) is to point to a property that the group R x SU(4)/K', where K' is generated by

$$(2\pi,\ -\mathbb{1}^{4\times 4}) \in R \times SO(4) \tag{111}$$

is lacking.

6) Apart from the abelian factor R in the Lie algebra only SU(N) -groups were used.

The work by N. Brene and myself[3)] is to "derive" these properties in what we could call a gauge glass model by an expression suggested to us by J. Greensite as an analogy to the concept of a spin glass. In the spin glass spins are interacting with random interactions. Our gauge glass is a lattice gauge theory with a random action or a random Hamiltonian.

That is to say we assume that the contribution to the hamiltonian (or to the action) from a plaquette in the lattice gauge theory taken to be fundamental has a randomly given functional form being defined on the set of conjugacy classes for the gauge group G say. It has to be a function of the conjugacy class of the plaquette variable U G only in order to ensure gauge invariance. Gauge invariance would in general be broken if we had postulated the plaquette energy contribution to be a random function U itself. Under a gauge transformation U transform into an equivalent element $\Lambda U \Lambda^{-1}$ and thus the conjugacy class of the plaquette variable U is gauge invariant. Taking the energy plaquette contribution to be a random function and different one for each plaquette we are assuming a non-translational invariant model. That is analogous to that a spin glass is not translational invariant except in a statistical sense.

We would then have to hope for that translational invariance would somehow reappear effectively. Actually the somewhat preliminary work by M. Lehto, M. Ninomiya and myself mentioned above under the example of "definitional way" (method I) is hoped to provide translational invariance because in gravitational theory translations are essentially a special type of reparametrizations.

But in the work by Brene and myself we simply postpone the problem

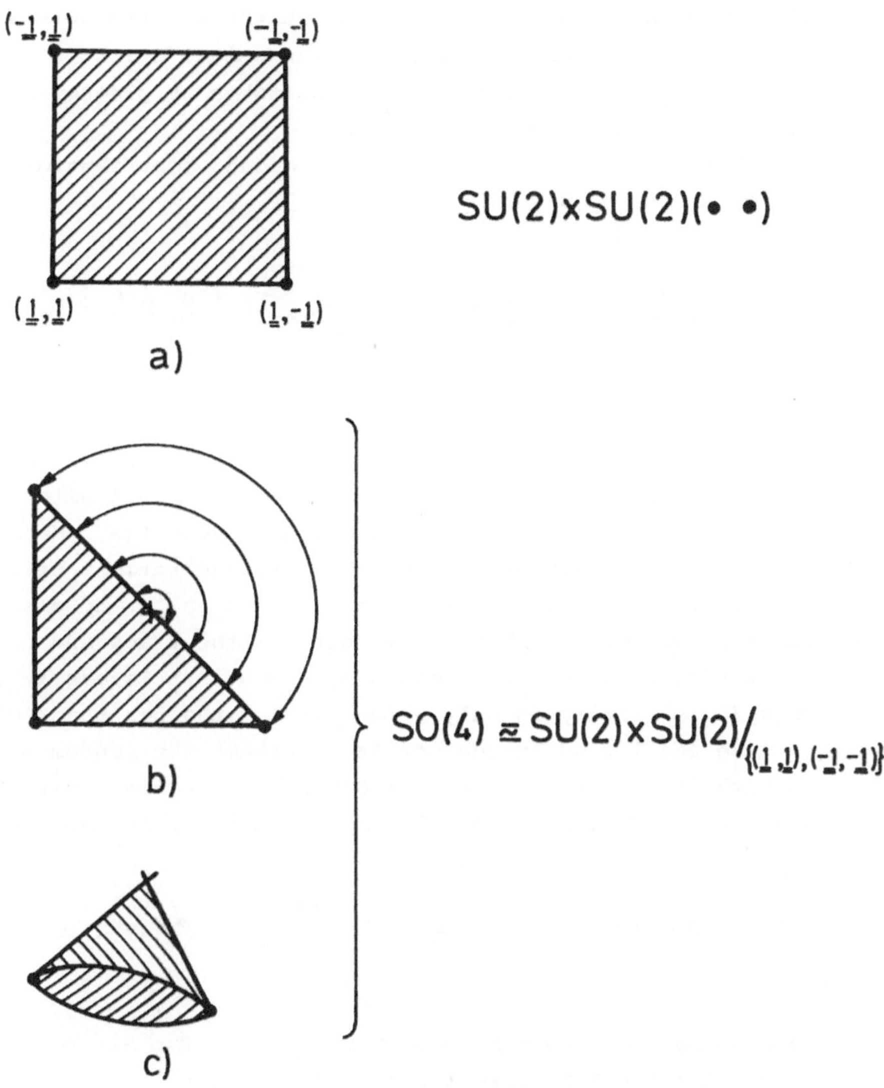

Fig. 8. Illustration of the construction of the space of conjugacy classes for $SO(4) = SU(2)xSU(2)/\{(\underline{1}, \underline{1}), (-\underline{1}, -\underline{1})\}$. First shown a) the class space for the covering group SU(2)xSU(2). The arrows on fig. b) ↶ denote identification of the points along the diagonal. Fig. c) is meant in perspective. The top of hat point is denoted by x.

of why we should get translational at the end and work with no model that have only translational invariance in the statistical way that the <u>distribution</u> of functional forms taken by plaquette energies is the same for different plaquettes.

For instance we may take hamiltonian contribution from a plaquette $\square$ to be given by

$$H_{\square}(class\,(U_{\square})) = \sum_{\substack{\text{represen-}\\ \text{tations } R}} \alpha_{\square R}\, Tr\,(\rho_R\,(U_{\square})) \tag{112}$$

where the coefficients $\alpha_{\square R}$ are stochastic (= random) variables with same distribution for different plaquettes $\square$ but with different actual values normally for the various plaquettes. Here the index R runs over the various inreducible representations of the group G and $\rho_R(U_{\square})$ means the matrix representing the group element $U_{\square}$ in the representation R. Further Tr denotes the trace and class $(U_{\square})$ means the conjugacy class represented by $U_{\square} \in G$. It is easily seen that $Tr(\rho_R(U_{\square}))$ really is only a function of the conjugacy class $(U_{\square})$.

We shall find it reasonable to assume the random variables $\alpha_{\square R}$ to have distributions that are Gaussian (= normal) with average zero and a width σ_R depending only on the inreducible representation R:

$$P(\alpha_{\square R})\, d\alpha_{\square R} = \frac{1}{\sqrt{2\pi}\,\sigma_R}\; e^{-\alpha_{\square R}^2/2\sigma_R}\; d\alpha_{\square R} \tag{113}$$

We also may need to take almost zero width for all representations R except for the very lowest nontrivial one(s).

The main point of our considerations now is to investigate in a classical approximation (and thus rather crudely) how the ground state of our gauge glass will look. Especially we are interested in if it is likely that all the plaquette variables $U_{\square}$ in the ground state (= vacuum) will take values in the center of the group Center (G). The point is that if some plaquette variables are in the ground state not taking values in the center Center (G) then this ground state is not invariant under global gauge transformations. By a global gauge transformation we understand a gauge transformation corresponding to an over space constant gauge transformation function $\Lambda(x) = \Lambda$.

The property of the vacuum not being invariant under the global gauge transformations may be considered the characteristic property of the Higgs-phase with $<\phi>_{vac} \neq 0$ in the Higgs-model distinguishing it from the "ordinary" charge scalar field phase in which $<\phi>_{vac} = 0$. Here $\phi(x)$ is the complex (or nonhermitean) charged scalar field in the Higgs model. It is therefore very reasonable to assume that if in the vacuum state the variables have their probabilities concentrated around a configuration not invariant under the global gauge transformations the situation is analogous to that of the Higgs phase and at least some of the gauge bosons aquire a nonzero mass. This main assumption of ours, that is to say we assume: spontaneous breakdown of the global gauge group leads to gauge boson mass(es). If the theory as our gauge glass one has neither translational nor (rotational) Lorentz invariance the concept of a mass is not so meaningful. We may take it a nonzero mass to mean that there will be no Goldstone boson like behavior of the particle in question (here the gauge boson). That is to that is got a nonzero mass means that it no longer holds that its energy goes to zero for zero (space) momentum.

With the philosophy that the fundamental energy scale is for example the Planck energy scale and least very large compared to experimentally accessible energy scales of today an "nonzero mass" comes to mean that the particle is unaccessible to present day experiments. Thus we can observe at experimental (i.e. low) energies only gauge bosons for which the vacuum state is invariant under the corresponding global gauge symmetry.

The main type of argument of our work (Brenes and mine) is to estimate in a classical approximation which gauge groups have the best chance of avoiding developing a vacuum state not invariant under the global gauge transformations. The better the group avoids in our gauge glass model a such spontaneous breakdown (collaps) the better should be the chance of us observing it (at low energy). A spontaneous breakdown obviously occurs (classically) if a plaquette variable $U_\square$ in the vacuum $U_{\square\, vac}$ takes on a value not belonging to the center of the group:

$$U_{\square vac} \notin \text{Center}(G) \Rightarrow \text{"collapse"}.$$

Whether a plaquette variable $U_\square$ belongs to the center or not is a gauge independent question while the link variables $U(\bullet\!\!\rightarrow\!\!\bullet)$ may be transformed in and out of the center under gauge transformations (sometimes).

It is rather easy to see that (at least classically) a group G with a trivial center Center (G) = $\{1\}$ has a very high chance of collapsing. In fact if the sign of the hamiltonian contribution from a plaquette happens to be opposite to that in a usual lattice gauge theory the unit element will be an energy maximum rather than an energy minimum as in a usual model, at least it may not be a minimum often. Thus the lowest energy state (= the vacuum) will not have such a plaquette variable equal to unity. But that means it cannot be in the center when the center is trivial. So the vacuum will not be globally gauge invariant and the group with trivial center will collaps. So it is not healthy for a group to have trivial center. Here "healthy" is used to denote what is improving the chance for the group avoiding collaps (= spontaneous breakdown for the global gauge) and other dangers like confinement preventing that the group be seen at low energy.

The most important property which we show to be healthy is for a group is to have a connected center. The argument for Center (G) disconnected being unhealthy i.e. leading to collaps hinges on the Bianchi identities.

From the definition of the plaquette variables (eq. 14) one can deduce that the product (by the group multiplication in G) of the six plaquette variables $U_{\square 1}, \ldots, U_{\square 6}$ around a unit cube must be equal to the unit element

$$\prod_{i=1}^{6} U_{\square i} = 1 \tag{115}$$

It is this relation we call the Bianchi identity. It corresponds to the Maxwell equations

$$\mathrm{div}\ \vec{B} = 0 \tag{116}$$

(and

$$\mathrm{rot}\ \vec{E} + \frac{\partial \vec{B}}{\partial t} = 0 \quad). \tag{117}$$

One may choose a gauge so that the Bianchi identity of this form (eq. 115) is true for any gauge group G, but since Brene and I are mainly concerned with the case of the $U_\square$ belonging to the center (or rather the question whether they can do that) we mainly need it for the center group Center (G) which is of course an abelian subgroup of G. For an abelian gauge group the Bianchi identity of the form (eq. 115) is true in any gauge and needs no specification of in what order the factors $U_{\square 1}$, $U_{\square 2}$, etc. should be multiplied together, as is needed in the non-abelian case.

Now because each plaquette - and especially each of the six plaquettes around a cube - has its individual randomly chosen hamiltonian form the plaquette variable value providing the lowest energy for one of the plaquettes is a random group element. It must, however, be an element of the center if the group shall have a chance of avoiding collaps (= spontaneous breakdown). So in cases of interest the six plaquettes around a cube would minimize their energy if they could take certain from the random dynamics determined center values. However, most likely the product of these six (from energy lowering) favoured centervalues will not be the unit element but rather a random center element in the group. But that is impossible to realize because of the Bianchi identities, there simply will not exist any choice of the fundamental link variables $U(\mapsto)$ that by means of the definition (eq. 14) can give the favoured $U_{\square}$ -values. Thus the plaquette variables in the vacuum state are forced to deviate somewhat from the favoured values that would have minimized the energy had it not been for the constraint from the Bianchi identities. Normally the energy cost by a small deviation of a plaquette variable $U_{\square}$ from a minimum (= favoured) value at a center element is approximately proportional to the square of this deviation. Thus it will often pay energetically to take the deviations between the actual vacuum plaquette variable and the favoured values be approximately equal in magnitude. The reason that an approximately equal distribution of the frustration for the cube over all plaquettes is that the sum of squares for a series (here 6) numbers with given sum is kept the lowest by taking the numbers equal.

Denoting the favoured values that would be obtained neglecting the Bianchi identity by $U_{\square fav\,1}$, $U_{\square fav\,2}$, ..., $U_{\square fav\,6}$ for the paluettes around a cube the frustration of the cube is

$$f = \prod_{i=1}^{6} U_{\square fav\, i}$$

and in general not equal to the unit element. Denoting further the corresponding true vacuum values $U_{\square vac1}$, $U_{\square vac2}$, $U_{\square vac6}$ we may construct a series of group elements

$$\begin{gathered} 1 \\ U_{\square vac1} \cdot U^{-1}_{\square fav1} \\ U_{\square vac1}\, U_{\square vac2} \cdot U^{-1}_{\square fav2} \cdot U^{-1}_{\square fav1}\,, \quad \ldots \end{gathered} \tag{118}$$

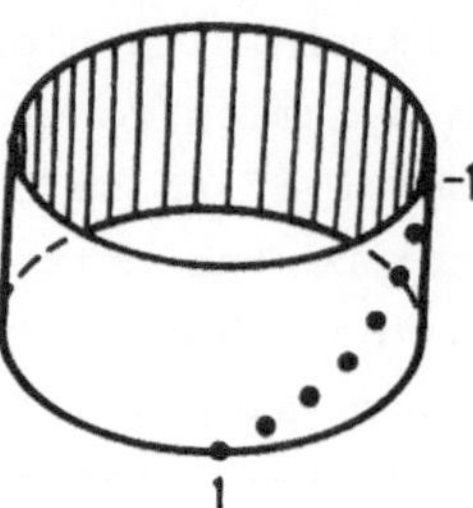

Fig. 11. Illustration of how in the example of the gauge group being U(1) x SU(2) with center consisting of two circles the chain of group elements (eq.(118)) may be represented by conjugacy classes. The conjugacy classes corresponding to the elements in chain (eq.(118)) are denoted by +'s. The first point in the chain is the unit element 1 the last one is f^{-1}.

$$U_{\square vac1} U_{\square vac2} \cdots U_{\square vac5} \cdot U^{-1}_{\square fav5} \cdots U^{-1}_{\square fav1}$$

$$U_{\square vac1} U_{\square vac2} \cdots U_{\square vac6} \cdot U^{-1}_{\square fav6} \cdots U^{-1}_{\square fav1}$$

$$= \left(\prod_{i=1}^{6} U_{\square favi} \right)^{-1} = f^{-1}$$

forming a chain in six about equal length steps through the group G from the unit element 1 to the inverse of the frustration f.

This cahin may be thought of as representing a curve in the group manifold G connecting the two elements 1 and f^{-1} of the center of G. If the center Center (G) is not topologically connected it will since the frustration f is a random center element be likely that for some cubes f^{-1} and 1 belong to different connected components of this center. In such a case there is no way in which the curve connecting 1 and f^{-1} can be kept inside the center. So it must pass elements of G outside the center. So we also expect that some elements of the chain (eq.118) being approximated by the curve will lie outside the center. Since we already consider the case that all the $U_{\square favi}$ $(i = 1,2,\ldots,6)$ are in the center we could conclude that some of the products

$$U_{\square vac1},\ U_{\square vac1} U_{\square vac2},\ \ldots,\ \prod_{i=1}^{6} U_{\square vaci}$$

must be non central, i.e. belong to G Center (G), then it would follow that some plaquette variables were noncentral.

So we see that the center Center (G) being not connected implies that $U_\square$'s take noncentral values again implying a spontaneous breakdown of the global gauge symmetry. So we expect a gauge group G with disconnected center to lead to massive gauge bosons at least. Thus only groups G with connected center will survive down to low energy avoiding collaps.

Explaining the other properties of S(U(2)xU(3))

The property 3) that the gauge group be not a cross product as a group (although perhaps as a Lie algebra) we cannot give any explanation for the simple 3+1 dimensional gauge glass model which we considered for the moment at least. However, if we imagine a model in which

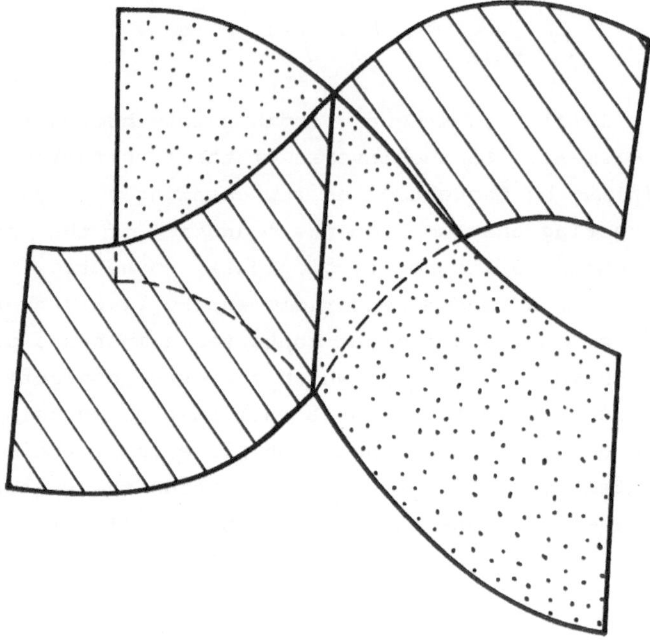

Fig. 12. The two surfaces here symbolize two 1+3 dimensional manifolds imbedded in a higher dimensional space. Along the [dotted] surface there is long range correlation in the field for group A, while there is for B inside [hatched] .

the space time at the fundamental have higher dimension than the experimental 3+1 there is a possibility for a speculative explanation for no full product gauge group G = AxB. The speculation is that our usual 3+1 dimensional space time lie as a four dimensional submanifold in the higher dimensional space-time and that this submanifold is characterized by there being long range correlations in the gauge fields along it while there should rather be a kind of confined or strong coupling like phase in the directions orthogonal to the submanifold(s). We are here talking about distances shorter compared to the scale of the confinement of Q.C.D. so that on the scales we are talking about there are long range correlations also in the $SU(3)_{Color}$. In fact we are deep inside the asymptotic freedom region. However, we imagine much shorter range correlations in the directions of the extra dimensions which of course only exist according to our speculations.

Now the argument that a gauge group of the form G = AxB should not be found is that it would be a strainge accident if both groups A and B should precisely have there long range correlations determine precisely the same 3+1 dimensional submanifold of the higher dimensional space. That is to say I would consider it more likely that there would be long range correlations in the gauge fields for group A along some 3+1 dimensional manifolds) and in the ones for B along other(s). These two types of manifold might then sometimes corss but without any connections between the groups A and B there seems no strong reason that they should exactly follow each other. Now we live in one such manifold and would not notice the other ones ans so we would see only one of the cross product factors i.e. either A or B. This is supposed to be an argument for why the group we should see as gauge group should not be factorizable into aetors i.e. not be of the form G = AxB.

If one does not like this argument with extra dimensions one may claim that there are such a factorization but that we only have interactions with the particles connected with one such factor.

The property 4) that the β-functions or rather the quadratic Cassimirs should be small is an immediate requirement to avoid confinement at higher than the experimental energy scale provided we make the assumption that all the coupling constants are roughly the same for all the simple sub Lie algebras at some fundamental scale (the Planck energy scale). Thus the running coupling constants that grow fastest towards the infrared become strong and confining at the highest energy scales. The Lie algebras surviving unconfined till the lowest scale in energy have smallest β-functions.

The property 5) of there being no top of hat point in the space of conjugacy classes is intuitively explainable by remarking that a singularity of this type would have a high chance of being the lowest energy state of a plaquette. Since the top of hat point does not belong to the center it would mean spontaneous symmetry breakdown of the global gauge symmetry if a plaquette variables U in vacuum takes on a value belonging to the (or a) conjugacy class at the top of hat singularity. Thus the existence of a top of hat singularity is expected to lead to an enhancement danger of collaps and so we should at experimental energy not see gauge groups with such singularities.

The argument for property 6) that only abelian group's and SU(N) groups are used the construction of the gauge group found experimentally is of a similar nature to the one for property 5) just given. In fact the SU(N) groups have a center element in every corner of their space conjugacy classes while the non SU(N) groups have corners in their class space which are not central. For example one might think of the rank two groups: For all the three simple Lie algebras of rank two A_2 = SU(3) (Dynkin diagram •—•), B_2 = spinor SO(5) ≅ C_2 = Sp(4) (Dynkin diagram ⚬⇒•) and the exceptional group G_2 (Dynkin diagram ⚬⇛•) the spaces of conjugacy classes for the covering groups (the unique simply connected group with the given Lie algebra) have the geometry of triangles. For A_2 = SU(3) all the three angles are not points representing center elements. For the covering group of the orthogonal group S$\tilde{O}$(5) (B_2 = spinor group of SO(5)) isomophic to the symplectic group C_2 = Sp(4) only two of the three "corners" of the triangle correspond to center elements so that there is one non central "corner". One would guess that there would be a high chance for the minimal energy of a plaquette to occur for a $U_\square$-value belonging to conjugacy class in the non central corner and that would lead to collapse. For the exceptional group G_2 the center is trivial and therefore the triangle of conjugacy classes has two non-central corners.

What groups are most fit to be seen at low energy

Now let us argue that the properties just argued for are suggestive for a group of the form

$$\hat{R} \times SU(2) \times SU(3) \times SU(5) \times SU(7) \times \cdots \times SU(P) / \hat{H}$$

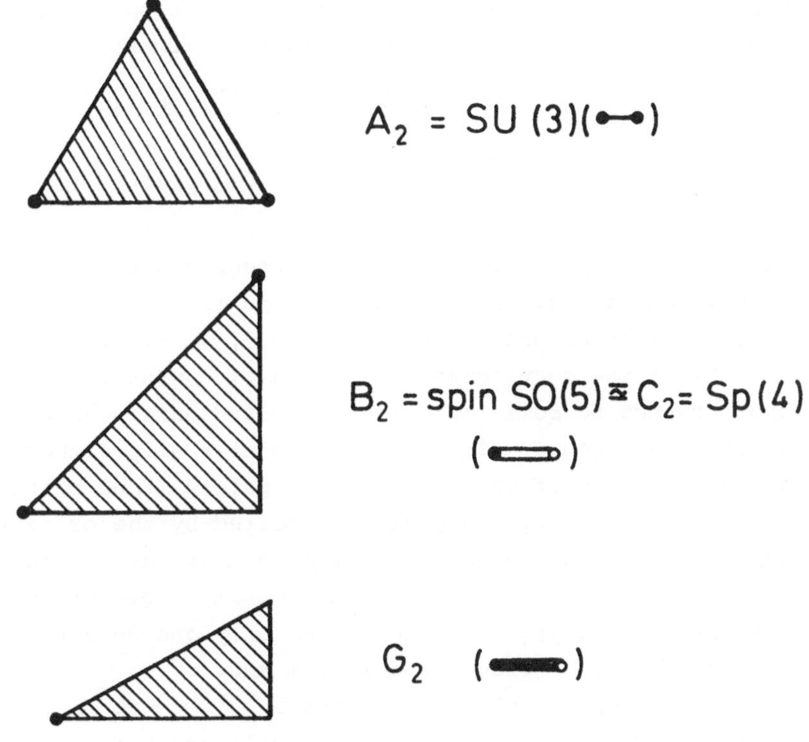

Fig. 10. Some conjugacy class spaces. Each point in these figures represent a class of mutually conjugate group elements. The dots • represent classes of a center element in each. The diagrams in bracket (...) are the Dynkin diagrams.

where $\hat{H}$ is constructed so as to make the group having a connected center isomorphic to U(1) and in fact is generated

$$\hat{h} = (2\pi\ ,\ -\underline{1}\ ,\ e^{i2\pi/3}\ \underline{1}\ ,\ e^{i2\pi p_5/5}\ \underline{1}\ ,\ \ldots\ ,\ e^{i2\pi p_P/P}\ \underline{1}\ ) \in$$

$$\grave{R} \times SU(2) \times SU(3) \times \cdots \times SU(P)$$

where p_5, p_7, p_{11},...,p_P are some integers not divisible by 5, 7, 11, ...P respectively. Here SU(N) -groups involved are those with N equal to the series of prime numbers 2, 3, 5, 7, 11, 13,...,P.

First the requirement of a connected center implies that the group cannot be semisimple i.e. it must contain an $\grave{R}$ (or U(1)) factor in the Lie algebra. Otherwise the center will either be trivial or it will consist of several isolated points (thus being disconnected).

For simplicity and partly supported by the property 3) that the gauge group G should not be of the form AxB as a group we shall assume that it is of the form of $\grave{R}$ x (a semisimple group)/(some invariant discrete subgroup). I.e. we assume only one $\grave{R}$-factor. Two or more may easily give a group of the form AxB.

From property 6) we only expect SU(N) -groups as the further factors. To keep the β-functions small according to property 4) we should take the N's of SU(N) so small as possible and that would favour a group of the form

$$\grave{R} \times SU(2) \times SU(2) \times \cdots \times SU(2)/K$$

with a series of SU(2)'s. However, with a suitable definition of the concept "top of hat singularities" we can show that if two SU(N)-groups occur among the factors SU(q_1) x SU(q_2) say, with q_1 and q_2 having a common integer factor ($\neq 1$) there will be such a singularity when the invariant discrete subgroup K is constructed so as to make the factor group have a connected center. The most important example of this is the case

$$\grave{R} \times SU(2) \times SU(2)/K \cong \grave{R} \times SO(4)/K'$$

where K has been constructed to make the center connected and it turns out that the group also can be written by means of the orthogonal group

$$SO(4) \cong SU(2) \times SU(2)/\{(\underline{1},\underline{1}),(\underline{1},\underline{1})\}$$

and K' generated by $(2\pi, -\underline{\underline{1}}^{(4x4)}) \in RxSO(4)$. But now SU(4) is the typical example of a group the space of conjugacy classes of which has a top of hat singularity. In fact the space of conjugacy classes for SO(4) has precisely the induced metric space structure of a paper hat.

To avoid N-values for the SU(N) having common factor and still have them as small as possible to make the β-functions small the best solution is just to take these N's to be a series of prime numbers starting 2, 3, 5,... At low energy we shall because of confinement only see the first few of these prime number SU(N)'s. For example we might only see the SU(2) and the SU(3) thus having arrived at the group

$$\dot{R} \times SU(2) \times SU(3) / \text{"}\dot{Z}\text{"} = S(U(2) \times U(3))$$

since a connected center is needed of course. Here "$\dot{Z}$" denotes the group generated by the center element (eq. 101).

So we have found a series of arguments not all watertight but still somewhat suggestive leading from a gauge glass model - i.e. a random dynamics gauge theory - to the "experimentally observed" low energy gauge group S(U(2)xU(3)). Since the latter also includes charge quantization among its predictions we can claim also to predict that.

Let me end this part of my talk by concluding that a series of results symmetries and even what group of symmetries are suggested from a random dynamics philosophy. So it might be worthwhile studying if further laws of nature than the ones considered could be understood from a fundamental chaotic law, i.e. from a so random fundamental law of nature that it is almost equivalent to no laws of nature at the fundamental level.

References

1) J.A. Wheeler, Physics and Austerity: Law without Law, Working paper Center of Theoretical Physics, University of Texas (1982); J.A. Wheeler, Beyond the Black Hole, Copyright © 1979 by John Archibald Wheeler, Center for Theoretical Physics, The University of Texas at Austin.

1a) C.H. Woo, Mission Impossible? A look at Past Setbacks in the Search for Elementary Matter and for Universal Symmetries, Zentrum für Interdisziplinäre Forschung, Univ. Bielefeld (W-Germany) preprint.

1b) R.P. Feynman, The Character of Physical Law (The M.I.T. Press, 1967).

1c) G.F. Chew, see for example Quark or Bootstrap: Triumph or Frustration for Hadron Physics? Proceedings of the ninth Course of the International School of Subnuclear Physics "E. Majorana", Properties of the Fundamental Interactions, Erice, July 8-26, 1971 - Edited by A. Zichichi - Published by Editrice Compositori, Bologna, Italy, 1973, or F. Chew, Hadron bootstrap triumph or frustration? Physics Today, October 1970 Vol. 23 no. 10, p. 23.

1d) S. Cuilli, Analytic Continuation and Prediction Theory, Part of an invited lecture and the closing address at the International Conference on Inverse Problems (Montpellier 1975) and of some lectures given in 1977 at Bern and in 1980 at Trinity College Dublin.

2) R. D. Peccei and H. R. Quinn, Phys. Rev. Lett. 38 (1977)1440; Phys. Rev. D16 (1977)1791; S. Weinberg, Phys. Rev. Lett. 40 (1978)223; F. Wilczek, Phys. Rev. Lett. 40 (1978)279.

3) N. Brene, H.B. Nielsen, Why the Standard Model Group should have a Connected Center, Niels Bohr Institute preprint NBI-HE-82-29.

4) H.B. Nielsen, "Fundamentals of Quark Models", Eds. J.M. Barbour and A.T. Davies (Univ. of Glascow, 1977) p. 528.

5) H.B. Nielsen, "Particle Physics 1980", Eds. I. Andric, I. Dadic and N. Zovkor (North Holland, 1981) p. 125.

6) H.B. Nielsen, Har vi brug for fundamentale naturlove? (Do we need fundamental laws of nature?), Gamma 36, p. 3 & Gamma 37, p. 35-46 (1978) (ed. at the Niels Bohr Institute).

6a) S. Chadha and H.B. Nielsen, Lorentz Invariance as a Low Energy Phenomenon, Niels Bohr Institute preprint HBI-HE-82-42.

7) D. Förster, H.B. Nielsen & M. Ninomiya, Phys. Lett. 94B (1980)135; S. Shenker, an unpublished research report.

8) C.D. Froggatt and H.B. Nielsen, Nucl. Phys. B147 (1979)277.

9) C.D. Froggatt and H.B. Nielsen, Nucl. Phys. B164(1979)114.

10) A shorter version of ref. 9) also mentioning a work by N. Brene (H. Bohr and myself) is in: Proc. Neutrino -79, Internat. Conf. on Neutrinos, Weak Interactions and Cosmology, Bergen, June 18-22,1979, vol. I, A. Haatuft and C. Jarlskog (eds.) (Astvedt Industrier A/S) 320.

11) M. Lehto, H.B. Nielsen and M. Ninomiya, Phys. Lett. 115B p. 129 (1982).

12) H.B. Nielsen and M. Ninomiya, Nucl. Phys. B141 (1979) 153.

13) H.B. Nielsen and I. Picek, Phys. Lett. 114B (1982) p. 141.

14) H.B. Nielsen and I. Picek, Lorentz non-invariance, Niels Bohr Institute preprint NBI-HE-82-30 (1982).

15) H.P. Dürr, "Heisenberg's einheitliche Feldtheorie der Elementarteilchen" in Heisenberg Gedenkbuch 1981, Deutsche Akademie der Naturforscher Leopoldina.

16) R. Peccei, these proceedings.

17) E.A. Ivanov and V.I. Ogievietsky, Pisma Zh. Eksp. Theor. Fiz. 23 (1976) 661.

18) E. Fradkin and S.H. Shenker, Phys. Rev. D19 (1979) 3682.

19) T. Banks and L. Rabinovici, Nucl. Phys. B160 (1979) 349.

20) M. Veltman, The infrared-ultraviolet connection, Acta Phys. Pol. B vol. B12 No. 5 p. 437-57 (May 1981), and talk given at the Hamburg meeting (1978).

21) J. Iliopoulos, D.V. Nanopoulos and T.N. Tamaros, Phys. Lett. 94B (1980) 141.
See also J. Iliopoulos, "Unification" in Particle Physics 1980, I. Andric, I. Dadic and N. Zovko (eds.) c 1981, North Holland Publishing Company. (The gauge symmetry explanation is on page 25 of the article beginning on page 1.)

22) T. Regge, Nuovo Cimento 19 (1961) 558.

23) R.M. Williams and M. Rocek, Phys. Lett. 104B, p. 31 (1981) and Calt-68-818 DOE Research and Development Report.

24) A.M. Polyakov, Phys. Lett. 103B (1981) 207.

25) L. Gross, Decay of Correlations in Classical Lattice Models at High Temperature, Commun. Math. Phys. 68, p. 9-27 (1979).

26) R. Jackiw, these proceedings.
S. Deser, R. Jackiw and S. Templeton, Annals of Physics 140 (1982) 372.

27) N.K. Nielsen, Nordita preprint and Nucl. Phys. B
G.t'Hooft, Nucl. Phys. B62 (1973) 444;
P.B. Gilkey, Proc. Symp. Pure Math. 27 (1975) 265;
J. Differential Geometry 10 (1975) 1975.

28) S. Weinberg in The second Workshop on Grand Unification, ed. J.P. Leveille, L.R. Sulak and D.G. Unger (Birkhäuser 1981), Phys. Rev. D26 (1982) 287.

29) L. Michel, Invariance in Quantum Mechanics; Group Theoretical Concepts and Methods in Elementary Particle Physics, Lectures at the Istanbul Summer School of Theoretical Physics, July 16 - August 4, 1962, Ed. by Feza Gürsey, Middle East Technical University, Ankara, Turkey, New York Gordon and Breach, 1964.

30) H. Georgi, B.L. Glashow, Phys. Rev. Lett. 32, 438 (1974).

31) See for example: Zou Zhenlong, Huang Peng et al. Some Research on Gauge Theories of Gravitation, Scientia Sinica Vol XXII No. 6 June 1979, p. 628.

32) See e.g. D.Z. Freedman and P.K. Townsend, Nucl. Phys. B177 (1981), 282-296, and N. Obukhov, Phys. Lett. 109B no. 3,p. 195, 18 Feb. 1982.

33) R. Pearson, Phase structure of antisymmetric tensor gauge fields, NSF-ITP-81-33 a preprint from Institute of Theoretical Physics, University of California, Santa Barbara, California 93106 (1981).

34) A. Brillet and J.L. Hall, Improved Laser Test of the Isotropy of Space, Phys. Rev. Lett. number 9, p. 549 (1979).

COMPOSITE MODELS OF QUARKS AND LEPTONS

R.D. Peccei

Max-Planck-Institut für Physik und Astrophysik
Werner-Heisenberg-Institut für Physik
Munich, Fed. Rep. Germany

Abstract

We motivate and review the desirability of, and necessary constraints for, having composite quarks and leptons. The dynamical requirement of having approximately massless fermion bound states and various ways of achieving this goal are discussed. Dynamical schemes for generating the correct fermionic mass patterns, as well as the subject of generations, are critically examined. The desirability of preventing proton decay at too fast a rate and yet generating enough baryon asymmetry in the universe leads us to consider a composite GUT scenario. Various features of this scenario, including the existence of natural mass hierarchies, are discussed. Failures and open questions regarding quark and lepton compositeness are addressed as a final topic.

I. Introduction

In these lecture notes I attempt to review various important issues facing composite models of quarks and leptons. This subject has received considerable attention in the last few years and, besides the many original papers, there exist already a number of fine reviews [1-4]. Although many separate issues in compositeness will be discussed, the subject matter will be addressed following a central theme. Namely, I shall ask, and try to answer, the question of what theoretical construct allows one to have computability for the observed parameters in nature? This will lead us, toward the end of these notes, to contemplate a rather complicated scenario of a composite GUT model where all masses are, in principle, computable from the Planck scale.

Focusing on computability, as the main issue and raison d'etre for compositeness, necessarily has slanted the material presented in a certain direction. It will be clear to the reader, however, that a great gap exists between asking for computability from a theory and actually being able to compute something in practice! Thus, and this is the most regrettable aspect of these notes, few models are really discussed and even then, they only illustrate certain aspects of a given issue [F1].

The material in these notes owes much to discussions which I had with my colleagues at the Max-Planck-Institut and elsewhere on the subject of compositeness. I am particularly grateful for the insights provided by my various collaborators, W. Buchmüller, S.T. Love, A. Masiero and last, but not least, T. Yanagida. Finally, I should also like to express my appreciation to J. Lindfors and R. Raito for having given me the opportunite to lecture in the splendid setting of Akäslompolo.

II. Motivations and Preliminary Constraints on Compositeness

The principal motivation for compositeness is the proliferation of apparently "elementary" excitations like quarks and leptons. Although the standard SU(3) x SU(2) x U(1) gauge theory, which we now believe correctly describes phenomenologically strong and electroweak interactions, can accommodate repetitive quark and lepton structures, it provides no guidelines for why they should exist. Indeed, although the standard SU(3) x SU(2) x U(1) model is a beautiful theoretical edifice, it fails to account for at least five attributes of nature - which I believe a reasonable theory should attempt to explain. Let me list and comment upon these deficiencies:

(1) No clues exist in the standard model for the apparent replication of fermions, with essentially identical properties (e.g., e, μ, τ, etc.). Indeed, in the model no constraints on the number of generations exist, although retention of asymptotic freedom for the SU(3) color group requires $N_f \leq 16$, with N_f being the number of flavors.

(2) Along the same lines, the masses of the fermions and bosons in the theory, as well as the various weak mixing angles, are theoretically beyond calculation. That is, these

parameters depend on, in principle, arbitrary coupling constants and vacuum expectation values, which cannot a priori be fixed.

(3) The symmetry breaking mechanism, via an elementary Higgs field, requires incredibly fine tuning of parameters, with respect to the scale of masses set by gravity, $M_{Planck} \simeq 10^{19}$ GeV. That is, there is no natural reason for having scalar excitations with mass much less than M_{Planck}.

(4) Because the charge Q is not a generator of SU(3) x SU(2) x U(1), there is no explanation why the charges of the quarks and leptons take the value they take. The absence of SU(2) x U(1) anomalies in the theory does provide a mild constraint in that one must have that $(TrQ)_{quarks + leptons} = 0$.

(5) The strengths of the SU(3), SU(2) and U(1) coupling constants do not bear any particular relation among themselves. The phenomenologically measured "running" coupling constants in the presently accessible range $\langle E \rangle \sim$ 10-20 GeV, appear to have rather disparate values.

I should emphasize that the above points are <u>not theoretical inconsistencies</u> of the standard model, <u>but physical limitations</u> of the model for describing nature. Because SU(3) x SU(2) x U(1) is a <u>renormalizable</u> theory, certain parameters, like masses, coupling constants, number of particles, etc., can take <u>arbitrary</u> values. Any values for these parameters give a consistent theory, and hence these parameters are <u>not</u> calculable. It could be that the world is such that, for example, the muon to electron mass ratio is incalculable and its value just turns out to be near 200. If this is so, then one needs look no further than the standard model. However, if one believes in the calculability of physical parameters, like the muon to electron mass ratio, then <u>one must go beyond the standard model</u>.

Four different, and in some sense complementary, ideas have emerged as ways to go beyond the standard model:

(a) Technicolor [5]

(b) Grand Unified Theories (GUTs) [6]

(c) Supersymmetric Theories [7]

(d) Compositeness

Each of these extensions of the standard model "explain" better some of the points in the "deficiency list" than others. On the other hand, certain points are left unexplained. For instance, GUTs typically are built so that the SU(3), SU(2) and U(1) coupling constants are related and so that the charge Q appears as a generator. Thus points (4) and (5) are understood. However, GUTs in general have no explanation for generations and little, or no, successful mass interrelations. Furthermore, the hierarchy problem - point (3) - is exacerbated in GUTs.

In these lectures, I shall try to address the open questions of the standard model from the point of view of compositeness. I will, however, along the way need to incorporate various features of GUTs, Technicolor and Supersymmetry to try to implement the goal of having a calculable theory. I should note that compositeness itself is a loose concept which takes different meanings depending on what objects one thinks are composite. In ascending order of what one may call radicalness one can contemplate:

(a) Composite Higgs bosons

(b) Composite quarks and leptons

(c) Composite heavy gauge bosons ($W^{\pm}$,Z)

(d) Composite massless gauge bosons (γ, gluons, graviton)

In these notes I shall stop at the second degree of radicalness, in which Higgs bosons, as well as quarks and leptons are composite [F2].

My stopping at composite quarks and leptons does not imply, necessarily, that there is no interesting physics beyond this level. Rather, I am afraid to speculate further. In the lectures of Fritzsch [8] at this school you will find some of the intriguing consequences of supposing that the $W^{\pm}$ and Z are composite and that the weak interactions are the Van der Waal forces of a yet more elementary interaction. An even more speculative idea, which is fraught with theoretical dangers, is that even the massless gauge fields - the photon, gluons and the graviton - are not elementary. There have been some recent attempts in this direction, mostly by Terazawa and his collaborators [9]. Somewhat different ideas, but still rather radical, are being pursued by the Munich school of H.P. Dürr, which supposes that all interactions and fermionic excitations emerge from an underlying nonlinear spinor field theory [10]. Their suggestion is that, although the underlying fundamental theory has only a small symmetry group - usually SU(2), the nonlinear binding generates effective approximate symmetries in a way analogous to how in atomic physics one generates large approximate symmetries from just angular momentum conservation. Although these attempts are all rather interesting, I will not pursue them further here.

III. Limits on the Scale of Compositeness

Any attempt of thinking of quarks and leptons as composite must face, immediately, the fact that these objects appear, by today's experimentation, as pretty elementary or point-like. If one lets the effective radius of quarks and leptons be $\langle r \rangle$, or perhaps, better, associates a momentum scale $\Lambda \sim \langle r \rangle^{-1}$ characteristic of the scale where the composite nature of quarks and leptons becomes manifest, then one can establish from experiment, respectively, an upper bound for $\langle r \rangle$ or a lower bound for Λ. It behooves us, as a first assay into compositeness, to see what these bounds are [1-4]. We remark, however, that one should not take these bounds too strictly, since many of them depend on particular model assumptions. They, nevertheless, serve to give one a benchmark for the distance below which, or the énergies above which, compositeness may become apparent.

Before examining bounds for the compositeness scale of quarks and leptons it will be useful, for future use, to establish a scale, Λ_{TC}, associated with having the elementary

Higgs excitation ϕ be replaced by interactions (Technicolor) of new fermionic degrees of freedom (technifermions, T). As is well known, in Technicolor theories [5] the technifermion condensates play the role of the Higgs vacuum expectation value:

$$\langle \phi \rangle \rightarrow \langle \bar{T} T \rangle \tag{III.1}$$

and no elementary Higgs fields exist.

The argument to obtain Λ_{TC} is rather standard [5], but it is worth repeating. It goes as follows: One knows that in QCD - the SU(3) of color part of the standard model - the chiral symmetry obtained by letting quark masses go to zero ($m_q \rightarrow 0$) is broken spontaneously, with the ensuing Goldstone bosons being the pions [F3]. Although QCD, in this limit, is a classically scale invariant theory, one can associate a typical scale for it, Λ_{QCD}, related to renormalization effects. This is the scale which appears in the running coupling constant, and which can be "measured" in deep inelastic scattering experiments. Alternatively, and more conveniently, from our point of view, Λ_{QCD} can be defined as the scale where $\alpha_3(q^2) = g_3^2(q^2)/4\pi$, with $g_3(q^2)$ being the SU(3) running coupling constant, goes to unity. In view of what happens when $m_q \rightarrow 0$, one arrives at the following formula for the pion mass:

$$m_\pi^2 \sim m_q \Lambda_{QCD} \tag{III.2}$$

with Λ_{QCD} appearing since the pion mass must depend in some way on the strong coupling constant α_3 and yet vanish in the chiral limit.

Eq. (III.2) is a more modern version of an old result by Gell-Mann, Oakes and Renner [11] which related m_π^2 to the pion decay constant $f_\pi \simeq 95$ Mev and the vacuum expectation value of the light quark densities, $\langle \bar{u}u \rangle$, whose existence in the chiral limit breaks chirality spontaneously:

$$m_\pi^2 = \frac{m_q \langle \bar{u} u \rangle}{f_\pi^2} \tag{III.3}$$

Since $\langle uu \rangle$ is nonvanishing as $m_q \rightarrow 0$, it follows from dimensional grounds, that

$$\langle \bar{u} u \rangle \sim \Lambda_{QCD}^3 \tag{III.4}$$

and thus

$$f_\pi \sim \Lambda_{QCD} \tag{III.5}$$

This result, which will be needed shortly, embodies the fact that f_π itself is perfectly well defined in the chiral limit. It correpsonds to the statement, in the σ-model of Gell-Mann and Levy [12], that $f_\pi = \langle \sigma \rangle$.

The key observation of Susskind [5], which was at the origin of Technicolor, is that pure QCD in the chiral limit ($m_q = 0$ and $\langle \bar{u}u \rangle \neq 0$) also breaks SU(2) x U(1) spontaneously.

The composite pion takes the role of the Higgs fields and gets absorbed to give the $W^{\pm}$ and Z masses. The Higgs expectation value is replaced, as in the σ-model, by f_π. Thus even without Higgs fields, the standard model suffers spontaneous breakdown of SU(2) x U(1) and one finds a W-mass [5]

$$M_W = \frac{e f_\pi}{2 \sin\theta_W} \simeq 35 \text{ MeV} \tag{III.6}$$

This formula is totally inadequate phenomenologically, but suggests that if there existed another QCD-like theory (Technicolor) among other fermionic excitations (Technifermions) with $\Lambda_{TC} \gg \Lambda_{QCD}$, then one could obtain a perfectly adequate value for $M_W \approx 80$ GeV. If T denotes the technifermion fields, one has in analogy to the $\langle \bar{u}u \rangle$ condensate the chiral breaking condensate

$$\langle \bar{T}T \rangle \sim \Lambda_{TC}^3 \tag{III.7}$$

and the technipion decay constant

$$F_\pi \sim \Lambda_{TC} \tag{III.8}$$

Whence, the scaling law

$$\frac{f_\pi}{\Lambda_{QCD}} \sim \frac{F_\pi}{\Lambda_{TC}} \tag{III.9}$$

along with the analog of (III.6):

$$M_W = \frac{e F_\pi}{2 \sin\theta_W} \simeq 80 \text{ GeV} \tag{III.10}$$

yields for Λ_{TC} the value

$$\Lambda_{TC} \simeq 10^3 \text{ GeV} \tag{III.11}$$

For the scale of quark and lepton compositeness, which we shall call Λ_{MC}, one cannot be as precise. It is pretty clear that Λ_{MC} must be quite large, since from all indications, leptons and, inferentially, quarks look pretty elementary. This observation can be made a bit more quantitative, and a variety of bounds for Λ_{MC} ensue which, however, depend sometimes on specific theoretical assumptions:

(1) Deviations from the electroweak theory at PETRA, parametrized by some cutoff parameter Λ_{MC}, give typical bounds [13]

$$\Lambda_{MC} \gtrsim 150 - 200 \text{ GeV} \tag{III.12}$$

(2) The values of (g - 2) for electrons and muons agree amazingly well with theoretical predictions from QED [14]

$$\delta a = \frac{1}{2}\left\{(g-2)_{exp} - (g-2)_{QED}\right\} = \begin{cases} 2.7\times 10^{-10} & (e) \\ 1.5\times 10^{-8} & (\mu) \end{cases} \tag{III.13}$$

One can interpret these tiny discrepancies as coming from an effective extra interaction, arising because the leptons are not pointlike:

$$\delta\mathcal{L} = \frac{e}{\Lambda_{MC}}\,\bar{\ell}_L\sigma_{\mu\nu}\ell_R F^{\mu\nu} \tag{III.14}$$

Then $\delta a \simeq (m_\ell/\Lambda_{MC})$, where m_ℓ is the lepton mass, and this gives the bound:

$$\Lambda_{MC} \gtrsim 2\times 10^6\ \text{GeV} \tag{III.15}$$

This bound, however, may be naive [15] It is rather likely, as we shall explain soon, that any chirally violating interaction, like that of Eq. (III.14), must contain factors proportional to the lepton mass. Thus Eq. (III.14) should perhaps be replaced by

$$\delta\mathcal{L} = \frac{e\,m_\ell}{\Lambda_{MC}^2}\,\bar{\ell}_L\sigma_{\mu\nu}\ell_R F^{\mu\nu} \tag{III.16}$$

and $\delta a \simeq (m_\ell/\Lambda_{MC})^2$. Then the bounds are very much weaker [15]:

$$\begin{aligned} \Lambda_{MC} &\gtrsim 50\ \text{GeV} \quad (e) \\ \Lambda_{MC} &\gtrsim 900\ \text{GeV} \quad (\mu) \end{aligned} \tag{III.17}$$

(3) The present limit on the decay $\mu\to e\gamma$, $B(\mu\to e\gamma) \lesssim 2\times 10^{-10}$ also can give a useful limit [16]. If one assumes that the transition $\mu\to e\gamma$ is induced by a term similar to (III.16) - including an explicit lepton mass term -

$$\mathcal{L}_{\mu\to e\gamma} = \frac{e\,m_\mu}{\Lambda_{MC}^2}\sin\theta_{e\mu}\,\bar{e}_L\sigma_{\mu\nu}\mu_R F^{\mu\nu} + \text{h.c.} \tag{III.18}$$

then one obtains the bound [16]

$$\Lambda_{MC} \gtrsim 2\times 10^5\,(\sin\theta_{e\mu})^{1/2}\ \text{GeV} \tag{III.19}$$

In the above, $\sin\theta_{e\mu}$ is an unknown intragenerational mixing angle. If this angle is of the order of the Cabibbo angle, then (III.19) is a rather strong bound. If, however, one assumes that $\mu\to e\gamma$ vanishes as both m_e and m_μ vanish, then $\sin\theta_{e\mu}\sim m_e/\Lambda_{MC}$ and the bound is correspondingly reduced:

$$\Lambda_{MC} \gtrsim 100 \text{ GeV} \tag{III.20}$$

In my opinion the second option above for $\sin\theta_{e\mu}$ appears to be the more sensible one.

(4) The strongest potential bound on Λ_{MC} comes from a possible direct contribution to the $K_L^o - K_S^o$ mass difference, arising from effective mixings between s and d quarks. Compositeness could induce a 4-fermion term

$$\delta\mathcal{L} = \frac{\sin^2\theta_{ds}}{\Lambda^2_{MC}} (\bar{d}_L\gamma^\mu s_L)(\bar{d}_L\gamma_\mu s_L) \tag{III.21}$$

with θ_{ds} being again an intragenerational angle. If θ_{ds} is of the order of the Cabibbo angle, then the observed $K_S^o - K_L^o$ mass difference gives [3]:

$$\Lambda_{MC} \gtrsim 10^5 - 10^6 \text{ GeV} \tag{III.22}$$

Because the operator in (III.21) does not flip chirality, there is no reason why θ_{ds} should vanish as the quark masses vanish. Thus we cannot, in this way, weaken the above bound. Nevertheless, there may be dynamical circumstances which could make θ_{ds} much smaller than θ_c, so that lower values of Λ_{MC} than the above can be contemplated. We will return to this point later.

IV. Dynamical Requirements

From the bounds on Λ_{MC} obtained in the preceeding section, it is clear that if one is to contemplate quarks and leptons as composite objects then the dynamics must be such that the masses of the bound state objects turn out to be much less than Λ_{MC}:

$$m_q, m_\ell \ll \Lambda_{MC} \tag{IV.1}$$

Such a constraint is rather a "funny" one to require of the dynamics. It means that the typical size of quarks and leptons $\langle r\rangle \sim 1/\Lambda_{MC}$ is much smaller than their own Compton wavelength! In atomic physics we are used to thinking of objects which are much bigger than the relevant Compton wavelengths. For instance for positronium $\langle r\rangle_{pos.} \simeq 1/m\alpha$ and

$$\langle r\rangle_{pos.}\, m \gg 1 \tag{IV.2}$$

In QCD, hadrons have sizes which are comparable to their Compton wavelength

$$\langle r\rangle_{prot.}\, M_{prot.} \sim 1 \tag{IV.3}$$

The constraint (IV.1) reverses the inequality in Eq. (IV.2).

I would like to argue that the dynamical requirement (IV.1) will only be sensibly satisfied by theories in which there is an (approximate) symmetry which forces the masses m_q, m_ℓ to (nearly) vanish. The argument goes as follows: because quarks are confined, whatever force binds their constituents (which I shall call, following general usage, preons) must also be a confining force. Confined quarks made up of unconfined preons make no sense! Thus the preon theory must also be some non-Abelian gauge theory. Furthermore, it is reasonable not to give any mass to the preons themselves. Doing so, would start anew the round of questions: what sets the scale of m_{preon}, etc. Now a non-Abelian gauge theory of massless preons is a classically scale invariant theory, which quantum mechanically acquires a scale through the running of its coupling constant. The "typical" scale Λ_{MC} of such a theory can be defined as the place where the running coupling constant $\alpha_{MC} = g^2_{MC}/4\pi$ goes through unity, as shown in Fig. IV.1

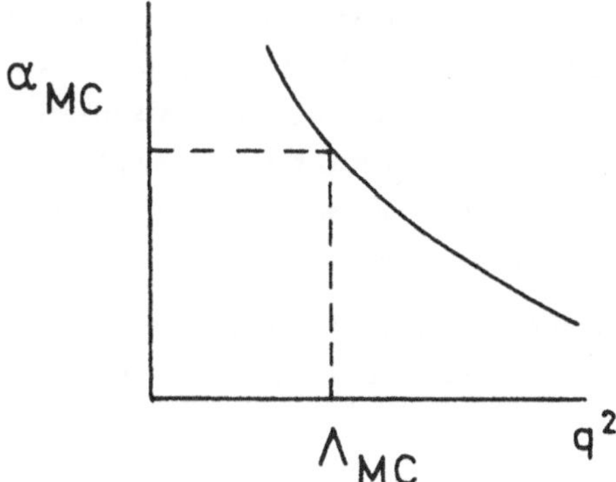

Fig. IV.1: Definition of Λ_{MC} from the behaviour of the running coupling constant

Because Λ_{MC} is the only scale of the preon theory, it is clear that all preon bound states must have masses $M \sim \Lambda_{MC}$. The only exception to this rule can be for states which by some symmetry reason are forced to have $m = 0$. Two flavor QCD in the chiral limit, $m_u = m_d = 0$, is precisely an example of such a theory. All states (proton, Δ-resonances, ρ mesons, etc.) have masses $M \sim \Lambda_{QCD}$, except for the pions, which being the Goldstone bosons of a spontaneously broken chiral symmetry, have $m_\pi = 0$. If this analogy applies to the preon theory, one would have a spectrum of preon bound states with masses $M \sim \Lambda_{MC}$ and then, forced by some yet unspecified symmetry, one would have in addition massless quarks and leptons.

Three questions immediately arise from these considerations:

(1) What symmetries force m_q, m_ℓ to vanish?

(2) How does one then, in actuality, get the small (with respect to Λ_{MC}) quark and lepton masses?

(3) What relation does Λ_{MC} have with other scales we know, like Λ_{QCD} and M_W?

Before I tackle these questions, it is worthwhile making an aside. There is a direct benefit of being forced, in first approximation, to have quark and lepton masses which vanish dynamically: then one may "understand" why no angular momentum repetitions of quarks and leptons appear. One is used in atomic, nuclear and hadronic physics to the idea that compositeness implies necessarily the appearance of angular momentum excitations. Why then, if quarks and leptons are composite do we not see spin 3/2, 5/2, etc. repetitions of these states?

The answer to this query is provided by a theorem, first enunciated in the early 60's by Case and Gasiorowicz [17] and rediscovered and amplified recently by Weinberg and Witten [18]. The theorem states that in a theory with a Lorentz covariant energy momentum tensor, $\theta_{\mu\nu}$, one can have no massless particles with $J > 1$. A very nice discussion of this theorem is contained in the review of Peskin [1]. One expects all composite models in which the binding is nongravitational (binding related to gravity is unlikely) to obey the conditions of the theorem. Thus, there is a clear dynamical reason for only observing (approximately) massless quarks and leptons and no higher spin excitations with low mass. Of course, higher spin states of mass $M \sim \Lambda_{MC}$ are perfectly well allowed.

I begin now by trying to address the first dynamical question posed above: What symmetries of the preon theory force the quark and lepton masses to vanish? To my knowledge there are only three possible options:

(a) m_q, $m_\ell = 0$ because the preon theory has an overall chiral symmetry. Furthermore, when composites form this chirality is not spontaneously broken, thereby requiring massless fermions in the theory.

(b) The quark and lepton masses vanish because the preon theory is a supersymmetric theory and these excitations are the Goldstinos arising from the spontaneous breakdown of the supersymmetry.

(c) m_q, $m_\ell = 0$ because the preon theory is a supersymmetric theory which possesses an overall global symmetry G. When composite bound states form, G is spontaneously broken although the supersymmetry remains exact. Then the quarks and leptons can be identified as the massless fermionic partners of the Goldstone bosons arising from the G-breakdown.

There are a variety of comments that can be made concerning the above possibilities. Option (a) was suggested originally by 't Hooft [19] and in many respects appears to be the most reasonable possibility. As we mentioned before, it is rather natural to assume that preon masses vanish. Hence the presence of an initial chiral symmetry appears to be a good supposition. What is harder, however, is to guarantee that in the binding the chiral symmetry is not spontaneously broken. I will return shortly to discuss constraints that need to be satisfied to guarantee that chirality is preserved. Here, let me just add as a further point that if chirality is preserved in the binding, one can understand, immediately, the appearance (or not) of the various factors m/Λ_{MC} in our discussion of the preceeding section on bounds on Λ_{MC}. If chirality is preserved in the binding, then any effective operators which cause transitions which break chirality (like those of Eq. (III.16)) must vanish as the quark and lepton masses vanish.

Option (b) was suggested by Bardeen and Visnjić [20] and, although it is very intriguing, it suffers in credibility because it requires that one should ascribe each known fermion to a distinct broken supersymmetric generator. Furthermore, because for $N > 8$ one has theories with $J > 2$, one is faced with severe theoretical problems, since there are no consistent ways to handle field theoretically such excitations [F4]. I will ignore this option for the remainder of these notes.

The last option (c), above, is an idea that grew out of some work on supersymmetric theories which I have done recently in collaboration with W. Buchmüller, S.T. Love and T. Yanagida [21]. One of the drawbacks of this option is that along with the massless fermions one necessarily has massless bosons. This, however, may not be all that serious since supersymmetry must eventually be broken and it could well be that the bosonic partners of quarks and leptons get heavy masses in this way. A more serious drawback of this idea is that the massless fermions one obtains are real. This follows directly since the massless fermions are supersymmetric partners of the Goldstone bosons arising from the breakdown of the global symmetry G to G'. Hence the fermions sit in the adjoint representation of G/G' and are either real or come in complex conjugate pairs. In fact, this may not be totally bad since by choosing G and G' appropriately one may get at least the massless fermions one wants, besides the inevitable doubling. A particularly nice example is provided by the breakdown of E_6 into SO(10). In this breakdown the adjoint of E_6 decomposes in terms of SO(10) representations as

$$78 = 45 + 16 + \overline{16} + 1 \tag{IV.4}$$

and a whole 16 of zero mass fermions appears. However, the problem remains of how, when masses are eventually generated, one guarantees that the $\overline{16}$ states get much heavier masses than the 16 states.

I will in what follows, when appropriate, indicate possible consequences of this last way of generating zero mass fermions. However, for the most part I shall assume that m_q and m_ℓ vanish, as a first approximation, because of an unbroken chiral symmetry. To be more specific, I imagine that the preon theory is some non-Abelian gauge theory based on a group G_{MC}, which, furthermore, has some overall chiral global symmetry G_f. Quark and leptons are then supposed to emerge as G_{MC} singlet bound states of preons, which have zero mass because the G_f chirality is unbroken.

This scenario suffers from an immediate problem. The only theory we know which has the same starting hypothesis - namely, QCD - is not realized in this way! QCD of n_f flavors of quarks, in the chiral limit when all the quark masses vanish, has an overall

$SU(N_f) \times SU(N_f) \times U(1)$ global symmetry, which reflects the ability one has to rotate independently the n_f right and left helicity quarks among each other. We know, however, that in the binding process of making hadrons the chiral symmetry does not survive, since chiral breaking condensates like $\langle\bar{u}u\rangle \neq 0$ form. In fact, $SU(N_f) \times SU(N_f) \times U(1)$ breaks down to $SU(N_f) \times U(1)$, with the consequence that protons are massive, but there exist $n_f^2 - 1$ Goldstone bosons.

For the preonic theory one would like to have the opposite happen. No condensates must form to destroy the chirality and no Goldstone bosons must ensue. One may honestly ask whether this ever obtains? We should remark, that, in this sense, the example of a supersymmetric theory with an internal symmetry G is closer in spirit to what happens in QCD. In this case we do want a G breakdown to occur to get Goldstone bosons and their $m = 0$ fermionic partners.

In his Cargese lectures [19] G. 't Hooft enunciated a set of necessary criteria which must be obeyed if one wants overall global chiral symmetries not to undergo spontaneous breakdown. These are the so-called 't Hooft consistency conditions. Before examining these conditions in detail, I want to make two qualitative remarks on which theories might expect to show different chiral behaviour than QCD:

(1) If the preon confining group $G_{MC} = SU(N)$ and the preons are in vector-like representations, leading to some overall G_f chiral global symmetry, it is clearly difficult to see why this theory should not behave as QCD. On the other hand, if the preons are in some chiral representation, which must be anomaly free with respect to SU(N), one can imagine things may be different.

(2) Even for the case of purely real representations, it may be possible that special circumstances exist which do not lead to the formation of any chirality breaking condensates. A nice example of this point was provided by Barbieri, Maiani and Petronzio [16]. Suppose $G_{MC} = O(N)$ and assume that the preons χ_i transform according to the adjoint representation of O(N). Then one can construct composite states ψ simply by screening the metacolor via an O(N) gluon, g_i. That is,

$$\psi \sim \chi_i g_i \tag{IV.5}$$

In this theory one can argue that, perhaps, no condensates of the type

$$\langle \chi_i \chi_i \rangle = \langle \sigma \rangle \neq 0 \tag{IV.6}$$

form, since the "bound" states σ actually would be unstable via $\sigma \to \psi\psi$, as indicated schematically in Fig. IV.2

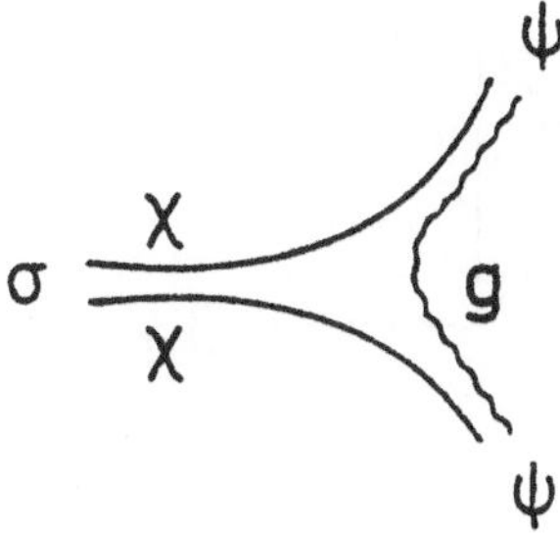

Fig. IV.2: $\sigma \to \psi\psi$ decay

Some confirmation of this idea has emerged from a lattice calculation by Banks and Kaplunovsky [22].

The 't Hooft consistency conditions are a remarkably simple, yet profound, requirement for preserving chirality in the binding process. They are based on the fact that theories which possess global chiral symmetries have Green's functions, involving the currents associated with these symmetries, which have anomalous divergences. One knows from the pioneering work of Adler and Bell and Jackiw [23] that even though chiral currents may be formally conserved, at the Lagrangian level, their divergence are anomalous in actual Feynman graph calculations. In particular, the anomaly arises because of singular short-distance behaviour in triangle graphs involving fermionic loops and has a universal character.

Let J_μ be a global symmetry current and define the three point function [F5]

$$(2\pi)^4 \delta^4(\Sigma q_i)\, \Gamma_{\alpha\beta\gamma}(q_1,q_2,q_3) = \int \prod_i d^4x_i\, e^{iq_i x_i} \langle 0|T(J_\alpha(x_1)\, J_\beta(x_2)\, J_\gamma(x_3))|0\rangle \tag{IV.7}$$

Then one can show that $\Gamma_{\alpha\beta\gamma}$ has an anomalous divergence:

$$q_3^{\ \gamma}\, \Gamma_{\alpha\beta\gamma}(q_1,q_2,q_3) = A_J\, \epsilon_{\alpha\beta\rho\tau}\, q_1^{\ \rho} q_2^{\ \tau} \tag{IV.8}$$

where the coefficient A_J is a constant. More precisely, A_J is determined entirely in terms of the representation content of the fermions to which J_μ couples in the triangle graph of Fig. IV.3.

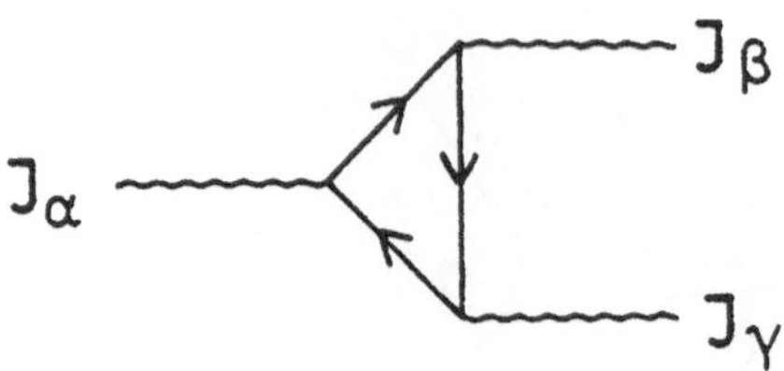

Fig. IV.3: Contribution to the anomaly of $\Gamma_{\alpha\beta\gamma}$

If we write the current J_μ in terms of the fermions of the theory as

$$J_\mu = \sum_i \bar{\psi}_i \left\{ \lambda_L^{(i)} \gamma_\mu \frac{(1-\gamma_5)}{2} + \lambda_R^{(i)} \gamma_\mu \frac{(1+\gamma_5)}{2} \right\} \psi_i \tag{IV.9}$$

where $\lambda^{(i)}_{L,R}$ are appropriate representation matrices, then [F6]

$$A_J = \frac{1}{8\pi^2} \sum_i \mathrm{Tr} \left\{ (\lambda_R^{(i)})^3 - (\lambda_L^{(i)})^3 \right\} \tag{IV.10}$$

The 't Hooft consistency conditions can now be stated very simply [19]. Compute in the preon theory A_J for all possible global symmetry currents. Then a necessary condition for chirality not to be spontaneously broken in the process of binding is that, for all currents J_μ,

$$(A_J)_{\text{bound states}} = (A_J)_{\text{preons}} \tag{IV.11}$$

Here $(A_J)_{\text{bound states}}$ is to be computed only in terms of the massless excitations which are presumed to result from the binding. Note that (IV.11) is only a <u>necessary</u> condition. There is no guarantee that even if (IV.11) is satisfied chirality will remain unbroken. However, if (IV.11) does not hold, then necessarily chirality is spontaneously broken.

The anomaly matching condition of 't Hooft has been elucidated by the work of Frishman, Schwimmer, Banks and Yankielowicz [25] and more recently by Coleman and Grossman [24]. Remarkably, the existence of the anomaly (IV.8), which is a short distance effect, implies something for the form of $\Gamma_{\alpha\beta\gamma}$ for <u>all values</u> of momentum. In particular, one can show [24,25]that Eq. (IV.8) implies, at the symmetric point $q_1^2 = q_2^2 = q_3^2 = q^2$, that

$$\Gamma_{\alpha\beta\gamma}(q_1,q_2,q_3)\Big|_{\substack{\text{sym.}\\ \text{pt.}}} = \frac{A_J}{q^2}\Big\{\epsilon_{\alpha\beta\rho\tau}q_1^\rho q_2^\tau q_{3\gamma} + \epsilon_{\beta\gamma\rho\tau}q_2^\rho q_3^\tau q_{1\alpha}$$

$$+\epsilon_{\gamma\alpha\rho\tau}q_3^\rho q_1^\tau q_{2\beta}\Big\} + \qquad \text{(IV.12)}$$

$$+ \text{ Non singular terms}$$

The $1/q^2$ singularity above is a direct consequence of the anomaly. Since $\Gamma_{\alpha\beta\gamma}$ is a perfectly respectable Green's function of the theory, this singularity must also be reproduced when one calculates this Green's function in terms of the bound states of the theory. Now such singularities can only be a result of massless particles in the bound state spectrum. There are two possibilities: If there is spontaneous breakdown of the symmetry associated with J_μ, then we expect to have a Goldstone boson in the theory. In this case Eq. (IV.12) relates A_J to the coupling of the Goldstone boson to the remaining other two currents. This is precisely what happens in QCD and through the anomaly one can compute the $\pi^0 \to 2\gamma$ rate [26]. If, on the other hand, the symmetry associated with J_μ is not spontaneously broken, then there must be massless fermions in the bound state spectrum. Their contribution to the singularity in (IV.12) must also have residue A_J. This is precisely the 't Hooft constraint (IV.11). Clearly if the bound state spectrum of massless excitations does not satisfy Eq. (IV.11), it cannot be correct and the global chiral symmetry must suffer spontaneous breakdown.

Various comments on the 't Hooft anomaly matching conditions deserve to be made:

(1) The 't Hooft conditions require that one knows how the bound states of the preon theory are formed. The usual assumption made is that the quarks and leptons are made up of the simplest combination of preons which are G_{MC} singlets. In practice, however, one can imagine that more complicated contributions are also allowed. Thus the anomaly matching of Eq. (IV.11) is not purely an algebraic exercise, but requires some dynamical input.

(2) The flavor group G_f whose currents one uses to do the matching must contain SU(3) x SU(2) x U(1). In fact, G_f is only a symmetry, in general, in the limit in which the SU(3) x SU(2) x U(1) coupling constants are set to zero. Thus one expects that a larger set of zero mass states than the quarks and leptons are generated, and it is with respect to this larger set that one must do the anomaly matching. Eventually, for nonvanishing α_i, all massless bound states not forbidden by SU(3) x SU(2) x U(1) from acquiring mass will get masses of $O(\alpha_i\Lambda_{MC})$, where Λ_{MC} is the G_{MC} characteristic scale.

(3) The anomaly matching condition is usually augmented by a logical, but very restrictive, condition - the, so-called, persistent mass condition [19,27]. This extra condition is easily explained. Imagine that one of the preons acquires a very large mass $M \to \infty$. Then one would expect that all the bound states containing this preon would similarly get very large mass, $M \to \infty$. The resulting theory has a smaller global symmetry group G_f' and also fewer zero mass bound states. The persistent mass condition requires that also for this group the anomaly matching of Eq. (IV.11) hold. Having to match anomalies for a sequence of flavor groups G_f, G_f', G_f'', etc. in practice is very difficult to do. Thus vector-like theories in which preons can acquire G_{MC} singlet mass terms are often in trouble with the persistent mass condition. On the other hand, purely chiral theories, in which no mass terms are allowed, are better candidate theories since then one can bypass this constraint [F7].

I will not discuss in detail how one obtains solutions to the 't Hooft equations. I refer the interested reader to the very nice paper of Bars [28], where a very general treatment of the problem is presented. The result of these analyses is rather disappointing. The allowed groups G_{MC} and G_f turn out to be very big, while the preon representations are repetitious and have not simple structures. Some models even have more preons than the quarks and leptons they finally produce! A not atypical illustration [28], which in fact has some features I shall exploit later, is provided by the following preon model. The metacolor group is G_{MC} = SU(4) and there are three different kinds of preons

$$\begin{aligned} P_{1L} &\sim \overline{10} \\ P_{2L} &\sim \bar{4} \\ P_{2R} &\sim \bar{4} \end{aligned} \qquad \text{(IV.13)}$$

with the transformation properties under SU(4) indicated above. However, these preons come in replicas of 1, 8 and 16, respectively, so that the flavor group is U(1) x SU(8) x SU(16). There are three types of SU(4) singlet bound states formed, and I give below their SU(8) x SU(16) properties:

$$\begin{aligned} A &= (\bar{8}, \overline{16})_R \sim P_{1L}\, \bar{P}_{2L}\, \bar{P}_{2R} \\ B &= (\overline{36}, 1)_L \sim P_{1L}\, \bar{P}_{2L}\, \bar{P}_{2L} \\ C &= (1, 120)_L \sim P_{1L}\, \bar{P}_{2R}\, \bar{P}_{2R} \end{aligned} \qquad \text{(IV.14)}$$

The G_f anomalies of (IV.13) and (IV.14) match and the persistent mass relations, obtained by successively making pairs of P_{2L}, P_{2R} massive, also lead to further anomaly matching.

It can be shown that SU(3) x SU(2) x U(1) can be embedded in the SU(16) and that the SU(8) can be taken as a family group. Then the states A represent eight families of 16 quarks and leptons, with the conventional charges. The states B and C, when the SU(3) x SU(2) x U(1) coupling constants are turned on, are unprotected, being real under this group, and can acquire masses of $O(\alpha_i \Lambda_{MC})$.

Although the above example has some nice features, it is clearly not simple. In fact, under certain aspects, at this stage the preon world looks more complicated than the one of quarks and leptons. Roughly speaking, the 8 P_{2L}'s act as family preons, while the 16 P_{2R}'s are quantum number preons. However, including their metacolor degree of freedom the "elementary states" (202) are more than the resulting 8 families of 16 quark and lepton states (128)! The only nontrivial result is that the anomaly matching has required the appearance of certain repetitive bound state structures. This may be perhaps the first dynamical glimpse on the origin of generations.

Given the somewhat unsatisfactory state of affairs with the chiral idea for obtaining massless fermions, the supersymmetric option (c) which I indicated earlier - even with its fermion doubling - may begin to look attractive. First of all, in this case one is dealing with spontaneously broken flavor groups G_f. Thus one does not need to worry about anomaly matching to preserve some chiral symmetry. Furthermore, even though one is considering supersymmetric theories, the metacolor group G_{MC} can still be confining and one can perfectly well consider quarks and leptons as preon bound states.

The dynamical issue involved in obtaining essentially massless composite quarks and leptons are clearly far from settled. I will nevertheless assume that someday one shall be able to overcome this hurdle. Then one will be faced with the next dynamical question: How do quarks and leptons acquire their small dynamical masses? I will present in the next section some ideas on this question developed in collaboration with T. Yanagida [29].

V. Mass Generation and Families

I believe it is important to distinguish two aspects related to the question of quark and lepton masses. Namely, one ought to separate the issue of the mass generation mechanism from the issue of family replication. Family replication may be purely mechanical, forced on the theory, say, by the anomaly matching or by, in the supersymmetric option, a particular breakdown of $G \rightarrow G'$. It is of course probable that there is a much deeper reason for families. Thus, it may well be that progress will be made only by really understanding what causes families to appear. Nevertheless, even if there were no families, one would still have to ask the question of what gave the electron and the up and down quarks mass. A reasonable approach to take, and one which I shall follow here, is to assume that some mechanism caused the existence of families and try to understand, at least qualitatively, how their rather bizarre mass spectra gets generated.

I will approach this question in two stages. First I shall worry about mass generation proper and then I shall worry about what causes the mass splittings. Ideas similar to what I shall present here have been developed independently by Bars [28] and Preskill [30] and are also contained in my paper with Yanagida [29]. Quark and lepton masses violate SU(2) x U(1). At Λ_{MC}, SU(2) x U(1) can be taken to be unbroken and so it is perfectly reasonable to assume that quarks and leptons are among the G_{MC} singlet massless bound states. If quarks and leptons are composite, it is also sensible that there should be no elementary Higgs bosons. Thus some sort of Technicolor mechanism should be responsible for the SU(2) x U(1) breakdown, by the formation of technifermion condensates.

The following scenario can be envisaged. The preon theory is such that the flavor group of the theory contains besides SU(3) x SU(2) x U(1) also a technicolor group G_{TC}:

$$G_f \supset G_{TC} \times SU(3) \times SU(2) \times U(1) \tag{V.1}$$

Then one expects at Λ_{MC} to form as preon bound states not only massless quarks and leptons, but also massless technifermions, T. When, at a scale Λ_{TC}, the Technicolor coupling constant becomes strong - thereby producing Technicolor singlet bound states of masses of $O(\Lambda_{TC})$ - one can assume that technifermion condensates form, which lead to a breakdown of SU(2) x U(1). The existence of condensates $\langle \bar{T}_L T_R \rangle$, with nonvanishing SU(2) x U(1) properties, generates a mass for the $W^{\pm}$ and Z^0 bosons ($M_W \sim \Lambda_{TC}$), as discussed in Section II. More importantly, once SU(2) x U(1) is spontaneously broken, also quark and lepton masses can be generated.

A mechanism for generating the quark and lepton masses in the above picture is as follows. When the metacolor group becomes strong, at scales of $O(\Lambda_{MC})$, there will be, besides massless preon bound states, also a full spectrum of metahadrons of masses of $O(\Lambda_{MC})$. One may write effective Lagrangians which describe the interactions of the zero mass states - quarks, leptons and technifermions - with the massive metahadron states in much the same way that in QCD one can write effective interactions between the (nearly massless) pions and the nucleons. Schematically, for the quarks, one has

$$\mathcal{L}_{eff} = g\,\bar{q}_L T_R \varphi + g'\,\bar{q}_R T_L \tilde{\varphi} + \text{h.c.} \tag{V.2}$$

Here φ and its charge conjugate field $\tilde{\varphi}$ are massive states ($M \sim \Lambda_{MC}$) which can couple to both quarks and technifermions. Similar equations would also apply for the leptons. I should note that the interaction in (V.2) cannot violate SU(2) x U(1) by assumption, since at Λ_{MC} this is a perfectly good symmetry. Whence the quantum numbers of the T and φ fields must be appropriate. (For example, $T_L \sim 2$, $T_R \sim 1$ and $\varphi \sim 2$ of SU(2).) When at Λ_{MC} $\langle \bar{T}_L T_R \rangle$ condensates form, SU(2) x U(1) is broken spontaneously and masses for the quarks can be generated through the diagram shown in Fig. V.1.

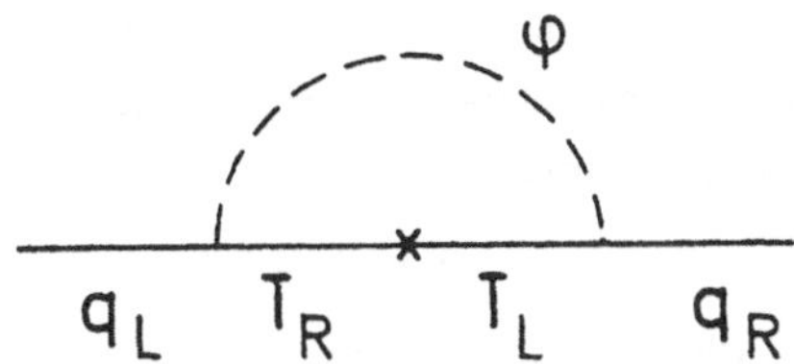

Fig. V.1: Mass generation for quarks via technifermion condensates

Clearly, from this diagram,

$$m_q \sim gg' \frac{\langle \bar{T}_L T_R \rangle}{M^2_\varphi} \sim gg' \frac{\Lambda^3_{TC}}{\Lambda^2_{MC}} \tag{V.3}$$

One notes from the above result that the dependence of m_q (and also of m_ℓ) on the SU(2) x U(1) symmetry breaking scale, Λ_{TC}, is different than that of $M_W \sim \Lambda_{TC}$. This is in contrast to the usual Higgs picture where $m_q \sim M_W \sim \langle \phi \rangle$. Although Eq. (V.3) is not precise, it is clear that if $\Lambda_{TC} \sim 10^3$ GeV, as our discussion in Section III indicated, then Λ_{MC} cannot be extremely large. Otherwise the factor $(\Lambda_{TC}/\Lambda_{MC})^2$ would soon push m_q below the MeV-GeV range. This is an interesting result in that it gives one the first dynamical hint that metacolor is reasonably "nearby". That is $\Lambda_{MC} \ll M_x \simeq 10^{15}$ GeV which is the typical scale of GUTs.

It should be pointed out that the mass generating mechanism just described has many of the characteristics of Extended Technicolor (ETC) [5]. However, it may suffer less from the problems of flavor changing neutral currents than ETC. The point is that ETC is a gauge theory and one cannot suppress transitions involving ETC gauge bosons among quarks, once one admits ETC interactions of quarks with techniquarks. In the present example, there is no gauge principle forcing intraquark transitions. Furthermore, even though the effective Lagrangian (V.2) will give 4-Fermi interactions among quarks, scaled by $1/\Lambda^2_{MC}$, in many instances there will be further suppression factors [F8]. Thus there may be no incompatibility here between having a relatively low value for Λ_{MC} ($\Lambda_{MC} \gtrsim 10^4$ GeV?), needed to generate reasonable quark and lepton masses, and constraints coming from flavor changing neutral currents (see Section III) which would like Λ_{MC} to be larger.

It is clear that Eq. (V.3) is just generic. It explains mass generation for quarks and leptons, but it gives no inkling of why there are such large intrafamily splittings. It does not explain either interfamily splittings, but I shall postpone discussing this interesting subject until later. In Ref. [29] Yanagida and I suggested a rough mechanism which might account for the large intrafamily splittings. We blamed the intrafamily splittings on two phenomena:

(1) The spectrum splittings of the heavy states with $M \sim \Lambda_{MC}$,

(2) the differing coupling of the higher massive states to the zero mass states.

Furthermore, we constructed a simple model to show how this might work.

Imagine that as the result of a preon theory n generations of quarks and leptons are formed but only one generation of technifermions is formed. (This, incidentally is precisely what happens in the Bars model of Section IV, if we embed an SU(2) of TC in the SU(8) family symmetry. The bound states are then 6 families of quarks and leptons plus a technifermion family.) Further, suppose that at Λ_{MC} not a single state φ is formed but an infinite tower of such states $\{\varphi_\alpha\}$. The obvious generalization of Eq. (V.2) is:

$$\mathcal{L}_{eff} = g_i^\alpha \bar{q}_{iL} T_R \varphi^\alpha + g_i'^\alpha \bar{q}_{iR} T_L \tilde{\varphi}^\alpha + h.c. \tag{V.4}$$

Calculation of the mass matrix by diagrams like those in Fig. V.1 yield

$$M_{ij} = \sum_\alpha g_i^\alpha g_j'^\alpha \frac{\Lambda_{TC}^3}{M_\alpha^2} \tag{V.5}$$

It is easy to see that if one bound state dominates - which naturally would be the lowest in the spectrum - then because of the structure of (V.5), the diagonalization of M_{ij} gives:

$$(M_{ij})_{diag} \sim \begin{pmatrix} 0 & & & \\ & 0 & & \\ & & 0 \ddots & \\ & & & 1 \end{pmatrix} \tag{V.6}$$

Clearly, including other states in the sum in (V.5) would alter this result. But if

$$\frac{g^{n+1} g'^{n+1}}{M_{n+1}^2} \ll \frac{g^n g'^n}{M_n^2} \tag{V.7}$$

then one would obtain

$$(M_{ij})_{diag} \sim \begin{pmatrix} \ddots & & & \\ & \delta & & \\ & & \epsilon & \\ & & & 1 \end{pmatrix} \tag{V.8}$$

with $1 \gg \epsilon \gg \delta \gg \ldots$ [F9].

We should remark that conditions like (V.7) are not unknown in QCD. For instance, the coupling of pions to the Roper resonance N' is very much smaller than their coupling to nucleons. One finds:

$$\left(\frac{g^2_{\pi NN'}}{M^2_{N'}}\right) \Big/ \left(\frac{g^2_{\pi NN}}{M^2_N}\right) \sim 1/30 \tag{V.9}$$

Our mechanism [29] for generating sizeable intragenerational splittings is obviously not unique. Preskill [30] suggests instead that these large splittings come from protective symmetries which force the lightest states to obtain mass only by multiple insertions of $\langle \bar{T}_L T_R \rangle$ condensates. Whence, for instance one would have

$$m_\tau \sim \frac{\Lambda^3_{TC}}{\Lambda^2_{MC}} \quad ; \quad m_\mu \sim \frac{\Lambda^6_{TC}}{\Lambda^5_{MC}} \quad ; \quad m_e \sim \frac{\Lambda^9_{TC}}{\Lambda^8_{MC}} \tag{V.10}$$

Before closing this section I want to make a final remark concerning neutrino masses in the above picture. Clearly if neutrinos have a mass, their mass in each family is substantially below the mean mass of the charged states. Such a disparity is difficult to generate naturally, unless right-handed neutrinos exist. Then, even though ν_R as well as ν_L could be forced to zero mass at Λ_{MC} by the preon binding, the turning on of the SU(3) x SU(2) x U(1) coupling constants would induce a natural Majorana mass for ν_R of order $\alpha_i \Lambda_{MC}$, because this state is not protected from getting a mass. Such a large Majorana mass would, by the see-saw mechanism [31], immediately generate neutrinos masses of order

$$m_\nu \sim \frac{\Lambda^6_{TC}}{\alpha_i \Lambda^5_{MC}} \ll \bar{m} \sim \frac{\Lambda^3_{TC}}{\Lambda^2_{MC}} \tag{V.11}$$

I have discussed elsewhere [32] the interesting consequences of having composite Majorana neutrinos in a theory.

VI. Constraints on Composite Models from Proton Decay and the Universe Baryon Asymmetry

In the preceeding section, we have argued that to generate quark and lepton masses the compositeness scale Λ_{MC} could not be much bigger than the technicolor scale $\Lambda_{TC} \simeq 10^3$ GeV. Say, $\Lambda_{MC} \sim 10^4$-10^5 GeV. This value is marginally close to some of the bounds we obtained in Section III. It might be acceptable, if dynamical factors could be invoked to suppress some of the flavor changing processes. However, in our discussion in

Section III we never worried about proton stability. If quarks and leptons are made of the same constituents, we expect to be able to write down an effective Lagrangian which would cause proton decay. Schematically, the simplest such Lagrangian is of the form

$$\mathcal{L}_{\Delta B \neq 0} = \frac{1}{\Lambda_{MC}^2}\, qqq\ell + h.c. \tag{VI.1}$$

where the compositeness scale Λ_{MC} is also the typical scale associated with the baryon violating processes. Since the proton lifetime from (VI.1) is of the order:

$$\tau_p \sim \Lambda_{MC}^4 / M_{prot.}^5 \tag{VI.2}$$

the present limit $\tau_p \lesssim 10^{30}$ years implies $\Lambda_{MC} \sim 10^{14}\text{-}10^{15}$ GeV. Such a value is many orders of magnitude above our estimate for Λ_{MC}! Indeed, turning the argument around, if $\Lambda_{MC} \sim 10^4\text{-}10^5$ GeV, then the present limit on proton decay tells us that no such effective interaction as that displayed in (VI.1) can be generated by the preon theory.

There are a variety of ways to bypass the proton decay conundrum. The interested reader can find a rather complete discussion of this problem in the papers in Ref. [33] of Casalbuoni and Gatto. The simplest way to guarantee that no fast proton decay occurs is to build in baryon conservation in the preon theory. In fact, the Bars example of a preon theory, discussed in Section IV, contains precisely such a custodial baryon conservation. The point is that, in this model, the quantum numbers of a single family of 16 are carried by just one kind of preon. Then preon number conservation is equivalent to baryon number conservation.

There are, however, other ways to prevent, but not forbid, proton decay. For instance, one can build models with discrete symmetries which do not allow writing down the simplest $\mathcal{L}_{\Delta B \neq 0}$ of Eq. (VI.1), but which allow the appearance of effective Lagrangians involving higher dimension interactions. An example of this sort has been discussed by Harari, Mohapatra and Seiberg [34]. In their model the lowest dimensional interaction which could cause proton decay had dimension nine and one has

$$\mathcal{L}_{\Delta B \neq 0} = \frac{1}{\Lambda_{MC}^5}\, qqq\ell\ell\ell + h.c. \tag{VI.3}$$

Then

$$\tau_p \sim \frac{\Lambda_{MC}^{10}}{M_{prot}^{11}} \tag{VI.4}$$

and $\Lambda_{MC} \gtrsim 10^6$ GeV gives an acceptable proton decay rate.

One should note, however, that even if one can bypass the proton decay problem - that is, get a rate for it which is either zero, or which is within the experimental bounds - there is a further constraint on preon theories coming from the universe's baryon asymmetry. I have studied this question with Masiero and Yanagida [35] and we concluded that a "pure" preon theory with $\Lambda_{MC} \ll 10^{15}$ GeV has in fact no chance of being able to generate the observed asymmetry $n_B/n_\gamma \sim 10^{-10}$. I do not want to repeat our arguments in detail. Essentially, preon theories with $\Lambda_{MC} \ll 10^{15}$ GeV lack some, or all, the necessary conditions needed to establish a baryon asymmetry in the early universe (violation of baryon number, violation of C and of CP, nonthermal equilibrium of baryon violating processes).

If one wishes to have a theory where this asymmetry can be possibly generated one must either:

(1) consider preon theories with $\Lambda_{MC} \sim 10^{15}$ GeV,

(2) consider non "pure" preon theories, with $\Lambda_{MC} \ll 10^{15}$ Gev, but ones which are embedded in a GUT.

In view of the rather compelling arguments for considering "low" values for the composite mass scale, the second alternative appears preferable. The idea then is that proton decay is possible, but is a consequence of the GUT theory and so its lifetime is set by $M_X \sim 10^{15}$ GeV. Similarly, the baryon asymmetry in the universe is established by the usual GUT scenario [6]. Below the scale of M_X, in the preon theory itself baryon number is conserved as a custodial symmetry, so that the low scale Λ_{MC} has no relevance for the proton lifetime.

VII. Hierarchical Scales and a Composite GUT Scenario

I have arrived at the need of a composite GUT scenario in a semiphenomenological way, by adducing reasons for both wanting $\Lambda_{MC} \ll M_X$ and yet generating a baryon asymmetry in the universe. There is, however, a different way to arrive at the same conclusion, by focusing on the issue of scales. In the scenario for generating fermion masses of Section V, the preon theory contained a number of gauge interactions. To be precise

$$G_{gauge} = G_{MC} \times G_{TC} \times SU(3) \times SU(2) \times U(1) \tag{VII.1}$$

Although each of the above non-Abelian gauge theories is scale invariant at the classical level, they are all characterized by scale parameters Λ_i, which one needs to determine from experiment [F10]. A perfectly logical question to ask is what determines the ratios Λ_i/Λ_j? This is the analog of the question we asked in the introduction, relating to the ratio m_μ/m_e, but at one level deeper! In principle, at the stage of theoretical development at which we have arrived, any ratio would do. I know of only one way to fix these ratios and that is by assuming that the separate groups in (VII.1) are unified into a truly grand unified theory.

There is a second reason for asking that the preon theory be grand unified and this is related to the issue of charge quantization. If the charge is not a generator of the group, but a sum of generators, then the assignment of charge to the preons - and thus to the quarks and leptons - is not fixed. In particular, as S. Nussinov has emphasized to me often [36], the choice of a charged and a zero-charged preon, as in the rishon model of Harari and Seiberg [37], is still a choice. Only if the $U(1)_{em}$ is part of a GUT will there be true charge quantization.

In collaboration with Yanagida [29], I have developed recently a scenario for a composite GUT, parts of which I have already described in Section V. I should emphasize that the key word here is scenario, for we do not have really any workable models, but just a set of ideas - which we, at least, think are compelling. It will be useful to discuss this scenario in steps. For this purpose, the rough sketch of the evolution of the various coupling constants in momentum space, given in Fig. VII.1, will be helpful. This figure will be modified and refined a little later on.

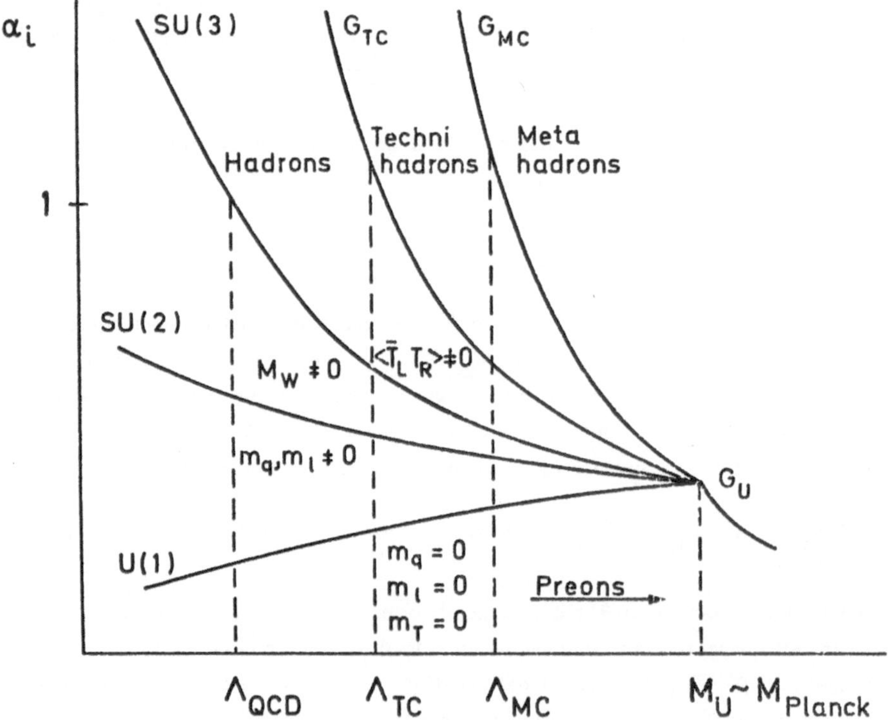

Fig. VII.1: Rough sketch of the evolution of the coupling constants α_i in the scenario

The scenario proceeds in 4 steps:

<u>Step 1</u>: At M_U, which is near the Planck mass $M_{Planck} \simeq 10^{19}$ GeV - the only natural scale in physics (?) - all the coupling constants of the gauge groups of Eq. (VII.1) unify. The picture may not be precisely as in Fig. VII.1 in that, for instance, SU(3) x SU(2) x U(1)

may unify first at $M_x \simeq 10^{15}$ GeV into SU(5), say, and later G_{TC} and G_{MC} unify with SU(5) at M_U. This first step is the one that we are the most unclear about. First of all, we really have no idea of what connections, if any, M_U has with M_{Planck}. Secondly, we do not know what causes the spontaneous breakdown at M_U. (We surely do not want to introduce Higgs fields to do this here!) Finally, the unifying group - if one exists at all - will be stupendously large. Why is this so? All these questions just expose our ignorance of the physics that might occur near the Planck mass. This ignorance notwithstanding, we proceed to the next step, down the energy ladder.

Step 2: Given the breakdown at M_U, all the remaining scales in the theory are determined by the renormalization group equations. The only input is the common value $\alpha_i(M_U)$, plus group theoretical information of how the fermions in the theory (preons and their composites) transform under the various groups. The metacolor group, G_{MC}, coupling constant α_{MC}, is assumed to run faster than the others. At a scale Λ_{MC}, α_{MC} will go through unity and the metacolor forces become strong. The dynamics is such that in the binding a symmetry allows the formation of massless quarks, leptons and technifermions, as well as massive ($M \sim \Lambda_{MC}$) metahadrons out of the preons. The ratio M_U/Λ_{MC} is a calculable number given the initial condition of $\alpha_{MC}(M_U)$. There is no ununderstandable hierarchy.

Step 3: The G_{TC} group is the one whose coupling constant α_{TC} becomes of 0(1) next, at a scale Λ_{TC}. Again this scale has a dynamical relation to M_U and there is no hierarchy. When the technicolor forces become strong, thereby binding the technifermions to form technihadrons, condensates of the type $\langle \bar{T}_L T_R \rangle$ form which break spontaneously SU(2) x U(1). The presence of these condensates, as discussed in Section V, generates both the masses for the $W^{\pm}$ and Z^o as well as the masses for the quarks and leptons. An interesting dynamical question is why the technifermion condensates do not break SU(3). A possible answer might be that no colored technifermions are bound. However, as I shall explain below, this is probably impossible. The more reasonable suggestion is to suppose that the technifermions are complex with respect to SU(2) x U(1) but real with respect to SU(3). In this case, one would expect the breaking to be along the complex direction. This kind of pattern is precisely what can emerge from the more detailed discussion to follow.

Step 4: Finally at Λ_{QCD} - again determined dynamically from M_U - the color group becomes strong and the quarks themselves bind to form hadrons.

The above picture may appear to be rather fanciful. However, it is the only scenario that I know, which tries to answer the dynamical question of what gives masses to the known "elementary" particles without introducing any other scale but M_U (which itself is at least near M_{Planck}). There are many dynamical assumptions in this scenario which I cannot begin to justify, but one can imagine that things might just work this way. There is, however, an important question which deserves an answer, since if this question cannot be answered, the scenario loses all its meaning. It is often stated that unification of the known forces in SU(3) x SU(2) x U(1) requires a desert from M_W to $M_X \simeq 10^{15}$. Our scenario is, instead, rather full of action in this energy range, as illustrated schematically in Fig. VII.2!

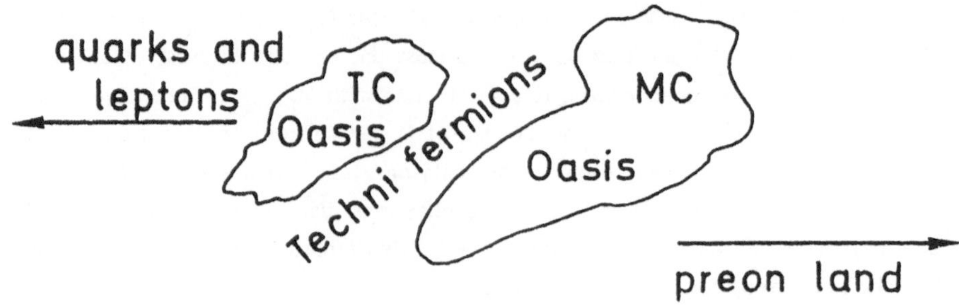

Fig. VII.2: Pictorial representation of the "populated desert" of the scenario of Ref. [29]

How can this scenario not destroy the unification of the SU(3) x SU(2) x U(1) coupling constants?

We know "experimentally" that if we take a reasonable value for $\sin^2\theta_W \simeq .2$ and $\Lambda_{QCD} \simeq 200$ Mev and include only the known quarks and leptons, then the evolution of the coupling constants α_3, α_2 and α_1 leads to a common meeting point at $M_x \simeq 10^{14}$-10^{15} GeV. With our scenario this evolution would begin being disturbed by the presence of the technifermions, which carry SU(3) and SU(2) x U(1) quantum numbers, above Λ_{TC}. Above Λ_{MC}, furthermore, the coupling constant evolution is governed, in its fermionic content, not by the quarks, leptons and technifermions, but by the preons! The crucial question then is, how can this so different dynamical scenario reproduce also the unification of α_3, α_2 and α_1 at M_x, which we know works for elementary quarks and leptons? The answer to this question is simple and yet very interesting.

I will dispose of the technifermion contribution to the evolution of α_3, α_2 and α_1 first. Because Λ_{TC} and Λ_{MC} are not that widely separated, it is clear that the technifermions can be at most a small perturbation in the evolution of the α_i up to Λ_{MC}. Hence, since at Λ_{MC} their role, as well as that of quarks and leptons, needs to be replaced by the preons, it is probably a good approximation to neglect the effect of technifermions altogether. The real issue is what is the evolution of the α_i in the preon region. (We will see, below, that once this latter problem is solved that this solution will also automatically fix the technifermion evolution, so as not to interfere with unification.)

Before discussing the evolution of the α_i in the preon region, it is useful to reproduce the formulas [6] which give the evolution of the α_i, to the lowest order, for the quark and lepton region:

$$\frac{d\alpha_1}{d\ln q^2} = -\frac{\alpha_1^2}{4\pi}\left\{0 - \frac{4}{3}N_g\right\} \qquad \text{(VII.2a)}$$

$$\frac{d\alpha_2}{d\ln q^2} = -\frac{\alpha_2^2}{4\pi}\left\{\frac{22}{3} - \frac{4}{3}N_g\right\} \qquad \text{(VII.2b)}$$

$$\frac{d\alpha_3}{d\ln q^2} = -\frac{\alpha_3^2}{4\pi}\left\{11 - \frac{4}{3}N_g\right\} \qquad \text{(VII.2c)}$$

In the above N_g is the number of generations of quarks and leptons. The appearance in all equations of the common factor $-4/3N_g$ has a simple explanation. One knows that the quarks and leptons of each generation transform according to some representation of a GUT group (e.g., $\bar{5}$ and 10 of SU(5)). Therefore, the fermionic contributions to the rate of change of the α_i must all be the same, since after unification there must be a common rate of change for the α_i and no more fermions need to be added to complete a GUT multiplet[F11]. Although N_g enters in Eqs. (VII.2), both the value of the unification scale, M_x, and of $\sin^2\theta_W$ are independent on the number of generations [F12]. Only the value of $\alpha(M_x)$, the common coupling constant at the unification point, depends on N_g.

The above remark makes it obvious how one can envisage having a composite GUT model. To guarantee that the values of M_x and $\sin^2\theta_W$ - which appear to be successfully phenomenologically - remain the same as that of an ordinary GUT, one must require that also the preons transform according to some representation of a group G_o which unifies SU(3) x SU(2) x U(1). Then the evolution equation in the preon sector of the theory for the coupling constant α_i will be precisely those of Eqs. (VII.2), with $4/3N_g$ replaced by some other constant K, which depends on the preon theory. Several remarks are in order:

(1) Because the preons transform according to some representation of G_o, when at Λ_{MC} preons bind to form massless quarks, leptons and technifermions plus massive states, all these bound states will also transform according to some G_o representation. In fact, in the limit of setting $\alpha_i = 0$, G_o is a good global symmetry. Clearly, since the technifermions transform according to some G_o representation, the evolution of the α_i in the region between Λ_{TC} and Λ_{MC} will not spoil unification.

(2) The appearance of a global symmetry G_o at Λ_{MC}, which is a precursor of the full local G_o at M_x, may have some consequences. For example, imagine G_o = SU(5), which is a pretty natural choice. Then if one assumes that the mass generating mechanism for, at least the heaviest, quarks and leptons is equivalent to having only breaking along

the 5 direction, one would predict at Λ_{MC}

$$m_b = m_\tau \tag{VII.3}$$

Whether similar relations hold for the lighter generations is unclear. Eq. (VII.3) would get altered away from Λ_{MC}. In a similar way to what one finds in ordinary GUT SU(5) [6], this would lead to the prediction:

$$\frac{m_b(q^2)}{m_\tau(q^2)} \simeq \left[\frac{\alpha_3(q^2)}{\alpha_3(\Lambda_{MC}^2)}\right]^{\frac{4}{11-\frac{4}{3}N_g}} \tag{VII.4}$$

To obtain the experimental number $m_b/m_\tau \simeq 2.8$ requires, for $\Lambda_{MC} \sim 10^4$-10^5 GeV, $N_g = 5$. Clearly, this prediction needs to be taken with a grain of salt! Obviously, however, it would be of interest to see whether any other consequences may be drawn from having a G_o global symmetry at Λ_{MC}.

(3) Because the fermionic contributions to the β-functions for SU(3) x SU(2) x U(1) - called K above - for the preons are unknown, it is possible that above Λ_{MC} the SU(2) and perhaps also the SU(3) coupling constant become stronger again. Thus a more realistic picture than that of Fig. VII.1 might be the one given below.

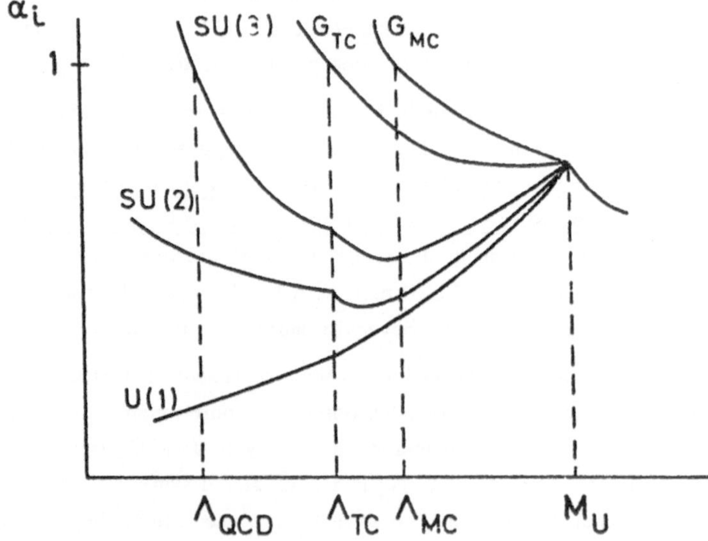

Fig. VII.3: A more realistic picture of the evolution of the α_i

Note that I have been very cavalier in our discussion of how coupling constants evolve in the presence of other strong forces. Obviously, in the neighborhood of Λ_{MC} and

Λ_{TC} when metahadrons and technihadrons form, the smooth evolution of the SU(3), SU(2) and U(1) coupling constants is just an approximation, which should hold on the average. Also I have never worried too much about how Λ_{TC} and Λ_{MC} evolve in detail, since what the groups G_{TC} and G_{MC} are, as well as what the relevant fermionic representations are, is very much an open question.

I have mentioned at the beginning of this section that there are really no complete models that fit this scenario. This should not be perhaps too surprising since the dynamica envisaged is very complicated. There are (at least!) three confining groups G_{MC}, G_{TC} and SU(3), which have running coupling constants which run in different ways and which are realized dynamically quite differently. In the case of G_{MC} we want massless fermionic excitations to exist, while G_{TC} and SU(3) have condensates and spontaneous breaking of chirality. Although it is at this stage impossible to provide answers to these more dynamical questions, one can at least construct toy models that satisfy some of the requirements. For instance, it is not too difficult - if one does not worry about the anomaly matching equations - to contruct models where indeed the metacolor coupling constant runs faster than the technicolor coupling constant, which in turn runs faster than the QCD coupling constant [29]. However, in general, the situation concerning the construction of a realistic composite GUT model, with the features described, appears even more grim than that discussed at the end of Section IV.

VIII. Concluding Remarks

I have tried in these lectures to discuss some of the issues involved in considering models of composite quarks and leptons. Although, it is quite clear that one is very far from being able to construct a realistic model, at least some of the principal dynamical constraints one must face have been exposed. In particular, the crucial problem remains of how one can generate essentially massless composites and then provide a mechanism for both inter- and intrafamily mass splittings. Models which obey theoretical constrainst, which allow one to have at least reasonable dynamical reasons for light composites are in general unrealistically complicated. Models, on the other hand, which describe quarks and leptons in terms of just a few constituents are in general not dynamically sound.

In view of these circumstances, and in view of the crucial importance of the question, any experimental information which can shed light on <u>any</u> scales beyond those of the standard model would be most welcomed. Experimentalists ought to be on the lookout for any possible deviations from the standard model orthodoxy. I list just a few: mirror fermions, which are partners of the ordinary quarks and leptons but with heavier mass and opposite weak couplings; "low mass" bosons, which could be remnants of the technicolor spectrum; multimuon events in very high energy cosmic ray processes, which could arise from interactions in which the compositeness radius has been broached, etc.

Perhaps, it is worthwhile ending by remarking that although compositeness might be very difficult to prove now, composite models of the type I discussed do predict spectacular things to happen at high energy (e.g., a technihadron and metahadron zoo of particles in the TeV range of energies). Two fundamental questions remain to be answered:

(1) How will we ever get there experimentally?

(2) How can we make these theories compelling?

Footnotes

[F1] Fortunately, the review of L. Lyons contains a very thorough discussion of most of the proposed model. Hence, the interested reader can partially remedy this deficiency in our presentation by consulting Ref. [4].

[F2] Theories in which leptons are composite but quarks are elementary, although a little weird are not totally out of the question. However, we shall not discuss them in these notes.

[F3] This is a little loose. If n different flavors of quark masses are sent to zero, then one expects $n^2 - 1$ Goldstone bosons, of which the pions really just correspond to the first three such states.

[F4] We should remark that in these theories the Case-Gasierowicz-Weinberg-Witten theorem does not apply.

[F5] As Coleman and Grossman [24] point out in their extremely clear analysis of the consistency condition, it suffices to consider the anomaly equations for identical currents.

[F6] Eq. (IV.10) explicitly shows that in a theory with no chiral current $(\lambda_R^{(i)} = \lambda_L^{(i)})$ there are no anomalies.

[F7] A very nice discussion of ways to avoid the persistent mass conditions and their implication is to found in Ref. [27].

[F8] We will discuss an example of this shortly - see [F9].

[F9] In this model the flavor changing neutral currents are also quite suppressed.

[F10] Actually for the SU(2) group, Λ_2 is irrelevant since technicolor breaks this group spontaneously.

[F11] At the unification scale there are of course heavy bosons contributing, which change the gauge field contributions in the curly brackets of Eqs. (VII.2) so that they are also all the same.

[F12] There are some small dependencies on N_g if we consider 2 loop corrections to Eqs. (VII.2).

References

[1] M.E. Peskin, Proceedings of the 1981 International Symposium on Lepton and Photon Interactions, Bonn, August 1981.

[2] H. Harari, Proceedings of the Canadian Summer School of Theoretical Physics, Banff, Alberta, August 1981.

[3] R. Barbieri, Proceedings of the X International Winter Meeting on Fundamental Physics and of the XIII GIFT International Seminar on Theoretical Physics, Masella, Alp, Girona, January 1982.

[4] L. Lyons, Oxford report Ref. 52/82, to appear in Prog. Part. and Nucl. Phys.

[5] L. Susskind, Phys. Rev. D20 (1979) 2619;
S. Weinberg, Phys. Rev. D13 (1976) 974, D19 (1979) 1277;
For a review see: M.A.B. Beg, Proceedings of the Lisbon International Conference on High Energy Physics, 1981.

[6] For reviews see: J. Ellis, "Gauge Theories and Experiments at High Energies", Scottish Universities Summer School in Physics, eds. N.C. Bowler and D.G. Sutherland,

Edinburgh, 1981, p. 201;
P. Langacker, Phys. Rep. 72C (1981) 185.

[7] For reviews see: K. Stelle and G. Farrar, these proceedings.

[8] H. Fritzsch, these proceedings.

[9] For some contributions along these lines, see for example: H. Terazawa, Proceedings of the INS Symposium, Tokyo, July 1981; K. Akama, ibid.

[10] For a review see for example: H.P. Dürr, Proceedings of the Heisenberg Symposium, Munich, July 1981.

[11] M. Gell-Mann, R. Oakes and B. Renner, Phys. Rev. 175 (1968) 2195.

[12] M. Gell-Mann and M. Levy, Nuov. Cim. 16 (1960) 705.

[13] J.G. Branson, Proceedings of the 1981 International Symposium on Lepton and Photon Interactions, Bonn, 1981.

[14] T. Kinoshita and W. Lindquist, Phys. Rev. Lett. 47 (1981) 1573;
G.P. Lepage and D.R. Yennie, Proceedings of the Second International Conference on Precision Measurements and Fundamental Constants.

[15] G.L. Shaw, D. Silverman and R. Slansky, Phys. Lett. 94B (1980) 343;
S.J. Brodsky and S.D. Drell, Phys. Rev. D22 (1980) 2236.

[16] R. Barbieri, L. Maiani and R. Petronzio, Phys. Lett. 96B (1980) 63.

[17] K.M. Case and S. Gasierowicz, Phys. Rev. 125 (1962) 1055.

[18] S. Weinberg and E. Witten, Phys. Lett. 96B (1980) 59.

[19] G. 't Hooft, "Recent Developments in Gauge Theories", Cargese Lectures 1979 (Plenum Press, New York, 1980) p. 135.

[20] W.A. Bardeen and V. Visnjic, Nucl. Phys. B194 (1982) 422.

[21] W. Buchmüller, S.T. Love, R.D. Peccei and T. Yanagida, Phys. Lett. 115B (1982) 233.

[22] T. Banks and V. Kaplunovsky, Nucl. Phys. B192 (1981) 270.

[23] S. Adler, Phys. Rev. 177 (1969) 2426;
J.S. Bell and R. Jackiw, Nuov. Cim. 60A (1969) 47.

[24] S. Coleman and B. Grossman, Nucl. Phys. B203 (1982) 205.

[25] Y. Frishman, A. Schwimmer, T. Banks and S. Yankielowicz, Nucl. Phys. B177 (1981) 157.

[26] S. Adler, "Lectures on Elementary Particles and Quantum Field Theory", Brandeis Lectures 1970 (MIT Press, Cambridge, 1970) p. 1.

[27] J. Preskill and S. Weinberg, Phys. Rev. D24 (1981) 1059.

[28] I. Bars, Phys. Lett. 106B (1981) 105, 109B (1982) 73, Yale preprint YTP 82-04, to be published in Nucl. Phys. B, Yale preprint YTP 82-12, to be published in the Proceedings of the Rencontre de Moriond 1982.

[29] R.D. Peccei and T. Yanagida, MPI-PAE/PTh 32/82, to be published in Nucl. Phys. B.

[30] J. Preskill, Harvard preprint HUTP-81/A051 and Proceedings of the Paris Conference 1982.

[31] T. Yanagida, Proceedings of the KEK Symposium 1979;
M. Gell-Mann, P. Ramond and R. Slansky, Proceedings of the Supergravity Workshop at Stony Brook, 1980.

[32] R.D. Peccei, MPI-PAE/PTh 72/82, to appear in the Proceedings of the VIII International Workshop on Weak Interactions and Neutrinos, Javea, Spain, 1982.

[33] R. Casalbuoni and R. Gatto, Phys. Lett. 108B (1982) 117, UGVA-DPT 1981/06-279, to appear in the Proceedings of the Johns Hopkins Workshop on Current Problems in Particle Theory;
see also I. Bars, Ref. 28 .

[34] H. Harari, R. Mohapatra and N. Seiberg, to be published in Nucl. Phys. B.

[35] A. Masiero, R.D. Peccei and T. Yanagida, Phys. Lett. 109B (1982) 357;
see also R.D. Peccei, MPI-PAE/PTh 81/81, to appear in the Proceedings of the 2nd Workshop on Unification of the Fundamental Particle Interactions, Erice, 1981.

[36] S. Nussinov, private communication.

[37] H. Harari and N. Seiberg, Phys. Lett. 96B (1981) 269.

LEPTONS AND QUARKS AS COMPOSITE OBJECTS+

Harald Fritzsch
Sektion Physik
and
Max-Planck-Institut für Physik and Astrophysik, München

The so-called standard model of particle physics is based on the idea that there exist besides gravity three types of forces in nature, the color forces, the electroweak forces and the X-boson forces (leading e.g. to the proton decay). Within the grand unified theories those forces turn out to be different manifestations of one fundamental gauge interaction, described e.g. by the SU(5) or SO(10) gauge theory[1)]. Thus the number of fundamental interactions is fairly small, which provides a justification for the belief that some day all of fundamental physics, including gravity, may be described within a single theory.

On the other hand it appears that the number of quarks and lepton taking part in these interactions is substantial. Three families of leptons and quarks are almost complete; further ones could be detected soon. This may be an indication that leptons and quarks are not fundamental entities unlike the gluons or the photon.

Recently a large number of authors has been interested in constructing composite models of leptons and quarks[2)]. The idea is that the leptons and quarks are composed of several constituents which are bound together by superstrong forces. There exist various constraints on the sizes of leptons and quarks, e.g. the agreement between theory and experiment of the anomalous magnetic moment of the electron, which imply that those sizes are less than about 10^{-17} cm.

If the leptons and quarks are composite objects, one may expect the same to be true for at least some of the intermediate (gauge) bosons. A possible point of view is to regard the fermions and the intermediate bosons as composite objects at a very high energy scale (of the order of the mass scale entering in the grand inification schemes ($\sim 10^{15}$ GeV)) or more. In this case the interactions at relatively low energies (QCD, flavor interactions) can be interpreted as effective gauge theories; essentially no deviations from the standard pattern of the QCD and QFG gauge theory framework are expected.

Here we should like to explore another possible road. We shall suppose that the W bosons and the fermions are bound states, while the massless bosons (photon, gluons) are elementary[3,4)]. The masses of the W bosons are generated dynamically by the binding forces in much the same way the ρ meson mass is generated in QCD. In this approach the weak interactions are indirect manifestations of the strong binding forces inside the W boson. One is reminded of the situation which has evolved during the last 20 years with respect to the strong interactions. Many years ago the strong interactions were interpreted as a gauge theory in which the baryons played the rôle of the elementary fermions while the vector mesons (ρ, ...) were considered to be gauge particles. Today the situation has changed considerably. Both the baryons and the vector mesons are composite objects, consisting of quarks. The strong interactions between the nucleons are regarded as residual effects of the gluonic forces between the quarks. Perhaps the weak interactions are of a similar nature, and the W bosons consist of constituents which are at the same time building blocks of the fermions. Of course, we realize that the weak interactions differ in their properties substantially from the strong interactions between nucleons. First of all, they violate parity. Moreover they are well discribed by the exchange of vector particles. No exchange forces generated by scalar or tensor particles are observed.

Moreover we have to find a way to understand why the masses of leptons and quarks are very small compared to the inverse radius of those objects (in case, they are composite objects at all). One is reminded of a corresponding situation in QCD. The π-meson plays a dual rôle in strong interaction physics. On the one hand it is a quark- antiquark bound state (the pseudoscalar partner of the ρ-meson), on the other hand it acts in a good approximation as a Goldstone boson of chiral symmetry. In nature the u- and d- quarks have an average mass of about 7 MeV, which implies that the π-meson has a mass of about 140 MeV. If the u- and d- quark masses would vanish, the π-meson would have zero mass. In some sort of "gedankenexperiment" one may consider the case $m_u = m_d = 1$ KeV. It is easy to estimate that in such a fictitious world the π-meson would have a mass of about 2 MeV. Nevertheless its radius, described by the inverse of the π-decay constant F_π, would be of the order of 10^{-13} cm, e.g. we have a situation where the mass of a particle is small compared to its inverse size. (In the real world the mass of the π-meson and its inverse size are of the same order of magnitude - an unfortunate accident.) If the ideas of a composite structure for the leptons and quarks make any sence, the leptons and quarks must behave in a similar way as the π-meson in the fictitious world described above - their masses must be small compared to their inverse sizes. Only one mechanism is known which may realize such a point of view - an approximate chiral symmetry, which is realized by composite fermions of nearly zero masses[5)].

We take the point of view that the leptons, quarks and W bosons are composed of constituents, while the photon and the gluons of QCD are elementary. In the absence of electromagnetism the global symmetry group of the weak interactions is SU(2) (weak isospin). It will be shown later that the observed structure of the neutral current is obtained if one takes into account the mixing between the photon and the neutral SU(2) boson W_3. This mixing arises dynamically, due to the electromagnetic annihilation of the W_3 constituents.

We shall assume that the underlying gauge symmetry is given by the group $SU(3)_c \times G_h \times U(1)_e$ (c: color, e: electric charge). The group G_h is the hypercolor gauge group describing the confining forces responsible for the binding of the hypercolored constituents.

The corresponding gauge theory is called QHD. For simplicity we shall use the hypercolor group SU(n), where n is yet unspecified. The extension to other groups is easily made.

If we regard the weak bosons as bound states, several possibilities are open. For example, the weak bosons could be composed of hypercolored fermions or bosons; these fermions or bosons could be colored or color singlets. The weak bosons could be composed of two or more constituents. For reasons which will become clear afterwards, models in which the weak bosons consist of more than two constituents cannot be accepted, likewise models in which the weak bosons are composed of scalar or pseudoscalar bosons, We shall concentrate on schemes in which the weak bosons are composed of spin 1/2 fermions. Furthermore we shall mainly consider the lightest family of leptons and quarks, consisting of the (u,d)-pair and the (ν_e, e^-) pair.

Two classes of models are possible:

a) The W-bosons consist of hypercolored fermions, which are color singlets.
b) The W-bosons consist of hypercolored and colored fermions.

In both classes the W-bosons are, of course, hypercolor and color singlet bound states. For illustration we consider the following schemes, in which the fermions are composed of (pseudo) scalar and fermionic constituents (haplons); the name is derived from the greek word "haplos" (simple). The underlying gauge group is

$$U(1)_e \times SU(3)_c \times G_H$$

(G_H: hypercolor gauge group).
The gauge theory QHD based on the group G_H (hypercolor gauge group) describes the dynamics of the haplon constituents bound together by the superstrong hypercolor forces. The QHD confinement parameter Λ_H is supposed to be of the order of a few hundred GeV ($\sim(G_F)^{-1/2}$). The gauge group G_H is not specified, but taken to be SU(n).

Case A (as far as the quantum numbers are concerned, this scheme has been considered by the authors of ref. (6)).

	spin	el. charge	color	hypercolor
α	1/2	1/2	1	n
β	1/2	-1/2	1	n
x	0	1/6	3	$\bar{n}$
y	0	-1/2	1	$\bar{n}$

The simplest QHD singlets one can form are fermions of electric charges $(2/3, -1/3) = [(\alpha x), (\beta x)]$ and $(0, -1) = [(\alpha y), (\beta y)]$ identified with (u,d), (ν_e, e^-) respectively.

Scheme B[7)]

	spin	el. charge	color	hypercolor
α	1/2	-1/2	3	n
β	1/2	+1/2	3	n
x	0	-1/6	3	$\bar{n}$
y	0	1/2	$\bar{3}$	$\bar{n}$

This scheme has the special feature that all haplons carry electric charge, color and hypercolor. The simplest QHD singlets are

$$\nu_e = (\bar{\alpha}\ \bar{y})_1 \qquad u = (\bar{\alpha}\ \bar{x})_3$$

$$e^- = (\bar{\beta}\ \bar{y})_1 \qquad d = (\bar{\beta}\ \bar{x})_3$$

(the index denotes the color of the state). Both in scheme A and B there exist vector bosons composed of the fermions ($\bar{\alpha}\beta\ \bar{\beta}\alpha$, ...). Those are interpreted as the carriers of the weak interactions. The global symmetry group SU(2) generated by the haplon doublet $\binom{\alpha}{\beta}$ is identified with the weak isospin.

The observed parity violation of the weak interaction can be accommodated in two different ways:

1. The bound state structure discussed above is assumed to be valid only for the lefthanded fermions.

2. Both the lefthanded and righthanded fermions are bound states, however the Λ_H parameter of the righthanded fermions is larger than the Λ_H parameter of the lefthanded ones, i.e. two QHD group are needed ($G_H^{(L)}$, $G_H^{(R)}$). In this case the global symmetry groups of weak interactions is $SU(2)_L \times SU(2)_R$; the W-bosons coupling to the righthanded fermions are heavier than those which couple to the lefthanded fermions.

Both in the schemes A and B the weak interaction is an effective interaction of the Van der Waals type, generated by the superstrong QHD force. The universality of the weak interaction between leptons and quarks follows from the global SU(2) symmetry in the α-β- space.

It is assumed that the spectral functions of the weak currents in QHD are qualitatively similar to the ones in QCD. At low energies they are dominated by the lowest lying pole, and at high energies (energies large compared to Λ_H) they can be described by a continuum of "weak quanta"(this term is borrowed from ref. (8)), i.e.a continuum of $\bar{\alpha}\alpha$, $\bar{\alpha}\beta$ $\bar{\beta}\alpha$, or $\bar{\beta}\beta$- pairs.

In the absence of mixing with the electromagnetic current there exist two neutral current channels, the isovector channel described by $(\bar{\alpha}\alpha-\bar{\beta}\beta)/\sqrt{2}$ and the isoscalar channel described by $(\bar{\alpha}\alpha + \bar{\beta}\beta)/\sqrt{2}$. Especially one expects that the isoscalar spectral function is dominated at low energies by the lowest lying meson with the quantum numbers $(\bar{\alpha}\alpha + \bar{\beta}\beta)/\sqrt{2}$. The observed neutral current is a pure isovector, if we neglect for a moment the mixing with the electromagnetic current. The experimental data allow at most a 10 % isoscalar contribution to the neutral current, i.e. the isoscalar boson W_0 must be much heavier than the isovector boson W_3 (or the Z-boson). The isoscalar boson has the internal quantum numbers of the vacuum, i.e. its mass term may receive contributions from the hypergluon annihilation channel. The situation is quite similar to the one in QCD. Here the η'-meson is about 6 times heavier than the π_0-meson. It is generally assumed that the source of the mass gap are the gluons of QCD. The isoscalar quark configuration $(uu + dd)/\sqrt{2}$ can mix with gluonic configurations. This implies that the isoscalar meson is heavier than the isovector meson.

Our approach makes sense only if something similar happens in QHD for the weak bosons. The hypercolor annihilation channels must lead to a rise of the mass of the isoscalar weak boson relative to the mass of the isovector boson. For our further discussion we shall assume that this is indeed the case.

The experimental data on the neutral current interaction require a mixing between the photon and the W_3 boson (the neutral, isovector partner of W^+ and W^-), which in the standard SU(2) x U(1) scheme is caused by the spontaneous symmetry breaking. Within our approach this mixing is due the W_3 - γ transitions, generated dynamically like the ρ-γ transitions, in QCD (for an early discussion, based on vector meson dominance, see ref. (8)). The magnitude of $\sin^2\theta_W$ is directly related to the strength of the γ-W_3 transition. The latter is determined by the electric charges of the W-constituents and by the W wave function near the origin.

We suppose that in the absence of electromagnetism the weak interactions are mediated by the triplet (W^+, W^-, W^3), where $M(W^+) = M(W^-) = M(W^3) = 0\ (\Lambda_H)$.

After the introduction of the electromagnetic interaction the photon and the W^3-boson mix. We denote the strength of this mixing by a parameter λ, following ref. (9), which is related to g (W-fermion coupling constant) and the effective value of $\sin^2\theta_W$

$$\sin^2\theta_W = \frac{e}{g} \cdot \lambda$$

Furthermore one has:

$$M_W = g \cdot 123 \text{ GeV}$$

$$M_Z^2 = \frac{M_W^2}{1-\lambda^2}$$

The mixing parameter λ is determined by the decay constant F_W (or f_W) of the W-boson, which we define in analogy to the decay constants of the ρ_0-meson (F_ρ, f_ρ respectively):

$$|W^3> = \frac{1}{\sqrt{2}} \frac{1}{\sqrt{n}} \frac{1}{\sqrt{3}} \sum_{j=1}^{n} \sum_{j=1}^{3} (\bar{\alpha}_{i,j}\alpha_{i,j} - \bar{\beta}_{i,j}\beta_{i,j}) \; \phi\,(x)$$

(i: color index). The current matrix element can be written as

$$<o|j_\mu{}^3|W^3> = \varepsilon_\mu \sqrt{n} \cdot \gamma \sqrt{2M_W} \cdot \phi\,(o) = \varepsilon_\mu \cdot M_W\, F_W$$

$$F_W = \sqrt{n} \cdot \gamma \cdot \sqrt{2/M_W}\; \phi\,(o)$$

where $\gamma = 1$ in case A and $\gamma = \sqrt{3} = \sqrt{n_c}$ in case B.

$$\sin^2\theta_W = \frac{e^2}{g} \sqrt{n} \cdot \gamma \sqrt{2/M_W^3}\; \phi\,(o),$$

$$= e^2/\,g \cdot F_W\,/\,M_W.$$

e.g. $\sin^2\theta_W$ is proportional to the coordinate space wave function of the W-boson at the origin. Taking for example $g = 0.65$ and $M_W = 79$ GeV, one obtains $F_W = 123$ GeV, a value which seems resonable for a bound state of the size 10^{-16} cm.

In the SU(2) x U(1) gauge theory the SU(2) coupling constant g is related to e by the relation $g = e/\sin\theta_W$. In bound state models of the weak interactions discussed here this relation need not be true in general. However it has been emphasized recently[10,11] that this relation is approximately fullfilled if the lowest lying W-pole dominates the weak spectral function at low energies. This leads to the relation

$$g = F_W\,/\,M_W \approx 0.65$$

(we have used $\sin^2\theta_W = 0.22$).

It is interesting that many aspects of the bound state models can be derived from a local current algebra of the weak currents[10].We observe that the left-handed leptons and quarks form doublets of the weak isospin. The weak isospin charges

F_i^W (i = 1, 2, 3) obey the isospin charge algebra

$$[F_i^W, F_j^W] = i\,\varepsilon_{ijk}\,F_k^W$$

We shall assume that these charges can be constructed as integrals over local charge densities $F_{oi}^W(x)$, i.e.,

$$F_i^W(x^o) = \int F_{oi}^W(x)d^3x.$$

Furthermore we assume that the charge densities obey at equal times the local current algebra

$$[F_{oi}^W(x), F_{oj}^W(y)]_{x^o = y^o} = i\varepsilon_{ijk}F_{o\kappa}^W\,\delta^3(\bar{x}-\bar{y}).$$

The local algebra is trivially fulfilled in a model in which leptons and quarks are pointlike objects and the weak currents are simply bilinear in the lepton and quark fields. However, if leptons and quarks are extended objects, the situation changes drastically. Currents, which are bilinear in the (composite) lepton and quark fields would not obey the local algebra, just like the currents, which are bilinear in nucleon fields, do not obey the local current algebra of QCD. Thus the local algebra becomes a highly non-trivial constraint. It is fulfilled in the haplon models discussed above, in which the currents are bilinear in α and β.

It is not known what the spectral functions of the weak isospin currents look like. It could be that they are dominated at low energies by a single pole (like the spectral functions of the $\bar{u}u$ or $\bar{d}d$ currents in hadron physics), by several poles (like the spectral functions of heavy quark currents, e.g., $\bar{c}c$), or by a continuum of states. We shall suppose that the first case is realized, and that the weak spectral functions at low frequencies are dominated by the lowest-lying pole (W dominance). Of course, at higher energies, higher excited states as well as the continuum will become relevant.

We consider matrix elements of the weak currents between the various fermion fields. In order to do so, we shall assume that the higher families composed of μ, τ, ... etc. are dynamical excitations of the first family (ν_e, e^-, u, d), without specifying in detail the dynamical structure of these states.

Let us look at the form factors of the left-handed weak neutral current $F_{\mu 3}^L(x)$, i.e., the matrix elements of this current between different lefthanded lepton or quark states, e.g., $\langle e_L^-|F_{\mu 3}(0)|e_L^-\rangle$. Denoting these form factors by $F_e(t)$, $F_\mu(t)$, $F_\tau(t)$, etc., the weak isospin algebra requires a universal normalization at t = 0, i.e.,

$F_e = F_{\nu_e} = F_\mu = F_{\nu_e} = \ldots = 1$. Assuming W dominance to be a reasonably good approximation, we may write for the dependence on the four-momentum transfer t,

$$F_f(t) = \frac{m_W^2}{f_W} \frac{f_W{}^{ff}}{m_W^2 - t} ,$$

where f denotes any one of the fermions e^-, ν_e, μ^-, ν_μ, etc. From $F_f(0) = 1$, we obtain the universality relation

$$f_W = f_W^{ff} \equiv g .$$

The neutral W and, because of the weak isospin algebra, also the charged W bosons couple universally to leptons and quarks, $f_{Wff} \equiv g$. Thus the universality of the weak interactions follows from the W-dominance.

Let us stress that W cominance is a dynamical approximation of the underlying hypercolour dynamics, which is expected to be valid in the low energy region. At high energies (~1 TeV) lepton-lepton (quark) scattering is expected to show completely new phenomena, like multiple lepton (quark) production, similar to hadron-hadron interactions at high energies. Since universality is valid, we expect W dominance to be a good approximation in the low energy region. As a consequence, deviations from the predicted charged and neutral boson masses may not be too large. Nevertheless, deviations of the order of $\pm 10\%$ to $\pm 20\%$, which cannot be accounted for by radiative corrections within the SU(2) x U(1) gauge theory, are rather expected than excluded. If the lowest neutral boson mass is close to the SU(2) x U(1) value of 37.3 GeV/$\sin\theta_W \cos\theta_W \cong 89$ GeV, it would mean that the first excited state is rather heavy. If the lowest mass deviates substantially from the SU(2) x U(1) value, the first excited state may well be below 200 GeV.

If the weak interactions are a manifestation of hypercolor dynamics, the question of SU(2) breaking is again open. One may wonder why the large violation of the weak isospin in the lepton - quark spectrum does not imply a large breaking of the symmetry in dynamical parameters like the W- fermion- coupling constants.

We should like to study the violation of the isospin in the pion dynamics. It is well - known that there exist two different sources of the violation of the isospin in strong interaction: the difference of the quark masses m_u and m_d, and the electromagnetic interaction. Only the second source contributes to the $\pi^+ - \pi^0$ mass difference. If we set $m_u = m_d = m$ and let m go to zero, the pion mass approaches zero as well, provided we neglegt the electromagnetic interaction.

Following the laws of chiral symmetry breaking, one finds:[12)]

$$M_\pi^2 = (m_u + m_d) \cdot B + O(m_q^2 \ln m_q)$$

$$B = - \frac{2}{F_\pi^2} <o|\bar{u}u|o>$$

where u, d denote the light quark flavors, m_u, m_d the quark masses, $|o>$ the QCD vacuum, F_π the pion decay constant. Typical values are $(m_u + m_d) = 14$ MeV, $B = 1300$ MeV[12].

The electromagnetic self energy of the π^0 vanishes in the chiral limit. The electromagnetic self energy of the charged pion can be calculated to order α in terms of the vector and axial vector spectral functions[12]:

$$(\Delta M_{\pi^+}^2)_{el.} = \frac{3\alpha}{4\pi \cdot F_\pi^2} \int_0^\infty ds \cdot s \ln \left(\frac{\mu^2}{s}\right) [\rho^V(s) - \rho^A(s)]$$

(ρ_V, ρ_A: vector - and axial vector spectral functions).

Saturating the integral above with the ρ- and A_1- poles and using the spectral function sum rules one obtains:

$$(\Delta M_{\pi^+}^2)_{el.} = \frac{3\alpha}{4\pi} \cdot M_\rho^2 \cdot \left(\frac{F_\rho}{F_\pi}\right)^2 \cdot \ln \left(\frac{F_\rho^2}{F_\rho^2 - F_\pi^2}\right).$$

Using the measured values $F_\pi = 132$ MeV and $F_\rho = 204$ MeV, one obtains $(\Delta M_{\pi^+}^2)_{el.} \cong (36.4 \text{ MeV})^2$, which is close to the observed mass difference $\Delta M_\pi^2 = (35.6 \text{ MeV})^2$.

Combining the two relations, denoted above, one finds:

$$M_{\pi^0}^2 = (m_u + m_d) \cdot B + \ldots$$

$$M_{\pi^+}^2 = (m_u + m_d) \cdot B + \alpha \cdot M_\rho^2 \cdot 0.31 + \ldots$$

In the chiral limit $m_u = m_d = 0$ we obtain $M_{\pi^0} = 0$, $M_{\pi^+} \approx 36$ MeV. As an illustration we consider the case $m_u = m_d = 1$ KeV. One finds $M_{\pi^0} = 1.6$ MeV, $M_{\pi^+} = 36.4$ MeV, i.e. the neutral and charged pion mass differ by a factor of about 23.

We have just found a situation in QCD, which resembles the one in the lepton-quark- spectrum, namely a large isospin breaking despite the fact that for $m_u = m_d$ the isospin is an exact symmetry of QCD. In the chiral limit the π-mesons are particles, which in the absence of electromagnetism have zero mass, but have a finite size. Their inverse size is of order Λ(Λ: QCD cutoff parameter).

Including the electromagnetic interaction has the effect of lifting the charged pion mass from zero to the finite value $M_{\pi^+} \approx 0.16 \cdot e \cdot M_\rho \approx 36$ MeV. The neutral pion stays massless. The charged pion mass is of order $e \cdot \Lambda[QCD]$, i.e. $e \cdot$ (inverse size of pion).

We note that the π^+ - mass is of electromagnetic origin. The self energy diagram consists of a charged pion emitting a virtual photon and turning itself into a massive state (ρ, A_1, ...). Due to the chiral symmetry the sum of all these contributions is finite and of order $e \cdot \Lambda(QCD)$.

With these preparations in mind, we are ready to consider the lepton quark spectrum. Let us assume that the leptons and quarks are massless bound states in the limit $e = 0$ like the pions in QCD in the limit $m_u = m_d = 0$ and $e = 0$. Introducing the QED interactions means in particular introducing self energy diagrams where a lepton and quark emits a virtual photon and turns itself into a massive fermion with a mass of the order of Λ_H (analogous to the ρ or A_1 mesons in the case of the pion self energy). The result will depend strongly on the mass spectrum of states at the energy of Λ_H, about which very little is known. In general one finds:

$$M(\text{fermion}) \cong \frac{3\alpha}{4\pi} \cdot Q^2(\text{fermion}) \cdot K \cdot \Lambda_h$$

where Q is the electric charge, and K is a constant depending on details of the intermediate states. Using as an illustrative example $\Lambda_H = 100$ GeV and $K = 1$ one finds the mass spectrum

M(neutral lepton) = 0 M(u-type quark) = 77 MeV

M(charged lepton) = 174 MeV M(d-type quark) = 19 MeV.

Of course this mass spectrum is not very realistic, however it displays a number of interesting features, which are also fulfilled for the real lepton and quark masses:

a) The neutrino remains massless (in the first order of α)
b) The up-type quark is heavier than the d- type quark.
c) The mass splitting inside the weak doublets is large compared to the lepton or quark masses.

Property b) is not fulfilled for the u-d system (the u-quark is lighter than the d-quark), but for the second and third family. Probably this is a consequence of the weak interaction mixing between the various families neglected here.

The example discussed above shows the possibility to interpret the lepton and quark masses as electrodynamic self energies. The self energies are finite since a real cut-off given by Λ_H enters in the calculations. Using definite bound state models for the leptons and quarks one may be able to develop an actual theory of the lepton and quark masses, and of the weak interaction mixing parameters.

Concluding this lecture, I would like to stress that there are reasonable prospects for a bound state structure of the leptons and quarks to show up at distances of the order of 10^{-17} cm. The W dominance, combined with the local algebra of weak currents, enforces relations between the coupling constants etc., which are identical to the ones predicted within the SU(2) x U(1) gauge theory. In fact, it may not be surprising that the latter works so well in the low energy region - it is an effective low energy theory. As soon as one reaches energies of the order of Λ_H, new phenomena will come in, which will give us essential insights into the internal dynamics of quarks and leptons.

It is my pleasure to thank Drs. R. Raitio and J. Lindfors for organizing this Arctic School in Lapland, one of the few remaining wilderness areas of Europe.

References

1. H. Georgi and S. Glashow, Phys. Rev. Lett. 32 (1974) 438.
   H. Fritzsch and P. Minkowski, Ann. Phys. 93 (1975) 193.
   H. Georgi, Particles and Fields, ed. C.E. Carlson (AJP, New York, 1975).
2. See e.g.:
   H. Harari, Phys. Lett. 86 B (1979) 83.
   M. A. Shupe, Phys. Lett. 86 B (1979) 87.
   O. W. Greenberg and C.A. Nelson, Phys. Rev. D 10 (1974) 256.
   R. Casalbuoni and R. Gatto, Phys. Lett. 103 B (1981) 113.
3. L. Abbott and E. Farhi, Phys. Lett. 101 B (1981) 69.
4. H. Fritzsch and G. Mandelbaum, Phys. Lett. 102 B (1981) 319;
   Phys. Lett. 109 B (1982) 224.
   R. Barbieri, A. Masiero and R. N. Mohapatra, Phys. Lett. 105 B (1981) 369.
5. See e.g.:
   G. 't' Hooft, in: Recent Developments in Gauge Theories,
   Plenum Press, N.Y. (1980), p. 135.
6. O.W. Greenberg and J. Sucher, Phys. Lett. 99 B (1981) 339.
   R. Casalbuoni and R. Gatto, Phys. Lett. 103 B (1981) 113.
7. H. Fritzsch and G. Mandelbaum, Ref. (4).
8. J. D. Bjorken, Phys. Rev. D 19 (1979) 335.
9. P. Hung and J. Sakurai, Nucl. Phys. B 143 (1978) 81.
   J. Sakurai, Proceedings of the Int. Neutrino Conference, Balatonfüred (June 1982)
10. H. Fritzsch, D. Schildknecht and R. Kögerler, Phys. Lett. 114 B (1982) 157.
11. R. Kögerler and D. Schildknecht, CERN preprint TH 3231 (1982).

MASSIVE DIRAC VS. MASSIVE MAJORANA NEUTRINOS

S. T. Petcov
Institute of Nuclear Research and Nuclear Energy
Bulgarian Academy of Sciences
Boul. Lenin 72, 1184 Sofia, Bulgaria

1. Introduction

The problem of the neutrino mass is as old |1,2| as the idea |1| of the existence of the neutrinos, but no convincing solution has been proposed yet. The neutrinos may well be massless. However, unlike the exact gauge invariance which ensures the masslessness of the photon in electrodynamics, no profound principle excluding the possibility of massive neutrinos has been discovered. Moreover, effects of a nonzero mass of the electron neutrino are claimed |3| to be observed by an experimental group from ITEP (Moscow), which studied the shape of the electror spectrum near the end point in the tritium β-decay[F1].

Finite neutrino masses arise naturally in the modern gauge theories of the electroweak interactions |6| and especially in the grand unified theories |7| (GUTs). In some GUTs as those based on the group SO(10) it is almost impossible to avoid them. At the same time the simplest versions of these theories, namely, the standard SU(2) x U(1) electroweak theory |8|, wherein the right-handed (RH) components of the neutrino fields $\nu_{\ell R}$ (ℓ = e, μ, ...) are not present, and its SU(5) grand unified generalization |9| predict massless neutrinos.

Nonzero neutrino masses and neutrino mixing imply an extremely wide spectrum of possible neutrino properties. For instance, several varieties of massive Dirac neutrinos differing, e.g. by their magnetic moments are possible |10|. Being electrically neutral, the massive neutrinos could be truely neutral objects |11,12|, i.e. Majorana particles, identical with their antiparticles. In this case the neutrinos with a definite mass would be fermionic analogs of the π^0-meson.

The properties of massive Dirac and massive Majorana neutrinos |13| and the physics they are associated with, as we imagine it today, are very different. The former arise in a rather natural way in the minimimally extended standard electroweak theory to include the RH neutrino field components as SU(2) singlets. Besides the nonzero neutrino masses, the only predicted new phenomena that might lead to observable effects in this case are |14|, in essence, the oscillations |11,15| between different neutrino flavours. In contrast, the massive Majorana neutrinos arise |6,7| usually in theories with considerable extentions of the standard theory which predict, as a rule, the existence of bizarre particles and processes, observable in practice. In GUTs these are the nucleon decays and the neutron-antineutron oscillations; in the SU(2) x U(1) theories containing no ν_R fields, these could be, e.g. relatively light charged, doubly charged and massless neutral Higgs particles, as well as a multitude of specific processes in which they might take part. For this reason it is generally believed today that the massive neutrinos of Majorana type could be the "visiting card" of some new physics, beyond that predicted by the standard model.

In the present lecture we shall discuss some aspects of the problem of massive neutrinos. A special attention will be paid to the differences between the neutrinos of Dirac and Majorana type. We shall recall first the well established special neutrino features. Then the varieties of neutrino mass terms, which determine the type of the massive neutrinos in gauge theories, as well as some examples of neutrino mass generation will be considered. A comparison between the properties of massive Dirac and massive Majorana neutrinos will follow. And finally, the possibilities to distinguish experimentally between the two possible types of massive neutrinos will be discussed.

Throughout the lecture we shall refer to the properties of the standard SU(2) x U(1) theory of electroweak interactions |8|, using them as a basis for our considerations.

2. Known peculiarities of the neutrinos

As in the case of charged leptons and quarks we can use four component Dirac fields $\nu_\ell(x)$, where ℓ ($\ell = e,\mu,\ldots$) labels the neutrino flavour, to describe the neutrinos which take part in the weak interactions. Since no intrinsic characteristics of the neutrinos such as mass, electric charge, magnetic moment etc. which could be used to distinguish between them have been observed we have to clarify what is meant by different neutrino flavours. By definition, neutrino (antineutrino) of the type ℓ, ν_ℓ $(\tilde{\nu}_\ell)$ is the particle that is produced together with the charged lepton $\ell^+(\ell^-)$ in the weak decays $(\pi^+ \to \mu^+\nu_\mu$, $K^- \to e^-\tilde{\nu}_e$, $F^+ \to \tau^+\nu_\tau$ etc.). So, unlike the case of the other known particles (charged leptons, mesons, baryons etc.), the flavour of a given neutrino is specified by the weak interactions. That is why the neutrinos ν_ℓ are often called weak interaction eigenstates.

The existing data |16| suggest that the six known neutrinos $\overset{(\sim)}{\nu}_e$, $\overset{(\sim)}{\nu}_\mu$ and $\overset{(\sim)}{\nu}_\tau$ are different particles[F2]. For example, it is established with a rather good accuracy that $\nu_\mu \not\equiv \overset{(\sim)}{\nu}_e$, $\overset{(\sim)}{\nu}_\tau$, $\tilde{\nu}_\mu$, i.e. ν_μ does not produce $e^\pm$, $\tau^\pm$ or μ^+ when interacting with the nucleons. This fact is in consonance with another experimental observation, namely, the unwillingness of the charged leptons to undergo transitions among themselves in which neutrinos are not involved. Stringent experimental upper limits for the branching ratios (BR) and the cross sections (σ) of a number of such processes exist |17-20|:

$$\begin{aligned} &BR(\mu \to e\gamma) < 1.9 \times 10^{-10} \\ &BR(\mu \to 3e) < 1.9 \times 10^{-9} \\ &\frac{\sigma(\mu^- + S \to e^- + S)}{\sigma(\mu^- + S \to \text{capture})} < 7 \times 10^{-11} \end{aligned} \tag{1}$$

All these are strict inequalities.

$$\frac{\sigma(\mu^- + S \to e^+ + S_i^*)}{\sigma(\mu^- + S \to \text{capture})} \lesssim 9 \times 10^{-9}$$

$$BR(\tau \to e\gamma) < 6.4 \times 10^{-4}$$

$$BR(\tau \to e\mu\mu) < 3.3 \times 10^{-4}$$

$$BR(\tau \to e\rho) < 3.7 \times 10^{-4}$$

The quoted limits and, in fact, all existing data on the weak processes suggest that the leptons of a given flavour may possess an additive quantum number L_ℓ ($\ell = e, \mu, \ldots$), called respectively electron, muon etc. lepton charge, which is conserved in all reactions and decays. The lepton charges usually assigned to the charged leptons and the neutrinos are the following:

$$L_\ell = \begin{cases} +1 & \text{for } \ell^- \text{ and } \nu_\ell \\ -1 & \text{for } \ell^+ \text{ and } \tilde{\nu}_\ell \\ 0 & \text{for all other particles} \end{cases} \tag{2}$$

The conservation of L_ℓ would imply that the weak interaction Lagrangian is invariant under the global transformations of the lepton fields:

$$\begin{aligned} \ell(x) &\to e^{i\alpha_\ell} \cdot \ell(x) \\ \nu_\ell(x) &\to e^{i\alpha_\ell} \cdot \nu_\ell(x) \end{aligned} \tag{3}$$

where α_ℓ are constant parameters. In this case the sum of the electron, muon etc. lepton charges, i.e. the lepton charge $L = \sum_\ell L_\ell$ would also be conserved.

From the studies of the properties of the weak interactions it is also known that only the left-handed (LH) components of the neutrino fields (i.e., only the fields $\nu_{\ell L}$ and $\tilde{\nu}_{\ell L}$ where $\nu_{\ell L} = \frac{1}{2}(1 + \gamma_5)\nu_\ell$) enter the standard weak interaction Lagrangian. There is no experimental evidence for the physical relevance of the RH components $\nu_{\ell R}$. For this reason the fields $\nu_{\ell R}$ are not necessary ingredients of the theory of weak interactions. In particular, $\nu_{\ell R}$ are not present in the standard electroweak theory, where the lepton doublets and singlets have the well-known form:

$$\begin{pmatrix} \nu_{\ell L} \\ \ell_L \end{pmatrix}, \quad \ell_R \qquad (\ell = e, \mu, \dots) \tag{4}$$

The neutrinos are known to be remarkably lighter than the fermions of the generation they belong to. The experimental limits on the values of the neutrino masses obtained so far are, in fact, not very stringent for ν_μ and ν_τ:

$$14 \text{ eV} < m_{\tilde{\nu}_e} < 46 \text{ eV} \qquad [3]$$

$$m_{\tilde{\nu}_e} < 55 \text{ eV} \qquad [21]$$

$$m_{\nu_\mu} < 550 \text{ keV} \qquad [22]$$

$$m_{\nu_\tau} < 250 \text{ MeV} \qquad [23]$$

However, we shall assume that if massive, the neutrinos are relatively light so that the sum of the neutrino masses satisfies the cosmological bound |24|:

$$\sum_i m_i \lesssim 100 \text{ eV} \tag{5}$$

And since in the present and planned experiments with neutrino beams the neutrino energy (E_ν) exceeds roughly 1 MeV, effects of the order of $(m_i/E_\nu)^2$ will not be considered here.

Finally, the neutrinos are electrically neutral. As we shall see, this most easily established property of the neutrinos may have far reaching implications for the neutrino physics.

3. Massive Dirac versus massive Majorana neutrinos

3.1 The neutrinó mass matrix

The type of massive neutrinos in a gauge theory is specified by the neutrino mass term, more precisely, by the symmetries it has. By definition, fermion mass term is any invariant under the proper Lorentz transformations piece of the Lagrangian, formed only by fermion fields and belinear in them. The neutrino mass matrix originates usually 6 in gauge theories of the electroweak interactions from Yukawa type coiplings of the lepton doublets and/or singlets with Higgs scalar fields, some components of which develop nonzero vacuum expectation values. In order not to spoil the renormalizability of the theory these couplings have to be gauge invariant.

Let us consider how the possible types of nuetrino mass terms. It is convenient to divide them into three categories. For simplicity, we shall discuss first the case of one neutrino flavour.

(i) Dirac mass term. Both components ν_L and ν_R of the neutrino field are needed to construct a mass term of Dirac type:

$$-\mathcal{L}_D^{\nu} = m\left(\bar{\nu}_L \nu_R + \bar{\nu}_R \nu_L\right) = m\bar{\nu}\nu \tag{6}$$

where m is real. $\mathcal{L}_D^{\nu}$ is invariant under the global transformation $\nu \to e^{i\alpha}\nu$. This implies that $\mathcal{L}_D^{\nu}$ conserves an additive quantum number, which is the fermion number F, coinciding with the lepton charge L in this case. The massive neutrino is a Dirac particle, distinguished from its antiparticle by the value of the fermion number F.

The mass term of Dirac type arises most naturally in the standard SU(2) x U(1) theory containing the RH components of the neutrino fields as SU(2) singlets. It is generated |14| by the Yukawa-type interaction

$$G\,\bar{\nu}_R\,\phi^c \begin{pmatrix} \nu_L \\ \ell_L \end{pmatrix} + h.c. \tag{7}$$

where $\phi^c = i\tau_2\,\phi^*$ is the charge conjugate of the standard Higgs doublet $\phi = \begin{pmatrix}\phi^+\\ \phi^o\end{pmatrix}$, the neutral component of which has a nonzero vacuum expectation value $\langle\phi^o\rangle_o \neq 0$, and we have assumed one lepton family. The coupling (7) gives rise to $\mathcal{L}_D^\nu$ (eq. (6)) with $m = G\langle\phi^o\rangle_o$. In this case neutrinos are treated on equal footing with the other fermions of the theory.

(ii) <u>Majorana mass term for ν_L</u>. Using the properties of the charge conjugate spinors we can form a Lorentz invariant bilinear of the neutrino fields, which contains only the component ν_L. Indeed, let us define the field

$$\nu_R^c \equiv C\,\bar{\nu}_L \tag{8}$$

Here C is the charge conjugation matrix, satisfying the conditions

$$C^{-1}\gamma_\mu C = -\gamma_\mu^T \quad , \quad C^T = -C \tag{9}$$

Like ν_L, the field ν_R^c describes LH neutrinos and RH antineutrinos (the former as antiparticles, the latter as particles). Note that ν_R^c transforms as ν_L under the proper Lorentz transformations. Therefore the Lagrangian 12

$$-\mathcal{L}_M^\nu = m\,\bar{\nu}_R^c\,\nu_L + h.c = m\left(C^{-1}\nu_L\nu_L + \bar{\nu}_L C\bar{\nu}_L\right) \tag{10}$$

(m is real, in essence) has all the properties of a mass term and indeed it is. Expressed in terms of the combination of ν_L and ν_R^c

$$\chi = (\nu_L + \nu_R^c)/\sqrt{2} \tag{11}$$

it takes the standard form

$$-\mathcal{L}_M^\nu = m\,\bar{\chi}\chi$$

o m is the mass of the field χ. It follows from eqs. (8) and (11) that χ satisfies the Majorana condition

$$\chi = C\bar{\chi} \tag{12}$$

i.e. χ is a massive Majorana field.

Note that the mass term (10) differs substantially from $\mathcal{L}_D^\nu$ as no charge carried by ν_L is conserved. Even the fermion number is not preserved and, e.g., the transition of a neutrino into an antineutrino in one space-time point becomes possible due to $\mathcal{L}_M^\nu$. Obviously, no charged particle can have such a mass term.

It is impossible to generate $\mathcal{L}_M^\nu$ in the minimal $SU(2) \times U(1)$ theory (no $\nu_{\ell R}$) in a gauge invariant manner as the product $C^{-1}\nu_L\nu_L$ changes the weak isospin by one unit and the only Higgs field available is isodoublet. However, if a triplet of Higgs particles

$$H = \begin{pmatrix} -H^+/\sqrt{2} & H^{++} \\ H^0 & H^+/\sqrt{2} \end{pmatrix} \tag{13}$$

whose neutral component has a nonzero vacuum expectation value, i.e. $\langle H^0\rangle_0 \neq 0$, is introduced, the gauge invariant coupling |6|

$$h(\bar{\nu}_L \ \bar{\ell}_L)\, H^+ i\tau_2 \begin{pmatrix} \nu_R^c \\ \ell_R^c \end{pmatrix} + \text{h.c.} \tag{14}$$

$\left(i\tau_2 \begin{pmatrix} \nu_R^c \\ \ell_R^c \end{pmatrix} = \begin{pmatrix} C\bar{\ell}_L \\ -C\bar{\nu}_L \end{pmatrix} \right)$ leads to (10) with $m = h\,\langle H^0\rangle_0$. Furthermore, if we assign a lepton charge to H ($L_H = 2$) and assume that the lepton charge is conserved by the Higgs potential of the theory (note that the coupling (14) conserves it), the global symmetry corresponding to this conservation law will be spontaneously broken |25| in case $\langle H^0\rangle_0 \neq 0$. The resulting model is due to Gelmini and Roncadelli |26| and has been widely discussed recently |27|. Its most remarkable feature is the presence of a physical massless neutral scalar particle (the Goldstone boson of the broken global symmetry) called Majoron, which couples extremely weakly to the charged leptons and quarks. This theory has an interesting phenomenology |27|.

An alternative mechanism for generating $\mathcal{L}_M^\nu$ within the SU(2)xU(1) theories with a minimal fermionic content and an enlarged Higgs sector, including several Higgs doublets ϕ_i ($i = 1,2,...$), was suggested by Zee |28|. This mechanism relies on the fact that there exist more than one lepton families. The left-handed neutrinos $\nu_{\ell L}$ acquire a radiatively induced Majorana mass as a result of the introduction of a SU(2) singlet charged Higgs field H'^+. It couples to SU(2) singlet combinations of two lepton doublets which are antisymmetric in the flavour indices:

$$\sum_{\ell\ell'} f^0_{\ell\ell'} (\bar{\nu}_{\ell L} \; \bar{\ell}_L) H'^+ i\tau_2 \begin{pmatrix} \nu^c_{\ell' R} \\ \ell'_R \end{pmatrix} + h.c$$
$$= 2 \sum_{\ell\ell'} f^0_{\ell\ell'} \; \bar{\ell}'_L H'^+ \nu^c_{\ell R} + h.c. \tag{15}$$

where $f^0_{\ell\ell'} = - f^0_{\ell'\ell}$ are, in general, complex constants. According to (15) H'^+ can be assigned two units of the lepton charge L. The lepton number violation effects originate then from trilinear couplings of H'^+ to ϕ_i, which have to be antisymmetric in the indices of the Higgs doublets in order to preserve the gauge symmetry:

$$\sum_{j,k} c_{jk} \; \phi_j^+ \phi_k^c \; H'^+ + h.c. \tag{16}$$

and $c_{jk} = - c_{kj}$ are constants (for the case of two Higgs doublets c_{12} is real). The interactions (15) and (16) together with the standard Yukawa couplings of the lepton doublets and ℓ_R with ϕ_i, which give rise to the charged lepton mass matrix, combine at one loop level to produce a finite Majorana mass term of the type given by eq. (10). The model of Zee can be accommodated within the SU(5) theory with an enlarged Higgs sector by adding a 10-plet of Higgs fields, in which H'^+ is the colour and SU(2) singlet. It is essentially the only mechanism of neutrino mass generation in the SU(5) theory that yields sizeable neutrino masses, mass differences and mixing angles |29|. Let us note also that the supersymmetric (susy) version of the corresponding SU(5) model avoids |30| the difficulties of the minimal susy SU(5) GUT (e.g. the raising of the predicted value of $\sin^2 \theta w$). It should be mentioned that the model of Zee is extremely rich in properties that make it interesting both from theoretical and experimental point of view |31-33|.

<u>(iii) Dirac and Majorana mass terms.</u> In the most general case the neutrino mass Lagrangian may include |34| both $\mathcal{L}_D^\nu$, $\mathcal{L}_M^\nu$ and a Majorana piece formed by ν_R:

$$-\mathcal{L}_{D+M}^{\nu} = \kappa_1 \bar{\nu}_R^c \nu_L + \kappa_2 \bar{\nu}_R \nu_L + \kappa_3 \bar{\nu}_R \nu_L^c + \kappa_4 \bar{\nu}_R^c \nu_L^c + h.c. \tag{17}$$

where $\nu_L^c = C\bar{\nu}_R$ and κ_i are, in general, complex constants. Obviously, $\mathcal{L}_{D+M}^\nu$ does not conserve any charge assigned to ν_L and/or ν_R.

Since $\bar{\nu}_R^c \nu_L^c = \bar{\nu}_R \nu_L$ we have $\kappa_4 = \kappa_2$. Introducing the column of fields

$$\eta = \begin{pmatrix} \nu \\ \nu^c \end{pmatrix}$$

one can rewrite $\mathcal{L}_{D+M}^\nu$ in the form

$$-\mathcal{L}_{D+M}^{\nu} = \bar{\eta}_R^c M \eta_L + h.c. \tag{18}$$

where

$$M = \begin{pmatrix} \kappa_1 & \kappa_2 \\ \kappa_2 & \kappa_3 \end{pmatrix} \tag{19}$$

is a complex symmetric matrix: $M^T = M$. Any complex symmetric matrix can be reduced to a diagonal matrix with real nonnegative elements via the transformation:

$$U^T M U = m \quad , (m)_{ik} = \delta_{ik} m_k \quad , m_k \geq 0 \quad (i,k=1,2) \tag{20}$$

where in our case U is a 2x2 unitary matrix. The neutrino fields (χ) possessibg definite masses (m_1) can easily be found:

$$\chi_i = U_{i\alpha}\eta_{\alpha L} + U^*_{i\alpha}\eta^c_{\alpha R} \qquad (i=1,2) \tag{21}$$

They satisfy the Majorana condition

$$\chi_i = C\bar{\chi}_i \qquad (i=1,2) \tag{22}$$

and hence describe massive Majorana neutrinos[F3].

Thus, we have started in this case with one Dirac field which has four independent components corresponding to the two spin states of a Dirac neutrino and its antiparticle. We have ended up with twice as many different Majorana fields each having two independent components describing the two possible spin states of a Majorana particle.

Neutrino mass Lagrangian of (Dirac + Majorana) type arises, e.g., in the SO(10) GUTs |7| and that is crucial for generating neutrino masses compatible with the observations. In these theories the neutrino of a given flavour (i.e.,generation) acquires at the three level a Dirac mass of the order of m_q, where m_q is the mass of one of the quarks in the same generation. So, $\kappa_2 \sim m_q$ and $\kappa_1 = 0$. As is well known, neutrino masses of this order are excluded experimentally. A possible solution to this problem is achieved by assuming that the RH components of the neutrino fields $\nu_{\ell R}$ are superheavy, having a Majorana mass, e.g., of the order of the unification scale of the electroweak and strong interactions $M_{GUT} \sim 10^{15}$ GeV. It turns out to be possible to generate $\kappa_3 \sim M_{GUT}$ and for the neutrino mass matrix one gets:

$$M^{\nu}_{SO(10)} = \begin{pmatrix} 0 & m_q \\ m_q & M_{GUT} \end{pmatrix} , \quad M_{GUT} \gg m_q$$

Since m_q and M_{GUT} are real, the neutrino mass term is CP conserving. The eigenvalues $m_{1,2}$ of $M^{\nu}_{SO(10)}$ and the corresponding eigenvectors $\chi_{1,2}$ can be easily found :

$$
\begin{aligned}
&m_1 \simeq -\frac{m_q^2}{M_{GUT}} \quad , \quad m_2 = M_{GUT} \\
&\chi_1 \simeq (\nu_L + \nu_R^c) + \frac{m_q}{M_{GUT}} (\nu_L^c + \nu_R) \\
&\chi_2 \simeq -\frac{m_q}{M_{GUT}} (\nu_L + \nu_R^c) + (\nu_L^c + \nu_R) \\
&\chi_1 = C\bar{\chi}_1 \quad , \quad \chi_2 = C\bar{\chi}_2
\end{aligned}
\tag{23}
$$

Obviously, the neutrino χ_1 takes part in the weak interactions (as $\chi_{1L} \simeq \nu_L$) and it is light enough to satisfy the existing limits on the neutrino masses. The second neutrino χ_2 is superheavy and it practically decouples at low energies. This is the famous mechanism for generation of small neutrino masses in SO(10) GUTs suggested by Gell-Mann, Ramond and Slansky, and independently by Yanagida |35|.

It is amusing to note that a mass term of the type of $\mathcal{L}^{\nu}_{D+M}$ (with $\kappa_1 = 0$) and, consequently, massive Majorana neutrinos may appear in the standard model containing the RH neutrino field components $\nu_{\ell R}$. Indeed, besides the Dirac piece, generated in the standard way (see eq. (7)), the most general neutrino mass Lagrangian includes in this case a Majorana term for $\nu_{\ell R}$. Since $\nu_{\ell R}$ are SU(2) and U(1) singlets, the latter is gauge invariant and does not spoil the renormalizability of the theory. However, the resulting neutrino mass spectrum is arbitrary which makes this possibility unattractive.

One final remark is in order. Diagonalizing $M^{\nu}_{SO(10)}$ we have obtained a negative value for the mass of the neutrino χ_1: $m_1 < 0$. The field with a positive mass $|m_1|$ is in fact χ_1' :

$$
\chi_1' = \gamma_5 \chi_1 = (\nu_L - \nu_R^c) + \frac{m_q}{M_{GUT}} (\nu_L^c - \nu_R) \tag{24}
$$

Note, however, that χ_1' satisfies the condition:

$$\chi_1' = -C\chi_1' \tag{25}$$

This simple example shows that when CP is conserved, the massive Majorana neutrinos can be of two varieties |11, 32|, differing by the sign in the Majorana condition their fields satisfy:

$$\chi = \eta C \bar{\chi} \quad , \quad \eta = +1 \text{ or } -1 \tag{26}$$

As noticed recently by Wolfenstein |32|, the product of the η-factors of two neutrinos is, in principle, an observable quantity and may play a crucial role in the analyses of the processes involving Majorana neutrinos. In the case of CP nonconservation the factor η can be absorbed in the mixing matrix U (see eq. (21)) by a redefinition of one of the CP violating phases in U.

This case completes our classification and discussion of the neutrino mass terms that might arise in the theories of the electroweak interactions which do not contain additional neutral fermions able to mix (i.e. to form mass terms) with the neutrinos |36|. One of the conclusions following from the above considerations is that the discovery of a non-zero neutrino mass would imply either that the lepton charge L is not conserved or that the RH components of the neutrino fields $\nu_{\ell R}$ are physically significant, or both.

3.2 Taking the flavour into account

Our analysis can be easily generalized to include the various flavours of the neutrinos. Omitting the technical details, let us give below the resulting relations between the weak and the mass eigenstate fields for n neutrino flavours in each of the three cases discussed above |37|:

(i) $$\nu_i = \sum_{l=e,\mu,\ldots} U^D_{il}\, \nu_{lL} + \sum_l U'^D_{il}\, \nu_{lR} \qquad (i=1,\ldots n) \qquad (27)$$

(ii) $$\chi_i = \sum_l U^M_{il}\, \nu_{lL} + \sum_l U^{M*}_{il}\, \nu^c_{lR} \qquad (i=1,\ldots n) \qquad (28).$$

(iii) $$\chi'_i = \sum_l \left(U_{il}\, \nu_{lL} + V_{il}\, \nu_{lR} + U^*_{il}\, \nu^c_{lR} + V_{il}\, \nu^c_{lL} \right) \quad (i=1,\ldots 2n) \qquad (29)$$

Here ν_i and $\chi_i^{(\sim)} = \eta_i\, C\, \chi_i^{(\sim)}$ correspond to massive Dirac and massive Majorana neutrinos, U^D , U'^D and U^M are $n \times n$ mixing matrices and $W = (UV)$ is a $2n \times 2n$ mixing matrix. If CP is not conserved by the neutrino mass matrix, written in terms of the weak interaction eigenstates, the mixing matrices are unitary and $\eta_i = +1$. In theories with CP-invariant lepton mass Langrangians the mixing matrices are orthogonal and $\eta_i = +1$ or -1.

Among the most important consequences of the connections between the neutrino mass eigenstates and the weak interaction eigenstates of the type we have considered is the possibility of transitions in flight between different flavours of neutrinos |11,15| (neutrino oscillations). Inverting, e.g., eqs. (27) and (28) in the cases (i) and (ii) respectively, and taking into account the time evolution of the states describing the free massive neutrinos, it is not difficult to obtain the probability amplitudes to find neutrino (antineutrino) of the type ℓ' at time t in a beam which at time t=0 consists of neutrinos (antineutrinos) of the type ℓ:

$$\begin{aligned} A(\nu_l \to \nu_{l'}) &= \sum_k e^{-iE_k t}\, U^{D,M}_{kl} \left(U^{D,M}_{kl'}\right)^* \\ A(\bar\nu_l \to \bar\nu_{l'}) &= \sum_k e^{-iE_k t} \left(U^{D,M}_{kl}\right)^* U^{D,M}_{kl'} \qquad (l,l'=e,\mu,\ldots) \end{aligned} \qquad (30)$$

where $E_k = \sqrt{p^2 + m_k^2}$ is the energy of the neutrino with mass m_k. Thus, if CP is not conserved, i.e. if $U^{D,M}$ are complex matrices, the $\nu_\ell \to \nu_{\ell'}$ and $\tilde\nu_\ell \to \tilde\nu_{\ell'}$ oscillation amplitudes and probabilities may be different for $\ell \neq \ell'$.

3.3 Massive Majorana and massive Dirac neutrinos: a comparison of the properties

Let us consider now in more detail the properties of a Majorana neutrino having a field χ,

$$\chi = \eta C \bar{\chi} \quad , \quad \eta = \begin{cases} +1 & \text{if CP is not conserved} \\ +1 \text{ or } -1 & \text{if CP is conserved} \end{cases} \tag{31}$$

and compare them with the properties of a Dirac neutrino ν. First, it is not difficult to show using eq. (31) that the particle and the antiparticle creation and annihilation operators in χ coincide. Therefore, χ describes a particle identical with its antiparticle, i.e. a truely neutral spin 1/2 object, which cannot carry additive quantum numbers as lepton charge, fermion number, weak hypercharge etc. Hence, in contrast to the Dirac neutrino fields, the Majorana fields cannot be subjected to continuous global phase transformations $\chi \to e^{i\alpha}\chi$ (they cannot "absorb" phases). Such transformations are incompatible with the Majorana condition (31). Secondly, besides the standard lepton number conserving propagator $\overbrace{\chi_\alpha(x)\bar{\chi}_\beta(y)}$ the Majorana neutrino can have a nontrivial lepton number nonconserving propagator:

$$\overbrace{\chi_\alpha(x)\,\chi_\beta(y)} = \eta\, C_{\alpha\beta}\, \overbrace{\chi_\alpha(x)\,\bar{\chi}_\beta(y)} \neq 0 \tag{32}$$

Obviously, $\overbrace{\nu_\alpha(x)\nu_\beta(y)} = 0$.

And, thirdly, it is easy to show using eq. (31) that

$$\begin{aligned} \bar{\chi}\, \sigma_{\mu\nu}\, \chi &= 0 \\ \bar{\chi}\, \sigma_{\mu\nu}\gamma_5\, \chi &= 0 \\ \bar{\chi}\, \gamma_\mu\, \chi &= 0 \end{aligned} \tag{33}$$

Eqs. (33) imply that massive Majorana neutrinos cannot possess intrinsic magnetic or electric dipole moments, which is not valid, in general, for the massive Dirac neutrinos.

3.4 Distinguishing between the two possible types of massive neutrinos

If neutrinos turn out to be massive, the question about the type of the neutrinos (Dirac or Majorana) will inevitably arise. Even if we leave aside our present theoretical prejudices this will be a fundamental question to answer as it will concern the very nature of the neutrinos. Let us first discuss the possibility to determine the type of massive neutrinos in experiments studying the neutrino oscillations |37|. The most interesting is the case when the number of the neutrino mass eigenstates coincides with the number of the neutrino flavours, i.e. when the corresponding neutrino mass matrix is either of Dirac type or Majorana type for $\nu_{\ell L}$. The expressions for the oscillation amplitudes in this case have been given earlier (see eq. (30)). They depend, in particular, on the lepton mixing matrix which enters the lepton charged weak current:

$$j_\mu^{lep} = \sum_\ell \bar{\ell}_L \gamma_\mu \nu_{\ell L} = \sum_\ell \bar{\ell}_L \gamma_\mu \begin{cases} U^{D+}_{\ell k} \nu_{kL} & \text{case (i)} \\ U^{M+}_{\ell k} \chi_{kL} & \text{case (ii)} \end{cases} \qquad (34)$$

If CP is not conserved the n x n unitary matrices $U^{D,M}$ will contain n^2 independent parameters, namely, $\frac{n(n-1)}{2}$ mixing angles and $\frac{n(n+1)}{2}$ the presence of which may imply CP-parity nonconservation. Not all phases in $U^{D,M}$ are observable. Indeed, using the phase arbitrariness in the definition of the Dirac fields (i.e., the invariance of any observable quantity with respect to the transformations $\ell(x)\to e^{i\alpha_\ell}\ell(x), \ell=e,\mu,..$ and $\nu_k(x)\to e^{i\beta_k}\nu_k(x)$, k=1,2,..,where α_ℓ and β_k are arbitrary real parameters) and recalling that the Majorana fields are not subject to such arbitrariness, it is possible to reduce the number of the phase parameters in U^D and U^M to $(n-1)(n-2)/2$ and $n(n-1)/2$, respectively. In particular, for three neutrino flavours there will be one CP-violating phase in the Dirac case and three CP-violating phases in the Majorana case. It is easy to notice, however, that the relevant oscillation amplitude $A(\overset{(\sim)}{\nu}_\ell \to \overset{(\sim)}{\nu}_{\ell'})$ does not change |37| under the transformation

$$U^M_{k\ell} \to e^{i\beta_k} U^M_{k\ell} \quad , k=1,\dots n \ , \ell = e,\mu,\dots \tag{35}$$

This implies that the additional CP-violating parameters in U^M do not lead to observable effects in the oscillations $\overset{(\sim)}{\nu}_\ell \rightleftarrows \overset{(\sim)}{\nu}_{\ell'}$. A more detailed investigation |38| reveals that these parameters are always associated with effects $\sim(m_i/E_\nu)^2$.

In fact, the existing large number of studies |13,39| allow to conclude that the majority of the effects typical only for massive Majorana neutrinos are very subtle, being of the order of $(m_i/E)^2$, where E is some characteristic energy scale for the given process involving real or virtual neutrinos. If neutrinos have masses smaller than 100 eV, most sensitive to such effects seem at present the experiments searching for neutrinoless double β-decay |40| ($(\beta\beta)_{o\nu}$) of certain nuclei:

$$(A,Z) \to (A,Z+2) + e^- + e^- \tag{36}$$

Let us discuss in more detail this process. If the electron neutrino is a Majorana particle ($\nu_e \equiv \chi$, $\chi = \eta C\bar{\chi}$) with nonzero mass (m), then by exchanging a virtual electron neutrino two neutrons in a nucleus may undergo transition into a pair of protons and a pair of free electrons, which leads to (36). In such a way, a nontrivial contribution to the $(\beta\beta)_{o\nu}$ -decay amplitude $A(\beta\beta)_{o\nu}$ may arise in the second order of perturbation theory in the Fermi coupling constant G_F. The element of the S-matrix which generates it can be written in the form:

$$S^{(2)}_{(\beta\beta)_{0\nu}} = -\frac{1}{2}\frac{G_F^2}{2}\int d^4x\, d^4y\ \bar{p}(x)\gamma_\alpha(1+\gamma_5)n(x)\cdot$$
$$\bar{e}(x)\gamma_\alpha(1+\gamma_5)\overbrace{\chi(x)\bar{e}(y)\gamma_\beta(1+\gamma_5)\chi}(y)\ \bar{p}(y)\gamma_\beta(1+\gamma_5)n(y) \tag{37}$$

where $p(x)$, $n(x)$ and $e(x)$ are the proton, neutron and electron fields and the other notations are obvious[F4]. It is evident that the process (36) cannot proceed via exchange of Dirac neutrinos for which the lepton number nonconserving propagator is identically equal to zero. Using the Majorana condition $\chi = \eta C \bar{\chi}$ and the properties of the charge conjugation matrix C (see eq. (9)) we can rewrite the product of the two lepton currents in $S^{(2)}_{(\beta\beta)_{0\nu}}$ as:

$$\bar{e}(x)\gamma_\alpha(1+\gamma_5)\overbrace{\chi(x)\bar{e}(y)\gamma_\beta(1+\gamma_5)\chi}(y) =$$
$$-\eta\,\bar{e}(x)\gamma_\alpha(1+\gamma_5)\overbrace{\chi(x)\bar{\chi}}(y)(1+\gamma_5)\gamma_\beta C\bar{e}(y) \tag{38}$$

It follows from eq. (38) that only the mass term in the numerator of the neutrino propagator in momentum space contributes to $A(\beta\beta)_{0\nu}$. An explicit evaluation of the nuclear matrix elements of $S^{(2)}_{(\beta\beta)_{0\nu}}$ shows |41| that, apart from negligible corrections, $A(\beta\beta)_{0\nu}$ depends on m linearly provided the neutrino mass is much smaller than a few MeV:

$$A(\beta\beta)_{0\nu} \sim \eta m\ , \quad m \ll \text{few MeV} \tag{39}$$

The experiments on $(\beta\beta)_{0\nu}$-decay which are in progress or should be performed in the near future are expected to be sensitive to values of m in the range from few tens to few eV. However, even if some of the neutrinos have masses of the order of, say, 30 eV, as is possibly indicated by the ITEP data |3|, experiments with much higher precision or entirely new methods may turn out to be required to determine the nature of neutrinos. Indeed, a straightforward generalization of eq.

(39) in the case of n neutrino flavours and nontrivial neutrino mixing leads to |32,41|:

$$A(\beta\beta)_{o\nu} \sim \sum_{k=1}^{n} \eta_k m_k \left(U^{M+}_{ek} \right)^2 , \quad m_k << \text{few MeV} \tag{40}$$

where for concreteness the charged lepton current has been assumed to have the form given by eq. (34) (case (ii)). In the case of CP non-conservation at least some of the elements U^M_{ek} (k = 1,...,n) should be complex and, as follows from eq. (40), there may be a partial cancellation of the different neutrino contributions to $A(\beta\beta)_{o\nu}$. If CP is conserved $(U^{M+}_{ek})^2 \geq 0$, but analogous cancellation may occur if not all Majorana neutrinos have identical η-factors |32|. These cancellations may even be complete.

A remarkable example of realization of the latter possibility is provided by one version of the model of Zee. In the case of three generations, flavour diagonal lepton-Higgs doublet couplings and weak flavour dependence of the H'^{+}-lepton coupling constants $f^o_{\ell\ell'}$ (see eq. (15)), the model of Zee contains |31| two almost degenerate in mass Majorana neutrinos, say, $\chi_{2,3}$, which are much heavier than the third one χ_1: $(m_2 - m_3) \sim m_1 \sim m_2 \frac{m_\mu^2}{m_\tau^2} << m_{2,3}$. In particular, values of $m_{2,3}$ as large as (30-40) eV are possible. Further, CP is conserved and $\eta_{1,3} = -1$, while $\eta_2 = +1$. The lepton mixing matrix has a special form which together with the mentioned relations between the neutrino masses and the values of the η-factors leads to exact mutual compensation |32| of the three terms in the sum that usually determines $A(\beta\beta)_{o\nu}$ to leading order:

$$\sum_{k=1}^{3} \eta_k m_k \left(U^{M+}_{ek} \right)^2 \Big|_{\text{Model of Zee}} = 0 \tag{41}$$

Tiny contributions from diagrams with exchange of virtual charged Higgs bosons make $A(\beta\beta)_{o\nu}$ different from zero but very much suppressed |33| in this theory. Similar cancellations were shown to take place in most of the SO(10) models as well |42|, although they are not as effective as in the case considered above.

This completes our discussion of the $(\beta\beta)_{o\nu}$-decay[F5] and of the possibilities to distinguish experimentally between massive Dirac and massive Majorana neutrinos. It indicates that the determination of the type of massive neutrinos might be a remarkably difficult problem.

4. Conclusion

It should have become clear from this lecture that although the idea of the existence of neutrinos has been proposed 52 years ago and the first neutrino induced reaction has been observed 37 years ago, we may still know rather little about these fascinating particles.

Acknowledgements

I would like to thank the organizers of this year Arctic School of Physics and especially R. Raitio and J. Lindfors for creating a pleasant working atmosphere at the School and for the kind hospitality extended to me during my stay in Finland. Thanks also go to C. N. Leung who read the manuscript and made numerous valuable remarks which lead to a considerable improvement of the initial text of this lecture. I am indebted to K. Enqvist, C.N. Leung, K. Mursula, R. Peccei, M. Roos and M. Schepkin for fruitful discussions. Finally, the hospitality of the Bartol Research Foundation of The Franklin Institute, where the preparation of the written version of this lecture was completed, is gratefully acknowledged.

Footnotes

F1. Indications of nonzero neutrino masses have been reported |4| to be observed in the experiments of Reines et al. with reactor antineutrinos, but they are not compatible with the most recent data |5| of Mössbauer et al.

F2. It can already be concluded on the basis of the available data that $\nu_\tau \neq \overset{(-)}{\nu}_\mu, \overset{(-)}{\nu}_e$, although the production of τ^- by ν_τ has not been observed yet.

F3. Obviously, under certain conditions $\mathcal{L}^{\nu}_{D+M}$ may lead to massive Dirac neutrinos (e.g. if $\kappa_{1,3} = 0$).

F4. The process (36) may be induced by interactions with right-handed currents as well |40,41|. This mechanism of $(\beta\beta)_{0\nu}$-decay can be distinguished experimentally |41| from that described by us and we shall assume for simplicity that it is not dominant.

F5. Further details will be given in the lecture of M. Schepkin.

References

1. W. Pauli, letter (4. December 1930) to a meeting of physicists in Tübingen. (See, for example, Brown, Phys. Today, September 1978).
2. F. Perrin, Comptes Rendus 197 (1933) 1625; E. Fermi, Nuovo Cim. 11 (1934) 1.
3. V.A. Lubimov et al., Phys. Lett. 94 B (1980) 266.
4. F. Raines et al., Phys. Rev. Lett. 45 (1980)1907.
5. J.L. Vuilleumier et al., SIN preprint PR-82-07, 1982.
6. See, e.g., T.P. Cheng and L.-F. Li, Phys. Rev. D22 (1980) 2860.
7. For a review, see, e.g.: J. Ellis, Lectures presented at Les Houches Summer School 1981 (CERN preprint, TH-3174, 1981).
8. S.L. Glashow, Nucl. Phys. 22 (1981) 579; S. Weinberg, Phys. Rev. Lett. 19 (1967) 1264; A. Salam, Proc. of the 8th Nobel Symposium, ed. N. Svartholm (Almquist and Wicksell, Stockholm, 1968).

9. H. Georgi and S.L. Glashow, Phys. Rev. Lett. 32 (1974) 438.
10. L. Wolfenstein,Nucl. Phys. B 186 (1981) 147.
11. B. Pontecorvo, JETP 34 (1958) 247.
12. V. Gribov and B. Pontecorvo, Phys. Lett. 28 B (1969) 49.
13. The properties of massive Majorana neutrinos have been widely discussed recently. See, e.g.: P.B. Pal and L. Wolfenstein, Phys. Rev. D 25 (1982) 766; J. Schechter and J.F.W. Valle, Phys. Rev. D24 (1981) 1883, errata ibid. D25 (1982) 283; R. Schrock, preprint ITP-SB-82-2, January 1982; B. Kayser, SLAC preprint 2879, January 1982; S.T. Petcov, Phys. Lett. 110B (1982) 245; J. Nieves, University of Puerto Rico preprint, 1981.
14. S.T. Petcov, Sov. J. Nucl. Phys. 25 (1977) 340, errata ibid. 25 (1977) 698.
15. A review of the subject with the literature therein up to 1978 is given in: S.M. Bilenky and B. Pontecorvo, Phys. Reports 41 (1978) 226.
16. G. Danby et al., Phys. Rev. Lett. 9 (1962) 36; S.E. Willis et al. Phys. Rev. Lett. 44 (1980) 522; A.M. Cnops et al., Phys. Rev. Lett. 40 (1978) 144; see also e.g. M. Perl, SLAC-PUB-2446 (1979).
17. J.D. Bowman et al., Phys. Rev. Lett. 45 (1979) 556.
18. S.M. Korenchenko et al., JETP 70 (1976) 3.
19. A. Baderstcher et al., Nuovo Cim. Lett. 28 (1980) 40.
20. K.G. Mayers et al., Phys. Rev. D25 (1982) 2869.
21. K.E. Bergkvist, Nucl. Phys. B 39 (1972) 371.
22. D. Daum et al., Phys. Lett. 74 B (1978) 126.
23. W. Bacino et al., Phys. Rev. Lett. 41 (1978) 13.
24. S.S. Gershtein and Ya.B. Zeldovich, JETP Lett. 4 (1966) 174; R. Cowsic and J. McClelland, Phys. Rev. Lett. 29 (1972) 669.
25. Y. Chikashige, R.N. Mohapatra and R.D. Peccei, Phys. Lett. 98 B (1981) 265.
26. G.B. Gelmini and M. Roncadelli, Phys. Lett. 99 B (1981) 411.
27. See e.g. H. Georgi, S.L. Glashow and S. Nussinov, Nucl. Phys. B 193 (1981) 297; V. Barger, W.Y. Keung and S. Pakvasa, Phys. Rev. D 25 (1982) 907; J. Schechter and J.F.W. Valle, Phys. Rev. D 25 (1982) 774.
28. A. Zee, Phys. Lett. 93 B (1980) 389.
29. J.F. Nieves, Nucl. Phys. B 189 (1981) 182.
30. A. Masiero et al., preprint MPI-PAE/PTh 29/1982.
31. L. Wolfenstein, Nucl. Phys. B 175 (1980) 93.
32. L. Wolfenstein, Phys. Lett. 107 B (1982) 77.
33. S.T. Petcov, Phys. Lett. 115 B (1982) 401.
34. S.M. Bilenky and B. Pontecorvo, Lett. Nuovo Cim. 17 (1976) 569.
35. M. Gell-Mann, P. Ramond and R. Slansky, unpublished; T. Yanagida, Proc. Workshop on the Unified Theory and the Baryon Number in the Universe, KEK, Japan 1979.
36. For a more general discussion see, e.g.: J. Schechter and J.F.W. Valle, Phys. Rev. D22 (1980) 2227; An example of a theory with neutral fermions which mix with the neutrinos was considered, e.g. by K. Enqvist and J. Maalampi, Phys. Lett. 97 B (1980) 217.
37. See, e.g.: S.M. Bilenky, J. Hosek and S.T. Petcov, Phys. Lett. 94 B (1980) 495,
38. J. Schechter and J.F.W. Valle, Phys. Rev. D 23 (1981) 1666.
39. See, e.g.: B. Kayser and R. Schrock, preprint SLAC-PUB-2815 (T), 1981; S.P. Rosen, preprint PURD-TH-81-8; M. Doi et al., Prog. Theor. Phys. 67 (1982) 281.
40. For reviews see e.g.: C.S. Wu, Proc. of Neutrino Mass Mini-conference, Telemark, Wisconsin, October 2-4, 1980 (ed. V. Barger and D.Cline), p. 97; S.P. Rosen, talk given in Orbis Scientiae 1981, Univ. of Miami, 19-22 January 1981.
41. M. Doi et al., Prog. Theor. Phys. 66 (1981) 1739 and 1765; errata Osaka Univ. Preprint, March 1982.
42. D. Chang and P.B. Pal, Phys. Rev. D26 (1982) 3113.

m_ν AND L-CONSERVATION

M.G. Schepkin
Institute for Theoretical and Experimental Physics
117259 Moscow, USSR

Contents:

1. Introduction

As is well known neutrino mass (if it exists) is much smaller than masses of quarks and charged leptons from the same generation[1-3]. This concerns at least known neutrinos, entering weak Lagrangian. As will be discussed below it is more natural to interpret such a small neutrino mass as Majorana mass. Therefore the question of neutrino mass is closely related now to the question of lepton conservation. The function performed by Majorana mass is twofold (as distinct from Dirac mass). The first (common for both types of masses) is kinematical function when the mass is considered as a quantity defining the position of the pole of neutrino propagator. The presence of Majorana and Dirac masses could equally well explain for example the deviation of Kurie spectrum from straight line in the vicinity of the end-point, $\nu_e \leftrightarrow \nu_\mu \leftrightarrow \nu_\tau$ oscillations[4], dark mass of the Universe[5]. Besides Majorana mass term $m_M \nu C \nu$ carries double lepton number L and thus should lead to the observable processes with $\Delta L = \pm 2$, the amplitude of such processes being proportional to m_M. On the other hand the discovery of $\Delta L = 2$ processes would not unambigously mean the existence of Majorana mass, as there could exist another lepton violating mechanisms. In this lecture we shall describe some of the widely discussed processes which could take place in the presence of Majorana masses and compare this situation with the case when Dirac masses are nonzero. Before that we shall give a short review of neutrino masses phenomenology (part 2) and neutrino masses generation in GUT (part 3). We shall see from the comparison of different phenomena sensitive to neutrino masses, that one of the most interesting processes to be searched for experimentally is neutrinoless double β -decay ($2\beta(0\nu)$ -decay).

2. Neutrino mass matrix. Phenomenology.

Let us consider for the beginning one generation of fermions, ψ describing neutrino field of definite flavour, say ν_e . Then in the most general case the Lagrangian may contain three types of mass terms[6,7]

$$-\mathcal{L} = m\bar{\psi}\psi + m_+\psi C\psi + m_-\psi C\gamma_5\psi + h.c. \quad (1)$$

It is convenient to work with chiral components of the field ψ :

$$\nu_{L,R} = \frac{1\pm\gamma_5}{2}\psi.$$

Indeces L and R describe the properties of two--component spinors $\nu_{L,R}$ with respect to γ_5 -transformations ($\gamma_5\nu_{L,R} = \pm\nu_{L,R}$) and not to the helicity once neutrino masses are nonzero. In terms of chiral fields ν_L and ν_R the Lagrangian (1) has the form

$$-\mathcal{L} = m_D\bar{\nu}_L\nu_R + m_L\nu_L C\nu_L + m_R\nu_R C\nu_R + h.c. \quad (2)$$

where

$$m_D = m, \qquad m_{\pm} = \frac{1}{2}(m_L \pm m_R). \tag{3}$$

When all the three terms in (2) are nonzero ν_L and ν_R are not anylonger eigenstates of the mass matrix. Eigenstates (let's denote them as φ_1 and φ_2) with masses μ_1 and μ_2. respectively are orthogonal superpositions of ν_L and ν_R

$$\begin{aligned} \varphi_{1L} &= \nu_L \cos\xi + \nu_L^c \sin\xi \, ; \quad \varphi_{1R} = \nu_R^c \cos\xi + \nu_R \sin\xi \, ; \\ \varphi_{2L} &= \nu_L \sin\xi - \nu_L^c \cos\xi \, ; \quad \varphi_{2R} = \nu_R^c \sin\xi - \nu_R \cos\xi . \end{aligned} \tag{4}$$

Mixing angle ξ and mass eigenvalues μ_1 and μ_2 when expressed in terms of m_D and $m_{\pm}$ has the form:

$$\begin{aligned} \mu_{1,2} &= m_+ \pm \sqrt{m_-^2 + m_D^2} \, , \\ tg\, 2\xi &= \frac{m_D}{m_-} . \end{aligned} \tag{5}$$

The inverse transformation looks as follows:

$$\begin{aligned} \nu_L &= \varphi_{1L} \cos\xi + \varphi_{2L} \sin\xi \, ; \\ \nu_R &= \varphi_{1R} \sin\xi - \varphi_{2R} \cos\xi . \end{aligned} \tag{6}$$

For the case of neutrinos of N flavours ν_i $(i = e, \mu, \tau, \ldots)$ the most general form of massive Lagrangian including all possible mixings between different flavors is

$$-\mathcal{L} = m^D_{iK}\,\bar{\nu}_{iL}\,\nu_{KR} + m^L_{iK}\,\nu_{iL} C\,\nu_{KL} + m^R_{iK}\,\nu_{iR} C\,\nu_{KR} + h.c. \tag{7}$$

Three $\mathcal{N}\times\mathcal{N}$ matricies in equation (7) can be presented as one $2\mathcal{N}\times 2\mathcal{N}$ matrix $\mathcal{M}$:

$$\mathcal{M} = \begin{pmatrix} \| m^L_{iK}\|^* & \frac{1}{2}\| m^D_{iK}\| \\ \frac{1}{2}\| m^D_{iK}\|^T & \| m^R_{iK}\| \end{pmatrix} \tag{8}$$

To diagonalize it one has to find $2\mathcal{N}\times 2\mathcal{N}$ matrix K expressing mass eigenstates φ_A in terms of neutrino of definite flavour

$$\varphi_{AL} = U_{Ai}\,\nu_{iL} + V_{AK}\,\nu^c_{KL},$$

$$\varphi_{AR} = U^*_{Ai}\,\nu^c_{iR} + V^*_{AK}\,\nu_{KR}, \tag{9}$$

where

$$K = \begin{pmatrix} U_{11} \ldots & U_{1\mathcal{N}} & V_{11} \ldots & V_{1\mathcal{N}} \\ - & - & - & - \\ U_{2\mathcal{N},1} \ldots & U_{2\mathcal{N},\mathcal{N}} & V_{2\mathcal{N},1} \ldots & V_{2\mathcal{N},\mathcal{N}} \end{pmatrix} \tag{10}$$

and $\nu^c_{R,L} = C\,\bar{\nu}_{L,R}$.

It is easy to see, that the number of eigenstates φ_A, satisfying condition

$$\varphi_A = C\,\bar{\varphi}_A \tag{11}$$

as well as the number of mass eigenvalues μ_A is equal to $2\mathcal{N}$.

For the unitary matrix K $(K^+K = 1)$

$$\nu_{iL} = K^+_{iA}\varphi_{AL} \equiv U^+_{iA}\varphi_{AL}; \quad \nu^c_{iR} = U^T_{iA}\varphi_{AR}; \tag{12}$$

$$\nu_{iR} = K^T_{i+\mathcal{N},A}\varphi_{AR} \equiv V^T_{iA}\varphi_{AR}; \quad \nu^c_{iL} = V^+_{iA}\varphi_{AL}.$$

It can be shown, that the **unitarity of K-matrix follows** unambigously from the properties of kinetic terms in the Lagrangian.

In terms of diagonal fields

$$-\mathcal{L} = m_{AB}\,\bar{\varphi}_{AL}\varphi_{BR} + h.c. \tag{13}$$

where m_{AB} is $2\mathcal{N}\times 2\mathcal{N}$ mass matrix

$$\| m_{AB} \| = K\mathcal{M}K^T = \mu_A \delta_{AB}. \tag{14}$$

Now let's consider the case when weak interactions **contain** only left-handed charged currents and is described by the Lagrangian

$$\mathcal{L}_L = g_L\,\bar{\nu}_{iL}\gamma_\mu l_{iL}\cdot W^{\mu +}_L + h.c. \tag{15}$$

In terms of mass eigenstates

$$\mathcal{L}_L = g_L\,\bar{\varphi}_{AL}U_{Ai}\gamma_\mu l_{iL}\cdot W^{\mu +}_L + h.c. \tag{16}$$

Thus for description of left-handed charged currents it is necessary to know only $2\mathcal{N}\times\mathcal{N}$ matrix U_{Ai} and $2\mathcal{N}$ mass eigenvalues. It can be shown that matrix U_{Ai} contains $3\mathcal{N}^2-\mathcal{N}$ physically relevant parameters: $(3\mathcal{N}^2-\mathcal{N})/2$ mixing angles and the same **number** of CP-odd phases. In the particular case, when $m_L \neq 0$, $m_R = m_D = 0$ there are $\mathcal{N}$ mass eigenstates, entering left-handed weak interactions [8]. Symmetrical mass matrix m^L_{ik} is diagonalized by $\mathcal{N}\times\mathcal{N}$ unitary matrix U_{Ai}, which contains now $\mathcal{N}(\mathcal{N}-1)/2$ mixing angles and the same number of physical relevant CP-odd phases. For $\mathcal{N}=3$ ($i = e, \mu, \tau$) the unitary matrix U_{Ai} describing charged currents can be written in the form [9]

$$U = \begin{pmatrix} c_1 & s_1 c_2 e^{i\varphi_2} & s_1 s_2 e^{i\varphi_3} \\ -s_1 c_3 & (c_1 c_2 c_3 - s_2 s_3 e^{i\varphi_1}) e^{i\varphi_2} & (c_1 s_2 c_3 + c_2 s_3 e^{i\varphi_1}) e^{i\varphi_3} \\ s_1 s_3 & (-c_1 c_2 s_3 - s_2 c_3 e^{i\varphi_1}) e^{i\varphi_2} & (-c_1 s_2 s_3 + c_2 c_3 e^{i\varphi_1}) e^{i\varphi_3} \end{pmatrix} \qquad (17)$$

where $c_i \equiv \cos\alpha_i$, $s_i \equiv \sin\alpha_i$, φ_i - CP-odd phases. If **the** mass matrix is of Majorana type, the number of CP-odd phases is equal to 3. Note, that for Kobayashy-Maskawa description for quarks [10] (where masses are of Dirac type) only one CP-odd phase is physically relevant. The comparison of some particular forms of mass matrix $\mathcal{M}$ (see eq. (8)) are given in the Table assuming weak interaction contains only left-handed neutrinos.

Table

Variant of the model	number of mass eigen-values	number of mixing angles	number of CP-odd phases
1. $m_L \neq 0,\ m_R = m_D = 0$	3	3	3
2. $m_L = m_R = 0,\ m_D \neq 0$	3	3	1
3. $m_L \neq 0,\ m_R \neq 0,\ m_D \neq 0$	6	12	12

3. Neutrino masses in gauge theories

In this Section we consider the appearance of neutrino masses as a result of spontaneous symmetry breaking. What kind of neutrino masses are expected in a theories unifying different interactions? Mechanisms leading to appearance of Dirac and Majorana mass are quite different. The difference is connected with the simple fact that Majorena mass does violate lepton number conservation law. Therefore soft generation of "left" Majorana masses means not only the breaking of SU(2)xU(1)-symmetry but also breaking of some additional symmetry associated with lepton conservation. This symmetry is also broken by "right" Majorana masses, which however leaves intact standard SU(2)xxU(1)-symmetry. This symmetry (let's call it L-symmetry) can be either global or local one. If L-symmetry is global (as in the standard model) then its spontaneous breaking should lead to the appearance of massless Goldstone boson. It is a real physical particle, which should bring a new physics at low energies.

Let's assume now that L-symmetry is gauged. If lepton number conservation is exact, there should exist massless gauge field, associated with unbroken L-symmetry. The absence of such a gauge field with a sensible soupling to the locally conserved current means that L-symmetry should be spontaneously broken. It's

spontaneous breaking naturally leads to the appearance of Majorana mass. Note, that in this case symmetry breaking is not accompanied by appearance of any massless Goldstone boson, as it is eaten by Higgs mechanism.

As for Dirac mass it can be generated in a way analogous to generation of quarks and charged lepton masses. Then the smallness of neutrino mass is due to the fact that corresponding Yukawa coupling constant is anomalously small.

Now we shall give a short review of some concrete models of neutrino mass generation, starting with the standard model. It is known that the appearance of Majorana mass terms of the form $m_L \nu_L C \nu_L$ or $m_R \nu_R C \nu_R$ requires the extension of Higgs and/or fermion sectors of the standard model. For simplicity we consider one generation of fermions (say with electron and up and down quarks). In the scheme proposed in Ref. [11] the existence of new Higgs singlet χ^0 was assumed, χ^0 interacting with Majorana-like combination of right neutrino:

$$h \, \nu_R C \nu_R \, (\chi^0)^+ . \tag{18}$$

Once ν_R **is incorporated in the scheme the appearance of Dirac** mass is expected, its value being of the order of (Dirac) masses of other fermions from the same generation.

Now let's assume that χ^0 field develops vacuum expectation value $\langle \chi^0 \rangle$. It will mean the appearance of "right" Majorana mass $m_R = h \langle \chi^0 \rangle$. If $m_R \gg m_D$ this situation **at low energies is equivalent** to the presence of small "left" Majorana mass of the order $m_L \sim m_D^2/m_R$. The diagram illustrating the last statement is shown on Fig. 1 .

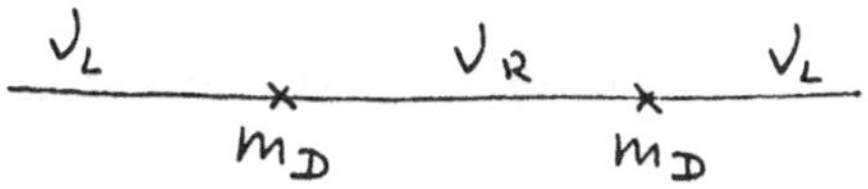

$$\nu_L \quad \overset{\times}{m_D} \quad \nu_R \quad \overset{\times}{m_D} \quad \nu_L$$

Fig. 1.

One has to remember that Majorana propagator of right neutrino includes C-matrix, $C = i\gamma_2\gamma_0$. The imaginary component of the field $\chi = \frac{1}{\sqrt{2}}\left[\langle\chi^0\rangle + i\xi + \rho\right]$ describes massless Goldstone boson, **usually called Majoron** [11]. It can transform into $\nu\nu$ pair with the coupling proportional to $h\,(m_D/m_R)^2$.

In Refs.[12,13] the scheme **for the generation of small left** Majorana mass **without introduction of right components was considered.** In this scheme left Majorana mass arises due to v.e.v. of neutral component of Higgs triplet H , interacting with Majorana-like combination of left leptons $\ell_L = \begin{pmatrix} e^- \\ \nu \end{pmatrix}_L$:

$$f \cdot \ell_L C \ell_L (H)^+ \tag{19}$$

Electric charges of the members of triplet H are **determined** by U(1) invariance:

$$H = \begin{pmatrix} H^0 & H^-/\sqrt{2} \\ H^-/\sqrt{2} & H^{--} \end{pmatrix}$$

The vacuum expectation value of the triplet $\langle H^0\rangle = v$ should be much smaller then standard v.e.v. of doublet $u = \langle\varphi^0\rangle \sim 250\,\text{GeV}$ **in order not** to change the canonical predictions of the model [13]. The Higgs content of the model is rather complicated. Besides massless boson (Majoron) there are the scalars which survive **after the action of Higgs mechanism:**

- two neutrals, one of them being very light with the mass of the order of v.e.v. $\langle H^0 \rangle$;
- singly charged field, described by linear combination $\frac{u H^- - \sqrt{2} v \varphi^-}{\sqrt{u^2 + 2v^2}}$;
- doubly charged field H^{--}.

For $v \ll u$ the coupling of Majorana to neutrino pair has the form

$$\frac{1}{\sqrt{2}} f \, \nu_L C \nu_L (M^0)^* + h.c. \tag{20}$$

Recently in Ref.[14] **the spontaneous breaking of B-L in SU(5) in connection with the origin of left** neutrino masses was considered. This is straightforward extension of the model proposed in Refs. [12,13] . It was shown that the low-energy predictions are the same as in the standard SU(2)xU(1) model.

In the schemes described so far lepton conservation was associated with the global symmetry. **Generation of Majorana masses connected with** the symmetry breaking was accompanied by appearance of massless Majoron, which is **weakly coupled** to **all** fermions except for neutrinos. In a model based on SO(10)-symmetry the spontaneous breaking of lepton conservation giving rise to neutrino Majorana masses does not lead to any massless scalars because B-L is gauged. There are two reasons for the appearance of Majorana masses in SO(10)-model:

1) As it was mentioned by Gell-Mann, Ramond and Slansky [15] the introduction of "right" Majorana mass m_R (say of the order of GU scale) allows to get rid of unreasonably large

Dirac mass (as compared to experimental limit) which could appear together with the masses of charged leptons and quarks and **would** have the same order of magnitude. Then at low energies neutrino, entering weak Lagrangian will behave like very light Majorana particle with the mass $m_L \sim m_D^2/m_R << m_D$.

2) The second reason was already mentioned: B-L cannot be an exact symmetry, for there does not exist a "second photon" - massless gauge field with a sensible coupling to locally conserved B-L current.

The question of neutrino mass in SO(10)-model was described in detail in Ref. [16]. It is possible to produce the large mass m_R **at the tree level if the existence of 126-plet of Higgs scalars is assumed. Then it's most natural to assume that** m_R **is of** order of superheavy mass, 10^{15} GeV. It is not necessary however to introduce 126-plet of scalars to give Majorana mass to "right" neutrino. As it was shown [16] in the absence of a Higgs 126 ν_R will automatically **acquire** mass at the two loop level, this contribution being much smaller than 10^{15} GeV. As a result the effective "left" Majorana mass $m_L \sim m_D^2/m_R$ is much larger than the estimate 10^{-4} eV based on an explicit Higgs 126. The numerical estimation of two loop diagram leads to the following result: $m_L \sim 10^{-7}\, m_q$, where m_q is the mass of the quark from the same generation. Thus for example for the first generation $m_L \sim 1$ eV.

It is easy to extend the schemes described here for N generations of fermions taking into account possible mixing between neutrinos of different flavour.

4. Dirac versus Majorana masses

Here we shall describe in some details widely discussed processes sensitive to neutrino masses and neutrino mixing, paying particular attention to the case of "left" Majorana masses.

<u>Neutrino oscillations.</u> To find amplitudes of transitions $\nu_i \to \nu_K$, $\nu_i \to \tilde{\nu}_K$ one has to fix "initial" conditions for the neutrino state. It can be done considering gedanken experiment [17], shown on Fig. 2.

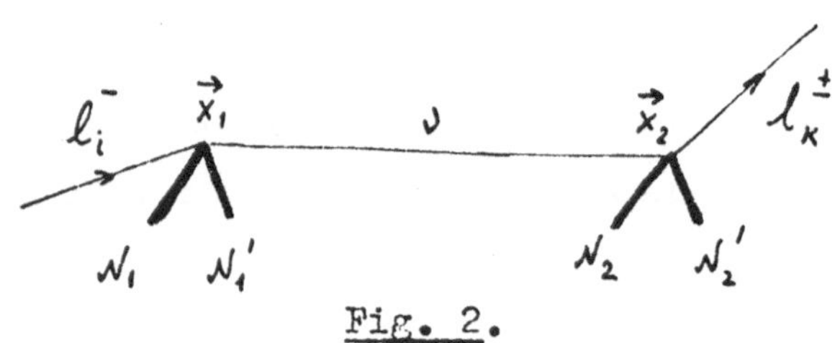

<u>Fig. 2.</u>

Here neutrino is produced by charged lepton ℓ_i^- , interacting with infinitely heavy nucleus **localized at point** $\vec{x}_1$. Due to the flavor-nondiagonal terms in the neutrino mass Lagrangian the state ν produced in this way is a superposition of mass eigenstates φ_A with the same energy, but with different momenta and helicities. At point $\vec{x}_2$ neutrino collides with another heavy nucleus and a charged lepton ℓ_K^- or an antilepton ℓ_K^+ are produced. This choice of quasiexperimental conditions allows to discuss the properties of oscillations in most transparent way. The adaptation of the results to the conditions of real experiment is rather obvious. Below we derive amplitudes and effective cross-sections for the above two-stage processes. The expressions we obtain do not assume that neutrino mass is small compared to its energy ε .

Consider at first the process $\ell_i^- \to \nu \to \ell_K^-$ **which doesn't violate the leptonic number.** S-matrix of this process is equal to

$$S = -\frac{1}{2!}\left(\frac{G}{\sqrt{2}}\right)^2$$

$$\int d^4x\, d^4y \langle \ell_K^-; N_1', N_2' | T\{ j_\mu(x) J_\mu^+(x) j_\lambda^+(y) J_\lambda(y) | N_1, N_2; \ell_i^- \rangle, \tag{21}$$

where N_1 , N_2 and N_1', N_2' characterize the initial and final nuclei states, $j_\mu(x)$ - charged leptonic current

$$j_\mu(x) = \sum_n \bar{\nu}_n(x) \gamma_\mu (1+\gamma_5) \ell_n(x),$$

(the sum is taken over leptonic flavours $n = e, \mu, \tau, \ldots$), $J_\mu(x)$ - charged hadronic current; for heavy spinless nuclei

$$J_\mu(x) = \delta_{0\mu}\, \delta(\vec{x} - \vec{x}_{1,2}).$$

T -product of leptonic currents is reduced to

$$\langle \ell_K^- | T j_\mu(x) j_\lambda^+(y) | \ell_i^- \rangle =$$

$$= \bar{u}(p_K) \hat{O}_\mu \langle 0 | T \varphi_A(x) \bar{\varphi}_B(y) | 0 \rangle \hat{O}_\lambda u(p_i) U_{iA}^+ U_{BK} e^{ip_K y - ip_i x} \tag{22}$$

where $\hat{O}_\mu = \gamma_\mu (1+\gamma_5)$ and

$$\langle 0 | T \varphi_A(x) \bar{\varphi}_B(y) | 0 \rangle = \int \frac{e^{-iq(x-y)}}{\hat{q} - m_A} \frac{d^4q}{(2\pi)^4} \delta_{AB} .$$

Substitution of these expressions into S-matrix and integration in 3-dimensional momentum space

$$\int e^{i\vec{q}\vec{L}} \frac{\hat{q}+\mu_A}{\varepsilon^2-\vec{q}^{\,2}-\mu_A^2} \frac{d^3q}{(2\pi)^3} = \frac{1}{4\pi L}\left[\varepsilon\gamma_0+\mu_A-\left(p_A+\frac{i}{L}\right)\frac{\vec{L}\vec{\gamma}}{L}\right]e^{ip_AL} \tag{23}$$

leads to the following expression for S-matrix

$$S = 2\pi\delta(\varepsilon_i-\varepsilon_K)\frac{(G/\sqrt{2})^2}{\sqrt{2\varepsilon_i}\sqrt{2\varepsilon_K}} M, \tag{24}$$

$$M = \sum_A \frac{e^{ip_AL}}{2\pi L} U_{iA}^{+} U_{AK}\, \bar{u}(p_K)(\varepsilon\gamma_0+p_A\vec{n}_L\vec{\gamma})(1+\gamma_5)u(p_i).$$

Here $p_A = \sqrt{\varepsilon^2-\mu_A^2}$, $\varepsilon_i = \varepsilon_K = \varepsilon$, $\vec{n}_L = \vec{L}/L$.
The cross section is

$$\frac{d\sigma}{d\Omega_K} = \frac{G^4}{64\pi^2}|M|^2\frac{v_K}{v_i}, \tag{25}$$

where $d\Omega_K$ is the solid angle of momentum $\vec{p}_K$; v_i and v_K are the velocities of leptons ℓ_i and ℓ_K, respectively.

S-matrix for the process $\ell_i^- \to \nu \to \ell_K^+$ is obtained from Eq. (21) by replacement $j_\lambda^+(y) \to j_\lambda(y)$. $J_\lambda(y) \to$ $\to J_\lambda^+(y)$. T -product of leptonic currents contains now the vacuum expectation $\langle 0|T\,\bar{\varphi}_A(x)\cdot\bar{\varphi}_B(y)|0\rangle$ that is nonzero only for Majorana neutrinos. In this case we have $\varphi_A = C\bar{\varphi}_A$ where $C = i\gamma_2\gamma_0$ and thus

$$\langle 0|T\,\bar{\varphi}_A(x)\bar{\varphi}_B(y)|0\rangle = -C\langle 0|T\,\varphi_A(x)\bar{\varphi}_B(y)|0\rangle . \tag{26}$$

The similar calculations gives the matrix element for the process $\ell_i^- \to \nu \to \ell_K^+$:

$$M(\ell_i^- \to \ell_K^+) = \sum_A \frac{e^{ip_A L}}{2\pi L} U_{iA}^T U_{AK} \mu_A \bar{u}(p_K)(1+\gamma_5)u(p_i). \qquad (28)$$

We have used **the relation** $u^T(-p_K)C = \bar{u}(p_K)$ **valid for the solution of Dirac equation.** For the case when the momentum of initial lepton $\vec{p}_i$ is parallel to $\vec{L}$ the cross sections are proportional to:

$$\sigma(\ell_i^- \to \ell_K^-) \sim \frac{G^4 \varepsilon^2}{L^2} \sum_{A,B} e^{i(p_A - p_B)L} U_{iA}^+ U_{AK} U_{KB}^+ U_{Bi} \cdot \qquad (29)$$

$$\{\varepsilon^2 + p_A p_B + \varepsilon(p_A + p_B)v_i\} \frac{v_K}{v_i} ;$$

$$\sigma(\ell_i^- \to \ell_K^+) \sim \frac{G^4 \varepsilon^2}{L^2} \sum_{A,B} e^{i(p_A - p_B)L} U_{iA}^T U_{AK} U_{KB}^+ U_{Bi}^* \mu_A \mu_B \frac{v_K}{v_i} . \qquad (30)$$

Eq. (29) is a general one, whereas Eq. (30) is valid only in the case when diagonal neutrinos are Majorana particles. (We assumed **that weak currents are left-handed).**

Now let's consider some particular cases of neutrino mass matrix.

1. Only left Majorana masses are nonzero:

$$-\mathcal{L} = m_{iK}^L \nu_{iL} C \nu_{KL} + h.c.$$

As it was shown in Section 2 symmetrical matrix m_{iK}^L is diagonalized by unitary $N \times N$ matrix U_{Ai} . Let us calculate the total neutrino and antineutrino fluxes at point $\vec{x}_2$, which are characterized by quantities $\sum_K \frac{1}{v_K} \sigma(\ell_i^- \to \ell_K^-)$ and $\sum_K \frac{1}{v_K} \sigma(\ell_i^- \to \ell_K^+)$ respectively.

Due to the unitarity of matrix U

$$\sum_K U_{AK} U^+_{KB} = \delta_{AB}$$

and we see that the total neutrino and antineutrino fluxes separatly do not oscillate:

$$\sum_K \frac{1}{v_K} \sigma(\ell_i^- \to \ell_K^-) \sim \varepsilon^4 \frac{G^4}{L^2} \sum_A U^+_{iA} U_{Ai} (1 + v_A^2 + 2 v_A v_i) \frac{1}{v_i} \; ; \quad (31)$$

$$\sum_K \frac{1}{v_K} \sigma(\ell_i^- \to \ell_K^+) \sim \varepsilon^4 \frac{G^4}{L^2} \sum_A U^T_{iA} U^*_{Ai} (1 - v_A^2) \frac{1}{v_i} \; ; \quad (32)$$

$$\frac{\sum_K \frac{1}{v_K} \sigma(\ell_i^- \to \ell_K^+)}{\sum_K \frac{1}{v_K} \sigma(\ell_i^- \to \ell_K^-)} = \frac{\sum_A |U_{iA}|^2 (1 - v_A^2)}{\sum_A |U_{iA}|^2 (1 + v_A^2 + 2 v_A v_i)} \; . \quad (33)$$

The sum rules (31) and (32) are valid for all energies $\varepsilon \geqslant m_K$, (m_K - mass of charged lepton ℓ_K). The fact that these sum rules are fulfilled separately for $\ell_i^- \to \ell_K^-$ and $\ell_i^- \to \ell_K^+$ transitions means that there are no oscillations between neutrinos and antineutrinos in Majorana case. The presence of antineutrino in the neutrino beam does not yet mean the appearance of $\nu_i \leftrightarrow \tilde{\nu}_K$ oscillations. Consider for example $\mathcal{N} = 2$ (ν_e and ν_μ). Let the initial lepton be electron: $\ell_i^- = e^-$. The production of μ^- means then that $\nu_e \leftrightarrow \nu_\mu$ oscillations take place, the increase of ν_μ being due to the decrease of ν_e . The sum of probabilities to find ν_e and ν_μ in the beam is constant, as follows from Eq. (31) (more precisely proportional to $1/L^2$). So the oscillations take place only between ν_e and ν_μ and independently between $\tilde{\nu}_e$ and

$\tilde{\nu}_\mu$. The picture looks so as if electron produces ν_e already with $\tilde{\nu}_e, \tilde{\nu}_\mu$ admixture and neutrinos and anti-neutrinos oscillate later separately. The amplitudes of $\tilde{\nu}_e$ and $\tilde{\nu}_\mu$ admixture are proportional at point $\vec{x}_1$ to the elements of the Majorana mass matrix m_{ee} and $m_{e\mu}$ respectively.

2. In the most general case neutrino mass Lagrangian includes both Dirac and Majorana masses and are given by Eq. (7). $2N \times N$ matrix U_{Ai} entering the amplitude for the processes $\ell_i^- \to \nu \to \ell_K^\pm$ is now a part of $2N \times 2N$ unitary matrix K (see Eq. (10)). Equations (29) and (30) for cross sections are valid in this case as well but unitary condition now has the form.

$$K K^+ = U_{AK} U_{KB}^+ + V_{AK} V_{KB}^+ = \delta_{AB} \tag{34}$$

The second term in Eq. (34) corresponds to the contribution of sterile states and consequently $\sum_K U_{AK} U_{KB}^+ \neq \delta_{AB}$. The cross sections of $\ell_i^- \to \ell_K^-$ and $\ell_i^- \to \ell_K^+$ summed over K can therefore oscillate as neutrino produced at point $\vec{x}_1$ can transform into sterile states.

3. Only Dirac masses are nonzero. The analysis of this possibility is rather simple. The cross section of the process $\ell_i^- \to \nu \to \ell_K^-$ is given by Eq. (29) and the process $\ell_i^- \to \nu \to \ell_K^+$ is forbidden. As in the case of left-handed Majorana neutrinos the cross sections satisfy the sum rule (31).

The results obtained here are valid for general form of complex matrix U allowing for CP-nonconservation. For

$\mathcal{N} = 3$ ($l_i = e, \mu, \tau$) there are three physically relevant CP-odd phases in Majorana case and one in Dirac case. However **by measuring** $\nu_i \leftrightarrow \nu_K$ oscillations it is possible to find only one CP-odd phase. This can be easily seen by substitution the explicit form of matrix U given by Eq. (17) into Eq. (24) characterizing the amplitude of $\nu_i \leftrightarrow \nu_K$ oscillations. To find the other CP-odd phases it **is necessary to search for** $\nu_i \to \tilde{\nu}_K$ **transitions** which are hardly possible in the nearest future because their amplitudes are proportional to m_ν / ε . Thus we see that measuring neutrino oscillations it is impossible to distinguish between Dirac and Majorana mass matrix.

β -<u>spectrum</u>. The influence of neutrino masses on the shape of β-spectrum is pure kinematical effect. Therefore it's impossible to distinguish between them by searching for the deviation of Kurie spectrum from the straight line. If electron neutrino is mixed with some others (say, neutrino of another flavour or/and their antineutrino including $\tilde{\nu}_e$) the β -spectrum has rather peculiar form. Consider the simplest example, when neutrino is the superposition of two mass eigenstates:

$$\nu_{eL} = \frac{1+\gamma_5}{2}\left(\varphi_1 \cos\alpha + \varphi_2 \sin\alpha \right).$$

The probability of the β -decay is

$$W(E_e) = \cos^2\alpha \cdot W_1(E_e) + \sin^2\alpha \cdot W_2(E_e) \tag{35}$$

where

$$W_A(E_e) \sim p_e E_e (E_0 - E_e)\left[(E_0 - E_e)^2 - \mu_A^2\right]^{1/2}, \quad A = 1;2. \tag{36}$$

μ_A - mass eigenvalues, $E_0 = E_e + E_\nu$, α - mixing angle. The shape of Kurie spectrum, defined by $F(E_e) = \left[\frac{W(E_e)}{p_e E_e} \right]^{1/2}$ is shown in Fig. 3.

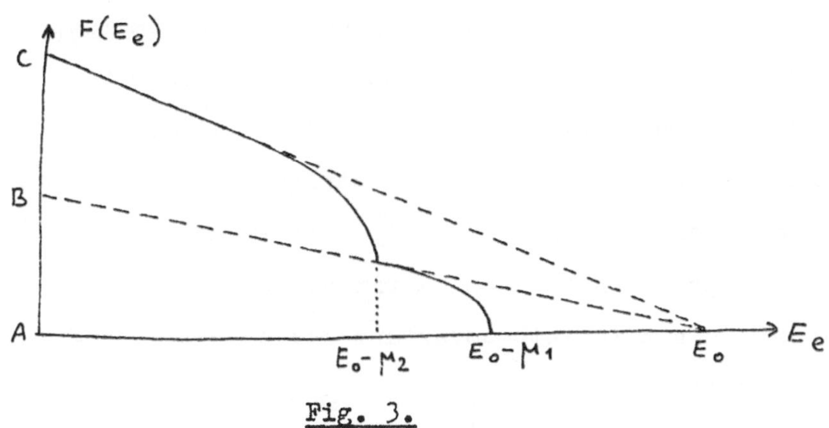

Fig. 3.

Note, that the ratio AB:AC (see Fig. 3) is equal to $|\cos\alpha|$ Therefore the precise measurement of Kurie spectrum can give rather rich information about neutrino mass matrix.

For $\mathcal{N}$ neutrino eigenstates

$$W(E_e) = \sum_A |U_{eA}|^2 W_A(E_e) \quad . \tag{37}$$

there are $\mathcal{N}$ thresholds in the spectrum instead of two shown in Fig. 3.

In the region $E_e < E_0 - \max\{\mu_A\}$ the spectrum is sensitive to some effective electron neutrino mass

$$m^\beta_{eff} = \left[\sum_A |U_{eA}|^2 \mu_A^2 \right]^{1/2} . \tag{38}$$

Neutrinoless double β -decay, $2\beta(0\nu)$.

This decay should appear if Majorana mass of electron neutrino is not equal to zero. The amplitude of the $2\beta(0\nu)$ - decay is proportional to m^L_{ee} -diagonal matrix element of Majorana mass matrix m^L_{ik} in nondiagonal representation. Later we shall discuss also some other mechanisms which can contribute to the amplitude of $2\beta(0\nu)$ -decay. For the beginning we shall find the contribution of Majorana mass. This was done in Refs. [18; 11] . It seems easier to derive the expression for the matrix element of decay $\mathcal{N} \to \mathcal{N}' e^- e^-$ in a Lorentz-covariant way starting from S-matrix in the form

$$S = -\frac{1}{2!}\left(\frac{G}{\sqrt{2}}\right)^2 \int d^4x\, d^4y \langle \mathcal{N}' e^- e^- | T\, \mathcal{L}(x)\, \mathcal{L}(y) | \mathcal{N} \rangle,$$

$$\mathcal{L}(x) = J_\mu(x) \cdot \bar{e}(x) \gamma_\mu (1+\gamma_5) \nu(x), \tag{39}$$

$J_\mu(x)$ - hadronic current. The effective neutrino propagator is proportional to m_{ee}/q^2 , and thus the forces acting between two nucleons (or two quarks) in diagram of Fig. 4 are Coulomb-like forces.

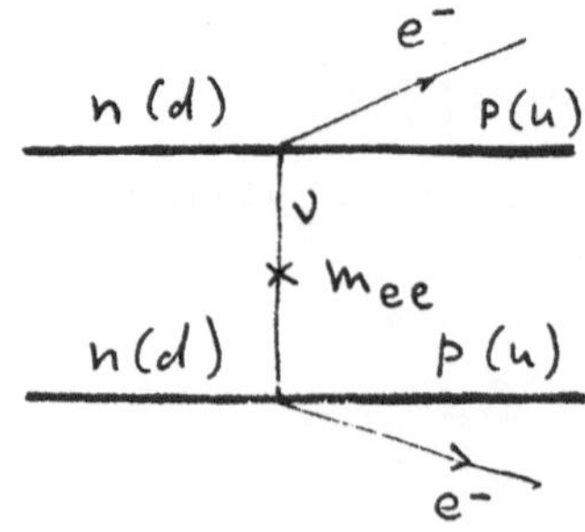

Fig. 4.

The dependence on spins and momenta of final electrons is defined by the factor

$$\bar{e}(k_1)\gamma_\mu\gamma_\lambda(1-\gamma_5)C\bar{e}(k_2).$$

After Pauli principle is taken into account this transforms into

$$\bar{e}(k_1)(1-\gamma_5)C\bar{e}(k_2)\cdot\delta_{\mu\lambda}.$$

The nuclear matrix element $\langle N'|J_\mu J_\lambda \frac{1}{r}|N\rangle\cdot\delta_{\mu\lambda}$ for two-nucleon mechanism in nonrelativistic limit has the from

$$\langle N'|\sum_{i\neq k}\frac{1}{r_{ik}}(1-g_A^2\vec{\sigma}_i\vec{\sigma}_k)|N\rangle \equiv \langle 1/r\rangle. \tag{40}$$

Indicies i, k refer to nucleons. Finally the matrix element of the decay can be written as

$$M = G^2\frac{(2m_{ee})}{\pi}\cdot\bar{e}(k_1)(1-\gamma_5)C\bar{e}(k_2)\cdot\langle 1/r\rangle. \tag{41}$$

The probability of the decay equals to

$$W = \frac{G^4(2m_{ee})^2}{15\cdot(2\pi)^5}\langle 1/r\rangle^2\Delta^5 F_c^2, \tag{42}$$

where F_c - Coulomb corrections:

$$F_c \approx \frac{2\pi\alpha Z}{1-\exp(-2\pi\alpha Z)},$$

Δ is the energy released. Here we neglected electron mass in comparison with Δ . The main uncertainties in numerical estimates of the decay probability are due to low accuracy in the estimations of nuclear matrix element. As an example let us put nuclear matrix element equal to $1/R$, where R is the radius of the nucleous. Then from experimental limit $T_{1/2} > 2 \cdot 10^{21}$ yr for ^{48}Ca [20] it follows that $2 m_{ee} < 50$ eV. (Note, that in the case when $m^L_{ik} = 0$ except for m^L_{ee} electron neutrino is the mass eigenstate with the mass $\mu_{\nu_e} = 2 m^L_{ee}$). In general case

$$2 m^L_{ee} = \sum_A U^2_{eA} \mu_A \tag{43}$$

where μ_A is mass eigenvalues. Note, that the operator responsible for nuclear transition (see Eq. (40)) has vacuum quantum numbers. Therefore $2\beta(0\nu)$-decay in case of nonzero Majorana mass occures without changing of spin and parity of nuclei. The $0^+ \longrightarrow 0^+$ nuclear transitions are of practical interest.

If Majoron M^0 exists there should take place also 2β - decay $\mathcal{N} \to \mathcal{N}' e^- e^- M^0$ [13], described by diagram of Fig.5.

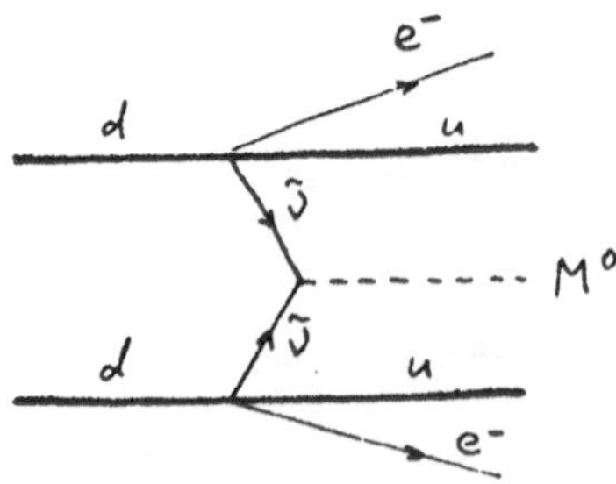

Fig. 5.

The corresponding matrix element can be calculated in a similar way and has the form

$$M(e^-e^-M^0)=\frac{G^2 f\sqrt{2}}{\pi}\bar{e}(k_1)(1-\gamma_5)C\bar{e}(k_2)\,\varphi^+_{M^0}\,\langle N'|\sum_{i\neq k}\frac{1}{r_{ik}}\left(1-g_A^2\vec{\sigma}_i\vec{\sigma}_k\right)|N\rangle, \quad (44)$$

φ_{M^0} - wave function of M^0 , $f/\sqrt{2}$ - coupling constant of Majoron to neutrinos. Note, that nuclear matrix element is just the same as for decay $2\beta(0\nu)$. This is because Fourier transformation of $[\hat{q}(\hat{q}-\hat{k})]^{-1}$ (k is Majoron momentum) gives the same result as $1/q^2$. The probability of the decay $N\to N'e^-e^-M^0$ is given by

$$W(e^-e^-M^0)=\frac{G^4 f^2}{315(2\pi)^7}\,\Delta^7\langle 1/r\rangle^2 F_c^2 . \quad (45)$$

The ratio of probabilities of decays $N\to N'e^-e^-M^0$ and $N\to N'e^-e^-$ equals to

$$\frac{W(e^-e^-M^0)}{W(e^-e^-)}=\frac{f^2}{84\pi^2}\left(\frac{\Delta}{2m_{ee}}\right)^2 \quad (46)$$

and does not depend on nuclear matrix element. In the model [12,13] neutrino mass $m_{ee}=\frac{1}{\sqrt{2}}f\langle H^0\rangle$, where H^0 is neutral component of triplet, and the ratio (46) can be written as

$$\frac{W(e^-e^-M^0)}{W(e^-e^-)}=\frac{1}{84\pi^2}\left(\frac{\Delta}{v}\right)^2, \quad v=\langle H^0\rangle\cdot\sqrt{2} . \quad (47)$$

As is known the main signature of neutrinoless 2β --decay $N\to N'e^-e^-$ is the fixed total energy $E=\varepsilon_1+\varepsilon_2$ carried by electrons. In case of decay $N\to N'e^-e^-M^0$ a part of the energy is carried away by Majoron. The distribution over $E = \varepsilon_1 + \varepsilon_2$ is described by

$$\frac{dW(e^-e^-M^0)}{dE} \sim E^5(\Delta - E)$$

and is shown in Fig. 6.

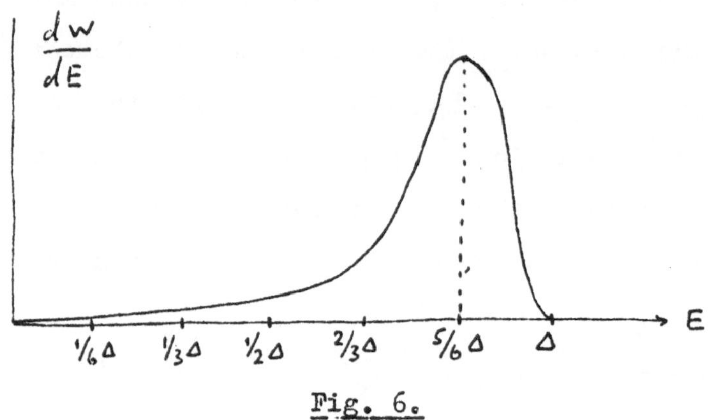

Fig. 6.

Here again we neglected m_e in comparison with Δ .

Majorana mass is not the only mechanism contributing to the amplitude of $2\beta(0\nu)$ -decay. The models [11-13] cited here provide mechanisms like those shown in Fig. 7.

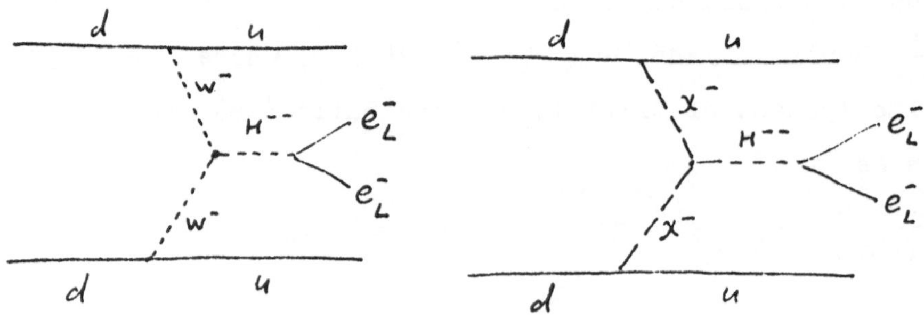

Fig. 7.

Recently Schechter and Valle [21] estimated their contribution and showed that it is negligible with respect to the contribution of Majorana mass. Furthermore, under some resonable assumptions it was shown [21] that the existence of $2\beta(0\nu)$--decay implies nonzero Majorana mass term of the form $m_L \nu_L C \nu_L$. These assumptions are:

1) Weak interactions are described by gauge theory;

2) There exist W's coupled to ($\bar{\ell}\, \nu_\ell$) currents as well as to ($\bar{u}$d)-current;

3) Crossing symmetry.

Consider the transition $e^+dd \longrightarrow uue^-$ which is the crossing process with respect to $2\beta(0\nu)$-decay dd ⟶ ⟶ uue^-e^-. Let the "black box" shown on Fig. 8 be some mechanism responsible for the process e^+dd --- uue^-

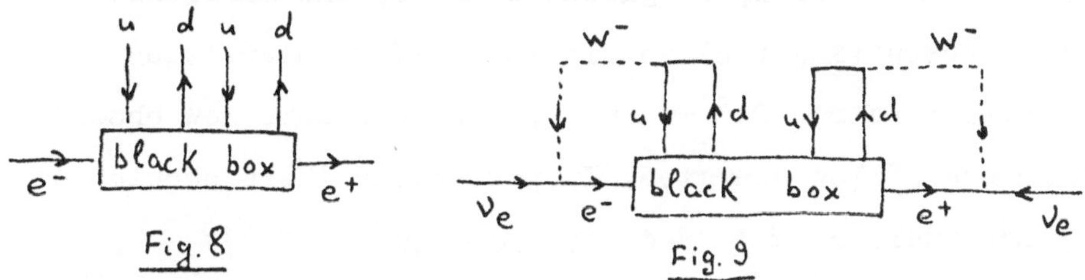

Fig. 8

Fig. 9

Closing quark lines and connecting them with leptonic lines by W-bosons we obtain the diagram (Fig. 9) which gives rise to "left" Majorana mass term of the form $m^L_{ee}\, \nu_{eL} C\, \nu_{eL}$.

So among the three processes considered above $2\beta(0\nu)$ -decay is the only one which enables to distinguish between Dirac and Majorana masses. Note, that due to the properties of matrix U the mass entering the amplitude of $2\beta(0\nu)$-decay $m_{2\beta} = \sum_A U^2_{eA}\, \mu_A$ is always smaller then the effective neutrino mass $m^\beta_{eff} = \left[\sum_A |U_{eA}|^2 \mu^2_A\right]^{1/2}$, measured by tritium β -spectrum: $m_{2\beta} \leqslant m^\beta_{eff}$.

5. Concluding remarks

From the phenomenological point of view neutrino Lagrangian may contain three types of mass terms (m_L, m_R and m_D). In the most general case there are $2\mathcal{N}$ mass eigenstates which are superpositions of two-component neutrinos ν_i and $\tilde{\nu}_K$ (i, K = 1, 2,... N). All types of masses can arise in a gauge models described **above**.It is more natural to expert that either $m_L \neq 0$, $m_R = m_D = 0$ or $m_L = 0$, $m_D << m_R$, m_D being of the order of typical mass of the generation. At low energies these two possibilities are indiscernible, because in the second case left neutrino will look like Majorana particle with the mass m_D^2 / m_R. In a model where Majorana mass arises due to spontaneous breaking of global symmetry the massless Goldstone boson (Majoron) should exist. Its interaction with neutrino pair, $M^0 \rightarrow \nu\nu$, provides some new observable phenomena at low energies. In particular there should take place nuclear 2β -decay $\mathcal{N} \rightarrow \mathcal{N}' e^- e^- M^0$, which can compete with the "usual" $2\beta(o\nu)$-decay $\mathcal{N} \rightarrow \mathcal{N}' e^- e^-$. Searching for double β -decay seems very important at present, as it is practically the only process which can help to distinguish between Dirac and Majorana masses. Furtheemore the discovery of $2\beta(o\nu)$ -decay would most probably mean that Majorana mass m^L_{ee} **isn't equal to zero.**

Neutrino oscillations are insensitive to the nature of neutrino mass matrix and can proceed independently from $2\beta(o\nu)$ -decay. Neutrino mass measured by tritium β - -spectrum must be nonzero if $2\beta(o\nu)$ -decay or/and ν -

-oscillations (including ν_e) take place. The inverse statement is obviously **not valid.** Fig. 10 illustrates the **connection** of the three phenomena discussed here.

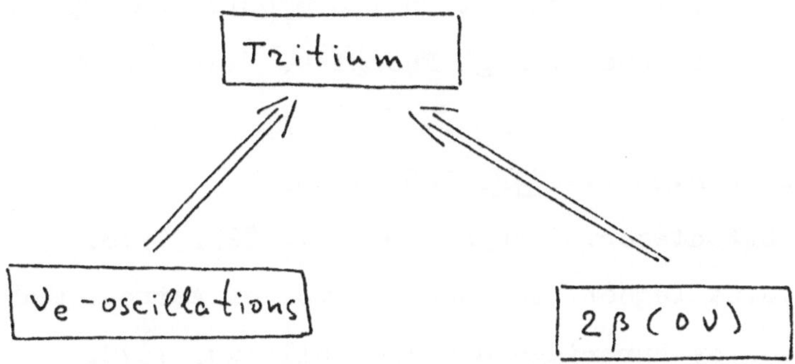

Fig. 10.

I am grateful to Prof. I.Yu.Kobzarev for helpful discussions.

References

1. V.A.Lyubimov et al. Phys.Lett., 94B, 266, 1980.
2. J.Kirkby. Proc. of the International Symposium on Lepton and Photon Interactions at High Energies. FNAL, Batavia, Illinois, 1979.
3. M.Daum et al. Phys.Lett., 74B, 126, 1978.
4. S.M.Bilenky, B.Pontecorvo. Phys.Rep., 41, 225, 1978.
5. S.M.Faber, J.S.Gallagher. In Annual Revs. of Astron. and Astrophys., ed. by Burbridge and J.G.Phillips, 1979. Ya.B.Zeldovich, M.Yu.Khlopov. Uspekhi Fiz.Nauk, v. 135, issue 1, p. 45, 1981.
6. V.Burger et al. Phys.Rev.Lett., 45, 692, 1980.
7. I.Yu. Kobzarev et al. Preprint ITEP-90, 1980.
8. V.Gribov, B.Pontecorvo. Phys.Lett., 28B, 493, 1969.
9. B.V.Martemjanov. Preprint ITEP-35, 1979.
10. M.Kobayashi, K.Maskawa. Prog. Theor. Phys., 49, 652, 1973.
11. V.Chikashige, R.N.Mohapatra, R.D.Peccei. Phys.Lett., 98B, 265, 1981.
12. G.B.Gelmini, M.Roncadelli, Phys.Lett., 99B, 441, 1981.
13. H.M.Georgi, S.Glashow, S.Nussinov. Nucl.Phys., B193, 297, 1981.
14. F.Buccella et al. Preprint MPI-PAE/PTh 21/82, 1982.
15. M.Gell-Mann, P.Ramond, R.Slansky, in "Supergravity", Proc. of the Supergravity Workshop at Stony Brook, 1979, p. 315.
16. E.Witten. Phys.Lett., 91B, 81, 1980.

17. I.Yu.Kobzarev et al. Preprint ITEP-153, 1981.

18. E.Greuling, R.C.Whitten. Ann.Phys., 11, 510, 1960.

19. M.Doi et al. Prog. Theor. Phys., 66, 1739 and 1765, 1981; 68, 347, 1981.

20. R.K.Bardin, P.J.Gollon, J.D.Ullman, C.S.Wu. Nucl.Phys., A158, 337, 1970.

21. J.Schechter, J.W.F.Valle. Phys.Rev., 25D, 2591, 1982.

INTRODUCTION TO SUPERSYMMETRY

K.S. STELLE
BLACKETT LABORATORY
IMPERIAL COLLEGE
LONDON SW7 2BZ

Abstract:

The basic features of supersymmetric field theories are presented in this article, emphasizing the structure of the representations of the supersymmetry algebra. In Chapter 1, the simple and N-extended supersymmetry algebras are presented, and the representations on massive and massless states derived using Wigner's method of induced representations. In Chapter 2, the representations of the supersymmetry algebra on fields are introduced, together with the superspace formalism, which is applied to a discussion of the basic supersymmetric model, the Wess-Zumino model. In Chapter 3, supersymmetric Yang-Mills theories are discussed, and the formalism of superspace applied to the quantization of the Wess-Zumino model and supersymmetric Yang-Mills theories, and also to the analysis of the quantum supercurrent for conformally invariant and non-conformally-invariant theories. In Chapter 4, the technique of superspace quantization is applied to the maximal supersymmetic gauge theory, the N=4 supersymmetric Yang-Mills theory. This theory is finite to all orders in perturbation theory, as can be shown using manifestly N=2 supersymmetric Feynman rules.

CHAPTER I

Supersymmetry Algebras and their Representations

Supersymmetry is the only known way to have a non-trivial unification of space-time and internal symmetries of the S-matrix in a relativistic particle theory. In the context of ordinary groups of symmetries for a relativistic non-trivial S-matrix, the theorem of Coleman and Mandula[(1)] showed that the only allowed groups were locally isomorphic to the direct product of an internal symmetry group and the Poincaré group, subject to some general assumptions on analyticity and finiteness of the number of particle types. This direct product structure earned these results the name 'no-go theorem', because an internal symmetry can change neither spin nor mass.

The way to avoid the strictures of the no-go theorem proved to be the generalization from groups of symmetries to graded groups. Graded Lie groups are characterized by Graded Lie algebras, whose composition rules contain both commutators and anticommutators. The graded Poincaré algebra was first considered by Gol'fand and Likhtman[(2)] in 1971. A four dimensional field theory with nonlinearly realized supersymmetry was constructed by Volkov and Akulov,[(3)] while the first four dimensional theory with linearly realized supersymmetry was constructed by Wess and Zumino,[(4)] generalizing the supergauge transformations of dual models.

The difficulties with internal symmetry currents carrying non-trivial Lorentz representations are codified by the Coleman-Mandula theorem.[(1)] The basic idea is that new conserved quantities with non-trivial Lorentz representations would force the scattering matrix to be unity except when certain kinematical conditions are met. For example, conservation of momentum P_μ and angular momentum $M_{\mu\nu}$ in a 2-body collision leaves only the scattering angle unknown. Additional conservation laws would allow only a discrete set of scattering angles, but then the analyticity of the S-matrix would rule out scattering at all angles.

Suppose there were a conserved symmetric traceless tensor change $Q\alpha\beta$. Lorentz invariance would then require that the matrix element $\langle p|Q\alpha\beta|p\rangle$ take the form $(p^2+m^2=0)$

$$\langle p|Q_{\alpha\beta}|p\rangle = \left(p_\alpha p_\beta + \tfrac{1}{4}\eta_{\alpha\beta}m^2\right) f(m^2) . \tag{1.1}$$

Conservation of this quantity in a two particle interaction with momenta $p_1{}^\mu$, $p_2{}^\mu$ scattering to $q_1{}^\mu$, $q_2{}^\mu$ would require

$$p_{1\alpha}p_{1\beta} + p_{2\alpha}p_{2\beta} = q_{1\alpha}q_{1\beta} + q_{2\alpha}q_{2\beta} , \tag{1.2}$$

but this can happen only for zero scattering angle. Then by analyticity, the scattering would have to be zero for all angles.

The above difficulty occurs with all bosonic symmetries carrying non-trivial Lorentz indices, beyond those already present in the Poincaré algebra. Supersymmetry escapes the requirements of the Coleman-Mandula theorem because the generators are <u>fermionic</u>. The simplest example is the Wess-Zumino model.[(4)] Starting with the free theory of a massless complex scalar and a Majorana spinor field $(\bar{\Psi}=\Psi^T C^*)$

$$I^{free} = \int d^4x \left(\partial_\mu \phi^* \partial^\mu \phi + i\bar{\Psi}\partial\!\!\!/\Psi\right) , \tag{1.3}$$

we find that there is a conserved spinorial current

$$Q^{free}_{\mu\alpha} = \left(\partial_\nu \phi^* \gamma^\nu \gamma_\mu \Psi\right)_\alpha , \tag{1.4}$$

as may be checked using the free field equations:

$$\begin{aligned} \partial^\mu Q^{free}_{\mu\alpha} &= \partial^\mu\partial_\nu \phi^* \gamma^\nu\gamma_\mu \Psi + \partial_\nu \phi^* \gamma^\nu \gamma^\mu \partial_\mu \Psi \\ &= \tfrac{1}{2}\partial_\mu\partial_\nu \phi^* [\gamma^\mu\gamma^\nu + \gamma^\nu\gamma^\mu]\Psi + \partial\!\!\!/\phi^* \partial\!\!\!/\Psi \\ &= \Box \phi^* \Psi + \partial\!\!\!/\phi^* \partial\!\!\!/\Psi \\ &= 0 . \end{aligned} \tag{1.5}$$

Of course, it is not surprizing to find such a conserved current in a free theory - in noninteracting purely bosonic theories there are always an infinite number of conserved 'Zilch' currents. What is truly remarkable about the Wess-Zumino model, however, is that it is possible to find a conserved supercharge if the interactions take a certain form:

$$I_{WZ} = \int d^4x \left[\partial_\mu \phi^* \partial^\mu \phi + i\bar{\Psi} \not{\partial} \Psi - g^2 (\phi^* \phi)^2 - g\left(\phi \bar{\Psi}\left(\frac{1+i\gamma_5}{2}\right)\Psi + \text{h.c.}\right) \right]. \tag{1.6}$$

In this interacting theory, the full current

$$Q^{WZ}_{\mu\alpha} = Q^{free}_{\mu\alpha} + g\gamma_\mu (\phi^*)^2 \left(\frac{1-i\gamma_5}{2}\right)\Psi \tag{1.7}$$

is conserved by virtue of the interacting field equations. The supercharge is given by the Majorana spinor

$$Q_\alpha = \int d^3x \; Q_{o\alpha} \;. \tag{1.8}$$

Since $Q_{o\alpha}$ is linear in Ψ, it will anticommute with itself at spacelike separations. More generally, we will need to know the algebra of anticommutators of the fermionic charge Q_α. Since Q_α is conserved, the anticommutator

$$\left\{ Q_\alpha, \bar{Q}_\beta \right\} \tag{1.9}$$

must be conserved as well. But this is a bosonic operator with non-trivial Lorentz structure. In order to be consistent with the Coleman-Mandula theorem, it must be a Poincaré generator, so we obtain the (flat space) supersymmetry algebra

$$\left\{ Q_\alpha, \bar{Q}_\beta \right\} = 2\gamma^\mu_{\alpha\beta} P_\mu \;. \tag{1.10}$$

Since the bosonic part of this algebra satisfies the Coleman-Mandula theorem, Q_α can be conserved even in an interacting theory. The Poincaré algebra together with its extension (1.10) is known as a graded Lie algebra.

The most general grading of the Poincaré algebra involves the addition of N spin ½ fermionic generators $Q_\alpha^i (i=1\ldots N)$ to the bosonic, or even part of the algebra. These fermionic generators are required to be irreducible under the Lorentz group in four dimensions. Accordingly, we impose upon them the Majorana constraint $C(\bar{Q}_\alpha^i)^T = Q_\alpha^i$, or equivalently we could use Weyl spinors. The bosonic part of the algebra is restricted by the Coleman-Mandula theorem to be a direct product $P \otimes (T \otimes Z)$, where P is the Poincaré algebra, T is a semisimple internal symmetry Lie group acting on the indices i, j, and Z is an Abelian Lie group, the centre of the algebra. The structure constants of the full algebra are then restricted via the Jaccobi identities by the required structure of the bosonic part of the algebra. It is convenient to use Van der Waerden notation and the Weyl representation for the spinors, splitting them up into complex two component spinors and their complex conjagates, $Q^i = (Q_\alpha{}^i, \bar{Q}^{\dot\alpha}{}_i)$, where $\bar{Q}^{\dot\alpha}{}_i = \varepsilon^{\dot\alpha\dot\beta} (Q_\beta{}^i)^*$ and $\varepsilon^{\alpha\beta} = \varepsilon^{\dot\alpha\dot\beta} = -\varepsilon_{\alpha\beta}$ ($\varepsilon^{12}=1$). The resulting most general grading is given by[5]

$$\{Q_\alpha{}^i, \bar{Q}_{\dot\beta j}\} = 2\,\sigma^\mu_{\alpha\dot\beta}\,\delta^i{}_j\,P_\mu$$

$$\{Q_\alpha{}^i, Q_\beta{}^j\} = 2\,\epsilon_{\alpha\beta}\,(\Omega^e)^{ij}\,Z_e$$

$$[Q_\alpha{}^i, T_a] = (f_a)^i{}_j\,Q_\alpha{}^j$$

$$[Q_\alpha{}^i, M_{\mu\nu}] = (\sigma_{\mu\nu})_\alpha{}^\beta\,Q_\beta{}^i$$

$$[Q_\alpha{}^i, P_\mu] = [Q_\alpha{}^i, Z_e] = [Z_e, P_\mu] = [Z_e, Z_k] = 0\,. \qquad (1.11)$$

The spinorial generators $Q_\alpha{}^i$ that form the odd part of the algebra transform according to the spin ½ representation matrices $f_a{}^i{}_j$ of the internal symmetry group, whose generators are the T_a. This internal symmetry group is an outer automorphism of the algebra of Q's, P's and Z's, since the T_a generators do not occur as the result of the (anti)commutation of these other generators. The Z_e do occur in the $\{Q,Q\}$ anti-commutator, and through the analysis of the Jaccobi identities it can be shown [5] that they must commute with all of the generators of the algebra, i.e. they form the centre of the algebra. The structure constants $(\Omega^e)^{ij}$ must be antisymmetric numerically invariant matrices under the action of the internal symmetry group.

It is clear from the algebra (1.11) that in the case $Z_e=0$ the algebra has a U(N) outer automorphism, and for this reason we have written the internal symmetry index in the lower position on $\bar{Q}_{\dot\alpha i}$. U(N) may or may not be a symmetry of a theory that is invariant under the Q's and the Poincaré generators. If U(N) is a symmetry of a given theory, the Q's and spinor fields must be taken as Weyl spinors, for a Majorana spinor mixes upper and lower i, j indices, $Q^i=(Q\alpha^i,\ \bar{Q}^{\dot\alpha}i)$. If the spinors are taken to be Majorana, then the maximal automorphism group is SO(N).

When the central charges are present, the requirement that the $(\Omega^e)ij$ be numerically invariant under the action of the internal symmetry group, i.e $(fa)^i j(\Omega^e)^{jk} + (fa)^k l(\Omega^e)^{il}=0$, reduces the internal symmetry group to some subgroup of U(N). In the case of just one central charge the maximal internal symmetry group is USP(N), for N even, in which case $(\Omega)^{ij}$ is the antisymmetric numerically invariant metric for that group. With more than one central charge, the internal symmetry group is further reduced.

We begin with the massive one-particle states in the absence of central changes ($Z_e=0$). Since $P^\mu P\mu$ is still a Casimir operator for the supersymmetry algebra, all the states in an irreducible representation must have the same mass M. We choose the Lorentz rest frame so that $P^\mu=(M,0)$. The little group is now SU(2), under which the α and $\dot\alpha$ indices transform in the same way.

The anticommutators of the Q's in the rest frame take the form

$$\{Q_\alpha^{\ i},\ \bar{Q}_{\dot\beta j}\} = M\,\delta_{\alpha\dot\beta}\,\delta^i_{\ j}$$
$$\{Q_\alpha^{\ i},\ Q_\beta^{\ j}\} = \{\bar{Q}_{\dot\alpha i},\ \bar{Q}_{\dot\beta j}\} = 0. \tag{1.12}$$

This is the algebra of 2N fermionic creation and annihilation operators. It has a maximal automorphism group SO(4N), for which it is in fact the Clifford algebra(7), as may more easily be seen by going over to a Majorana representation for the four component spinors, in which case the stability subalgebra becomes

$$\{Q_\gamma^{\ i},\ Q_\lambda^{\ j}\} = M\,\delta_{\gamma\lambda}\,\delta^{ij};\quad \gamma,\lambda = 1\cdots4\ ;\ i,j = 1\cdots N. \tag{1.13}$$

The unique irreducible representation of the Clifford algebra

(1.12) has dimension 2^{2N}, and can be derived by starting from a singlet Clifford vacuum Ξ satisfying

$$Q_\alpha{}^i \Xi = 0 \qquad \forall \alpha, i \tag{1.14}$$

and building up the 2^{2N} states

$$\begin{array}{l} \Xi, \ \bar{Q}_{\dot{\alpha} i}\Xi, \ \bar{Q}_{\dot{\alpha}_1 i_1}\bar{Q}_{\dot{\alpha}_2 i_2}\Xi, \cdots, \ \bar{Q}_{\dot{\alpha}_1 i_1}\bar{Q}_{\dot{\alpha}_2 i_2}\cdots\bar{Q}_{\dot{\alpha}_k i_k}\Xi, \cdots \\ 1 + 2N \qquad + \cdots + \qquad \frac{2N(2N-1)}{2} \quad = \quad (1+1)^{2N} \end{array} \tag{1.15}$$

These states span the spinorial representation of SO(4N), which can be split into the + and - eigenstates of γ^{4N+1} each of which have dimension 2^{2N-1}, containing the bosons and fermions respectively.

The SO(4N) symmetry that classifies the one-particle states is not a symmetry of a supersymmetric Lagrangian, for it refers only to the rest-frame stability subalgebra. Another classification symmetry applies to the states of a given spin. If we define the 2N component spinors(7)

$$\begin{aligned} Q_\alpha^a &= Q_\alpha{}^i \quad , \qquad a = i = 1,2,\cdots,N \\ &= \bar{Q}^{\dot{\alpha}}{}_i = \epsilon^{\dot{\alpha}\dot{\beta}}\bar{Q}_{\dot{\beta} i} \ , \quad a = N+i = N+1,\cdots,2N \end{aligned} \tag{1.16}$$

we find that they satisfy the reality condition

$$(Q_\alpha{}^a)^* = \epsilon^{\alpha\beta}\Omega_{ab}\, Q_\beta{}^b \tag{1.17}$$

with

$$\Omega_{ab} = -\Omega_{ba} = \begin{pmatrix} 0 & 1 \\ -1 & 0 \end{pmatrix}_{ab} . \tag{1.18}$$

In this notation, the anticommutator of the Q's becomes, for $Z_e = 0$,

$$\{ Q_\alpha^a , Q_\beta^b \} = M \epsilon_{\alpha\beta}\Omega^{ab} \quad , \quad \Omega^{ac}\Omega_{cb} = \delta^a{}_b . \tag{1.19}$$

In this form, the stability subalgebra has manifest SU(2) USP(2N) symmetry, under which the spinor charges Q still transform irreducibly in the vector representation: $4N \to (2,2N)$. Thus we see that the states of a given spin are classified by representations of USP(2N), and the 2^{2N} states of the whole supermultiplet break up into SU(2) and USP(2N) representations as

$$2^{2N} = (N+1,1) + (N,2N) + \cdots + (N+1-k,[2N]_k) + \cdots + (1,[2N]_N) , \qquad (1.20)$$

where the first label is the dimension of the SU(2) representations ($J=(N-k)/_2$) and $[2N]k$ indicates the k index totally antisymmetric and traceless (with Ωab) representation of USP(2N).

In the above we have assumed that the Clifford vacuum carries no spinor internal symmetry representation. The general irreducible massive representations of supersymmetry on one particle states are obtained by allowing Ω to carry spin and some representation of the internal symmetry group for the theory of interest, in general U(N). For the spin states, we just multiply the Clifford vacuum spin into the SU(2) representations given in (1.20). If the Clifford vacuum is not a singlet under the internal symmetry then the USP(N) representations given in (1.20) must be reduced into internal symmetry representations, and then the product taken with the representation of the Clifford vacuum Ξ . The total dimension of the supermultiplet is then

$$D = 2^{2N} \times d_{\Xi} , \qquad (1.21)$$

where d_{Ξ} is the dimension of the spin $\otimes$ internal symmetry representation carried by the Clifford vacuum.

For example, we have the following irreducible massive supermultiplets, with Clifford vacuums of spin J_Ξ (all examples are singlets under internal symmetry):

TABLE 1

N \ J	2	3/2	1	½	0	J_Ξ
1				1	2	0
			1	2	1	½
		1	2	1		1
	1	2	1			3/2
2			1	4	5	0
		1	4	5⊕1	4	½
	1	4	5⊕1	4	1	1
3		1	6	14	14	0
	1	6	14⊕1	14⊕6	14	½
4	1	8	27	48	42	0

The quantity J_Ξ is known as the superspin of the multiplet. As can be seen, the lowest dimensional massive representation has spin states from 0 to $N/2$. In general, the range is $\max(0, J_\Xi - N/2)$ to $J_\Xi + N/2$.

Irreducible supersymmetry representations on massless states can be derived in the same fashion as above, only now choosing the standard four-momentum $p^\mu=(k,o,o,k)$. In this case, the stability subalgebra is

$$\begin{aligned} \{ Q_\alpha^i , \bar{Q}_{\dot\beta j} \} &= k\,(\mathbb{1}+\sigma_3)_{\alpha\dot\beta}\,\delta^i{}_j \\ \{ Q_\alpha^i , Q_\beta{}^j \} &= 0 \quad , \end{aligned} \tag{1.22}$$

from which it can be seen that $Q_2{}^i, \bar{Q}_{2j}$ are no longer creation and annihilation operators, so we set $Q_2{}^i=0$ and have a Clifford algebra for only N creation and annihilation operators, whose irreducible representation contains 2^N states. If we rescale $Q_1{}^i = \frac{1}{2}Q^i$, we have

$$\begin{aligned} \{ Q^i , \bar{Q}_j \} &= \delta^i{}_j \\ \{ Q^i , Q^j \} &= \{ \bar{Q}_i , \bar{Q}_j \} = 0 \, . \end{aligned} \tag{1.23}$$

We can then define

$$Q^a = \left[\frac{Q^i + \bar{Q}_i}{\sqrt{2}} , \; i\frac{(Q^i - \bar{Q}_i)}{\sqrt{2}} \right] \quad a=1\cdots 2N \tag{1.24}$$

and the anticommutation relations for the Q^a become the Clifford algebra of SO(2N). The 2^N states in the fundamental massless multiplet span the spinor representation of the classification symmetry SO(2N), and can again be separated into bosonic and fermionic states using the $\pm$ eigenvalues of $\gamma^{(2N+1)}$.

In the massless case, the states of the multiplet can be classified by helicity, with the classification symmetry for states of the same helicity now reduced to U(N), as can be seen since with $Q_2{}^i=0$, the anticommutators (1.19) just reduce to (1.23), with manifest $[U(1)]_{\text{helicity}} \otimes U(N)$ symmetry. We are thus interested in the decomposition of the 2^N states in the spinor representation of SO(2N) into $[U(1)]_{\text{helicity}} \times U(N)$:

$$\begin{aligned} 2^N = (\lambda, 1) + (\lambda-\tfrac{1}{2}, \bar{N}) + \cdots + (\lambda - \tfrac{k}{2}[\bar{N}]_k) \\ + \cdots + (\lambda - \tfrac{N}{2}, 1) \end{aligned} \tag{1.25}$$

where λ is the helicity of the Clifford vacuum and $[\bar{N}]_k$ is here the totally antisymmetric representation of U(N) with k indices.

In a field theory, we must have a PCT conjugate state for every helicity, and these are not in general contained in (1.25). Thus we generally have to double the multiplet (1.25) by adding the PCT conjugate multiplet $(-\lambda, 1), \cdots, (-\lambda+{}^{N}\!/_{2}, 1)$. If in addition the Clifford vacuum carries a representation of some internal symmetry group $\subset$ U(N), then the PCT conjugate multiplet must transform in the conjugate representation of the internal symmetry group.

The only case where the addition of PCT conjugate states is not necessary is when $\lambda={}^{N}\!/_{4}$, when the supersymmetry multiplet (1.25) already contains the PCT conjugates. This also gives the minimum helicity range for a massless multiplet: $\lambda=0$ to $\lambda=\pm{}^{N}\!/_{4}$.

Just as we may decompose the massive spin state representations of the Poincaré group into massless helicity states, in supersymmetry, we may decompose the massive multiplets into massless ones. The pattern is given by the decomposition of the 2^{2N} states of the fundamental massive representation (1.15) into massless multiplets. The massless multiplets can be denoted $\{\lambda, [N]_k\}$ where λ is the maximum helicity in the massless multiplet, and this state (the Clifford vacuum of the massless multiplet) also carries the representation $[N]_k$ of U(N), i.e.

$$\{\lambda,[N]_k\} = (\lambda,[N]_k) + (\lambda-\tfrac{1}{2}, [N]_k\otimes\bar{N}) + \cdots + (\lambda-\tfrac{\ell}{2}, [N]_k\otimes[\bar{N}]_\ell) + \cdots + (\lambda-\tfrac{N}{2}, [N]_k) . \tag{1.26}$$

The decomposition of the fundamental massive multiplet is then

$$2^{2N} = \left\{\frac{N}{2}, 1\right\} + \left\{\frac{N-1}{2}, N\right\} + \cdots + \left\{\frac{N-k}{2}, [N]_k\right\} + \cdots + \{0,1\} . \tag{1.27}$$

In the massive supermultiplets discussed above, we have set the central charges to zero. If the central changes are active, the multiplet structure may still be analysed by the method of induced representations.[8] As we have pointed out above, the presence of central charges also affects the internal symmetry structure that can be given to the supermultiplets. The simplest case is that of a single central charge occurring as

$$\{Q_\alpha^i, Q_\beta^j\} = 2\epsilon_{\alpha\beta}\Omega^{ij}Z \tag{1.28}$$

with Ω^{ij} the numerically invariant antisymmetric metric of USP(N), for N even. In this case, the maximal internal symmetry group is USP(N), and the massive particle states are also classified by this symmetry.

The most striking feature of supermultiplets with central charges is the reduction in the range of spins. In the simplest case, as given above, the lowest dimensional multiplet has maximum spin $J={}^N/4$, half that of the massive multiplet without central charges. The structure of the multiplets is just that of the massive multiplets without central charges for $^N/_2$ extended supersymmetry (again, for N even), but doubled. The central charge just rotates every state into its double.

CHAPTER II

Superactions in Superspace

In order to formulate supersymmetric field theories, we need to know the representations of the supersymmetry algebra on fields. In fact, we can already discover much about these representations from the results of the last chapter, even before we discuss the details of the supersymmetry transformations. As with the representations of the Poincaré group, we can learn most of the structure of the irreducible representations on fields from the irreducible representations on massive particle states.

For example, a massive spin zero particle corresponds to a scalar field ϕ satisfying

$$(\Box - m^2)\,\phi = 0. \tag{2.1}$$

A more revealing example is that of a massive spin one particle, which can be represented by a vector field A_μ satisfying the Proca field equation

$$\partial_\mu F^{\mu\nu}(A) - m^2 A^\nu = 0. \tag{2.2}$$

Taking the divergence of this equation, we obtain

$$m^2\, \partial_\nu A^\nu = 0\,, \tag{2.3}$$

so $A^\nu = A^{\nu T}$, i.e. it is a Poincaré irreducible transverse vector field. The remaining content of (2.2) is then just

$$(\Box - m^2)\, A^{\nu T} = 0\,. \tag{2.4}$$

Thus a massive spin one field is described by a vector satisfying the Klein-Gordon equation (2.4) together with the auxiliary condition (2.3) which ensures Poincaré irreducibility. This pattern persists for all massive fields:

Poincaré irreducible fields which satisfy the massive Klein-Gordon equation describe particle states with spin determined by the Lorentz representation of the field.

The above correspondance between massive particles and Poincaré irreducible fields is always one-to-one in terms of the number of degrees of freedom of the particles and the number of components of the fields. Spinor representations of the Lorentz group have numbers of components equal to integral multiples of four. Spinor particles, however, have numbers of degrees of freedom equal to multiples of two. Thus, a Majorana or Weyl spinor field must correspond to two massive spin ½ particles when we impose the massive Klein-Gordon equation, and similarly for higher spins. In the case of spin ½ particles, we may separate the Klein-Gordon equation for a Majorana field

$$(\Box - m^2)\Psi = 0 \tag{2.5}$$

into two first order equations,

$$\begin{aligned} (\lambda - \partial\!\!\!/\Psi) &= 0 \\ (\partial\!\!\!/\lambda - m^2\Psi) &= 0 . \end{aligned} \tag{2.6}$$

If we now redefine

$$\begin{aligned} \lambda &= \chi + \varphi \\ m\Psi &= \chi - \varphi \end{aligned} \tag{2.7}$$

we obtain the system of equations

$$\begin{aligned} (\partial\!\!\!/ + m)\varphi &= 0 \\ (\partial\!\!\!/ - m)\chi &= 0 \end{aligned} \tag{2.8}$$

and if one wishes to change the sign on the mass term in the second of these, it is sufficient to define $\chi=\gamma_5\chi'$ Thus we see explicitly that the massive Klein-Gordon equation for a Majorana spinor field describes two massive spin ½ particles.

The supersymmetry algebra without central charges is a direct extension of the Poincaré algebra, and we find that the above pattern extends to cover full irreducible representations of supersymmetry on fields. For example, take the simplest massive particle representation given in Table 1, containing 1 spin ½ particle and 2 spin 0 particles. Because of the 2 - 1 correspondence between spinor particles and fields explained above, we must double the particle representation in order to have the same number of spinor components as a Majorana (or Weyl) spinor field. The correspondence then gives us the simplest representation on fields,

$$\left(A(x),\, B(x),\, \psi_\alpha(x)\left(\bar{\psi}_{\dot\alpha}(x)\right),\, F(x),\, G(x)\right) \quad , J_{\mathrm{I}} = 0 \tag{2.9}$$

i.e., four scalars and one complex two-component spinor field. This multiplet is known as the chiral multiplet of N=1 supersymmetry. The relative dimensions of the component fields are not fixed by the correspondence to massive particle states, and we shall return to this point later. The parity of the above fields is also not determined.

The next particle representation in Table 1 can be taken over without doubling, as there are already four spin ½ degrees of freedom:

$$\left(C(x),\, \lambda_\alpha(x)\left(\bar{\lambda}_{\dot\alpha}(x)\right),\, V_\mu(x)\right) \quad , \quad J_{\mathrm{I}} = \frac{1}{2} \tag{2.10}$$

where the vector field V_μ must be transverse to ensure Poincaré irreducibility:

$$\partial^\mu V_\mu = 0. \tag{2.11}$$

The need for the condition (2.11) in the irreducible supermultiplet (2.10) is to ensure that there are the same number of effective bosonic as there are fermionic field components, as we always must have in representations of supersymmetry, either on fields or on particles. Thus the vector V_μ counts for only three bosonic components, and in total in (2.10) we have 4 fermionic plus 4 bosonic components. The multiplet (2.10) is known as the N=1 linear multiplet.

We may of course wish to consider reducible representations of supersymmetry as well, particularly in order to avoid having to impose differential subsidiary conditions like (2.11) on some component fields. Just as with ordinary fields we may combine a scalar and a transverse vector to obtain an unconstrained 4-vector, $V_\mu = V^T_\mu + \partial_\mu \phi$, so in supersymmetry the multiplets (2.9) and (2.10) can be combined to form the N=1 general scalar multiplet:

$$\left(C(x)\, ,\, \lambda_\alpha(x)\, ,\, H(x),\, K(x),\, V_\mu(x)\, ,\, \chi_\alpha(x)\, ,\, D(x) \right) \tag{2.12}$$

The irreducible submultiplets of the general scalar multiplet are the linear multiplet $(C,\ \lambda_\alpha,\ V^T_\mu)$ and the chiral multiplet $(H,\ K,\ \lambda_\alpha,\ D,\ \partial^\mu V\mu)$ containing the longitudinal part of V^μ.

Now we consider the detailed form of the supersymmetry transformations. The basic structure can be established by dimensional considerations. From the supersymmetry algebra (1.10) we see that the supersymmetry generator has dimensions of $(\text{mass})^{\frac{1}{2}}$, so the parameter ε^α of supersymmetry must have dimensions of $(\text{length})^{\frac{1}{2}}$. This is in accord with the canonical dimensions of scalar and spinor fields, so that in the chiral multiplet with a scalar field A, a pseudoscalar field B and a four component Majorana spinor $\Psi=(\Psi\alpha, \bar{\Psi}^{\dot{\alpha}})$ of canonical dimensions (respectively 1, $^3/_2$), we have

$$\begin{aligned} \delta A &= i\bar{\varepsilon}\Psi \\ \delta B &= i\bar{\varepsilon}\gamma_5\Psi \end{aligned} \tag{2.13}$$

In order to be dimensionally consistent, the spinor Ψ in turn must transform into fields of dimension two or into derivatives of fields of dimension one. It does both:

$$\delta\Psi = \partial_\mu (A - \gamma_5 B)\gamma^\mu \epsilon \; + \; (F + \gamma_5 G)\,\epsilon \; . \tag{2.14}$$

Since we have now exhausted the fields in the multiplet (2.9), the dimension two fields can only transform into derivatives of the spinor Ψ:

$$\begin{aligned} \delta F &= i\bar{\epsilon}\gamma^{\mu}\partial_{\mu}\Psi \\ \delta G &= i\bar{\epsilon}\gamma_5\gamma^{\mu}\partial_{\mu}\Psi \, . \end{aligned} \tag{2.15}$$

Due to the presence of the derivatives in the transformations (2.14, 2.15), the commutator of two supersymmetry transformations with parameters ε_1 and ε_2 on any field gives a translation (generator $P\mu=i\partial_{\mu}$) with parameter $(\bar{\varepsilon}_2\gamma_{\mu}\varepsilon_1 - \bar{\varepsilon}_1\gamma_{\mu}\varepsilon_2)$. In order for this not to vanish, the Majorana spinors ε_1, ε_2 must be taken to be anticommuting objects. This is in accord with the classical limit of quantized spinor fields, which must also be taken to be anticommuting. Before continuing with the structure of the supersymmetry transformations, we present a discussion due to Valuyev[(9)] on the nature of anticommuting C- numbers.

In order to realize the Grassman algebra of anticommuting C- numbers concretely, we take the Clifford algebra for SO(2*N*) where *N* denotes some very large integer. The γ-matrices for this group satisfy

$$\{\gamma_i , \gamma_j\} = 2\,\delta_{ij} \, . \tag{2.16}$$

We can define the basis for the Grassman algebra by

$$\Theta_n = \frac{1}{\sqrt{2}}\left(\gamma_{2n} + i\,\gamma_{2n+1}\right) \tag{2.17}$$

$$d\Theta_n = \frac{1}{\sqrt{2}}\left(\gamma_{2n} - i\,\gamma_{2n+1}\right) , \tag{2.18}$$

where the Θn anticommute among themselves and the $d\Theta_m$ anticommute among themselves, but give δmn when mixed anticommutators are taken:

$$\{\Theta_m, \Theta_n\} = \{d\Theta_m , d\Theta_n\} = 0 \tag{2.19}$$

$$\{d\Theta_m, \Theta_n\} = 2\,\delta_{mn} \tag{2.20}$$

Thus, we can define the SO(2N) invariant integral

$$\int d\Theta\, f(\Theta) = 2^{-N} \mathrm{tr}[d\Theta\, f(\Theta)] , \tag{2.21}$$

where dΘ and Θ are now general elements of the Grassman algebra formed by taking linear combinations of the basis (2.17, 2.18). In particular, for the integral over a single Θ the tracelessness of the γ matrices implies

$$\int d\Theta = 0 \tag{2.22}$$

while the integral

$$\int d\Theta\, \Theta = 1 \tag{2.23}$$

follows from (2.20) and a construction of dΘ from the basis (2.18) conjugate to that of Θ from the basis (2.17). The rules (2.22, 2.23) are just Berezin's rules for integrating over anticommuting variables(10).

Requiring that the spinor fields and the spinor parameter ε of the supersymmetry transformations be anticommuting elements of a Grassman algebra as above, we can now see that the transformations (2.13-2.15) do indeed realize the supersymmetry algebra. This is trivially seen in the transformations on the scalar A and pseudoscalar B. More interesting is the check of the algebra's closure on the spinor Ψ. In order to check the algebra, we take the commutator of two supersymmetry transformations with parameters ε_1 and ε_2, since $[\delta\varepsilon_2 , \delta\varepsilon_1]\psi = -\bar{\varepsilon}_2{}^{\alpha}\varepsilon_1{}^{\beta}\{Q_\alpha, Q_\beta\}\psi$
Thus,

$$\delta_{\epsilon_1}\Psi = \partial_\mu (A - \gamma_5 B)\gamma^\mu \epsilon_1 + (F + \gamma_5 G)\epsilon_1 \tag{2.24}$$

so
$$[\delta_{\epsilon_2}, \delta_{\epsilon_1}]\Psi = i\,\partial^\mu(\bar{\epsilon}_2\Psi - \bar{\epsilon}_2\gamma_5\Psi\gamma_5)\gamma^\mu\epsilon_1 + i(\bar{\epsilon}_2\gamma^\mu\partial_\mu\Psi + \bar{\epsilon}_2\gamma_5\gamma^\mu\partial_\mu\Psi\gamma_5)\epsilon_1 - (1\leftrightarrow 2) ,$$

and in order to bring the two ε's together we must make a Fierz transformation:

$$
\begin{aligned}
[\delta_{\epsilon_2}, \delta_{\epsilon_1}]\Psi = & -\tfrac{i}{4}\,\bar{\epsilon}_2\gamma_\nu\epsilon_1\left(\gamma^\mu\gamma^\nu\partial_\mu\Psi - \gamma_5\gamma^\mu\gamma^\nu\gamma_5\partial_\mu\Psi\right) \\
& +\tfrac{i}{2}\,\bar{\epsilon}_2\sigma_{\rho\tau}\epsilon_1\left(\cancel{\gamma^\mu\sigma^{\rho\tau}\partial_\mu\Psi} - \cancel{\gamma_5\gamma^\mu\sigma^{\rho\tau}\gamma_5\partial_\mu\Psi}\right) \\
& -\tfrac{i}{4}\,\bar{\epsilon}_2\gamma_\nu\epsilon_1\left(\gamma^\nu\gamma^\mu\partial_\mu\Psi + \gamma_5\gamma^\nu\gamma_5\gamma^\mu\partial_\mu\Psi\right) \\
& +\tfrac{i}{2}\,\bar{\epsilon}_2\sigma_{\rho\tau}\epsilon_1\left(\cancel{\sigma^{\rho\tau}\gamma^\mu\partial_\mu\Psi} + \cancel{\gamma_5\sigma^{\rho\tau}\gamma_5\gamma^\mu\partial_\mu\Psi}\right) \\
= & -i\,\bar{\epsilon}_2\gamma_\nu\epsilon_1\left(\gamma^\mu\gamma^\nu\partial_\mu\Psi + \gamma^\nu\gamma^\mu\partial_\mu\Psi\right) \\
= & -2i\,\bar{\epsilon}_2\gamma_\nu\epsilon_1\,\partial^\nu\Psi
\end{aligned}
\qquad (2.25)
$$

as required.

Thus, all the properties of anticommuting Majorana spinors come into play when verifying the supersymmetry algebra. If we were to repeat the above derivation using the two-component Van der Waerden notation, the Fierz transformation step would correspond to symmetizing and antisymmetizing on the two component indices to be joined. Note also that there is a consistent restriction on the chiral multiplet which sets F and G to zero and requires the fields A,B and Ψ to satisfy free field equations. In that case, the restricted multiplet (A,B,Ψ) still forms a realization of the supersymmetry algebra, but the closure calculation (2.25) is then valid only subject to the imposition of these field equations. The dimension two fields F and G, while they do not contribute physical degrees of freedom, are necessary in order to have a complete linear representation of supersymmetry without imposing the field equations. These non-dynamical fields are called auxiliary fields.

The transformations (2.13 - 2.15) cause the following combinations of fields to transform by total derivatives:

$$\mathcal{L}_{\text{kinetic}} = -\tfrac{1}{2}(\partial_\mu A)^2 - \tfrac{1}{2}(\partial_\mu B)^2 - \tfrac{i}{2}\bar{\Psi}\partial\!\!\!/\Psi + \tfrac{1}{2}F^2 + \tfrac{1}{2}G^2 \qquad (2.26)$$

$$\mathcal{L}_{\text{mass}} = m\left(FA + GB - \tfrac{i}{2}\bar{\Psi}\Psi\right) \qquad (2.27)$$

$$\mathcal{L}_{\text{interaction}} = g\left(FA^2 - FB^2 + 2GAB - i\bar{\Psi}(A-\gamma_5 B)\Psi\right). \qquad (2.28)$$

The integrals of these three terms over $\int d^4x$ are thus separately supersymmetrically invariant. The action for the interacting Wess-Zumino model(4) is the sum of these three integrals. The auxiliary fields F and G then have algebraic equations of motion:

$$F + mA + g(A^2 - B^2) = 0 \tag{2.29}$$

$$G + mB + 2gAB = 0 . \tag{2.30}$$

Since these equations of motion are algebraic, the dynamical consequences of the action are unchanged if we substitute for F and G back into the rest of the action. For m=o, the result is the same as we had in e.g. (1.6) for $\phi = \frac{A+iB}{\sqrt{2}}$
With $m \neq 0$, the mass terms are accompanied by trilinear scalar interaction terms: after elimination of F and G,

$$I_{WZ} = \int d^4x \Big(-\tfrac{1}{2}(\partial_\mu A)^2 - \tfrac{1}{2}(\partial_\mu B)^2 - \tfrac{i}{2}\bar{\Psi}\not{\partial}\Psi - \tfrac{1}{2}m^2A^2 - \tfrac{1}{2}m^2B^2 - \tfrac{i}{2}m\bar{\Psi}\Psi - gmA(A^2+B^2) - \tfrac{1}{2}g^2(A^2+B^2)^2 - ig\bar{\Psi}(A-\gamma_5 B\Psi)\Big). \tag{2.31}$$

The supersymmetry invariance of the Wess-Zumino model gives rise to the existence of a conserved vector-spinor current

$$\sqrt{2}\, Q^\mu = \gamma^\lambda \partial_\lambda (A - \gamma_5 B)\gamma^\mu \Psi - (F + \gamma_5 G)\gamma^\mu \Psi . \tag{2.32}$$

Upon elimination of the auxiliary fields F and G using their equations of motion, this current coincides with the one given in eg. (1.7).

The potential for the scalar fields in the supersymmetric model (2.31) is positive definite, as can clearly be seen since it is just the sum of the squares of the auxiliary fields,

$$V = \tfrac{1}{2}(F^2 + G^2) . \tag{2.33}$$

In addition to allowing the supersymmetry algebra to close without use of the equations of motion, the auxiliary fields are necessary in order to have a linear representation of supersymmetry:

the action (2.31) is still supersymmetric, but under nonlinear transformations obtained from (2.13 - 2.15) by substituting (2.29 - 2.30) for F and G.

In order to exploit the linear realization of supersymmetry afforded by the chiral representation and others to be discussed, we need a convenient notation that makes the linear realization manifest. Such a notation is provided by following the analogy of constructing the induced representations of the Poincaré group. Just as Minkowski space can be viewed as the space of left cosets of the Lorentz group H within the Poincaré group $\mathcal{P}$, so can one construct superspace(3,6) as GP/H, where GP is the graded Poincaré group, whose algebra is the supersymmetry algebra given in (1.10).

The co-ordinates of superspace label the above cosets. From the supersymmetry algebra, we can see that these co-ordinates will carry the same Lorentz indices as the generators P_μ and Q_α, $\bar{Q}_{\dot\alpha}$. Thus we have a space of vectorial bosonic co-ordinates x^μ and spinorial fermionic co-ordinates θ_α, $\bar\theta_{\dot\alpha}$ (or equivalently, a four-component Majorana spinor). In order to define the action of the supersymmetry group on these co-ordinates, we start with a corresponding group element

$$G(x, \theta_\alpha, \bar\theta_{\dot\alpha}) = e^{i[-x^\mu P_\mu + \theta^\alpha Q_\alpha + \bar\theta_{\dot\alpha}\bar{Q}^{\dot\alpha}]} \quad . \tag{2.34}$$

In writing the group element this way, we are treating the Q's as if they were Lie algebra generators, but with anticommuting parameters. Thus the θ_α, $\bar\theta_{\dot\alpha}$ must be taken to be anticommuting elements of a Grassman algebra, as discussed above.

Two group elements like (2.34) can be multiplied together using Hausdorff's formula.

$$e^A e^B = e^{A+B+\frac{1}{2}[A,B]+\cdots} \quad . \tag{2.35}$$

The higher order commutators in this formula are not needed, since they vanish for the supersymmetry algebra. The action of a translation $e^{-i(\delta x^\mu p_\mu)}$ works as in the Poincaré group: $x^\mu \to x^\mu + \delta x^\mu$. The supersymmetry transformation of $x^\mu, \theta_\alpha, \bar\theta_{\dot\alpha}$ is given by computing

$$G(0, \epsilon_\alpha, \bar\epsilon_{\dot\alpha})\, G(x^\mu, \theta_\alpha, \bar\theta_{\dot\alpha}) = G(x^\mu + i\theta\sigma^\mu\bar\epsilon - i\epsilon\sigma^\mu\bar\theta,\ \theta+\epsilon,\ \bar\theta+\bar\epsilon). \tag{2.36}$$

Thus, group multiplication induces a motion in the coset parameter space

$$(x^\mu, \theta_\alpha, \bar\theta_{\dot\alpha}) \to (x^\mu + i\theta\sigma^\mu\bar\epsilon - i\epsilon\sigma^\mu\bar\theta,\ \theta_\alpha+\epsilon_\alpha,\ \bar\theta_{\dot\alpha}+\bar\epsilon_{\dot\alpha}) \tag{2.37}$$

Just as the Poincaré shift in x^μ is generated by $i\,\partial/\partial x^\mu$, so the shifts induced by a supersymmetry transformation are generated by

$$Q_\alpha = \frac{\partial}{\partial\theta^\alpha} - i\,\sigma^\mu_{\alpha\dot\alpha}\,\bar\theta^{\dot\alpha}\,\partial_\mu \tag{2.38}$$

$$\bar Q^{\dot\alpha} = \frac{\partial}{\partial\bar\theta_{\dot\alpha}} + i\,\theta^\alpha\sigma^\mu_{\alpha\dot\beta}\,\epsilon^{\dot\alpha\dot\beta}\,\partial_\mu \tag{2.39}$$

which give the required algebra

$$\{Q_\alpha, \bar Q_{\dot\alpha}\} = 2i\,\sigma^\mu_{\alpha\dot\alpha}\,\partial_\mu \tag{2.40}$$

$$\{Q_\alpha, Q_\beta\} = \{\bar Q^{\dot\alpha}, \bar Q^{\dot\beta}\} = 0. \tag{2.41}$$

The representations of supersymmetry that follow from the above construction are superfields $F(x,\theta,\bar\theta)$, a priori general functions of the commuting and anticommuting co-ordinates of superspace. In order to relate them to the fields used in our ordinary formulations of physical theories, we may expand a superfield in a Taylor series in $\theta,\bar\theta$. Due to the anticommuting property of $\theta,\bar\theta$ this series will terminate since the highest possible product is $\theta^\alpha\,\theta_\alpha\,\bar\theta_{\dot\alpha}\,\bar\theta^{\dot\alpha}$. Thus we obtain a set of coefficient functions of x^μ only, carrying various Lorentz representations:

$$\begin{aligned} F(x,\theta,\bar\theta) = {} & f(x) + \theta^\alpha\chi_\alpha(x) + \bar\theta_{\dot\alpha}\bar\varphi^{\dot\alpha}(x) + \theta\theta\, m(x) + \bar\theta\bar\theta\, n(x) \\ & + \theta\sigma^\mu\bar\theta\, V_\mu(x) + \theta\theta\,\bar\theta^{\dot\alpha}\bar\lambda_{\dot\alpha}(x) + \bar\theta\bar\theta\,\theta_\alpha\psi^\alpha(x) \\ & + \theta\theta\bar\theta\bar\theta\, d(x). \end{aligned} \tag{2.42}$$

While indicating correctly the number and types of "component" ordinary fields contained within a general complex superfield , the expansion (2.42) is not in fact the most convenient one. Just as we define the coefficients in a Taylor expansion by taking repeated derivatives of the function to be expanded and evaluating the result at the zero value of the expansion co-ordinate, we would like to do the same for superfields in terms of repeated fermionic derivatives. However, we run into the problem that the ordinary partial derivative $\partial/\partial\theta^\alpha$ is not covariant:

$$\left\{\frac{\partial}{\partial\theta^\alpha}, \bar{Q}_{\dot\beta}\right\} = i\,\sigma^\mu_{\alpha\dot\beta}\,\partial_\mu \, . \tag{2.43}$$

Because of this non-covariance, the derivative $\partial/\partial\theta^\alpha$ of a superfield does not transform according to the same rule as a superfield, ie by operation with $Q_\alpha, \bar{Q}_{\dot\beta}$. In order to cure this problem, we introduce fermionic covariant derivatives:

$$D_\alpha = \frac{\partial}{\partial\theta^\alpha} + i\,\sigma^\mu_{\alpha\dot\alpha}\,\bar\theta^{\dot\alpha}\,\partial_\mu \tag{2.44}$$

$$\bar{D}_{\dot\alpha} = -\frac{\partial}{\partial\bar\theta^{\dot\alpha}} - i\,\theta^\alpha\sigma^\mu_{\alpha\dot\alpha}\,\partial_\mu \, . \tag{2.45}$$

These covariant derivatives preserve the superfield character of that which they operate upon, since

$$\{D_\alpha, Q_\beta\} = \{D_\alpha, \bar{Q}_{\dot\beta}\} = \{\bar{D}_{\dot\alpha}, Q_\beta\} = \{\bar{D}_{\dot\alpha}, \bar{Q}_{\dot\beta}\} = 0 \, . \tag{2.46}$$

Among themselves, they satisfy an algebra similar to that of the Q's,

$$\{D_\alpha, \bar{D}_{\dot\alpha}\} = -2i\,\sigma^\mu_{\alpha\dot\alpha}\,\partial_\mu \tag{2.47}$$

$$\{D_\alpha, D_\beta\} = \{\bar{D}_{\dot\alpha}, \bar{D}_{\dot\beta}\} = 0 . \tag{2.48}$$

In fact, it may be checked that D_α and $\bar{D}_{\dot\alpha}$ can also be viewed as the generators of parameter shifts induced by <u>right</u> multiplication, analogously to (2.36). Their definitions (2.44, 2.45) and anti-commutation relations (2.47) differ from those of the Q, $\bar{Q}$ only by the signs of the x_μ shifting terms.

Using the covariant derivatives, a convenient expansion of a superfield is to take the $\theta=\bar{\theta}=0$ components of a sequence of superfields obtained from the original one by differention with $D_\alpha, \bar{D}_{\dot{\alpha}}$. Thus we may define

$$
\begin{aligned}
f(x) &\equiv F|_{\theta=\bar{\theta}=0} \\
\chi_\alpha(x) &\equiv (D_\alpha F)|_{\theta=\bar{\theta}=0} \quad , \text{etc.}
\end{aligned}
\tag{2.49}
$$

The reason that this expansion into components is more convenient than the straight expansion in θ ,$\bar{\theta}$ of (2.42) is that the supersymmetry transformations of the components may be obtained by a simple rule. The supersymmetry transformation of any superfield $F(x\,,\theta,\bar{\theta}\,)$ is given by

$$
\delta F = (\epsilon^\alpha Q_\alpha + \bar{\epsilon}_{\dot{\alpha}} \bar{Q}^{\dot{\alpha}}) F \tag{2.50}
$$

so in particular, the transformation of the $\theta=\bar{\theta}=0$ component of F is

$$
\delta f = [(\epsilon^\alpha Q_\alpha + \bar{\epsilon}_{\dot{\alpha}} \bar{Q}^{\dot{\alpha}}) F]|_{\theta=\bar{\theta}=0} \,. \tag{2.51}
$$

Moreover, since the expression on the right is evaluated at $\theta=\bar{\theta}=0$, we obtain the same result if we operate with D_α and $\bar{D}_{\dot{\alpha}}$ instead of the Q_α, $\bar{Q}_{\dot{\alpha}}$ since the difference is set to zero with $\theta=\bar{\theta}=0$:

$$
\begin{aligned}
\delta f &= [(\epsilon^\alpha D_\alpha + \bar{\epsilon}_{\dot{\alpha}} \bar{D}^{\dot{\alpha}}) F]|_{\theta=\bar{\theta}=0} \\
&= \epsilon^\alpha \chi_\alpha + \bar{\epsilon}_{\dot{\alpha}} \bar{\varphi}^{\dot{\alpha}} \,.
\end{aligned}
\tag{2.52}
$$

This same rule applies to the transformations of all the higher components of the multiplet, since each of them is obtained from the $\theta=\bar{\theta}=0$ component of some superfield. Thus, for example,

$$
\delta \chi_\alpha = [(\epsilon^\alpha D_\alpha + \bar{\epsilon}_{\dot{\alpha}} \bar{D}^{\dot{\alpha}}) D_\alpha F]|_{\theta=\bar{\theta}=0} \,, \tag{2.53}
$$

and so on: one needs to use only the algebra of the D's and $\bar{D}$'s together with the definitions of the various components.

So far, we have considered our basic superfield F to be a general function of superspace, with an expansion into components given as in (2.42), or the more convenient expansion in terms of D's and $\bar{D}$'s given above.

However, not every scalar superfield need contain as many components as given in (2.42). For example, we could impose a reality condition F = F*. In imposing such a condition, it is conventional to invent the order of strings of Θ's $\bar{\Theta}$'s:

$$(\Theta_\alpha \bar{\Theta}_{\dot\beta})^* = \Theta_\beta \bar{\Theta}_{\dot\alpha} . \tag{2.54}$$

Supersymmetry allows another type of restriction, however, due to the fact that the superspace co-ordinates $(x^\mu, \Theta_\alpha, \bar{\Theta}_{\dot\alpha})$ fall into a reducible representation of the Lorentz group. As a consequence it is possible to impose a fermionic differential constraint without constraining the entire superspace dependence of a superfield. The fundamental example is a chiral superfield $\phi(x, \Theta, \bar{\Theta})$ satisfying

$$\bar{D}_{\dot\alpha} \phi = 0 \tag{2.55}$$

Using the constraint (2.55), we find that the independent component fields of ϕ are just

$$a = \frac{A+iB}{\sqrt{2}} \equiv \phi|_{\theta=\bar\theta=0} \tag{2.56}$$

$$\chi_\alpha \equiv (D_\alpha \phi)|_{\theta=\bar\theta=0} \tag{2.57}$$

$$h = \frac{F-iG}{\sqrt{2}} \equiv (D^2\phi)|_{\theta=\bar\theta=0} \quad , \quad D^2\phi = D^\alpha D_\alpha \phi . \tag{2.58}$$

Note that a chiral superfield cannot be made real without constraining it to be a constant. Evaluating the supersymmetry transformations of the multiplet according to the rule given above, we find

$$\begin{aligned} \delta a &= [(\epsilon^\alpha D_\alpha + \bar\epsilon_{\dot\alpha} \bar D^{\dot\alpha})\phi]|_{\theta=\bar\theta=0} \\ &= \epsilon^\alpha \chi_\alpha \end{aligned} \tag{2.59}$$

$$\begin{aligned} \delta \chi_\alpha &= [(\epsilon^\beta D_\beta + \bar\epsilon_{\dot\beta} \bar D^{\dot\beta}) D_\alpha \phi]|_{\theta=\bar\theta=0} \\ &= [-\tfrac{1}{2}\epsilon_\alpha D^2\phi + 2i\,\sigma^\mu_{\alpha\dot\beta}\bar\epsilon^{\dot\beta}\partial_\mu \phi]|_{\theta=\bar\theta=0} \\ &= -\tfrac{1}{2} h\,\epsilon_\alpha + 2i\,\partial_\mu a\,\sigma^\mu_{\alpha\dot\beta}\bar\epsilon^{\dot\beta} \end{aligned} \tag{2.60}$$

$$\delta h = [(\epsilon^{\beta} D_{\beta} + \bar{\epsilon}_{\dot{\beta}} \bar{D}^{\dot{\beta}}) D^2 \phi]_{|\theta=\bar{\theta}=0}$$
$$= -2i\, \bar{\epsilon}_{\dot{\beta}}\, \sigma_{\mu}{}^{\dot{\beta}\alpha}\, \partial^{\mu} \chi_{\alpha} \,, \tag{2.61}$$

where the constraint on ϕ has been used as well as the property

$$D_{\alpha} D_{\beta} D_{\gamma} = 0 \tag{2.62}$$

following from the anticommutation relations of the 2-component D's.

Due to the dimensionality $(\text{length})^{\frac{1}{2}}$ of the supersymmetry parameter ε, the component of highest dimension of any supermultiplet must transform into a derivative of some component with one half unit of dimension less than the highest. This means that the integral over all of spacetime of such a component is a supersymmetric invariant. The existence of both chiral and general scalar supermultiplets thus implies the existence of two types of supersymmetric invariant. The multiplet whose highest component is to be taken as a supersymmetric density is generally a product of the basic field multiplets and their derivativesin a given model.

The superspace formalism provides a very compact notation for supersymmetric actions when account is taken of the Berezin superspace integration rules (2.22, 2.23). Because of these rules, it is apparent that integration in superspace is equivalent to differentiation

$$\int d\theta = \frac{\partial}{\partial\theta} \,. \tag{2.63}$$

Furthermore, since in an action we also integrate over $\int d^4x$, we may replace an overall $\partial/\partial\theta^{\alpha}$ by D_{α}, which differs from it only by an overall spacetime derivative. Thus, the supersymmetric invariant action formula for a superfield Lagrangian F of the general scalar type, such as the superfield in (2.42), is given by the full superspace integral

$$I_{GS} = \int d^4x\, d^4\theta\, F(x,\theta,\bar{\theta}) \tag{2.64}$$

where $d^4\theta = d\theta^{\alpha}\, d\theta_{\alpha}\, d\bar{\theta}_{\dot{\beta}} d\bar{\theta}^{\dot{\beta}}$. The full superspace integral just picks out the highest component of the multiplet. Since we may add total space-time derivatives at will under the spacetime integral, this action is equivalent to

$$I_{GS} = \int d^4x\, D^2\bar{D}^2 F\,. \tag{2.65}$$

We can check the invariance by evaluating the supersymmetry transformation of the integral as discussed above,

$$\delta(D^2\bar{D}^2F) = \left[(\epsilon^{\alpha}D_{\alpha} + \bar{\epsilon}_{\dot{\alpha}}\bar{D}^{\dot{\alpha}})\, D^2\bar{D}^2F\right]_{|\theta=\bar{\theta}=0}\,. \tag{2.66}$$

In fact, due to the fact that we are integrating over all spacetime, the restriction to $\theta=\bar{\theta}=0$ is unnecessary here since the θ, $\bar{\theta}$ dependent terms in the derivatives give total spacetime derivatives. Thus,

$$\begin{aligned} D_{\alpha}(D^2\bar{D}^2F) &= 0 \\ \bar{D}_{\dot{\alpha}}(D^2\bar{D}^2F) &= [\bar{D}_{\dot{\alpha}}, D^2]\bar{D}^2F + D^2\bar{D}_{\dot{\alpha}}\bar{D}^2F \\ &= 4i\,\partial_{\mu}\sigma^{\mu}_{\alpha\dot{\alpha}}\, D^{\alpha}\bar{D}^2F\,, \end{aligned} \tag{2.67}$$

so the integral in (2.65) varies by a total spacetime derivative as required for supersymmetric invariance.

The chiral multiplet (2.59 - 2.61) gives rise to an action formula with integration only over $d^2\theta$:

$$I_{CH} = \int d^4x\, d^2\theta\, \mathcal{L}\,. \tag{2.68}$$

This is invariant subject to the chiral constraint $\bar{D}_{\dot{\alpha}}\mathcal{L}=0$ on the chiral superspace density $\mathcal{L}$, as may again be checked

$$I_{ch} = \int d^4x\, D^2\mathcal{L} \tag{2.69}$$

and

$$D_\alpha D^2\mathcal{L} = 0 \tag{2.70}$$

$$\begin{aligned} \bar{D}_{\dot{\alpha}} D^2\mathcal{L} &= [\bar{D}_{\dot{\alpha}}, D^2]\mathcal{L} + D^2\bar{D}_{\dot{\alpha}}\mathcal{L} \\ &= 4i\,\partial_\mu \sigma^\mu_{\alpha\dot{\alpha}} D^\alpha \mathcal{L} \end{aligned} \tag{2.71}$$

where it has been necessary to use the chiral constraint to obtain (2.71).

Examples of the two types of action formulas are provided by the three terms (2.26 - 2.28) which make up the action for the Wess-Zumino model. The basic fields of this model are contained in a chiral superfield ϕ. The kinetic term may be written using either of the two action formulas:

$$I_{\text{kinetic}} = \int d^4x\, d^4\theta\, \bar{\Phi}\Phi = \int d^4x\, d^2\theta\, \Phi\bar{D}^2\bar{\Phi}\ , \tag{2.72}$$

where the second form of the action uses the fact that $\bar{D}_{\dot{\beta}}\bar{D}^2\bar{\Theta} = 0$, so $\phi\bar{D}^2\bar{\phi}$ is chiral. Although it might appear that the second form is complex and one should take the real part, this is in fact unnecessary since the imaginary part is the integral of a total spacetime derivative. We can evaluate this kinetic action as follows:

$$\begin{aligned} I_{\text{kinetic}} &= \int d^4x\, D^2(\Phi\bar{D}^2\bar{\Phi}) \\ &= \int d^4x\, D^2\Phi\bar{D}^2\bar{\Phi} + 2\int d^4x\, D^\alpha\Phi(D_\alpha\bar{D}^2\bar{\Phi}) + \int d^4x\, \Phi D^2\bar{D}^2\bar{\Phi} \\ &= \int d^4x \left[h h^* + 4i\,\chi^\alpha \partial_\mu \sigma^\mu_{\alpha\dot{\alpha}} \bar{\chi}^{\dot{\alpha}} + 8\, a \Box a^*\right] \end{aligned} \tag{2.73}$$

which, after some rescalings just gives (2.26).

The mass term of the Wess-Zumino model can only be written using the chiral action formula (2.68):

$$I_{mass} = \mathrm{Re}\left[m \int d^4x \, d^2\theta \, \Phi^2\right] . \tag{2.74}$$

This can also be evaluated using the rules given above,

$$\begin{aligned} m \int d^4x \, d^2\theta \, \phi^2 &= m \int d^4x \, D^2 \phi^2 \\ &= 2m \int d^4x \, (D^2\phi)\phi + 2m \int d^4x \, D^\alpha\phi \, D_\alpha \phi \\ &= 2m \int d^4x \, [h a + \chi^\alpha \chi_\alpha] , \end{aligned} \tag{2.75}$$

which gives (2.27) after taking the real part.

Finally, the interaction term (2.28) must also be written as a chiral superspace integral:

$$I_{interaction} = g \int d^4x \, d^2\theta \, \phi^3 . \tag{2.76}$$

If we generalize the Wess-Zumino model to include a number of chiral superfields ϕ^a, then the most general renormalizable Lagrangian is

$$I_{matter} = \int d^4x \, d^4\theta \, \bar{\Phi}^a \phi^a + \mathrm{Re} \int d^4x \, d^2\theta \, f(\Phi^a) , \tag{2.77}$$

where the function f, known as the superpotential, must be at most trilinear for renormalizability:

$$f(\Phi^a) = \eta_a \Phi^a + m_{ab} \Phi^a \Phi^b + g_{abc} \phi^a \phi^b \phi^c . \tag{2.78}$$

In the case that this matter model possesses some rigid symmetry in which the ϕ^a transform according to the representation matrices $T_k{}^a{}_b$, then the superpotential f must satisfy

$$f_{,a}(\phi) \, T_k{}^a{}_b \, \Phi^b = 0 \tag{2.79}$$

where $f_{,a} = \partial f / \partial \phi^a$.

Evaluation of the interaction for the general matter model (2.77) gives

$$\mathrm{Re} \int d^4x \left(h^a f_{,a} + \chi^{\alpha a} \chi_\alpha{}^b f_{,ab} \right). \tag{2.80}$$

Taking this together with the $h^a \bar{h}^a$ term from the kinetic term, h^a can be eliminated to give a general form to the potential

$$V = \sum_a | f_{,a} |^2 \tag{2.81}$$

whose manifest positivity is an essential feature of supersymmetric theories.

CHAPTER III

Quantum Supersymmetric Gauge Theories

In order to describe a massless supersymmetric gauge theory(11), we must start from the irreducible representation of supersymmetry containing a vector V^μ and its spin $\frac{1}{2}$ partner $\lambda\alpha$, i.e. the superspin $\frac{1}{2}$ multiplet given in (2.10). The condition $\partial^\mu V_\mu = 0$, which is necessary for the irreducibility of this multiplet, indicates that in order to find an action for the multiplet in terms of unconstrained fields, it is necessary to add another supermultiplet containing the divergence of the vector field. In the final action, this extra multiplet must be eliminable by gauge transformations, so we can see that a supersymmetric gauge theory will have a whole multiplet's worth of gauge invariances.

The only other irreducible multiplet that can be used as the multiplet of gauge components is the $J_{\Xi}= 0$ chiral multiplet (2.9). The combination of the "physical" and the "gauge" multiplets may be described by a real general scalar superfield $V(x,\theta,\bar{\theta}) = V^*(x, \Theta, \bar{\Theta})$. Since the gauge variation of this superfield must itself be real, we conclude that for a chiral superfield of gauge transformation parameters Λ, in an Abelian theory we must have

$$\delta V = i(\Lambda - \Lambda^*) \quad , \qquad \bar{D}_{\dot\alpha}\Lambda = 0 \tag{3.1}$$

where the choice of the imaginary part of Λ is necessary for the vector V^μ to be a parity even object. If the components of Λ are denoted $(A+iB,\zeta,E-iF)$, then the components of the real multiplet (3.1) are $(B,\zeta\ E,F,\partial_\mu A,0,0)$, as can be worked out using the rules given in the last chapter. If the components of the superfield V are denoted $(C,\chi,H,K,V_\mu,\lambda,D)$, then the transformation (3.1) allows us to gauge away C,χ,H, K and $\partial^\mu V_\mu$. The remaining gauge invariant quantities are described by a multiplet whose lowest component is the Abelian gauge invariant spinor λ_α. In accordance with our definition of the components of a superfield given in the last chapter, we see that the gauge invariant superfield is given by

$$W_\alpha = \bar{D}^2 D_\alpha V \tag{3.2}$$

It is easy to check the gauge invariance of W_α:

$$\delta W_\alpha = i\,\bar{D}^2 D_\alpha (\Lambda - \Lambda^*) = i\bar{D}^2 D_\alpha \Lambda = -2i\partial_\mu \sigma^\mu_{\alpha\dot{\beta}} \bar{D}^{\dot{\beta}} \Lambda = 0 . \tag{3.3}$$

Due to the derivative factor $\bar{D}^2$, W_α is a chiral spinor superfield,

$$\bar{D}_{\dot{\beta}} W_\alpha = \bar{D}_{\dot{\beta}} \bar{D}^2 D_\alpha V = 0 , \tag{3.4}$$

and in addition the reality of V implies the reality of the scalar component of W_α:

$$D \equiv (D^\alpha W_\alpha)_{|\theta=\bar{\theta}=0} = (D^\alpha \bar{D}^2 D_\alpha V)_{|\theta=\bar{\theta}=0} = (\bar{D}_{\dot{\alpha}} D^2 \bar{D}^{\dot{\alpha}} V)_{|\theta=\bar{\theta}=0} = D^* . \tag{3.5}$$

Conditions (3.4) and (3.5) define an irreducible superfield, whose components are $(\lambda_\alpha, \partial_\mu V_\gamma - \partial_\nu V_\mu, D, \sigma^\mu_{\alpha\dot{\beta}} \partial_\mu \lambda^{\dot{\beta}})$. This multiplet is known as the supersymmetric field strength multiplet, since it contains the field strength $F_{\mu\nu} = \partial_\mu V_\nu - \partial_\nu V_\mu$ of the vector field. It possesses in a local expression the same content as the superspin $\frac{1}{2}$ multiplet $(\frac{1}{\Box} D, \frac{\partial\!\!\!/}{\Box}\lambda, V_\mu^T)$.

Since the superfield W_α is chiral, we may form the gauge invariant action

$$I = \mathrm{Re} \int d^4x\, d^2\theta\, W^\alpha W_\alpha . \tag{3.6}$$

Expressing this in components, we have after some rescalings the action for the vector V_μ, spinor λ and auxiliary field D:

$$I = \int d^4x \left(-\frac{1}{4} F_{\mu\nu} F^{\mu\nu} + \frac{i}{2} \bar{\lambda} \partial\!\!\!/ \lambda + \frac{1}{2} D^2 \right) . \tag{3.7}$$

If we had taken the imaginary part of the integral in (3.6) instead of the real part, we would have obtained a total divergence whose spin 1 part is the Pontryagin index $\int d^4x\, F_{\mu\nu}{}^*F^{\mu\nu}$.

In order to describe non-Abelian supersymmetric gauge theories, we must generalize the above discussion. The clue that tells us how to do this is provided by the gauge transformation of some doublet of chiral matter superfields ϕ^i covariantly coupled to the gauge superfield V. The natural generalization of the ordinary gauge transformation is

$$\phi^i \rightarrow [e^{-i\Lambda T}]^{ij}\, \phi^j \quad , \quad T = \begin{bmatrix} 0 & 1 \\ -1 & 0 \end{bmatrix} \quad . \qquad (3.8)$$

The chirality of the gauge parameter superfield Λ is essential for the consistency of (3.8). If we let $\delta V = i(2g)^{-1}(V - V^*)$, we can write the gauge invariant kinetic term for the ϕ^i as

$$I_{\text{kinetic}} = \int d^4x\, d^4\theta\, \bar{\Phi}^i [e^{2gVT}]^{ij}\, \Phi^j \quad . \qquad (3.9)$$

This may now be expanded into components using the rules given in the last chapter. Unlike the kinetic term for the gauge multiplet (3.6), however, the gauge coupled matter kinetic term depends upon all of the components of the gauge multiplet, not just the gauge invariant parts contributing to W_α. In order to simplify the component expression, it is convenient to make a non-supersymmetric gauge choice known as a Wess-Zumino gauge choice, setting the low dimension components of V to zero: $C=\chi=H=K=0$. The only remaining unfixed gauge transformation is then the ordinary Maxwell gauge transformation for the vector field V_μ . In this gauge, the action (3.9) becomes

$$I_{\text{Kinetic}} = \int d^4x \Big[-8(\nabla^\mu a^*)^i (\nabla_\mu a)^i + 4i\, \chi^{\alpha i} \sigma^\mu_{\alpha\dot\beta} (\nabla_\mu \bar\chi^{\dot\beta})^i + h^{*i} h^i + \mathrm{Re}\,(4g\, \lambda^\alpha \chi^i_\alpha T^{ij} a^{*j}) + 2g\, a^i T^{ij} a^{*j} D \Big] \qquad (3.10)$$

where $\nabla^\mu = \partial^\mu - igV^\mu T$ is the usual covariant derivative.

The generalization to non-Abelian matter couplings is now straightforward, for matter fields ϕ belonging to some gauge group representation with generator matrices T_r must now transform as

$$\phi \rightarrow e^{-i\Lambda}\phi \quad , \quad \begin{array}{l}\Lambda = \Lambda^r T_r \\ \bar{D}_{\dot\alpha}\Lambda = 0\end{array} \tag{3.11}$$

which still preserves the chirality of $\phi, \bar{D}_{\dot\alpha}\phi=0$. In the matter kinetic term, we need to cancel a factor $e^{i\Lambda^*}$ coming from ϕ and a factor $e^{-i\Lambda}$ coming from ϕ. Thus if we again write the gauge coupled kinetic action as

$$I_{\text{kinetic}} = \int d^4x\, d^4\theta\, \bar{\Phi} e^{2gV} \Phi \quad , \tag{3.12}$$

then the V^r must undergo non-linear gauge transformations such that

$$e^{2gV} \longrightarrow e^{-i\Lambda^*} e^{2gV} e^{i\Lambda} . \tag{3.13}$$

Taking the Abelian limit of this expression, we find that it agrees with (3.1).

In the non-Abelian case, there is a generalization of the field strength superfield (3.2):

$$W_\alpha = \frac{1}{2g}\bar{D}^2\left(e^{-2gV} D_\alpha e^{2gV}\right) . \tag{3.14}$$

This is still a chiral superfield: $\bar{D}_{\dot\beta}W_\alpha=0$, and it transforms covariantly in the adjoint representation of the gauge group:

$$\begin{aligned} W_\alpha \rightarrow & \frac{1}{2g}\bar{D}^2\left[e^{-i\Lambda}e^{-2gV}e^{i\Lambda^*} D_\alpha\left(e^{-i\Lambda^*}e^{2gV}e^{i\Lambda}\right)\right] \\ &= \frac{1}{2g} e^{-i\Lambda}\left(\bar{D}^2 e^{-2gV} D_\alpha e^{2gV}\right)e^{i\Lambda} + \frac{1}{2g} e^{-i\Lambda} \cancel{\bar{D}^2 D_\alpha} e^{i\Lambda} \\ &= e^{-i\Lambda} W_\alpha e^{i\Lambda} \end{aligned} \tag{3.15}$$

In order to expand the field strength multiplet into its components, we take repeated fermionic covariant derivatives as before. In this case, however, we want to preserve the gauge covariance as well as the supersymmetric covariance, Thus, we need a gauge covariantized fermionic derivative ∇_α. As one can see from the calculation (3.15), such a derivative is provided by

$$\nabla_\alpha \equiv e^{-2gV} D_\alpha e^{2gV} \equiv D_\alpha - g A_\alpha^r T_r \quad , \tag{3.16}$$

where the D_α is taken to act upon everything to the right. The superfield A_α^r is known as the superspace connection. Note that $D_{\dot\alpha}$ is still covariant without a connection since Λ is chiral. Expanding W_α using ∇_α, we find the set of gauge-covariant components

$$W_\alpha \leftrightarrow \left(\lambda_\alpha, F_{\mu\nu} = \partial_\mu V_\nu - \partial_\nu V_\mu + ig[V_\mu, V_\nu], D, \sigma^\mu_{\alpha\dot\alpha} \nabla_\mu \bar\lambda^{\dot\alpha}\right), \tag{3.17}$$

where ∇_μ is the ordinary spacetime covariant derivative. Every term in (3.17) is actually a contraction of a set of gauge component fields into the Hermitean generators $T_r^\cdot$, thus $D = D^r T_r$ is a Hermitean matrix. Note that in terms of the gauge covariantized ∇_α and $\nabla_{\dot\alpha} = D_{\dot\alpha}$, the algebra of covariant derivatives becomes

$$\{\nabla_\alpha, \bar\nabla_{\dot\alpha}\} = -2i\sigma^\mu_{\alpha\dot\alpha} \nabla_\mu \tag{3.18}$$

$$\{\nabla_\alpha, \nabla_\beta\} = \{\bar\nabla_{\dot\alpha}, \bar\nabla_{\dot\beta}\} = 0 \tag{3.19}$$

$$[\nabla_\mu, \bar\nabla_{\dot\beta}] = i\sigma_\mu{}^\beta{}_{\dot\beta} W_\beta \tag{3.20}$$

where W_α appears as a superspace field strength.

The action for the supersymmetric Yang-Mills theory may be written down as before, since W_α is chiral as before:

$$I_{S.Y.M.} = \mathrm{Re}\int d^4x\, d^2\theta\, \mathrm{tr}\, W^\alpha W_\alpha \quad , \tag{3.21}$$

where the trace is over the gauge group indices. Expanding (3.21) into components, we find after some rescalings

$$I_{S.Y.M.} = \mathrm{tr}\int d^4x \left(-\tfrac{1}{4}F_{\mu\nu}F^{\mu\nu} - \tfrac{i}{2}\bar{\lambda}\not{\nabla}\lambda + \tfrac{1}{2}D^2\right). \tag{3.22}$$

In writing down the form (3.22) it is important that we expanded W^α into components using ∇_α. If we had performed a non-gauge-covariant expansion with D_α, we would have obtained the form (3.22) only after going into a Wess-Zumino gauge. As in the Abelian case, if we had taken the imaginary part of the integral in (3.21), we would have obtained a total divergence, with spin one part now the Pontryagin index for the Yang-Mills field.

Combining the supersymmetric Yang-Mills theory with a general gauge coupled matter sector, the general supersymmetric matter plus Yang-Mills theory can be written.

$$I = \int d^4x\, d^2\theta\, \bar{\Phi}\, e^{2gV}\phi + \mathrm{Re}\int d^4x\, d^2\theta \left(\mathrm{tr}\, W^\alpha W_\alpha + f(\phi)\right), \tag{3.23}$$

where $f(\phi)$ is a group invariant superpotential in the sense of equation (2.79). If the gauge group contains some Abelian U(1) factors, then it is possible to add also the Fayet-Iliopoulos terms

$$I_{FI} = \int d^4x\, d^4\theta\, C^{\hat{k}}\, V_{\hat{k}} \tag{3.24}$$

where $C^{\hat{k}}$ are a set of coefficients, with $\hat{k}$ running over the number of Abelian U(1) factors. These terms play an important role in the study of spontaneous gauge symmetry and supersymmetry breaking.

So far, we have described supersymmetric field theories at the level of classical physics. In order to develop such theories into full quantum field theories, one could of course just explicitly write a given theory out in components and then proceed via the standard methods of perturbative quantum field theory. However, many of the surprising features of supersymmetry have to do with its control over the higher order quantum corrections and renormalizations, and in order to discuss these, we need to keep the supersymmetry manifest at the quantum level. For this reason, we need to be able to quantize directly in the superfield formalism. It is not within the purview of this article to give an exhaustive treatment of superfield quantization techniques, for which we refer the reader to the series of articles by Grisaru, Siegel and Rocek [12, 13]. Here we shall indicate only the main features.

We return to the Wess-Zumino model for a single chiral multiplet with a general renormalizable superpotential and add a source term in order to develop propagators and vertices:

$$I_{WZ} = \int d^4x\, d^4\theta\, \bar{\Phi}\Phi - \mathrm{Re}\int d^4x\, d^2\theta\,(m\Phi^2 + \tfrac{1}{3}g\Phi^3) + 2\,\mathrm{Re}\int d^4x\, d^2\theta\, J\Phi \tag{3.25}$$

where the source J is also a chiral superfield.

The rules for developing propagators and vertices in this theory are derived directly from those for ordinary field theory. If one uses the functional integral formalism, an integral over a superfield is defined to be the integral over its component fields.

Following the standard procedure, we are led to invert the free part of (3.25). Since in (3.25) we have chiral as well as full superspace integrals, it is convenient to rewrite the free part as a full superspace integral, so that we may do fermionic integration by parts. This can be done if we allow non-local expressions to appear, and since the propagator we are deriving will itself be non-local, there is no reason not to do so at this stage. The starting action (3.25) is of course local regardless of how we re-write it. In order to re-write (3.25), we selectfor example the source term and re-write it as

$$\int d^4x\, d^2\Theta\, J\phi = -\tfrac{1}{4}\int d^4x\, d^4\Theta\, J\, \frac{\bar{D}^2}{\Box}\,\phi\ . \tag{3.26}$$

This equality can be checked by re-writing the right hand side again as a chiral integral with an extra $-\frac{1}{4}D^2$:

$$\begin{aligned}-\tfrac{1}{4}\int d^4x\, d^4\Theta\, J\,\frac{\bar{D}^2}{\Box}\phi &= \tfrac{1}{16}\int d^4x\, d^2\Theta\, D^2\left(J\,\frac{\bar{D}^2}{\Box}\phi\right)\\ &= \tfrac{1}{16}\int d^4x\, d^2\Theta\, J\,\frac{D^2\bar{D}^2}{\Box}\phi\\ &= \int d^4x\, d^2\Theta\, J\phi\ .\end{aligned} \tag{3.27}$$

The relative factor of $-\frac{1}{4}$ that enters in the first line is a conventional one introduced to make $\int d^2\Theta\, \Theta^2 = 1$. Doing the same for the other bilinear chiral integrals, we can write the free part of the action as

$$I_{(2)} = \int d^4x\, d^4\Theta\, \left[\bar{\Phi}\phi - \mathrm{Re}\left(m\phi\left(-\frac{\bar{D}^2}{4\Box}\right)\phi - 2J\left(-\frac{\bar{D}^2}{4\Box}\right)\phi\right)\right]. \tag{3.28}$$

Performing the Euclidean functional integral, we obtain

$$Z_0(J) = \int [d\phi][d\bar{\phi}]\, e^{I_{(2)}} = e^{-\frac{1}{2}\int d^4x\, d^4\Theta\, B^T A^{-1} B} \tag{3.29}$$

where

$$B = \begin{pmatrix} -\frac{1}{4}\left(\frac{\bar{D}^2}{\Box}\right)J \\ -\frac{1}{4}\left(\frac{D^2}{\Box}\right)\bar{J}\end{pmatrix} \tag{3.30}$$

and the propagators are given by

$$A^{-1} = \begin{pmatrix} -\frac{1}{4}\frac{m\bar{D}^2}{\Box - m^2} & 1 + \frac{m^2\bar{D}^2D^2}{16\Box(\Box - m^2)} \\ 1 + \frac{m^2 D^2\bar{D}^2}{16\Box(\Box - m^2)} & -\frac{1}{4}\frac{mD^2}{\Box - m^2}\end{pmatrix} \tag{3.31}$$

The vertices are given by the usual functional formula

$$Z(J) = e^{I_{int}\left(\frac{\delta}{\delta J}\right)} Z_0(J) . \tag{3.32}$$

In order to evaluate this, we need to know how to take variational derivatives of superfields. This requires the notion of a superspace δ-function. The rules (2.22), 2.23) for Berezin integration in superspace show that the product $(\Theta_1-\Theta_2)^2(\bar\Theta_1-\bar\Theta_2)^2$ works like a δ function for the fermionic coordinates. Multiplying this by $\delta^4(x_1 - x_2)$ we get the full superspace δ- function

$$\delta^8(z_{12}) = \delta^4(x_1 - x_2)(\Theta_1 - \Theta_2)^2(\bar\Theta_1 - \bar\Theta_2)^2 . \tag{3.33}$$

The variational derivative of one superfield with respect to itself is then

$$\frac{\delta V(z_1)}{\delta V(z_2)} = \delta^8(z_{12}) \tag{3.34}$$

When we take variational derivatives of chiral superfields, it is necessary to account for the fact that an integral over functions of superfields of the same chirality is taken only over the chiral superspace, $\int d^4x d^2\Theta$. This requires removing two $\bar\Theta$'s from the δ-function (3.33). The required expression can be written covariantly as

$$\frac{\delta J(z_1)}{\delta J(z_2)} = -\frac{1}{4}\bar D_1^2\, \delta^8(z_{12}). \tag{3.35}$$

Note that the derivatives can equivalently be considered as acting on z_2.

The vertices can now be written as

$$I_{int}\left(\frac{\delta}{\delta J}\right) J(z_1)J(z_2)J(z_3) = \frac{-g}{6}\left[\int d^4x_4\, d^2\Theta_4 \left(\delta/\delta J(z_4)\right)^3\right] J(z_1)J(z_2)J(z_3) \tag{3.36}$$

$$= -g\int d^8z_4\, \delta^8(z_{14})\left[-\tfrac{1}{4}\bar D_2^2\delta^8(z_{24})\right]\left[-\tfrac{1}{4}\bar D_3^2\delta^8(z_{34})\right] ,$$

so that two of the three propagators entering the vertex will be acted upon by factors of $-\frac{1}{4}D^2$.

For general scalar superfields, the Feynman rules are also straightforward to develop: the action must be expanded into a series of interactions, after supplying a gauge-breaking term and adding Faddeev-Popov fhost terms with chiral ghost superfields C,C':

$$I_{GB} = -\frac{1}{16}\, tr \int d^4x\, d^4\Theta\, (D^2 V)(\bar{D}^2 V) \qquad (3.37)$$

$$I_{FP} = tr \int d^4x\, d^4\Theta\, (\bar{C}' - C')\, L_{gV/2} \left[(\bar{C}+C) + (\coth L_{gV/2})(C-\bar{C}) \right] \qquad (3.38)$$

where $L_X[Y] = [X, Y]$. For more details, see reference (12). For our present purposes, it is sufficient to quote the resulting Feynman rules from reference (12):

(i) Massless propagators are $\pm(1/p^2)\delta^4(\Theta_{12})$, with + for the $\phi\bar{\phi}$ propagators of chiral superfields (both physical and ghost chiral multiplets) and - for real superfields (gauge multiplets). In addition, the massive chiral multiplet has $+(p^2 + m^2)^{-1}\delta^4(\Theta_{12})$ for the $\phi\bar{\phi}$ propagator and a $\phi\phi$ propagator $m\{p^2(p^2 + m^2)\}^{-1}\frac{1}{4}D^2(p,\Theta_1) \times \delta^4(\Theta_{12})$.

(ii) Vertices are read directly from I, with an extra($-\frac{1}{4}\bar{D}^2$ for each antichiral one), but omitting one factor of $-\frac{1}{4}\bar{D}^2$ for converting $d^2\Theta$ into $d^4\Theta$ (a factor of $-\frac{1}{4}D^2$ for antichiral superfields).

(iii) There are the usual combinatorial factors, and -1 for each ghost loop.

(iv) The amputated one-particle-irreducible graphs should have each amputated external line multiplied by the appropriate superfield. For each external chiral superfield, a factor of $-\frac{1}{4}D^2$ should be omitted from a vertex (a factor $-\frac{1}{4}D^2$ for an antichiral external superfield).

(v) The diagram is to be integrated over

$$\int \left[\prod_{\text{loops}} \frac{d^4k}{(2\pi)^4}\right]\left[\prod_{\text{ext.}} \frac{d^4p}{(2\pi)^4}\right](2\pi)^4\, \delta^4\Big(\sum_{\text{ext}} p\Big)\int \prod_{\text{vertices}} d^4\Theta$$

The superspace Feynman rules have an immediate important consequence. Since all vertices are to be integrated over $\int d^4\Theta$, all contributions to the Green's functions must be written as full superspace integrals. Moreover, one can integrate by parts to move fermionic covariant derivatives on any propagator off from the corresponding $\delta^4(\Theta_{mn})$. One can then perform the Berezin integration $\int d^4\Theta_n$ on one connecting vertex, thus setting $\Theta_n = \Theta_m$. This can be performed within any loop until only one fermionic integration is left, with a factor $D\ldots D\delta^4(\Theta_{n1})_{\Theta_n=\Theta_1}$ The only such factor that is nonvanishing is $D^2\bar{D}^2\delta^4(\Theta_{n1})_{\Theta_n=\Theta_1}=16$ Repeating this procedure for all loops in a diagram, the result is an expression with a single remaining $\int d^4\Theta$ integral: all contributions to the effective action have the form

$$\int d^4x_1 \cdots d^4x_n\; F_1(x_1,\Theta)\cdots F_n(x_n,\Theta)\; G(x_1\cdots x_n) \tag{3.39}$$

where the F_n are products of superfields and their derivatives and G is translationally invariant, so that (3.39) is invariant under the supersymmetry transformations

$$\Theta^\alpha \rightarrow \Theta^\alpha + \epsilon^\alpha\,,\quad \bar{\Theta}^{\dot\alpha} \rightarrow \bar{\Theta}^{\dot\alpha} + \bar{\epsilon}^{\dot\alpha}\,,\quad x_n^\mu \rightarrow x_n^\mu - i(\epsilon^\alpha\bar{\Theta}^{\dot\beta} + \bar{\epsilon}^{\dot\beta}\Theta^\alpha)\,\sigma^\mu_{\alpha\dot\beta}\,. \tag{3.40}$$

In calculating the quantum corrections to Green's functions, it is also necessary to regularize the diagrams. While this is still a subject of some controversy, the technique of regularization by dimensional reduction[14] works in practice, at least at low orders in the loop expansion. In this technique, all the algebraic manipulations of the covariant derivatives are to be carried out first in 4-dimensional space, while the momentum integration is carried out subsequently in d-dimensional space.

Since the overall infinite parts of the contributions to the effective action are local expressions, i.e. involve only one overall $\int d^4 x$ integration, their structure is constrained in an important way by the result (3.39). Divergences due to infinite subdiagrams are non-local in structure, but are removed by the appropriate renormalizations at the corresponding lower loop level. Counterterms that cannot be rewritten locally as full superspace integrals according to (3.39) cannot arise, even if they are supersymmetric in structure.

Thus we see that in the Wess-Zumino model, mass and interaction terms like (2.74) and (2.76) cannot occur as counterterms, i.e. these terms in the classical action are not renormalized. For this reason, the result (3.39) is known as the non-renormalization theorem of N=1 supersymmetry. The only independent renormalization in the Wess-Zumino model is the wavefunction renormalization coming from the allowed kinetic counterterm (2.72).

The behaviour of a quantum field theory at high momenta is determined by the trajectories of the coupling constants as given by the solution to the renormalization group equations. In supersymmetric theories, there are important relations between the various renormalization constants and the associated β- and γ- functions. For example, as we have just seen, the interaction term (2.76) in the Wess-Zumino model cannot appear as a counterterm.

This imposes the constraint

$$Z_\phi^{3/2} Z_g = 1 \tag{3.41}$$

between the wavefunction and coupling constant renormalization constants, where

$$\phi_0 = Z_\phi^{1/2}\ \phi \tag{3.42}$$

$$g_0 = Z_g\ g \tag{3.43}$$

give the unrenormalized (ϕ_0, g_0) in terms of the renormalized (ϕ, g) field and coupling constant. In the massless Wess-Zumino model, this has the consequence that if there were a non-trivial fixed point g^∞ with $\beta(g^\infty) = 0$, then the γ function would have to vanish at that value as well, and all the fields would have their canonical dimensions, so conformal invariance would be preserved[15]. In such a model without gauge fields, it has been shown that the preservation of conformal invariance is only possible in a free theory [16], so $g^\infty = 0$. Whether such conclusions can be drawn is a gauge theory with $\beta(g) \neq 0$ is unknown.

Further restrictions on the quantum corrections to a theory are given by the constraints of supersymmetry, and gauge invariance in a gauge theory, upon the quantum supermultiplet containing the stress tensor. The stress tensor $T_{\mu\nu}$ is a conserved symmetric tensor. From our analysis of representation in Chapters 1 and 2, it can be seen that there is only one irreducible multiplet of fields containing a single spin 2 object like $T_{\mu\nu}$. This irreducible multiplet contains also a spin 3/2 object (Q_μ^{α}, $\bar{Q}_\mu^{\dot\alpha}$) and an axial spin 1 object J_μ^5. Irreducibility requires the following conditions on the components of the multiplet:

$$\partial^\mu T_{\mu\nu} = \partial^\mu Q_\mu^{\alpha} = 0 \tag{3.44}$$

$$T_\mu{}^\mu = \sigma^\mu_{\alpha\dot\alpha} Q_\mu^{\alpha} = \partial^\mu J_\mu^5 = 0 . \tag{3.45}$$

In the case of a conformally invariant theory, supersymmetry thus requires the existence of a conserved vector-spinor current Q_μ^{α} which is " σ-traceless" and a conserved axial vector current J_μ^5. The vector-spinor current is none other than the super-symmetry current which we have already met in Chapters 1 and 2. The axial current is conserved due to the existence of a symmetry which rotates spinors by γ_5 transformations and rotates scalars into pseudoscalars. The supermultiplet containing these three currents was found by Ferrara and Zumino (17), who gave the multiplet the name "supercurrent".

In superfields, we may describe the conformal super-current multiplet by an axial vector superfield $V_\mu(x,\theta,\theta) = \sigma_\mu{}^{\alpha\dot\alpha} V_{\alpha\dot\alpha}$, satisfying the condition

$$D^\alpha V_{\alpha\dot\alpha} = \bar{D}^{\dot\alpha} V_{\alpha\dot\alpha} = 0 \tag{3.46}$$

It may easily be checked using the rules given in Chapter 2 that this superfield contains the components

$$J_\mu^5 \equiv \sigma_\mu^{\alpha\dot\alpha} V_{\alpha\dot\alpha}\big|_{\theta=\bar\theta=0} \tag{3.47}$$

$$Q_\mu^{\alpha} \equiv \sigma_\mu^{\beta\dot\beta} Q_{\beta\dot\beta}{}^{\alpha} \equiv \sigma_\mu^{\beta\dot\beta} \left(D_\beta V^{\alpha}{}_{\dot\beta} \right)\big|_{\theta=\bar\theta=0} \tag{3.48}$$

$$\begin{aligned} T_{\mu\nu} &= \sigma_\mu^{\alpha\dot\alpha} \sigma_\nu^{\beta\dot\beta} T_{\alpha\beta\dot\alpha\dot\beta} \\ T_{\alpha\beta\dot\alpha\dot\beta} &= \tfrac{1}{2} \left([\bar{D}_{\dot\alpha}, D_\beta] V_{\alpha\dot\beta} + [\bar{D}_{\dot\beta}, D_\alpha] V_{\beta\dot\alpha} \right)\big|_{\theta=\bar\theta=0} \end{aligned} \tag{3.49}$$

satisfying the conditions (3.44, 3.45).

In non-conformally invariant theories, the conditions (3.45) must be relaxed so as to allow a trace for the stress tensor. In superfields, this means relaxing the condition (3.46), but not totally, for (3.46) is also responsible for the conservation conditions (3.44) which we must keep. Since we are trying to add lower spin projectors to the components $T_{\mu\nu}$, $Q_\mu{}^\alpha$ and $J_\mu{}^5$, we need to combine the conformal supercurrent with a multiplet containing lower spins. There are two basic ones available: the chiral multiplet (2.9) and the linear multiplet (2.10).

If we use a chiral superfield C to relax (3.46), we write

$$\bar{D}^{\dot\alpha} V_{\alpha\dot\alpha} = D_\alpha C \quad , \qquad \bar{D}_{\dot\alpha} C = 0 . \tag{3.50}$$

Expanding the superfield in components, we find that $\sigma^\mu_{\alpha\dot\alpha}\, Q_\mu{}^\alpha$, $T_\mu{}^\mu$ and $\partial^\mu J_\mu{}^5$ are no longer zero, but are given by the spinor and by the real and imaginary parts of the highest dimension complex scalar in C, respectively. In addition, there are now the real and imaginary parts of the lowest dimension scalar in C, which are not related to conserved currents. The multiplet described by (2.50) is an example of a reducible supermultiplet that is not given by a single superfield, since the lowest dimension components of C are not found in $V_{\alpha\dot\alpha}$ (although the gradients of these components are). This is not an unusual feature in supersymmetry. Irreducible representations can always be described by single superfields subject to constraints, while reducible representations may be described by single superfields with weakened constraints (such as the general scalar (2.42) with no constraints) or by several superfields linked by constraints (like (3.50)). The sense in which (3.50) describes a reducible representation is the same as that in which the non-conformal $T_{\mu\nu}$ is a reducible Poincare representation. The trace may obviously be extracted by a local operation, but the traceless part may not, since the conserved traceless part is given by the non-local expression

$$\left(\eta_{\mu\alpha} - \frac{\partial_\mu \partial_\alpha}{\Box}\right)\left(\eta_{\nu\beta} - \frac{\partial_\nu \partial_\beta}{\Box}\right)\left(T^{\alpha\beta} - \tfrac{1}{3}\eta^{\alpha\beta}\, T_\rho{}^\rho\right) \tag{3.51}$$

Similarly, (3.50) is not fully locally reducible, but it is non-locally reducible.

If we use the real linear multiplet to relax (3.46), we write

$$\bar{D}^{\dot{\alpha}} V_{\alpha\dot{\alpha}} = \bar{D}^2 D_\alpha L \tag{3.52}$$

where in order for the real superfield L to be linear, $D^2 L = \bar{D}^2 L = 0$. These latter conditions on L do not actually need to be applied in order to have a viable supercurrent, since the operator $\bar{D}^2 D_\alpha$ projects out only the linear multiplet ($J_\Xi = \frac{1}{2}$) part of a real general scalar superfield, so L may be taken to be a real general scalar without affecting the currents in $V_{\alpha\dot{\alpha}}$. The main difference between the form (3.50) and the form (3.52) is that in (3.52) the axial current $J_\mu{}^5$ remains conserved, due to the reality of L and the identity

$$D^\alpha \bar{D}^2 D_\alpha L = \bar{D}_{\dot{\alpha}} D^2 \bar{D}^{\dot{\alpha}} L \,. \tag{3.53}$$

Thus the tracelessness of $T_{\mu\nu}$ and $Q_\mu{}^\alpha$ is always broken in non-conformal theories, but the axial vector current can in some cases be preserved.

In order to make the above discussion more concrete, we give some examples of the two types of supercurrent at the level of classical physics. In the free massless Wess-Zumino model, the "improved" supercurrent is

$$V^{cl.}_{\alpha\dot{\alpha}} = D_\alpha \phi \bar{D}_{\dot{\alpha}} \bar{\phi} + 2i(\phi \partial_{\alpha\dot{\alpha}} \bar{\phi} - \partial_{\alpha\dot{\alpha}} \phi \bar{\phi}) \,. \tag{3.54}$$

This form of the supercurrent contains the "improved" form of the massless scalar stress tensor, and satisfies (3.46) using the field equations. "Non-improved" forms of the supercurrent also exist, both of type (3.50) and of type (3.52), but for a more definite example of a non-conformal supercurrent we consider the massive Wess-Zumino model. In that case, the free field equations are

$$\bar{D}^2 \Phi = m \Phi \tag{3.55}$$

and the condition (3.46) becomes

$$\bar{D}^{\dot{\alpha}} V^{cl.}_{\alpha\dot{\alpha}} = \frac{m}{2} D_\alpha (\Phi^2) , \tag{3.56}$$

which is clearly of the chiral type (3.50).

An example of the second type of non-conformal supercurrent (3.52) is the massless model with the same spin content as the Wess-Zumino model, but with the pseudoscalar represented by a gauge antisymmetric tensor(18). The gauge invariant fields of this non-conformal model are given by a real linear superfield G, $D^2 G = \bar{D}^2 G = 0$. As we saw in (2.10), this multiplet contains a conserved vector. In the present application, this is the field strength for the antisymmetric tensor,

$$G_\mu = \partial^\nu A_{\mu\nu} . \tag{3.57}$$

The other two fields in the multiplet are the physical scalar and spinor; there are no auxiliary fields in this model. The supercurrent for the model is

$$V^{cl.}_{\alpha\dot{\alpha}} = D_\alpha G \, \bar{D}_{\dot{\alpha}} G \tag{3.58}$$

Using the field equations

$$\partial^{\alpha\dot{\beta}} \bar{D}_{\dot{\beta}} G = 0 \tag{3.59}$$

we find that instead of (3.46), we have

$$\bar{D}^{\dot{\alpha}} V^{cl.}_{\alpha\dot{\alpha}} = -\tfrac{1}{2}\bar{D}^2 D_\alpha (G^2), \tag{3.60}$$

which is of the type 3.52 with $L = -\frac{1}{2}G^2$, a real general scalar superfield.

The supercurrent for a gauge theory is restricted by the additional requirement of gauge invariance. In the supersymmetric Yang-Mills theory (3.21), the only choice for a classical level supercurrent is

$$V^{cl.}_{\alpha\dot{\alpha}} = \mathrm{tr}\, W_\alpha \bar{W}_{\dot{\alpha}} \tag{3.61}$$

Using the field equations

$$\nabla^\alpha W_\alpha = \bar{\nabla}_{\dot{\alpha}} \bar{W}^{\dot{\alpha}} = 0, \tag{3.62}$$

we obtain

$$\bar{D}^{\dot{\alpha}} V^{cl.}_{\alpha\dot{\alpha}} = -\mathrm{tr}\, W_\alpha \bar{\nabla}^{\dot{\alpha}} \bar{W}_{\dot{\alpha}} \tag{3.63}$$

as is required of a conformally invariant theory. Note that when acting upon the antichiral $\bar{W}_{\dot{\alpha}}$, $\bar{\nabla}_{\dot{\alpha}} = \bar{D}_{\dot{\alpha}} - g(A^r_\alpha)^* T_r$.

In the quantum theories derived from classically conformally invariant actions, the scale dependence of the renormalized coupling constants upon the renormalization point causes a violation of conformal invariance, the well-known conformal anomaly. In supersymmetric theories, the conformal anomaly is given by a relaxation of (3.46) for the quantum supercurrent. The particular form this relaxtion takes may be determined by details of the quantization procedure, in particular by the various subtraction schemes employed in renormalization. The particularities of a given subtraction scheme are frequently implied by the regularization method chosen. Thus, in the Wess-Zumino model if one uses Pauli-Villars regularization[19], conformal invariance is broken by the regularization together with σ-tracelessness of $Q_\mu{}^\alpha$ and conservation of the current $J_\mu{}^5$

for the axial U(1) symmetry of the model. In this case, the anomaly equation takes the form (3.50)

$$\bar{D}^{\dot{\alpha}} V_{\alpha\dot{\alpha}}^{qu.} = c\, D_\alpha \left(\Phi \bar{D}^2 \bar{\Phi} \right), \tag{3.64}$$

where c is the anomaly coefficient.

On the other hand, if one uses higher derivative regularization, the kinetic term (2.72) is supplemented by the regulator term

$$I_{reg.} = -\frac{1}{m^2} \int d^4x \, d^4\theta \, \bar{\Phi} \Box \Phi \, . \tag{3.65}$$

The regulator term(3.65) is invariant under the axial U(1) transformations just like the kinetic term (2.72), from which it differs only by the extra, axially inert, d'Alembertian. Thus, higher derivative regularization preserves the axial U(1) invariance and the natural subtraction scheme which is suggested by it does too. Thus with higher derivative regularization, there can be no $J_\mu{}^5$ anomaly, and the anomaly equation takes the form (3.52)

$$\bar{D}^{\dot{\alpha}} V_{\alpha\dot{\alpha}}^{qu.} = c\, \bar{D}^2 D_\alpha \left(\Phi \bar{\Phi} \right). \tag{3.66}$$

In both (3.64) and (3.66), the anomaly multiplet is given by the kinetic part of the superspace Lagrangian multiplet, but in forms differing by a total derivative, corresponding to the two ways of writing the kinetic action given in (2.72). Thus the $J_\mu{}^5$ anomaly that seems to be present in (3.64) is in fact illusory, for it can be eliminated by a supersymmetric redefinition of the quantum supercurrent $V_{\alpha\dot{\alpha}}{}^{qu}$ similar to that involved in going from a "non-improved" to an "improved" form of the supercurrent in the classical theory.

In the supersymmetric Yang-Mills theory, there is no arbitrariness in the non-conformal part of the supercurrent such as there is in the Wess-Zumino model, for we have the additional constraint of gauge invariance. Again, the anomaly multiplet is the multiplet containing the kinetic superspace Lagrangian, which must be written as the chiral superspace integral (3.21) if manifest gauge invariance is to be preserved.

Thus, the only possible form for the quantum supercurrent anomaly is the form (3.50),

$$\bar{D}^{\dot{\alpha}} V_{\alpha\dot{\alpha}}^{qu} = d\, D_\alpha \left(\mathrm{tr}\, W^\beta W_\beta\right). \tag{3.67}$$

In this case, there is no way consistent with gauge invariance to redefine $V_{\alpha\dot{\alpha}}^{qu}$ so as to eliminate the $J_\mu{}^5$ anomaly. The attempt to derive the anomaly by using higher derivative regularization and subtraction does not allow us to circumvent this conclusion, for the requirements of gauge invariance now force us to add higher derivative vertices as well as kinetic-type terms, and these prevent the higher derivative technique from regularizing at the one loop level of perturbation theory. Only in a gauge theory which turns out to be finite at the one loop level can higher derivative regularization be used.

CHAPTER IV

The Finiteness of N = 4 Supersymmetric Yang-Mills Theory

Field theories based upon the extended supersymmetry algebras given in the first chapter of this article have even stronger constraints upon their quantum corrections than those provided by N = 1 supersymmetry. The most dramatic instance of this control is provided by the maximally extended supersymmetric theory consistent with having only spins 1 and less, the N = 4 supersymmetric Yang-Mills theory.

In components, the N = 4 theory contains the following propagating fields: one spin 1 field A_μ for an arbitrary gauge group, four spin ½ fields $\lambda_{\alpha i}$ in the adjoint representation of the gauge group and in the $\underset{\sim}{4}$ of SU(4) plus six scalar fields satisfying the self-conjugacy condition $\phi_{ij} = \frac{1}{2}\varepsilon_{ijkl}\bar{\phi}^{kl}$, also in the adjoint representation of the gauge group and in the real $\underset{\sim}{6}$ of SU(4). In terms of these fields, the action [20] is the manifestly SU(4) invariant expression.

$$I_{YM4} = \frac{1}{g^2}\int d^4x \, \mathrm{tr}\Big\{-\frac{1}{4}F_{\mu\nu}F^{\mu\nu} + \frac{1}{2}\nabla^\mu\bar{\Phi}^{ij}\nabla_\mu\Phi_{ij} - i\lambda_i\sigma_\mu\nabla^\mu\bar{\lambda}^i - \frac{i}{2}\bar{\lambda}^i_{\dot\beta}[\bar{\lambda}^{\dot\beta j},\Phi_{ij}] - \frac{i}{2}\lambda^\alpha{}_i[\lambda_{\alpha j},\bar{\Phi}^{ij}] + \frac{1}{4}[\Phi_{ij},\Phi_{kl}][\bar{\Phi}^{ij},\bar{\Phi}^{kl}]\Big\} \tag{4.1}$$

where i,j,k,l: 1→4, and where the trace is over the suppressed gauge group indices. In addition, the action (4.1) is conformally invariant, since it contains Yang-Mills fields minimally coupled to massless scalars and spinors with Yukawa and quartic scalar interactions.

The action (4.1) is also supersymmetric, although it lacks the auxiliary fields that we have had before in linear realizations of supersymmetry. In eliminating auxiliary fields from the action and supersymmetry transformations of an interacting theory, the transformations become nonlinear. This is also the case with the transformations for (4.1). In addition, as we have seen, the supersymmetry transformations without auxiliary fields form a closed algebra only subject to the equations of motion. Unfortunately, no set of auxiliary fields is known which linearizes the four Lorentz covariant supersymmetry transformations of (4.1), and there are strong arguments indicating that no such set of auxiliary fields exists[21]. Thus if we wish to quantize the action

(4.1) and retain control over some of the supersymmetry at the quantum level, we are forced to use a formalism which makes manifest only some part the supersymmetry. This requires a corresponding reduction in control over the other internal symmetries of (4.1), such as SU(4).

Probably the most practical way to handle the quantization of (4.1) is to introduce auxiliary fields for N = 1 supersymmetry and quantize it using the superspace Feynman rules for N=1 supersymmetry of Chapter 3. The fields then organise themselves into four superfields, with a manifest SU(3) x U(1) invariance: a singlet general scalar superfield V containing the Yang-Mills vector field and one of the four spinors, plus a triplet of chiral superfields ϕ_i containing the three remaining spinors and the six scalars. The N=1 superspace action [22,12] for the theory is the manifestly SU(3) x U(1) invariant expression

$$I_{YM4} = \mathrm{tr}\left\{ \frac{1}{32g^2} \mathrm{Re}\int d^4x\, d^2\theta\, W^\alpha W_\alpha + \int d^4x\, d^4\theta\, e^{-gV}\bar{\Phi}^i e^{gV}\Phi_i \right.$$
$$\left. + \frac{ig}{3} \mathrm{Re}\int d^4x\, d^2\theta\, \epsilon^{ijk}\Phi_i[\Phi_j,\Phi_k] \right\}, \qquad (4.2)$$

where now i,j,k : $1 \to 3$. The different appearance of the gauge superfield's coupling to the chiral superfields in (4.2) from the expression (3.23) is due to the fact that the adjoint representation coupling is here effected by contracting all fields into the group generators and taking the trace of group invariant products.

The prospect of dramatic ultraviolet cancellations in the N = 4 theory is apparent from a number of points of view. The theory's spin ½ and spin 0 fields saturate the bound for asymptotic freedom. Thus, the one-loop gauge coupling β-function vanishes. In addition, at the one-loop order the chiral multiplet does not require a wavefunction renormalization. Moreover, these same properties are maintained to at least the three-loop level, as has been found by explicit calculation[23-25].

The prospect of these cancellations continuing to all orders is strongly suggested by the consequences that any non-vanishing β-function would have for the trace and chiral anomalies of the theory[26]. As can be seen from (4.1), the theory has an SU(4) symmetry but not U(4) due to the self-conjugacy of the real $\underset{\sim}{6}$ of scalars, which thus cannot be given an extra U(1) charge. This self-conjugacy is a reflection of the self-conjugacy of the theory as a whole, as is characteristic of maximal supersymmetric theories, which already contain all the required PCT conjugate states within the basic supermultiplet of states. Thus, for example, the U(1) part of the SU(3) x U(1) manifest symmetry in (4.2) is really part of the SU(4) symmetry which is present although not fully manifest in (4.2). This U(1) part of SU(3) x U(1) is also the symmetry associated to the axial current $J_\mu{}^5$ in the supercurrent multiplet (3.47 - 3.49). The requirement of gauge invariance of the anomalous part of the supercurrent then links the anomaly in this axial U(1) current to the trace anomaly, as we saw in Chapter 3. The spin 1 part of the trace anomaly is determined by the gauge coupling constant β- function. Thus we conclude that if there is a non-vanishing β-function in this theory, then SU(4) invariance must be broken due to the associated axial anomaly. This seems most unlikely, because semi-simple groups like SU(4) have been found to be anomaly free in all other contexts.

Another reason to anticipate cancellation of the β-function to all orders in the N=4 Yang-Mills theory is provided by the non-renormalization theorem of N=1 supersymmetry, which was discussed in Chapter 3. This argument [24,27] uses the relation between the wavefunction renormalization constant for the chiral fields Z_ϕ and the coupling constant renormalization Z_g, arising from the non-renormalization of the trilinear chiral superfield coupling in (4.2). If, in addition, at least the non-axial SO(4) internal symmetry of the theory is preserved by the quantum corrections, then the coupling constant of the trilinear chiral superfield coupling must remain equal to the gauge coupling constant. This can be seen clearly since some Yukawa couplings to spinors come from the gauge covariantization of the chiral superfield kinetic term in (4.2), while others come from the trilinear chiral superfield interaction. Moreover, the wavefunction renormalization for the spinor in the gauge general scalar superfield V must equal that for the three spinors in the chiral superfields ϕ_i if SO(4) is to be preserved. If, in addition, one uses the background field method in N=1 superspace [12,13], the wave-function renormalization of the gauge superfield V is restricted by

$$Z_V^{1/2} Z_g = 1 , \tag{4.3}$$

as is required to keep the product gV unrenormalized and to have counterterms which are manifestly background gauge invariant. Since $Z_V = Z_\phi$ by SO(4) symmetry, as we have already noted, the only possibility consistent with both (4.3) and (3.41) is

$$Z_g = Z_\phi = Z_V = 1 . \tag{4.4}$$

Thus, in order to have a non-vanishing β-function there would have to be an anomaly in the SO(4) symmetry of the N=4 theory. This would seem to be impossible since there are no axial generators in this SO(4).

In order to definitively establish that the β-function for the N=4 Yang-Mills theory vanishes to all orders, it is necessary to use a formalism that keeps some of the extended supersymmetry manifestly and linearly realized in the Feynman rules. Although it does not appear to be possible to linearly realize all four of the theory's supersymmetries, it is possible to construct a Lorentz covariant, manifestly N=2 supersymmetric formulation of the theory[28]. In this formalism, the theory's finiteness follows using the extention to N = 2 supersymmetry of the background field method and the non-renormalization theorem derived for N=1 supersymmetry[13]. The full details of this N=2 background field method analysis will not be given here. For further details, we refer the reader to a forthcoming paper [29].

An alternative approach, which abandons manifest Lorentz covariance but keeps the SU(4) symmetry manifest using a light-cone gauge formalism, has recently been proposed[30]. In the light-cone-gauge approach, effectively half of the full supersymmetry is linearly realized, the same amount as in the Lorentz covariant N=2 superfields, and with the same degree of control over the infinites. Unfortunately, neither the N=2 superspace Feynman rules nor the light-cone-gauge formalism are particularly well suited to practical calculations; the former because the covariant derivative algebra in Feynman diagrams is complicated, the latter because it is complicated and also because Lorentz covariance is not manifest. The most manageable formalism for practical calculations of Feynman diagrams seems to be the N=1 superspace formulation with the action (4.2).

When decomposed into representations of N=2 supersymmetry, the propagating fields of the N=4 Yang-Mills theory fall into just two N=2 massless multiplets, each carrying the adjoint representation of the gauge group. These are the N=2 Yang-Mills multiplet (31,32) (A_μ, λ_i ϕ , ϕ^*) where now i=1,2 only, and the N=2 matter representation, the "hypermultiplet" (ψ_i, $L^{(ij)} = \varepsilon^{ik}\varepsilon^{jm}L_{km}$, $S = S^*$). In the N=2 supersymmetric formalism the manifest internal symmetry, apart from the Lorentz group, is SU(2) x U(1).

The superspace for N=2 supersymmetry has as fermionic coordinates $\Theta_\alpha{}^i$, where i is now an SU(2) doublet index. The Abelian N=2 gauge theory was described in N=2 superspace by Mezincescu (33). The Abelian action is given in terms of a prepotential $V_{(ij)}$(dimension-2) carrying the real $\underset{\sim}{3}$ representation of SU(2),

$$I_{YM2} = \mathrm{Re}\int d^4\theta\, W^2 \tag{4.5}$$

where the $d^4\theta$ integral is a chiral integral for N=2 supersymmetry and W is the chiral N=2 field strength superfield

$$\bar{D}_{\dot{\alpha}i} W = 0 \tag{4.6}$$

$$\bar{D}_{\dot{\alpha}i} = -\frac{\partial}{\partial\bar{\Theta}^{\dot{\alpha}i}} - i\,\Theta^{\alpha}{}_i\,\sigma^{\mu}_{\alpha\dot{\alpha}}\,\partial_\mu\ . \tag{4.7}$$

In terms of the prepotential V_{ij}, the field strength superfield W is given by

$$W = \bar{D}^4 D^{ij} V_{ij} \tag{4.8}$$

where $D^{ij} = D^{\alpha i} D_\alpha{}^i$. W is also expressible in terms of a connection superfield $\bar{A}^{\dot{\alpha} i}$

$$W = \bar{D}_{\dot{\alpha} i} \bar{A}^{\dot{\alpha} i} \tag{4.9}$$

where

$$A_\alpha{}^i = D^3_\alpha{}^i \bar{D}^{jk} V_{jk} - \frac{1}{2} D_\alpha{}^i \{ D^j{}_k , \bar{D}^{kl} \} V_{jk} \tag{4.10}$$

$$D^3_\alpha{}^i = \frac{1}{3} D_{\alpha j} D^{ji} , \tag{4.11}$$

and where SU(2) indices are raised and lowered with ε^{ij}. The construction of the vertices required to extend the Abelian action (4.5) to a full non-Abelian Yang-Mills theory will not be given here; the reader is referred to the report of the joint work of the author together with P.S. Howe and P.K. Townsend in reference (34).

The N=2 hypermultiplet matter sector has been discussed in superspace formulations with central charges in references (35). Unfortunately, these formulations are not sufficient for the establishment of N=2 supersymmetric Feynman rules, because in an interacting theory, the central charge transformations become non-linear. Thus, we require a formulation of the massless hypermultiplet in ordinary N=2 superfields without a central charge. This was given in reference (28); the formulation requires two prepotentials $\rho^\alpha{}_i$ (dimension -3/2) and $X^{(ijkl)}$ (dimension-1). Defining the quantities

$$L^{ij} = \mathrm{Re} \left(3 \bar{D}^{ij} D^3_\alpha{}^k \rho^\alpha{}_k + \bar{D}_k{}^{(i} D^{j)k} D_\alpha{}^l \rho^\alpha{}_l \right) \tag{4.12}$$

$$L^{ijkl} = \frac{2}{5} D^{(ij} \bar{D}^{kl)} \left(D_\alpha{}^m \rho^\alpha{}_m + \bar{D}^{\dot{\alpha}}{}_m \bar{\rho}_{\dot{\alpha}}{}^m \right) \tag{4.13}$$

$$V = D^{ij}\bar{D}^{kl} X_{ijkl} , \tag{4.14}$$

the action for the hypermultiplet can be written as the full N=2 superspace integral

$$I_{hm2} = \mathrm{Re} \int d^4x\, d^8\theta \left(\rho^{\alpha}{}_i D_{\alpha j} L^{ij} + X_{ijkl} L^{ijkl} \right) . \tag{4.15}$$

For the full coupling of this multiplet to the N=2 Yang-Mills multiplet, we refer the reader to references (28, 34, 29)

The finiteness of the N=4 Yang-Mills theory follows from the non-renormalization theorem for extended supersymmetry[13]. This requires the use of the background field method of reference (13) together with the construction of an appropriate set of extended superspace Feynman rules, such as that derived from the N=2 formalism sketched above for the linearized case and given in further detail in references (34, 29).

The extended supersymmetry non-renormalization theorem requires that, for a theory which can be quantized in superfields of N- extended supersymmetry, the quantum corrections to the effective action be expressed as a single full superspace integral $\int d^{4N}\theta$. Together with the requirement of locality, this restricts the allowable counterterms as we saw for N=1 superfields in Chapter 3. Moreover, there are restrictions on what combinations of superfields can appear in the superspace integrand of a quantum correction to the effective action.

For matter multiplets, such as the hypermultiplet given above, only certain components of the prepotentials actually appear in the action of the theory when written as an ordinary spacetime integral. In the free hypermultiplet, these components are those contained in the superfields L^{ij}, L^{ijkl} and V of equations (4.12 - 4.14). Since these superfields are constrained,

$$D_\alpha^{(i} L^{jk)} = D_{\alpha l} L^{ijkl}$$
$$D_{\alpha\beta} V = [D_\alpha{}^i , \bar{D}_{\dot{\alpha} i}] V = 0 , \tag{4.16}$$

it is necessary to introduce the unconstrained prepotentials

$\rho^{\alpha}{}_{i}$ and X_{ijke} for quantization. However, the fact that only certain components of the prepotentials appear in the action indicates that there is a gauge invariance, called a pregauge invariance, since it involves transformations of the prepotential. The non-renormalization theorem requires that only the pregauge invariant superfields appear in the integrand of the full superspace integral. Thus the hypermultiplet kinetic action (4.15) is ruled out as a counterterm, since the prepotentials $\rho^{\alpha}{}_{i}$ and X_{ijkl} appear in it "naked" i.e. without the derivatives necessary to form the allowed combinations L^{ij}, L^{ijkl} and V.

For gauge superfields, a similar restriction applies, but only at the order of 2 loops and higher. This is because the process of fixing the pregauge invariances requires special Faddeev-Popov ghost diagrams that only appear at one loop (13,29). At the 2-loop order and higher, the counterterms must be written as full superspace integrals of expressions written in terms of the superspace corrections. In the N=2 theory, the linearized connection $A_{\alpha}{}^{i}$ is given in (4.10); for the full Yang-Mills connection, see references (34, 29). Again, we see by inspection that the action (4.5) fails the requirements of the theorem, for it is not possible to rewrite (4.5) as a full superspace connection involving $A_{\alpha}{}^{i}$ and $\bar{A}_{\dot{\alpha}i}$.

At the 1 loop level of the quantum theory, a separate check needs to be made to determine whether the action (4.5) occurs as a counterterm. In the case of the N=4 Yang-Mills theory, an explicit calculation reveals no 1 loop gauge action counterterm, so the above theorem establishes finiteness to all orders. Note that in the background field method, the gauge dependent wavefunction renormalization of the gauge prepotential V_{ij} is tied to the coupling constant renormalization, so all the Green's functions are finite.

For gauge multiplets, the non-renormalization theorem

works like the Adler-Bardeen theorem(36) for ordinary axial anomalies, only here for <u>trace</u> anomalies. It is possible for a theory to have a non-vanishing contribution to the β-function at one loop, but not at higher orders. An example of this situation is provided by the N=2 Yang-Mills theory, which is protected by the non-renormalization theorem at 2 loops and higher, but which does have an infinite counterterm proportional to the action at the 1 loop order. This fact is borne out by explicit calculation of the 1,2 and 3 loop contributions to the β-function(37).

A final question about the quantization of the N=4 Yang-Mills theory concerns the problem of regularization. Since the theory turns out to be finite, in principal no regularization is needed. However, it is worth noting that the finiteness of this theory at the one-loop order allows a higher derivative regularization to be used, as we noted at the end of Chapter 3. It can be checked that the theory remains finite at the 1-loop order with the higher derivative regulator included in the action. At higher loop orders, the regularization comes into effect. Higher derivative regularization has the property of preserving all the axial internal symmetries of a theory. In particular, this includes the U(1) part of SU(2) x U(1), whose anomaly is linked to the trace anomaly by supersymmetry and gauge invariance, as we discussed in Chapter 3. Having for the N=4 theory a regularization procedure which respects this symmetry, we are guaranteed that there will be no $J_\mu{}^5$ anomaly, and hence no trace anomaly either. Thus we can complete the earlier arguments of reference (26) by providing a regularization which guarantees the absence of a $J_\mu 5$ anomaly, and so ruling out the trace anomaly too.

References

1. S. Coleman and J. Mandula, Phys.Rev. 159 (1967), 1251.
2. Y.A. Gol'fand and E.P. Likhtman, J.E.T.P.Lett. 13 (1971), 323.
3. D.V. Volkov and V.P. Akulov, Phys.Lett. 46B (1973), 109.
4. J. Wess and B. Zumino, Nucl.Phys. B70 (1974), 39; Nucl.Phys. B78 (1974), 1.
5. R. Haag, J.T. Lopuszanski and M.F. Sohnius, Nucl.Phys. B88 (1975), 257.
6. A. Salam and J. Strathdee, Nucl.Phys. B76 (1974), 4778; Nucl.Phys. B84 (1975), 127.
7. S. Ferrara, in Proceedings of the 9th International Conference on General Relativity and Gravitation, Jena, G.D.R. (198); CERN preprint TH-2957 (1980).
8. S. Ferrara, C.A. Savoy and B. Zumino, Phys.Lett. 100B (1981), 393.
9. B.N. Valuyev, JINR Report P2-11638 (1978) (in Russian).
10. F.A. Berezin, The Method of Second Quantization (Academic Press, New York, 1966).
11. S. Ferrara and B. Zumino, Nucl.Phys. B79 (1974), 413.
    A. Salam and J. Strathdee, Phys.Lett. 51B (1974), 353.
    B. de Wit and D. Freedman, Phys.Rev. D12 (1975), 2286.
12. M.T. Grisaru, W. Siegel and M. Roček, Nucl.Phys. B159 (1979), 429; M.T. Grisaru and W. Siegel, Nucl.Phys. B187 (1981), 149.
13. M.T. Grisaru and W. Siegel, Nucl.Phys. B201 (1982), 292.
14. W. Siegel, Phys.Lett. 84B (1979), 193.
    W. Siegel, P.K. Townsend and P. van Nieuwenhuizen, in Superspace and Supergravity, eds. S.W. Hawking and M. Rocek (Cambridge University Press, 1981). p.165.
15. S. Ferrara, J. Iliopoulos and B. Zumino, Nucl.Phys. B77 (1974), 413.
16. K. Pohlmeyer, Comm.Math.Phys. 27 (1972), 247.
17. S. Ferrara and B. Zumino, Nucl.Phys. B87 (1975), 207.
18. V. Ogievelsky and V. Polubarinov, Sov.J. Nucl.Phys. 4 (1966), 216.
    M. Kalb and P. Ramond, Phys.Rev. D9 (1974), 2273.
19. M.T.Grisaru in Recent Developments in Gravitation, eds.M. Levy and S. Deser (Plenum, New York, 1979) p. 577.
20. F. Gliozzi, J. Scherk and D. Olive, Nucl.Phys. B133 (1978), 253.
    L. Brink, J. Schwarz and J. Scherk, Nucl.Phys. B121 (1977), 77.
21. M. Roček and W. Siegel, Phys.Lett. 105B (1981), 275.
    V. Rivelles and J.G. Taylor, King's College preprint, London (May 1982).
22. P. Fayet, Nucl.Phys. B149 (1979), 137.

23. O.V. Tarasov and A.A. Vladimirov, Phys.Lett. 96B (1980), 94

24. M.T. Grisaru, M. Rocek and W. Siegel, Phys.Rev.Lett. 45 (1980), 1063.

25. W. Caswell and D. Zanon, Phys.Lett. 100B (1980), 152.

26. M.F. Sohnius and P.C. West, Phys.Lett. 100B (1981), 245.
S. Ferrara and B. Zumino, unpublished.

27. K.S. Stelle in Quantum Structure of Space and Time, eds. M.J. Duff and C.J. Isham (Cambridge University Press, 1982), p. 337.

28. P.S. Howe, K.S. Stelle and P.K. Townsend, CERN preprint TH- 3271 (1982) (Nucl.Phys. B, in press).
K.S. Stelle, in Proc. 21st Int.Conf. on High Energy Physics, eds. P. Petiau and M. Porneuf, Journal de Physique, Colloque C3 supp. au No 12 (1982), p. 326.

29. P.S. Howe, K.S. Stelle and P.K. Townsend, to appear.

30. S. Mandelstam, in Proc. 21st Int.Conf.on High Energy Physics, eds. P. Petiau and M. Porneuf, Journal de Physique, Colloque C3 Supp. au No 12 (1982), p. 331.
L. Brink, O. Lindgren and B.E.W. Nilsson, Goteborg preprint UTTG-1-82 (November 1982).

31. See the first paper of reference 11.

32. P. Fayet, Nucl.Phys. B113 (1976), 135.

33. L. Mezincescu, JINR Report P2-12572 (1979).

34. K.S. Stelle in Proc.Conf. on "Group Theoretical Methods in Physics", Zvenigorod, USSR (November 1982).

35. M.F. Sohnius, Nucl.Phys. B138 (1978), 109.
A. Galperin, E.A. Ivanov and V.I. Ogievetsky, JETP Lett. 33 (1981), 176; JINR Report E2-81-482 (1981).

36. S.L. Adler and W.A. Bardeen, Phys.Rev. 182 (1969), 182.

37. L.V. Avdeev and O.V. Tarasov, Phys.Lett. 112B (1982), 356.

SUPERGRAVITY AND GRAND UNIFICATIONS

E. Sokatchev

Joint Institute for Nuclear Research, Dubna USSR

I. Introduction

I.1. The Idea of Unification in Physics

For a very long time one of the main objectives of physics has been to find a unifying, common explanation for various, apparently different phenomena. One can recall such celebrated examples as the theory of gravitation of Newton, who found a common reason for celestial and terrestrial movements, or the greatest achievement of the 19th century physics: the electromagnetic field theory of Faraday and Maxwell, who unified electricity, magnetism and light in a beautiful scheme. More recently, Einstein in his Special Relativity theory unified the concepts of Space and Time, and later he put together geometry and dynamics to create the General Relativity theory, one of the most impressive achievements of modern physics. Nowadays, a lot of effort is concentrated on attempts to unify the three fundamental interactions in elementary particle physics: electromagnetic, weak and strong ones. The first success in this direction is the electroweak theory of Glashow, Salam and Weinberg and hopes exist of including also the strong interactions within the framework of a Grand Unified Theory. However, the fourth and most universal force - the gravitational one -seems to be ignored by elementary particle physicists. Why? Well, until recently one could present two main objections against gravity. First, it is too weak (its typical energy scale is given by the Planck mass $M_p \sim 10^{19}$ GeV), so it should not contribute significantly to processes at present or even foreseeable energies. Second, any reasonable theory of elementary particles is supposed to be quantized and gravity is very difficult to quantize. The problem is that the coupling constant $\varkappa$ in gravity has non-zero dimension (of length), so in every order of perturbation theory there are infinities different in form from the preceding ones. Thus, quantum gravity can either be finite or non-renormalizable. All known evidence seemed to indicate the latter.

Today, however, both those counter-arguments are not valid any more. Indeed, human fantasy is now reaching energies close to M_p.

For instance, in the widely discussed GUT models the unification of strong and electroweak interactions is expected to take place at energies of the order 10^{15} GeV; also, near the (hypothetical) black holes, as well as in the early Universe (Big Bang) huge gravitational fields are supposed to cause non-trivial quantum effects. On the other hand, the newly developed supergravity provides hopes that quantum gravity can be in fact a finite theory. Therefore, it is not justified to forget gravity any more and one may say that unification with it is really on the agenda.

There is another, related question which arises in the current research in QFT. We unify various kinds of matter particles, we also find a common origin for various forces. However, we never try to mix these two basic constituents of our world. Indeed, there is a sharp distinction between fundamental forces, the carriers of which are believed to be bosonic gauge fields, and the matter particles which are described by fermionic non-gauge fields. So, the question is: is not there a symmetry between them which really unifies all kinds of fields in QFT?

I.2. Supersymmetry as a Basis for Unification with Gravity

Here we come to the cardinal question of symmetries and their role in particle physics. The dominant philosophy nowadays is that the four fundamental forces are associated with gauge symmetries. Indeed, the electromagnetic field is the gauge field of local U(1) symmetry, the gauge group SU(2)xU(1) is the basis for the electroweak models, the gluon (strong interaction) fields are the gauge fields of local SU(3). Finally, the gravitational interaction is believed to be carried by the metric field of general coordinate transformations(also interpreted as the gauge theory of Poincare group). Moreover, all recent attempts at constructing grand unified theories are based on gauge symmetries (e.g., SU(5), SO(10), etc.). One can say that the choice of gauge group is the starting point for building unified field theories: one first chooses the symmetry, then develops the corresponding gauge theory and finally derives phenomenological consequences.

However, trying to unify gravity with the other interactions one has problems with the symmetries involved. The point is that gravity is based on a space-time gauge symmetry whereas the remaining interactions have to do with internal gauge symmetries. Therefore, a new symmetry combining space-time and internal space is required. Again, some years ago such symmetries were believed not to exist ("no-go" theorems). Today, the situation has changed with the discovery of supersymmetry/1/.

Since an extensive discussion of supersymmetry is contained in the lectures of K.S.Stelle in these Proceedings, I shall only recall the basic anticommutation relation of an -extended supersymmetry algebra:

$$\{ Q^i_\alpha , \bar{Q}_{\dot{\alpha} j} \} = 2i\, \delta^i_j \, (\sigma^\mu)_{\alpha\dot{\alpha}} P_\mu \, , \quad i = 1, \dots, N \, . \tag{1}$$

We see how space-time translations P_μ are mixed with internal symmetry rotations (the index i of Q belongs to some representation of some internal group $G \subseteq U(N)$).

I.3. Local Supersymmetry and Gravity

There are strong hints for gravity in eq. (1). Indeed, if supersymmetry is to be the gauge symmetry of a future interaction theory, then local supersymmetry transformations $(Q, \bar{Q})$ will lead to local translations (P_μ). The latter are general coordinate transformations, i.e. the gauge basis for gravity. This argument can be made more precise. Namely, the constant parameters ε_α of global (say, $N=1$) supersymmetry become local $(\varepsilon_\alpha(x))$ in gauge supersymmetry. To make the theory gauge invariant one needs a spin-vector gauge field, $\psi_{\alpha\mu}(x)$. The free equation of motion for such a field is the Rarita-Schwinger equation/3/

$$\varepsilon_{\mu\nu\lambda\rho} (\gamma^\nu \gamma_5 \partial^\lambda \psi^\rho)_\alpha = 0 \tag{2}$$

admitting gauge invariance

$$\delta \psi_{\alpha\mu} = \partial_\mu \varepsilon_\alpha \, . \tag{3}$$

In an interacting theory a source should appear in the r.h.s. of eq. (2)

$$\varepsilon_{\mu\nu\lambda\rho} (\gamma^\nu \gamma_5 \partial^\lambda \psi^\rho)_\alpha = \varkappa \, J_{\alpha\mu} \, . \tag{4}$$

Owing to the gauge invariance (3) it must be conserved

$$\partial^\mu J_{\alpha\mu} = 0 \, . \tag{5}$$

The obvious candidate for $J_{\alpha\mu}$ is the Noether current of supersymmetry related to the supersymmetry generators

$$Q_\alpha = \int d^3x \, J_{\alpha 0} \quad . \tag{6}$$

Comparing eqs. (4) and (6) one sees that the coupling constant $\varkappa$ in eq. (4)must have dimension of length, i.e. just as the gravitational constant. So, the conclusion is: local supersymmetry leads to gravity. In the extended case ($N>1$) eq.(1) suggests also an internal symmetry which may serve as a GUT group in such a theory.

The above argument can also be reversed. Suppose one would like to supersymmetrize gravity. A good reason for that could be the hope that supersymmetry, which is known to cause drastic cancelations of infinities in QFT, may help in the renormalizability problem of gravity. So, the first thing to do is to place the graviton (massless particle with helicity 2) in a supermultiplet. In the simplest case N=1 the two possible multiplets are (2,5/2) and (2,3/2). The first one can be discarded because it is believed that no consistent field theory for helicity 5/2 particles exists/2/. The second one contains a massless helicity 3/2 particle (called "gravitino"). Now, such a particle is described by the Rarita-Schwinger field $\psi_{\alpha\mu}$ with its gauge invariance (3). The local spinor parameter $\varepsilon_\alpha(x)$ must obviously be the parameter of localized supersymmetry. So, now the argument says that gravity plus supersymmetry lead to local supersymmetry.

I.4. Supergravity: Successes and Problems

Following similar ideas people have been able to construct the theory of gauge supersymmetry, or supergravity/3/. Soon the first big success came: N=1 pure supergravity (i.e. without matter) was shown to be one-and two-loop finite/4/. It was an impressive achievement compared to the disastrous renormalizability situation in pure gravity. However, it was also realized that adding (super) matter coupled to supergravity destroys the finiteness of the theory. Indeed, the infinity cancelation "miracle" is only due to the specific arrangement of interactions and coupling constants within a supersymmetry multiplet. If a matter supermultiplet with its own independent coupling constant is added, the miracle does not work. The way out is to use larger multiplets of a larger symmetry including all the fields of interest. Thus one turns to extended supergravity as the possible unification scheme of all interactions. One has to admit that not too much has been achieved in this direction yet. It is true that the one- and two-loop finiteness persists in extended supergravity too, but what happens at three loops and higher is not clear. Some "miracles" of the type apparently observed in N=4 supersymmetric Yang-Mills theory[5] may also help here. Further, even the largest, N=8 supergravity scheme does not seem (at first sight) to have enough room for all the known elementary particles. However, some sophisticated preon--type mechanism may provide the key to this problem. In a word, supergravity is a very complicated and not very well understood by now subject. A lot of work is still needed to be able to tell whether it has something to do with reality.

In introductory lectures like these it is hard to choose among the various topics in supergravity, many of which are frighteningly technical. I have decided to follow a "constructive" line. It reflects the situation now, people being more experienced in constructing supergravity theories rather than interpreting them. So I shall describe different methods for building various supergravity models, at the same time pointing out some of their basic properties. Particular attention will be paid to a geometric fromulation of N=1 supergravity and to the most interesting and at the same time most complicated model of N=8 supergravity. I shall try to skip as many technical details as possible, thus sometimes deliberately oversimplifying things, just to give an overall idea of what supergravity is about. My aim is to attract possible future "believers" in supergravity, and not to scare them right from the beginning with the awful lot of calculations that supergravitationists have to do.

II. Simple (N=1) Supergravity

II. 1. N=1 Supergravity in Terms of Fields

The Noether coupling method. The most straightforward method of constructing supergravity theories is to take the fields describing the various particles in the gravitational supermultiplet and try to arrange a supersymmetric interaction among them order by order in the coupling constant $\varkappa$. Thus, in the N=1 case one simply deals with the vierbein field $e_\mu^{\ a}$ (the vierbein formulation of gravity is more convenient here, in the presence of spinors) for the graviton and the Rarita-Schwinger field $\psi_{\alpha\mu}$ for the gravitino. The first step is to write down together the Einstein Lagrangian for $e_\mu^{\ a}$ and the general coordinate transformation covariantized Rarita-Schwinger Lagrangian

$$L = -\frac{1}{2\varkappa^2} e R(e,\omega) - \frac{1}{2} \varepsilon^{\mu\nu\rho\sigma} \bar{\psi}_\mu \gamma_5 \gamma_\nu D_\rho \psi_\sigma \quad . \tag{7}$$

Here $e = \det(e_\mu^a)$, ω_μ^{ab} is the Lorentz connection (a function of e_μ^a) and

$$D_\rho \psi_\sigma = \left(\partial_\rho + \tfrac{1}{2}\omega_\rho{}^{ab}\sigma_{ab}\right)\psi_\sigma \tag{8}$$

is the covariant derivative. The next step is to couple the gravitino field to the Noether current of global supersymmetry (see eq. (4)). Then some corrections to this Noether coupling and to the local supersymmetry transformations (a generalization of eq. (3)) are established order by order in $\varkappa$. The result /3/ of all that effort ammounts to the replacement of the Lorentz connection

$$\begin{aligned}\omega_{\mu ab}(e,\psi) &= \omega_{\mu ab}(e) + \\ &\quad + \frac{\varkappa^2}{4}\left(\bar\psi_\mu \gamma_a \psi_b - \bar\psi_\mu \gamma_b \psi_a + \bar\psi_a \gamma_\mu \psi_b\right)\end{aligned} \tag{9}$$

in eq. (7). The Lagrangian obtained is invariant under the following local supersymmetry transformations

$$\delta e_\mu^a = \tfrac{1}{2}\bar\varepsilon \gamma^a \psi_\mu , \quad \delta\psi_\mu = D_\mu(\omega(e,\psi))\,\varepsilon \ . \tag{10}$$

Auxiliary fields. The peculiarity of the transformations (10) is that their algebra closes only on-shell (i.e. using the equations of motion). In other words, supersymmetry is not manifest off-shell. There is a simple reason for that. One of the main properties of supersymmetry is that it mixes equal numbers of Fermi and Bose degrees of freedom. Off-shell the vierbein field e_μ^a has 16 components, 6 of which can be gauged away by local Lorentz transformations and further 4 by general coordinate transformations, so one has altogether 6 bosonic degrees of freedom. On the other hand, of the 16 components of the gravitino field $\psi_{\alpha\mu}$ only 4 can be gauged away by local supersymmetry transformations, so 12 fermionic degrees of freedom are left off-shell. To restore the equality one needs 6 additional bosonic fields which must vanish on-shell. These are the so called auxiliary fields

$$L_{aux} = -\frac{e}{3}\left(S^2 + P^2 - A_\mu^2\right) . \tag{11}$$

This is the minimal set of auxiliary fields for N=1 supergravity /6/, others exist as well. There are two main reasons why they are so important. First, without them supersymmetry is not manifest which makes quantization, investigation of inifinities, etc. much more difficult. Second, one of the known mechanisms of spontaneous breaking of supersymmetry (a major point for phenomenological applications) has to do with giving some auxiliary fields non-zero vacuum expectation values.

As simple as the auxiliary piece of Lagrangian (11) may look, it took about two years to discover it. In general, finding auxiliary fields is a major problem in supergravity. At present they are known only for N=1 and N=2 supergravities.

II. 2. Geometry of Superspace

General relativity and geometry. The Einstein theory of gravity is essentially a geometrical theory and this is what makes it so beautiful. It is natural to think that supergravity is also based on some kind of geometry. Here we shall see how N=1 supergravity can be formulated starting from simple geometric ideas.

Let us first recall some basic facts about general relativity viewed as a geometric as well as a gauge field theory. If one were to construct it now, one could write down the following five-step program:

a) choose a manifold. This is the four dimensional space-time

$$\mathbb{R}^4 = \{x^\mu\} ;$$

b) introduce a symmetry (gauge) group. This is the group of general coordinate transformations (GCT) in $\mathbb{R}^4$

$$x'^\mu = x'^\mu(x)$$

together with a local Lorentz group (if the vierbein formalism is to be used);

c) introduce a gauge field. It can be the metric field defined by the invariant interval

$$ds^2 = g_{\mu\nu}(x)\, dx^\mu dx^\nu$$

or the vierbein field $e_\mu{}^a$

$$g_{\mu\nu} = e_\mu{}^a e_{a\nu} .$$

The reason why we call $g_{\mu\nu}$ a gauge field is that its non-flat part $h_{\mu\nu}(x) = g_{\mu\nu}(x) - \eta_{\mu\nu}$ is subject to GCT with infinitesimal parameters

$$\delta h_{\mu\nu} = \partial_\mu \lambda_\nu + \partial_\nu \lambda_\mu + h_{\mu\rho}\partial_\nu \lambda^\rho + h_{\nu\rho}\partial_\mu \lambda^\rho$$

which look like, e.g., the gauge transformations of the electromagnetic field $\delta A_\mu = \partial_\mu \lambda$.

d) construct gauge covariant objects. The metric $g_{\mu\nu}$ is gauge dependent in the sense that 4 of its 10 degrees of freedom can be gauged away. The information about the structure of space-time which does not depend on the choice of coordinate frame (or gauge) is contained in the gauge covariant object, the Riemann tensor

$$R_{\mu\nu\lambda\rho}(g)$$

which is a function of $g_{\mu\nu}$.

Steps a)-d) outline the geometric framework for general relativity. But we still have to specify the geometry, i.e. to make a particular choice of $R_{\mu\nu\lambda\rho}$. Here comes the last, dynamical step of our program:

e) formulate a variational principle. The Einstein action is

$$S = \frac{1}{2\varkappa^2} \int d^4x \sqrt{-g}\, R(g)$$

and its variation with respect to $g_{\mu\nu}$ produces the equation of motion

$$R_{\mu\nu} - \frac{1}{2} g_{\mu\nu} R = 0$$

which determines the geometry dynamically.

In supergravity we are going to follow the same five-step procedure [7].

A) Superspace Manifold

From the lectures of K.S.Stelle we learned that the supersymmetry algebra can be realized as transformations in a superspace involving Grassmann coordinates:

$$\mathbb{R}^{4,4} = \{(x^\mu, \theta^\alpha, \bar\theta^{\dot\alpha})\} \;;$$

$$\begin{cases} \delta x^\mu = i\theta\sigma^\mu\bar\varepsilon - i\varepsilon\sigma^\mu\bar\theta, \\ \delta\theta^\alpha = \varepsilon^\alpha, \\ \delta\bar\theta^{\dot\alpha} = \bar\varepsilon^{\dot\alpha}. \end{cases} \tag{12}$$

It can be called a real superspace because the bosonic variable x^μ is real and the fermionic ones θ^α, $\bar\theta^{\dot\alpha}$ form a Majorana spinor. It is possible to consider the geometry of curved $\mathbb{R}^{4,4}$ as the basis for supergravity [8] but it turns out that there are too many superfluous degrees of freedom there which have to be eliminated by constraints.

A more adequate framework is provided by the so called complex (or chiral) superspace

$$\mathbb{C}^{4,2} = \{(z^\mu, \theta^\alpha)\} . \tag{13}$$

Here z^μ is a complex four-vector and θ^α is a complex two-component spinor. Again, the supersymmetry transformations are easily realized in $\mathbb{C}^{4,2}$:

$$\begin{cases} \delta z^\mu = 2i\theta\sigma^\mu\bar{\varepsilon} , \\ \delta\theta^\alpha = \varepsilon^\alpha . \end{cases} \tag{13'}$$

There is also an alternative (conjugate) parametrization of $\mathbb{C}^{4,2}$: $(\bar{z}^\mu, \bar{\theta}^{\dot{\alpha}})$.

$\mathbb{C}^{4,2}$ is most suitable for describing chiral superfields. Indeed, the chirality condition $\bar{D}_{\dot{\alpha}}\varphi = 0$ in a certain basis in $\mathbb{R}^{4,4}$ becomes $\frac{\partial}{\partial\bar{\theta}^{\dot{\alpha}}}\varphi = 0$.

The solution to it is just $\varphi(z^\mu, \theta^\alpha)$, i.e. a function on $\mathbb{C}^{4,2}$. The chiral representations are the fundamental ones in supersymmetry and one would like to preserve them in the curved case. Thus, $\mathbb{C}^{4,2}$ turns out to be a better framework for supergravity.

B) Supergravity Group

This is chosen to be the group of general analytic coordinate transformations in $\mathbb{C}^{4,2}$ (a generalization of eq. (13'))

$$\begin{cases} \delta z^\mu = \lambda^\mu(z,\theta) , \\ \delta\theta^\alpha = \lambda^\alpha(z,\theta) . \end{cases} \tag{14}$$

The superfunctions-parameters λ are functions of z and θ but not of their conjugates (analytic, or chiral). They contain a number of ordinary functions-parameters in their decomposition in powers of θ :

$$\begin{aligned} \lambda^\mu(z,\theta) &= a^\mu(z) \quad + \text{pure gauge} \\ \lambda^\alpha(z,\theta) &= \varepsilon^\alpha(z) + \theta_\beta\,\omega^{(\beta\alpha)}(z) + \\ &+ \theta^\alpha[a(z) + ib(z)] + \theta^\beta\theta_\beta\,\eta^\alpha(z) . \end{aligned} \tag{15}$$

The parameter a^μ in λ^μ is the one of general coordinate transformations in x-space. The remaining parameters in λ^μ correspond to pure gauge degrees of freedom. The parameter ε^α in λ^α is the one of local supersymmetry, $\omega^{\alpha\beta}$- of local Lorentz. The other three, a , b and η^α are the parameters of Weyl, local U(1) and local conformal supersymmetry; they are unwanted in physical (Einstein) supergravity. Therefore, a smaller subgroup should be considered. It is obtained by imposing the condition that <u>the supervolume of $\mathbb{C}^{4,2}$ should be preserved</u>. The supervolume element of $\mathbb{C}^{4,2}$ transforms as follows

$$(d^4z\, d^2\theta)' = d^4z\, d^2\theta \,.\, \mathrm{Ber}\left(\frac{\partial(z',\theta')}{\partial(z,\theta)}\right) \,.$$

Here Ber denotes the Berezinian (superdeterminant) of the transformations. So, the subgroup is defined by

$$\mathrm{Ber}\left(\frac{\partial(z',\theta')}{\partial(z,\theta)}\right) = 1 \tag{16}$$

or, infinitesimally,

$$\frac{\partial}{\partial z^\mu}\lambda^\mu - \frac{\partial}{\partial\theta^\alpha}\lambda^\alpha = 0$$

which eliminates the unwanted parameters.

Finally, another subgroup is also relevant:

$$\left|\mathrm{Ber}\left(\frac{\partial(z',\theta')}{\partial(z,\theta)}\right)\right| = 1 \,. \tag{17}$$

It allows for additional global U(1) transformations compared to the group (16).

C) Supergravity Gauge Superfield

So far we considered the complex superspace $\mathbb{C}^{4,2}$. The real one, $\mathbb{R}^{4,4}$, is embedded as a hypersurface

$$\mathbb{R}^{4,4} : \{(x^\mu = \tfrac{1}{2}(z^\mu + \bar z^\mu),\ \theta^\alpha,\ \bar\theta^{\dot\alpha})\} \,, \tag{18.a}$$

$$Y^\mu(x,\theta,\bar\theta) = \frac{1}{2i}(z^\mu - \bar z^\mu) \,. \tag{18.b}$$

Eq. (18.b) defines the hypersurface, it also introduces the vector superfield Y^μ which is going to be the main gauge (and dynamical) object in the theory. So, the geometric trick is the following: one takes a large superspace $(\mathbb{C}^{4,2})$, embeds in it the physical one $(\mathbb{R}^{4,4})$, thus making some coordinates $(Im\, z^\mu)$ physical fields.

The gauge group (14), (16) defined in $\mathbb{C}^{4,2}$ is realized non-linearly in $\mathbb{R}^{4,4}$ and on Y^μ:

$$\delta x^\mu = \frac{1}{2}[\lambda^\mu(z,\theta) + \bar{\lambda}^\mu(\bar{z},\bar{\theta})] \ ,$$

$$\delta\theta^\alpha = \lambda^\alpha(z,\theta) \ , \ \delta\bar{\theta}^{\dot{\alpha}} = \bar{\lambda}^{\dot{\alpha}}(\bar{z},\bar{\theta}) \ ; \qquad (19)$$

$$\delta Y^\mu(x,\theta,\bar{\theta}) = \frac{1}{2i}[\lambda^\mu(z,\theta) - \bar{\lambda}^\mu(\bar{z},\bar{\theta})] +$$

+ coordinate translation terms .

Here $z^\mu = x^\mu + iY^\mu(x,\theta,\bar{\theta})$ is not an independent coordinate any more.

The transformations (19) are gauge transformations for the superfield Y^μ. To see this and to find out its physical content let us consider its θ-decomposition.

$$Y^\mu(x,\theta,\bar{\theta}) = \theta\sigma_a\bar{\theta}\, e^{a\mu}(x) + \bar{\theta}\bar{\theta}.\theta^\alpha\psi_\alpha{}^\mu(x) +$$

$$+\theta\theta.\bar{\theta}_{\dot{\alpha}}\bar{\psi}^{\dot{\alpha}\mu}(x) + \theta\theta\frac{\partial^\mu}{\Box}(S(x)+iP(x)) +$$

$$+\bar{\theta}\bar{\theta}\cdot\frac{\partial^\mu}{\Box}(S(x)-iP(x)) + \theta\theta.\bar{\theta}\bar{\theta}\,A^\mu(x). \qquad (20)$$

The first few terms (θ^0, θ^1) are gauged away using parameters from $\lambda^\mu(z,\theta)$ (15) (Wess-Zumino gauge). The field $e^{a\mu}$ is the vierbein (graviton) field, $\psi_\alpha{}^\mu$ is gravitino, and S, P, A^μ are the auxiliary fields. The gauge freedom left after the partial gauge fixing corresponds to just general coordinate transformations in x-space, local Lorentz and local supersymmetry transformations. So, the superfield Y^μ introduced above is indeed the right gauge object for N=1 supergravity.

D) Supertensors

The hypersurface $\mathbb{R}^{4,4}$ is, in general, a curved superspace. To describe its geometry in gauge-independent terms one needs tensor (i.e. gauge covariant) objects. A standard way to do that is to develop the formalism of covariant differentiation (differential geometry). One introduces a tangent space Lorentz group (induced by the world superspace group (14), (16)) acting on Lorentz tensor indices $A = a, \alpha, \dot{\alpha}$. The covariant derivatives

$$D_A = E_A{}^{\mu} \frac{\partial}{\partial x^{\mu}} + E_A{}^{\alpha} \frac{\partial}{\partial \theta^{\alpha}} + E_A{}^{\dot{\alpha}} \frac{\partial}{\partial \bar{\theta}^{\dot{\alpha}}} + \omega_A{}^{BC} L_{BC}$$

involve vielbeins E_A and Lorentz connections ω (L_{BC} is the Lorentz generator). Both of them are expressed in terms of our single superfield Y^{μ}.

(Anti) commuting two such derivatives one finds

$$[D_A, D_B]_{\pm} = T_{AB}{}^{C} D_C + R_{AB}{}^{CD} L_{CD}$$

where T and R are torsion and curvature tensors. They are the objects containing all the covariant information about the geometry. They are subject to Bianchi (i.e. Jacobi) identities which in our case are automatically fulfilled because T and R are manifest functions of our <u>prepotential</u> Y^{μ}. It can be shown that T and R have just a few independent tensor components. Namely, they are reduced to a complex chiral superfield R ($\bar{D}_{\dot{\alpha}} R = 0$), a real vector superfield G^{μ}, a totally symmetric chiral superfield $W_{\alpha\beta\gamma}$ ($\bar{D}_{\dot{\alpha}} W_{\alpha\beta\gamma} = 0$) and their derivatives. There are certain relations among the latter (Bianchi identities), e.g.

$$D^{\alpha} W_{\alpha\beta\gamma} = D_{\beta\dot{\beta}} G^{\dot{\beta}}_{\gamma} + D_{\gamma\dot{\beta}} G^{\dot{\beta}}_{\beta} , \quad etc. \tag{21}$$

The superfields R, G^{μ} have dimension 1, $W_{\alpha\beta\gamma}$ has $\frac{3}{2}$.

E) Dynamics

As in general relativity, steps A)–D) give the geometric framework for the theory. Compared to the most general geometry of $\mathbb{R}^{4,4}$ (i.e., arbitrary torsion and curvature) our embedding of $\mathbb{R}^{4,4}$ in $\mathbb{C}^{4,2}$ produces a rather specific geometry. However, we still have to further specify it. This is done by the following simple dynamical (variational) principle /9/: minimize the invariant supervolume of $\mathbb{R}^{4,4}$. The latter is given by the integral

$$S_{SG} = \frac{1}{\kappa^2} \int d^4x \, d^2\theta \, d^2\bar{\theta} . \ E(x, \theta, \bar{\theta}) \tag{22}$$

where E = Ber $E_M^{\;A}$ is the Berezinian of vielbeins (the analog of $\sqrt{-g}$ in general relativity). In our case E is a function of our (now dynamical) field variable Y^μ:

$$E = \left[\det([\Delta_\alpha, \bar{\Delta}_{\dot{\alpha}}] Y^\mu)\right]^{-\frac{1}{3}} \left[\det(\delta_\mu^{\;\nu} + \partial_\mu Y^\lambda \partial_\lambda Y^\nu)\right]^{\frac{2}{3}},$$

$$\Delta_\alpha = \frac{\partial}{\partial\theta^\alpha} + i \frac{\partial}{\partial\theta^\alpha} Y^\mu . (\delta_\mu^{\;\nu} - i\partial_\mu Y^\nu)^{-1} \frac{\partial}{\partial x^\nu} .$$

The variation of Y^μ in S'_{SG} (22) produces the equation of motion

$$G_\mu = 0 \tag{23}$$

which has a consequence $R = \bar{R} = 0$. If supergravity interacts with supersymmetric matter one has to add the matter action with covariant derivatives

$$S = S_{SG} + \int d^4x d^2\theta d^2\bar{\theta} . L(\varphi, D\varphi).$$

In this case the equation of motion becomes

$$G_\mu = \text{æ}^2 V_\mu \tag{24}$$

where V_μ in the r.h.s. is the source of Y^μ, the so called supercurrent. Written out in terms of ordinary fields, eq. (24) reduces to the Einstein equation

$$R_{\mu\nu} - \frac{1}{2} g_{\mu\nu} R = \text{æ}^2 T_{\mu\nu},$$

the covariant Rarita-Schwinger equation

$$\varepsilon_{\mu\nu\lambda\rho} (\gamma_5 \gamma^\nu D^\lambda \psi^\rho)_\alpha = \text{æ} J_{\alpha\mu}$$

and equations for the auxiliary fields. The sources $T_{\mu\nu}$ (stress tensor) and $J_{\alpha\mu}$ (supersymmetry current) are components of the supercurrent V_μ.

This completes the superspace formulation of N=1 supergravity. Now one is prepared to investigate the interesting problem of renormalizability.

II.3. One- and Two-Loop Finiteness

In quantum supergravity one has to look for possible counterterms in every loop order

$$S_{QSG} = \frac{1}{\varkappa^2}\int d^4x\, d^2\theta\, d^2\bar\theta\, E +$$

$$+\int d^4x\, d^2\theta\, d^2\bar\theta\, E\, \Delta L^{I} + \varkappa^2 \int d^4x\, d^2\theta\, d^2\bar\theta\, E\, \Delta L^{II} + \dots$$

Here ΔL^{I} has dimension 2 , ΔL^{II} has 4, etc. On mass-shell, however, nothing much is left of the covariant tensors:

$$R = G_\mu = 0, \quad \bar D_{\dot\alpha} W_{\alpha\beta\gamma} = D^\alpha W_{\alpha\beta\gamma} = 0 \tag{25}$$

(see eqs. (21), (23)). There is only one candidate for a one-loop counterterm:

$$\Delta S^{I} = \int d^4z\, d^2\theta\, W^{\alpha\beta\gamma} W_{\alpha\beta\gamma}$$

($W^{\alpha\beta\gamma}$ is chiral, therefore a $\mathbb{C}^{4,2}$ description for it is natural). It can be shown to vanish which is a generalization of the Gauss-Bonet theorem

$$\int d^4x \left(R^{\mu\nu\lambda\rho} R_{\mu\nu\lambda\rho} - 4R^{\mu\nu}R_{\mu\nu} + R^2\right) = 0.$$

At two loops the only candidate is

$$\Delta S^{II} = \varkappa^2 \int d^4x\, d^2\theta\, d^2\bar\theta\, E\, (D^\alpha W^{\beta\gamma\delta})(D_\alpha W_{\beta\gamma\delta}) =$$

$$= \varkappa^2 \int d^4x\, d^2\theta\, d^2\bar\theta\, E\, W^{\beta\gamma\delta} D^\alpha D_\alpha W_{\beta\gamma\delta} = 0$$

(see eq. (25); integration by parts is possible in the curved case). There is also a simpler reason why ΔS^{II} vanishes: S_{SG} is U(1) globally invariant (i.e. under the wider subgroup (17)) but $\Delta S'^{II}$ is not. This U(1) invariance helps to eliminate a number of higher counterterms.

At three loops there are possible counterterms even on-shell (although nobody has calculated their coefficients, they may vanish as well). Thus, one can claim that N=1 supergravity is one- and two-loop finite on-shell (provided a regularization procedure consistent with supersymmetry is applied, anomalies do not spoil supersymmetry and all those subtleties of quantization are overcome).

What happens in the presence of matter? As we saw above, the equation of motion changes (24), G_μ is not zero any more and there is a possible one-loop counterterm

$$\Delta S^{I} = \int d^4x\, d^2\theta\, d^2\bar{\theta}\; E\, G^\mu G_\mu \qquad , \text{ etc.}$$

So, arbitrary matter interactions destroy the finiteness of supergravity. The only way out is to try to choose specific interactions involving a larger symmetry. In other words, one should try to put all matter together with gravity into a single large multiplet of extended supergravity.

III. Extended Supergravity

III.1. Limits on N and Particle Content

When looking for a wider supergravity multiplet we come across a natural upper limit on N. Indeed, the massless multiplets of N-extended supersymmetry starting with a helicity 2 state have the following structure

$$|2\rangle,\ |3/2\rangle,\ |1\rangle,\ \dots,\ |2-\tfrac{N}{2}\rangle\ .$$

It is obtained by applying the N different helicity-lowering (by steps of 1/2) operators constructed of the Q's. From here one immediately sees the limit $N \leq 8$, otherwise one would have helicities greater than 2 and one does not know how to deal with them. Notice that the case N=8 is special: in that case one gets the PCT complete multiplet $|2\rangle, \dots, |-2\rangle$. Recall the N=4 Yang-Mills case where one also has a PCT self-conjugate multiplet $|1\rangle, \dots, |-1\rangle$. It is believed that the exceptional renormalizability properties of the latter theory have to do with this feature and it is hoped that the same may apply to N=8 supergravity.

In Table 1 we can see the particle content of some extended supergravity multiplets (λ is helicity)

Table 1

N \ $\lvert\lambda\rvert$	2	3/2	1	1/2	0
1	1	1			
2	1	2	1		
⋮					
8	1	8	28	56	70

In the simplest case (N=2) a massless spin 1 particle appears. In the theory it becomes the photon field, so this is a consistent Maxwell-Einstein system. We see that the dream of Einstein can be realized only in supergravity, with the help of two intermediate particles, the gravitini.

We skip all other values of N and jump over to the most sophisticated case: N=8. Although rather rich, the particle spectrum even in this case does not seem very realistic/10/. First, let us look at the vectors. There are 28 of them, just enough to gauge the global SO(8) symmetry present in the theory. However, SO(8) does not contain the "standard model" subgroup

$$SO(8) \not\supset SU(3) \times SU(2) \times U(1).$$

At most it contains

$$SO(8) \supset SU(3) \times U(1) \times U(1).$$

Here there is place for Z^0 but not for $W^{\pm}$. Another problem is that SO(8) has only real representations, so the fermions will be real and that means they can acquire huge masses of the order of M_P. Further, a careful analysis of the representations of SO(8) shows that there is no room for such particles as μ, τ, b.

All that made some people think that supergravity could not be a realistic physical theory. However, later on some subtle hidden symmetries of the N=8 model were discovered which led to speculations about possible composite particle structures based upon N=8 supergravity. More about that in the final part of these lectures.

III.2. The N=2 Theory

Let us now discuss in some more detail the simplest model, the N=2 one, where we shall see some of the basic features of extended supergravity.

From Table 1 we know the particle content of the theory, i.e. spins $2, 2\times 3/2, 1$. The cprresponding fields are the vierbein e^a_μ, <u>two</u> gravitino fields $\psi^i_{\alpha\mu} (i=1,2)$ and a "photon" field A_μ. The Lagrangian can be obtained by the Noether method described in Section II.1, although it requires more work now. The result is/11/ ($\varkappa$ is not shown explicitly)

$$L = -\frac{e}{2} R - \frac{1}{2} \varepsilon^{\mu\nu\rho\sigma} \bar{\psi}^i_\mu \gamma_5 \gamma_\nu D_\rho \psi_{\sigma i} - \frac{e}{4} F_{\mu\nu}^2 +$$

$$+ \frac{1}{4\sqrt{2}} \bar{\psi}^i_\mu \left[e(F^{\mu\nu} + \hat{F}^{\mu\nu}) + \frac{1}{2} \gamma_5 (\tilde{F}^{\mu\nu} + \hat{\tilde{F}}^{\mu\nu}) \right] \psi_{\nu i} \quad . \tag{26}$$

Here $F_{\mu\nu}$ is the curl of A_μ , $\tilde{F}_{\mu\nu}$ is its dual tensor and

$$\hat{F}_{\mu\nu} = F_{\mu\nu} - \frac{1}{2\sqrt{2}} \left(\bar{\Psi}^i_\mu \Psi_{\nu i} - \bar{\Psi}^i_\nu \Psi_{\mu i} \right) .$$

The Lagrangian (26) is invariant under N=2 local supersymmetry transformations.

The first three terms in eq. (26) are natural: the Einstein, Rarita-Schwinger and electromagnetic kynetic terms. Notice, however, that the interaction between A_μ and Ψ_μ is only through $F_{\mu\nu}$, i.e. it is non-minimal. In other words, A_μ is not a true photon field and Ψ_μ has no electric charge.

In fact one can achieve minimal coupling in eq. (26)[12]. However, the supersymmetry forces one to introduce, alongside the electric charge g , a mass for the gravitino of the order $g\varkappa^{-1}$ and a cosmological constant of the order $g^2\varkappa^{-4}$. The value of the latter turns out to be huge compared to present estimates. This may not be a serious problem either because by adding some matter multiplets one can cancel the cosmological constant or even because having a large cosmological constant is not too bad according to Hawking et al.[13]. The appearance of a cosmological constant when gauging the global SO(N) symmetry (i.e. minimally coupling the corresponding gauge fields) is common in all extended supergravity theories.

Another feature of the N=2 model is the presence of a "hidden" symmetry. Besides a manifest SU(2) symmetry of the Lagrangian (26) the equations of motion have an additional U(1) symmetry realized as duality-chirality transformations[14]

$$\delta \Psi_\mu = -i\gamma_5 \Psi_\mu , \quad \delta \hat{F}_{\mu\nu} = i e \varepsilon_{\mu\nu\rho\sigma} \hat{F}^{\rho\sigma} . \tag{27}$$

Off-shell it disappears. Similar hidden on-shell symmetries are found in the other extended supergravity models too.

The auxiliary fields for the theory have been found[15]. Also, there exists a superspace formulation although it involves constrained superfields. The idea of complex superspace explained in Section II helps in this case too: it reduces the number of constraints and makes it possible to write down a superspace action formula[16]. However, it is not still a truely unconstrained formulation. Some new geometric ideas are needed.

III.3. The N=8 Theory

Let us now concentrate on the most interesting and complicated case of N=8 supergravity. Looking at Table 1 one realizes that trying to work out an interaction of so many fields using the straightforward Noether method is hopeless. Superspace is not of much help either. The problem with extended superfields is that they have too many components: 2^{4N} in N-extended supersymmetry. In the case N=8 there are 2^{32} of them, the majority of which are redundant and have to be eliminated by constraints which are not known.

Fortunately, an elegant method for deriving such complicated theories has been invented.

Dimensional reduction. The idea of the method is rather simple /17/. In a d-dimensional space the spinors of the corresponding $SO(d-1, 1)$ group have $2^{\frac{d}{2}}$ (d even) or $2^{\frac{d-1}{2}}$ (d odd) components. When reduced to 4 dimensions they split into a number of 4-dimensional spinors. So do the generators Q_α of simple (N=1) supersymmetry in d dimensions: they give rise to N generators $Q_{\alpha i}$ of an extended supersymmetry. Table 2 illustrates this

Table 2

d (N=1)	4	6	10	(11)
N (d=4)	1	2	8	(8)

In fact, some additional conditions (Majorana and/or Weyl) may reduce in some cases the number N obtained from a given d. Thus, from d=10 one can also obtain N=4 supergravity or N=4 Yang-Mills.

It turns out that for N=8 supergravity d=11 is most suitable /17/. So, the program is to first construct N=1 supergravity in 11 dimensions and reduce it to 4 dimensions to obtain N=8 supergravity.

Supergravity in 11 dimensions. Let us consider the fields and degrees of freedom of N=1 11 dimensional supergravity. First, there is the elfbein e_μ^a. It is an 11x11 matrix but in fact it can be made symmetric (using the local SO(10,1) Lorentz invariance), transversal and traceless (fixing the general coordinate transformation gauge), so it describes

$$\frac{1}{2}(9\times 10) - 1 = 44$$

degrees of freedom. Second, the gravitino field $\psi_{\mu\alpha}$ has 11x32 components, but it can also be made transversal and Γ-traceless, $\Gamma^\mu \psi_\mu = 0$

(local supersymmetry gauge) and it is real, so it describes

$$\frac{1}{2}(9\times 32-32)=128$$

degrees of freedom. The remaining 84 bosonic degrees of freedom can be supplied by an antisymmetric 3 index tensor $A_{\mu\nu\lambda}$. Fixing its gauge invariance $\delta A_{\mu\nu\lambda}=\partial_\mu A_{\nu\lambda}+cycl$ one can make it transversal and get exactly

$$\binom{9}{3}=84$$

components.

The Lagrangian for these three fields (together with the local supersymmetry transformation laws) can be worked out directly. The result is

$$\begin{aligned} L= & -\frac{e}{2}R-\frac{e}{2}\bar{\Psi}_\mu\Gamma^{\mu\nu\rho}D_\nu\Psi_\rho-\frac{e}{48}F_{\mu\nu\rho\sigma}F^{\mu\nu\rho\sigma}+ \\ & +\frac{e}{192}\bar{\Psi}_\mu\Gamma^{\mu\nu\lambda\rho\sigma\tau}\Psi_\nu\left(F_{\lambda\rho\sigma\tau}+\hat{F}_{\lambda\rho\sigma\tau}\right)+ \\ & +\frac{2}{(12)^4}\varepsilon^{\mu_1\ldots\mu_{11}}F_{\mu_1\ldots\mu_4}F_{\mu_5\ldots\mu_8}A_{\mu_9\mu_{10}\mu_{11}}\ . \end{aligned} \tag{28}$$

Again, the first three terms have an obvious meaning, the rest are non-minimal couplings.

<u>Reduction from 11 to 4 dimensions</u>/18/. It is achieved by compactification of the last seven dimensions. In other words, given a function $\varphi(x,y)$, $x=x_1,\ldots x_4$, $y=x_5,\ldots,x_{11}$, one makes the space shrink in all the y -directions into little circles (a torus-like form) with radii going to zero. Then in the Fourier transform the y -dependent modes acquire masses going to infinity and can be discarded. Effectively $\varphi(x,y)$ becomes $\varphi(x)$. The 11-dimensional vector and spinor indices split as follows

$$\mu(1\ldots 11)\rightarrow\mu(1\ldots 4)\ ,\ m(5\ldots 11)\ ;$$

$$\alpha(1\ldots 32)\rightarrow\alpha i(\alpha=1\ldots 4\ ,\ i=1\ldots 8)\ .$$

The elfbein field becomes now

$$e_\mu^a(11)\longrightarrow\begin{pmatrix} e_\mu^a & e_\mu^k \\ 0 & \varphi_m^k \end{pmatrix}\ . \tag{29}$$

The vanishing left-lower corner is obtained by breaking the local Lorentz group SO(10,1) down to

$$SO(10,1) \longrightarrow SO(3,1) \times SO(7).$$

The residual SO(3,1) is the 4-dimensional Lorentz group, SO(7) can be further fixed to make φ_m^k symmetric. On the other hand, this SO(7) can be extended (see below) to a local SU(8) invariance of the action.

Now, in 4 dimension the elfbein field describes

e_μ^a - 1 tensor (spin2) ,
e_μ^k - 7 vectors (spin 1) ,
φ_m^k - $\frac{1}{2}$ 7x8=28 scalars (spin 0) .

The gravitino field reduces to

$\psi_\mu^{\alpha i}$ - 8 gravitini (spin 3/2) ,
$\psi_m^{\alpha i}$ - 7x8=56 spinors (spin 1/2) .

Finally, the antisymmetric tensor.
It produces four types of 4-dimensional tensors:

i) $A_{\mu\nu\rho}$ - it is antisymmetric, so its curl is

$$F_{\lambda\mu\nu\rho} = \partial_{[\lambda} A_{\mu\nu\rho]} = \varepsilon_{\lambda\mu\nu\rho}\, \varphi$$

and the equation of motion $\partial^\lambda F_{\lambda\mu\nu\rho} = 0$ gives $\varphi =$ const, i.e. $A_{\mu\nu\rho}$ is pure gauge;

ii) $A_{\mu\nu k}$ - these are 7 spins 0. Indeed, now

$$F_{\lambda\mu\nu} = \partial_{[\lambda} A_{\mu\nu]} = \varepsilon_{\lambda\mu\nu\rho}\, \varphi^\rho$$

and the equation of motion yields $\varphi^\rho = \partial^\rho \varphi$ where φ is a scalar field

iii) $A_{\mu k l}$ - $\frac{1}{2}$6x7=21 axial vectors (spin 1) ,
iv) A_{klm} - $\binom{7}{3}$=35 pseudoscalars (spin 0) .

Putting together all the numbers above one finds 1 spin 2, 8 spins 3/2, 28 spins 1, 56 spins 1/2, 70 spins 0 which is exactly the particle content of N=8 supergravity (see Table 1).

It should be pointed out that the method of dimensional reduction produces the N=8 theory <u>without manifest supersymmetry off-shell</u>. The auxiliary fields are not known in 11 as well as 4 dimensions. This is one of the major open problems of the theory.

<u>Symmetries of the N=8 theory</u> /18/. The 11-dimensional Lagrangian was invariant under local SO(10,1) and general coordinate transformations. The former reduces to local SO(3,1) and local SO(7) (provided we keep e_m^k in eq. (29) non-symmetric). The latter give rise to general coordinate transformations in 4 dimensions plus global $GL(7,R)$

invariance acting on 7-dimensional world indices (like m in e_m^k, $\psi_m^{\alpha i}$, etc.).

As we saw in the N=2 case, there may be concealed symmetries of the theory on-shell. This is the case here where the global $GL(7,R)$ symmetry can be extended to a global non-compact E_7 symmetry. The latter acts only on the scalars and the curls of the vectors and is an <u>on-shell</u> symmetry.

At the same time the fake SO(7) local symmetry can be extended to an SU(8) local symmetry acting on the scalars and spinors (this time off-shell). It should be well understood that this SU(8) is an artificial (i.e. added by hand) symmetry. The precise relation between E_7 and SU(8) is not so simple. Here I shall try to give only a rough idea. Recall the analogy with the vierbein field in general realativity, e_μ^a. There are local SO(3,1) and global $GL(4,R)$ transformations acting on the indices a and μ respectively ($GL(4,R)$ is the global subgroup of the general coordinate transformation group). Now, one can fix the local SO(3,1) gauging away 6 of the 16 components of e_μ^a. Thus one is left with the metric tensor $g_{\mu\nu}$ corresponding to the coset

$$g_{\mu\nu} \sim GL(4,R)/SO(3,1).$$

In the N=8 case one introduces 63 scalars by hand in order to have local SU(8). The remaining 70 physical scalars of the theory parametrize the coset $E_7/SU(8)$:

$$\begin{aligned}&63 \text{ gauge scalars} \sim SU(8),\\&70 \text{ physical scalars} \sim E_7/SU(8).\end{aligned}$$

<u>Dynamical generation of bound states.</u> The analogy with the vierbein goes even further. To make general relativity invariant under the artificial local SO(3,1) one needs connections. They do not have to be independent dynamical fields; instead, they are composed of the vierbeins:

$$\omega_{\mu a b}(e) = -\tfrac{1}{2} e_a^\lambda e_b^\rho (\partial_\lambda e_{c\rho} - \partial_\rho e_{c\lambda}) e_\mu^c + \\ +[\tfrac{1}{2} e_a^\lambda (\partial_\mu e_{b\lambda} - \partial_\lambda e_{b\mu}) - (a \leftrightarrow b)] .$$

Similarly, in N=8 supergravity one can achieve local SU(8) invariance with the help of composite SU(8) connections

$$V_\mu \in (\partial_\mu V) V^{-1} \tag{30}$$

where V is a 56x56 matrix belonging to E_7 and containing the scalars. Thus, A_μ are not physical gauge fields.

However, now the optimist recalls the CP^{n-1} model in 2 dimensions[19]. There one has n complex scalar fields

$$A_i \,,\ i = 1,\dots,n \,,\ \sum_{i=1}^{n} |A_i|^2 = 1 \,.$$

One can write down a Lagrangian for them

$$L = \sum_{i=1}^{n} (\partial_\mu - i v_\mu) A_i^* (\partial_\mu + i v_\mu) A_i \tag{31}$$

which is invariant under local U(1) transformations

$$A_i \to e^{i\Lambda} A_i \,,\ v_\mu \to v_\mu - \partial_\mu \Lambda \,.$$

Again, the gauge field v_μ in eq. (31) is not independent. Its equation of motion expresses it in terms of the scalar fields

$$v_\mu = \frac{i}{2} \sum_{i=1}^{n} A_i^* \overleftrightarrow{\partial_\mu} A_i \,.$$

However, investigations using non-perturbative methods show that the propagator for v_μ develops a pole and v_μ becomes a physical field (bound state).

The only thing the optimist has to add is: maybe(?) some mysterious mechanism can render the composite gauge fields (30) poles dynamically and thus make SU(8) a physical gauge symmetry. Really, not too much can support this hypothesis, but let us be optimists and try to figure out what the consequences might be[20].

The vector gauge fields should belong to a whole supermultiplet of bound states. Their classification group should be SU(8). Supposingly, it is an exact symmetry only at energies of the order of M_P. One could expect that at energies of the order of M_{GUT} SU(8) should break down to SU(5) or some other GUT group and, further, to SU(3)xSU(2)xU(1) at energies $\sim$100 GeV. This promotes the philosophy that the fields of N=8 supergravity are preonic fields and the "elementary" particles observed at accelerator energies are their bound states.

The first question one has to ask is what kind of supermultiplet can the composite SU(8) gauge fields belong to? One guess is that it may resemble the stress tensor multiplets known in other supersymmetric theories. In Table 3 the helicity and SU(8) content of such a multiplet is shown.

Table 3

Helicity	-3/2	-1	-1/2	0	+1/2	+1	+3/2	+2	+5/2
SU(8)	$\overline{8}$	63	216	420	504	$\overline{378}$	$\overline{168}$	$\overline{36}$	$\overline{8}$
content		+1	+8	+28	+56	+70	$+\overline{56}$	$+\overline{28}$	

The representations in the first row are of the type $R^A{}_{[BC\ldots F]}$ with antisymmetrized lower indices and in the second row $R^A{}_{[AC\ldots F]}$ (SU(8) traces).

Most of those numerous SU(8) multiplets are certainly redundant. One has to resort to a long chain of not entirely convincing arguments to discard:

i) all the SU(8) traces (a point disputed by other authors /21/);

ii) all the helicity 5/2 and 3/2 states and those helicity 1 states which are not in the adjoint representation of SU(8). The reason is that they would lead to non-renormalizable theories at low energies which is not acceptable according to a theorem by Veltman;

iii) all the helicity 1/2 states that lead to non-zero anomalies (again the Veltman theorem);

iv) all the vector-like particles which can acquire masses $\sim M_p$.

The positive outcome of this struggle with the proliferation of particle states in N=8 supergravity are two encouraging indications: SU(5) is the only possible (within the given framework) GUT group and the fermions in nature form three SU(5) families: 3(10+5) . Unfortunately, these nice results are based on too many assumptions which are not easy to verify. Clearly, a lot more work is required to start understanding the extremely rich structure and the beauty of supergravity which may, indeed, turn out to be the ultimate theory of elementary particle physics (Hawking)!

References

1. D.Volkov, V.Akulov. Phys.Lett. 46B (1973) 109;
   J.Wess, B.Zumino. Nucl.Phys. B70 (1974) 39.
2. P. van Nieuwenhuizen. J.Phys. A13 (1980) 1643.
3. D.Freedman, P. van Nieuwenhuizen, S.Ferrara. Phys.Rev. D13 (1976) 3214;
   S.Deser, B.Zumino. Phys.Lett. 62B (1976) 335.
4. Review on renormalizability of supergravity: M.Grisaru, P. van Nieuwenhuizen in New Pathways in Theoretical Physics (Coral Gables, 1977), eds. B.Kursunoglu and A.Perlmutter.

5. See the last lecture by K.S.Stelle in these Proceedings.
6. K.S.Stelle, P.C.West. Phys.Lett. 74B (1978) 330.
P. van Nieuwenhuizen, S.Ferrara. Phys.Lett. 74B (1978) 333.
7. V.Ogievetsky, E.Sokatchev. Phys.Lett. 79B (1978) 222; Yadernaya Fiz. 31 (1980) 205, 821.
8. For a review see: B.Zumino, in Proc. 1979 Cargese school (Plenum Press).
9. J.Wess, B.Zumino. Phys.Lett. 74B (1978) 51.
10. M.Gell-Mann, Talk at the 1977 Washington Meeting of the American Physical Sciety (unpublished).
11. S.Ferrara, P. van Nieuwenhuizen. Phys.Rev.Lett. 37 (1976) 1669.
12. A.Das, D.Freedman. Nucl.Phys. B120 (1977) 221;
E.Fradkin, M.Vasiliev. Lebedev Institut Preprint N°197 (1976).
13. S.Hawking, D.Page, C.Pope. Phys.Lett. 86B (1979) 175.
14. J.Scherk, S.Ferrara, B.Zumino. Nucl.Phys. B121 (1977) 393.
15. E.Fradkin, M.Vasiliev. Phys.Lett. 85B (1979) 47.
B. de Wit, J. van Holten. Nucl.Phys. B155 (1979) 530.
16. E.Sokatchev. Phys.Lett. 100B (1981) 466.
17. E.Cremmer, B.Julia, J.Scherk. Phys.Lett. 76B (1978) 409.
18. E.Cremmer, B.Julia. Nucl.Phys. B159 (1979) 141.
19. A.D'Adda, P. DiVecchia, M.Lüscher. Nucl.Phys. B146 (1978) 63.
20. See J.Ellis, M.Gaillard, B.Zumino. CERN preprint TH-3152 (1981) and references therein.
21. J.Derendinger, S.Ferrara, C.Savoy. CERN preprint TH-3025 (1981); Nucl.Phys. B188 (1981) 77.

Supersymmetry Applied to Particle Physics - Phenomenological Problems

Glennys R. Farrar
Dept. of Physics and Astronomy
Rutgers University
New Brunswick, New Jersey 08903

In these two lectures I want to try to present the crux of the dilemma in making a useful, acceptable supersymmetry theory relevant to particle physics. By the end of the second lecture I will have shown that the most naive version of how supersymmetry could be helpful in providing an attractive GUT does not work; hopefully this will leave the students here with a clear idea of what directions may still be productive to pursue. In order to cover a large subject in two lectures I will assume the student has learned already the rudiments of supersymmetry and I will furthermore save myself some effort by not bothering with ± signs, factors of 2, etc. where not crucial for whatever point I am making. Thus formulae with "∿" rather than "=" are likely to be missing numerical factors and the student is forewarned not to use them out of context.

In the first lecture I will describe spontaneous supersymmetry breaking, and in the second I will describe a feature which is common to the most aesthetically appealing models and which is unacceptable phenomenologically, namely R-invariance. New kinds of models are being developed so rapidly that I cannot cover the most recent ones here; the student should bear in mind that some of the difficulties I will mention may not be present in all types of models. Unfortunately I am unable to competently discuss supergravity so my remarks are restricted to global supersymmetry. Nor do I discuss explicit supersymmetry breaking.

I. Spontaneous Supersymmetry Breaking

A. General remarks

Recall from K. Stelle's lectures that an N=1 supersymmetric action can be obtained[1] by integrating an arbitrary superfield or product of superfields over superspace

$$\int d^4x \int d^2\theta d^2\bar{\theta} \text{ [arbitrary superfield]} \tag{1}$$

In addition,

$$\int d^2\theta \text{ [chiral superfield]} \tag{2}$$

gives a supersymmetric lagrangian, where a chiral superfield ϕ is defined to satisfy

$$\bar{D}_{\dot{\alpha}}\phi = 0 \tag{3}$$

Here, as throughout, I use the notation of Wess's "Lectures in Supersymmetry"[2].

In terms of the gauge (V_k) and chiral (ϕ_i) superfields we can thus write

$$\mathcal{L} = \left[\phi_i^+ e^{V_k}\phi_i\right]_{\theta\theta\bar{\theta}\bar{\theta}} + \left[\mathrm{tr} W_k{}^{\alpha} W_{k\alpha}\right]_{\theta\theta} + \left\{g_{ijk}\left[\phi_i\phi_j\phi_k\right]_{\theta\theta} + m_{ij}\left[\phi_i\phi_j\right]_{\theta\theta} + \lambda_i\left[\phi_i\right]_{\theta\theta} + hc\right\} + \xi\left[V\right]_{\theta\theta\bar{\theta}\bar{\theta}} \tag{4}$$

where

$$W_\alpha = \bar{D}\bar{D}e^{-V}D_\alpha e^{+V} \tag{5}$$

The first kind of term describes gauge interactions and kinetic terms of matter, in addition to specific interactions among the matter fields which we will look at later. The second kind of term contains the gauge kinetic energy, including the interaction of gauginos with the vector gauge bosons. The third type of term is the super-Yukawa interaction including Yukawa and $A_\phi{}^4$ interactions among the matter fields. The remaining terms include intrinsic masses for the matter fields and terms relevant for supersymmetry breaking.

There are a number of observations I want to make about (4):
i) θ has dimension $\ell^{\frac{1}{2}}$, the vector superfield [V] is dimensionless, and $\phi \sim \ell^{-1}$. It is easy then to see that e^V and $W^\alpha W_\alpha$ are dimensionally correct, and that a cubic interaction among chiral superfields is the highest power allowed in a renormalizable theory. Note that in the superfield expansions

$$\phi \sim A_\phi + \theta\psi + \theta\theta F + \ldots \tag{6}$$

$$V \sim \bar{\theta}\bar{\theta}\theta\lambda + \theta\theta\bar{\theta}\bar{\lambda} + \theta\theta\bar{\theta}\bar{\theta}\, D + \ldots \tag{7}$$

the auxilliary fields F and D both $\sim \ell^{-2}$.
ii) Bose statistics for the ϕ superfields requires g_{ijk} and m_{ij} to be totally symmetric.
iii) When the gauge group is non-abelian, V is a matrix $= V_{ij} = V^a T^a{}_{ij}$ and under the gauge transformation

$$\phi \to \phi' = e^{-i\Lambda}\,\phi;\ \phi^+ \to \phi'^+ = \phi^+ e^{i\Lambda^+}$$

and (8)

$$e^V \to e^{V'} = e^{-i\Lambda^+}\, e^V e^{i\Lambda}$$

where Λ and Λ^+ are chiral and antichiral respectively, i.e.,

$$\bar{D}_{\dot{\alpha}}\Lambda = D_\alpha \Lambda^+ = 0 \tag{9}$$

It is interesting to note that the superspace "covariant derivative" appearing in the chiral constraint (9) and the definition of W^α does not contain a gauge field:

$$D_\alpha = \frac{\partial}{\partial\theta^\alpha} + i\sigma^m_{\alpha\dot{\alpha}}\,\bar{\theta}^{\dot{\alpha}}\partial_m \tag{10}$$

and the gauge invariance of $W^\alpha W_\alpha$ is assured because

$$e^{-V}D_\alpha e^V = D_\alpha V - \tfrac{1}{2}[V, D_\alpha V]. \tag{11}$$

Gauge invariance is consequently assured for the first two terms in (4), while it places severe restrictions on the form of the remaining terms. For instance in order for there to be a linear term, ξV, the gauge group must contain a U(1) factor and for there to be a linear $\lambda\phi$ term the theory must include a matter field which is a total gauge singlet. One of the prettiest features of supersymmetry in my opinion is the strongly constrained form of the scalar potential due to the fact that it arises from bilinear $[\phi^+ e^V \phi]_{\theta\theta\bar{\theta}\bar{\theta}}$, and trilinear, $[\phi^3]_{\theta\theta}$ terms in the superfields. For instance denoting the left-chiral quark isodoublet by Q, the left-chiral isosinglet by U^c and D^c, and the (left-chiral) higgs isodoublet by H, then $[QU^cH]_{\theta\theta}$ is supersymmetric and SU(2) x U(1) gauge invariant. However $[QD^cH]$ has the wrong hypercharge and $[QD^cH^+]_{\theta\theta}$ is not supersymmetric. Thus a second higgs superfield H' must be introduced. (As usual in SU(2) x U(1), no "explicit" mass term, here from $[\phi^2]_{\theta\theta}$, is possible for quarks.) When we discuss making realistic models we will see more interesting examples of the restrictiveness of the absence of quadrilinear terms in superfields.

iv) It has been shown[3)] that higher order perturbation theory corrections generate terms which have the form (1) and not (2). This is referred to as the non-renormalization theorem since a consequence of it is that the parameters g_{ijk}, n_{ij} and λ_i are not changed by renormalization of the theory, and if $\xi_i = 0$ at tree level, it is not generated at any order in perturbation theory.[4)] This feature has been exploited in trying to solve the "naturalness" problem which plagues GUT's: if a mass parameter is zero as a result of some symmetry, even if that symmetry is broken spontaneously at the unification scale, that parameter receives no radiative corrections at all - unlike the case without supersymmetry in which radiative corrections would typically be $O(M_{GUT})$. A less satisfactory case is when no symmetry rules out the unwanted term, so that a fine-tuning is involved in setting the relevant mass parameter to zero. Some people have argued[5)] that the supersymmetry case is still an improvement since the fine-tuning "only has to be done once" and need not be imposed order by order in perturbative theory. This seems to me to be a distinction only relevant as a result of how we calculate in perturbation theory and of no fundamental significance. Witten[4)] has noted that since the non-renormalization theorems are only true order by order in perturbation theory, they might be violated by non-perturbative effects. He conjectured that the origin of the mass hierarchy could then be that the supersymmetry breaking scale $M_{SS} \sim (e^{-1/g^2})M_{GUT}$ which can give about the right scales, because SU(2) x U(1) breaking would come about by supersymmetry breaking, causing the higgs scalars to get a negative $(\text{mass})^2 \sim M_{ss}^2$.

Now let us look at the components in the superfield expansions of $\mathcal{L}$ in (4),

$$[\phi^+ e^{gV} \phi]_{\theta\theta\bar{\theta}\bar{\theta}} \sim g A_\phi^* A_\phi D_V + \bar{\psi} \not{D} \psi + F_\phi^* F_\phi + \ldots \tag{12}$$

$$[W^\alpha W_\alpha]_{\theta\theta} \sim \bar{\lambda} \not{D} \lambda + D^2 + \ldots \tag{13}$$

$$g_H [Q U^c H]_{\theta\theta} \sim g_H \{\psi_Q \psi_{U^c} A_H + \psi_Q A_{U^c} \psi_H + A_Q \psi_{U^c} \psi_H + A_Q A_{U^c} F_H + A_Q F_{U^c} A_H + \ldots\} \tag{14}$$

When convenient we may wish to eliminate the auxilliary fields by using their equations of motion, giving, with just the terms above,

$$F_H^* = -g_H A_Q A_{U^c}, \quad F_Q^* = -g_H A_H A_{U^c}, \quad F_{U^c}^* = -g_H A_H A_Q \tag{15}$$

and

$$D = -g A_\phi^* A_\phi . \tag{16}$$

One can then see that the vacuum energy has the form

$$V = \sum_i |\langle F_i \rangle|^2 + \sum_j |\langle D_j \rangle|^2 . \tag{17}$$

B. Chiral Masses

What about supersymmetry breaking? Since

$$H = \tfrac{1}{4} \sum Q_i^* Q_i \tag{18}$$

In order to have $Q_i|0\rangle \neq 0$ for some i, it is necessary that $\langle 0|H|0\rangle > 0$. It is then apparent that in order to break supersymmetry at least one of the auxiliary fields must take a non-zero VEV. The picture of the potential as a function of scalar field is no longer the familiar one shown in 1a, but becomes

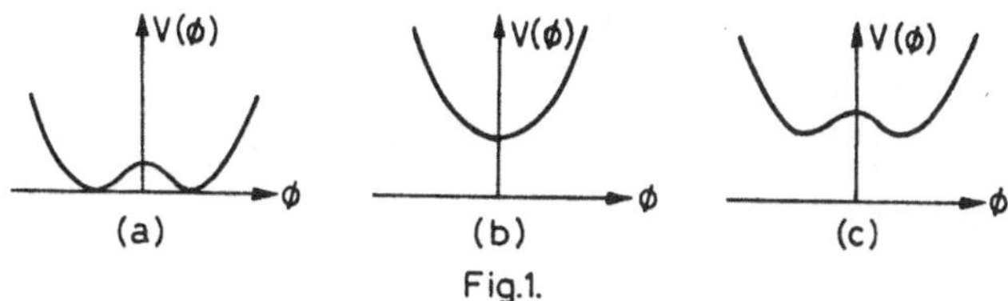

Fig.1.

that shown in 1b (or 1c, if some internal symmetry is spontaneously broken as well). For the moment I will put aside the issue of arranging that an $\langle F\rangle$ or $\langle D\rangle$ be non-zero and discuss the effect of a non-zero $\langle F\rangle$ or $\langle D\rangle$ on the mass-splittings within matter (chiral) supermultiplets.

Using the expressions (15) to eliminate the F's from the super-Yukawa term (14), we see that $\langle A_H\rangle \neq 0$ generates equal masses $g_H\langle A_H\rangle$ for the u quark and its scalar and pseudoscalar partners. (Recall that a massive spin ½ particle with its two helicity states has two bosonic partners which in general may not be degenerate when supersymmetry is broken.) This is the only source of mass generation for spin ½ quarks and is nothing more than the supersymmetric generalization of the Higgs mechanism for fermion masses.

What if one of the $\langle F\rangle$'s is non-zero? First observe that this must not happen in the simple case here with no other super-Yukawa terms since by (15) at least one of the A_Q or A_U would have to have non-zero VEV, breaking color, charge and baryon number. However the addition, for instance, of a new isosinglet chiral superfield X could permit a term $g_X[HH'X]_{\theta\theta}$ which would change the formula (15) for F_H to

$$F_H^* = -g_H A_Q A_{U^c} - g_X A_{H'} A_X. \tag{19}$$

Reinstating the h.c. terms for (14) and using CP so $\langle F_H\rangle$ is real, we have the spin-0 mass terms

$$(A_\phi^* A_\phi^* + A_\phi A_\phi)\langle F_H\rangle \sim [(\mathrm{ReA})^2 - (\mathrm{ImA})^2]\langle F_H\rangle \tag{20}$$

Thus the spin-0 mass eigenstates are ReA and ImA. Supersymmetry is broken but in a way which is unacceptable for quark mass-splittings since one of the spin-0 particles moves to lower $(\mathrm{mass})^2$ with the average $(\mathrm{mass})^2$ equal to that of the fermions. In fact even a small splitting of this <F>-type is a problem since it leads to parity violations in the strong interaction arising, e.g., from diagrams like that of Fig. 2.[6]

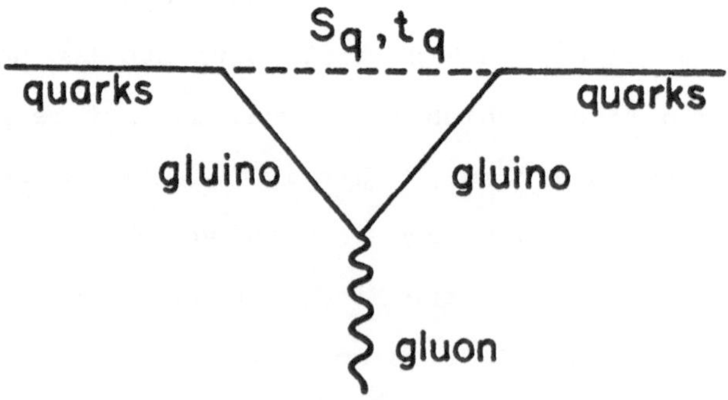

Fig. 2

In fact, even though an "X" field is often added, its scalar component usually does not get a non-zero VEV, so this problem may not arise.

What about $<D> \neq 0$ to raise the spin-0 quark masses? Since (12) $\sim g_Q A_Q^* A_Q <D> \sim [(\mathrm{Re}A_Q)^2 + (\mathrm{Im}A_Q)^2] g_Q <D>$ we see that the spin-0 masses are split from the spin-½ mass (which gets no analog contribution) but are degenerate with each other so parity violation is not a problem. Thus if the sign of g_Q can be properly chosen this term has the desired property of raising both scalar and pseudoscalar quark masses. The only D's which can be given non-zero VEV at tree level are in U(1) gauge supermultiplets. However $<D_{QED}> \neq 0$ does not give satisfactory masses to spin 0 quarks: The u and d quarks have opposite signs for their electric charge so that if the spin-0 partners of one are raised those of the other are lowered. This led Fayet[7)] to introduce an additional gauged $\tilde{U}(1)$, with all quarks and leptons having $\tilde{Y} = +1$ (so the H and H' have $\tilde{Y} = -2$ to permit the QU^cH and QD^cH' terms) and enabled him to construct the first supersymmetric model of particle physics which was not obviously in contradiction with experiment.

Before proceeding to describe how $<F>$ or $<D> \neq 0$ can be arranged at tree level I should point out an inherent difficulty in giving the spin-0 quark masses by letting them have $\tilde{Y} = +1$ under a new $\tilde{U}(1)$ with non-vanishing $<\tilde{D}>$. It is difficult, even perhaps impossible, to do that while insisting that the $\tilde{U}(1)$ be anomaly free (as required by the renormalizability of the theory) and that QCD not be spontaneously broken. Fayet's model[7)] is not anomaly-free, although supersymmetry is spontaneously broken and QCD is exact. Weinberg proposed a way to cancel the anomalies, however the addition of the required fields turns out to not allow supersymmetry breaking.[8] The origin of the difficulty is that to cancel the $\tilde{U}(1)$ color anomaly of the quarks, additional colored chiral fields with negative $\tilde{Y}$ must be added to the theory. However the $A_\phi^* A_\phi \tilde{g}\langle\tilde{D}\rangle$ contribution to their spin-0 $(\text{mass})^2$ is negative since for the quarks by construction it was positive. Unless this is cancelled, these spin-0 (colored) fields will get non-zero VEV, breaking color. No model is known which evades this problem and it may be possible to prove a no-go theorem here.

C. Forcing supersymmetry to break

Leaving aside dynamical supersymmetry breaking which is essentially completely conjectural up to now, and dimensional reduction which is too extensive a subject in its own right to explain in this lecture, there are two standard types of spontaneous supersymmetry breaking, known as the Fayet-Iliopoulos and O'Raifeartaigh mechanisms. In the first, addition of a $\tilde{V}$ term, when relative signs of ξ and the corresponding gauge coupling $\tilde{g}$ are appropriately chosen, leads to $\langle\tilde{D}\rangle \neq 0$; in the second, a $\mu^2[\phi]$ term, with other chiral interaction terms carefully chosen, leads to $\langle F_\phi\rangle \neq 0$.

For simplicity of illustration of the Fayet-Iliopoulos mechanism consider a single chiral field coupled to $\tilde{V}$ through

$$\sim [\phi^+ e^{\tilde{g}\tilde{V}}\phi]_{\theta\theta\bar{\theta}\bar{\theta}} + [\tilde{W}^\alpha \tilde{W}_\alpha]_{\theta\theta} + \xi[\tilde{V}]_{\theta\theta\bar{\theta}\bar{\theta}} \,.$$

Eliminating the auxiliary fields leaves as the potential

$$<V> \sim [\xi + \tilde{g}|<A_\phi>|^2]^2$$

By choosing $\tilde{g}/\xi>0$, $<V> \geq |\xi|^2$ and supersymmetry is broken. When there are several gauge groups there are several similar contributions to $<V>$.

The simplest $\mathcal{L}$ illustrating the O'Raifeartaigh mechanism requires three different chiral superfields, for instance

$$\mathcal{L} \sim [X^+X + Y^+Y + Z^+Z]_{\theta\theta\bar{\theta}\bar{\theta}} + [gXY^2 + X + mYZ]_{\theta\theta} \quad (21)$$

This yields

$$\begin{aligned} -F^*_X &= gY^2 + \lambda \\ -F^*_Y &= 2gXY + mZ \\ -F^*_Z &= mY \end{aligned} \quad (22)$$

These cannot simultaneously vanish so that supersymmetry is broken. (With only two chiral fields, there is always a supersymmetry conserving solution although it may require one of the scalar fields to have an infinite VEV.) It is typical of supersymmetric theories that only products or ratios of the scalar superfields are fixed at the potential minimum. Witten proposed[9)] the stragegy of using F-type breaking to "drive" supersymmetry breaking, letting radiative corrections supply spin-0 quark and lepton masses and lift the degeneracy of the vacuum. Recent supersymmetric GUTs incorporating this strategy are given in Ref. [5]. Note that the presence of a term $[XYZ]_{\theta\theta}$ in (21) would guarantee supersymmetry remains unbroken.

Such a term cannot be excluded by gauge invariance or discrete symmetries since the presence of X and YZ in $\mathcal{L}$ implies they are singlets. Thus unless it is excluded by an R invariance (see below) its absence is unnatural.

II. R-Invariance

A. General Remarks

The parts of the supersymmetry $\mathcal{L}$ (4) containing the gauge fields, V, are automatically invariant under a global U(1) rotation for which θ is assigned charge $R_o(\theta) = +1$ and $R_o(V) = 0$. Furthermore in the absence of dimensional parameters, m_i and λ_i^o, the $\mathcal{L}$ is automatically R_o invariant with $R_o(\phi) = 2/3$.[10] When there are dimensional parameters as well R invariance is not automatic, but is in fact present in many interesting cases. For instance in the example (21), an R-invariance is present under which $R_o(X) = +2$, $R_o(Y) = 0$ and $R_o(Z) = 2$. It seems generally to be the case, although I do not know if it is inevitable, that supersymmetric models with spontaneous F-type supersymmetry breaking will have an R-invariance if they are natural.

The higgs VEV's break R_o since they have $R_o = 2/3$, however if there is an additional gauged $\tilde{U}(1)$ in the theory under which both of the higgs have the same value of $\tilde{Y}$, say $\tilde{Y} = -2$, the higgs VEV's leave $R = R_o + 1/3\ \tilde{Y}$ unbroken. $\tilde{R}$ can be unbroken, or broken spontaneously by, e.g., $\langle X\rangle \neq 0$, or QCD condensates such as $\langle\lambda\lambda\rangle$, or by QCD anomalies. The analysis of the various possibilities can be found in Ref. [10] so I will only summarize the conclusions. Spontaneous $\tilde{R}$ breaking leads to unacceptable goldstone bosons. (In some cases this goldstone boson is axion-like and axion searches with negative results exclude it[10]; in the other cases beam-dump results are decisive[11]. Completely unbroken $\tilde{R}$, with the QCD $\tilde{R}$-anomaly of gluinos cancelled by that of $\tilde{R} < 0$ colored fermions (which need to be in the theory in any case to cancel the $\tilde{U}(1)$ anomaly), is a possibility. It would have to be realized through parity doubling among particles having $\tilde{R} \neq 0$, which is okay since only $\tilde{R} = 0$ particles have been observed so far. 't Hooft's anomaly-matching condition normally excludes strictly massive composite parity doublets, but not in this case when the anomaly vanishes. However the anomaly in the divergence of the $\tilde{R}$ current is, at one loop, the same by supersymmetry as that of θ^{μ}_{μ}. Hence a vanishing $\tilde{R}$ QCD anomaly is possible only if the QCD β-function vanishes to one loop; explicit calculation at two loops reveals a β-function such that QCD is not asymptotically free. For some people this is too hard a pill to swallow and they would prefer to avoid having the $\mathcal{L}$ be R invariant.

B. Gluino and Photino Masses

At tree level the fermionic partners of gauge bosons of an unbroken gauge group are massless even when supersymmetry is spontaneously broken: Since the photons and gluons must be massless by gauge invariance an explicit mass term is not acceptable. Furthermore a mass can be generated by the higgs mechanism only when a charged or colored scalar field is given a non-zero VEV, since the members of a gauge supermultiplet only couple to fields charged under the gauge group, and that is excluded by the requirement of unbroken U(1) and SU(3) gauge invariance. What about mass generation by radiative corrections?

Since gauginos have R = +1, a $\lambda\lambda$ mass-term is only possible when R-invariance is broken. If there is a suitable chiral multiplet with the same quantum numbers as the gaugino, then in principle an off-diagonal mass term $\lambda\psi$, mixing the gaugino and the fermionic member of that chiral multiplet may be possible. However, in fact, unless exotic chiral multiplets are added to the model[10,12], which may prevent supersymmetry breaking in realistic models, discrete symmetries are present which prevent off-diagonal R-conserving mass terms. Again the student should refer to ref. 10 for details.

If one sets aside the evidence against spontaneously broken R - coming from the absence of the required goldstone boson - one can ask what masses one would expect for photinos and gluinos[10),13)]. Remarkably strong bounds can be given[10)] in the case of $\tilde{D}$-type supersymmetry breaking - for instance, with maximal R-invariance breaking the gluino mass must be less than $\sim$1 GeV and the photino mass about a factor of one hundred smaller than that. The experimental limits are close to ruling out a massless gluino but it is not yet completely excluded (see below). On the other hand, a photino mass of $\sim$15 eV would be very nuce to solve astrophysical problems such as the galactic and larger scale missing mass[15)]. F-type models seem to give large gluino and photino masses quite naturally.[5)]

C. Other Phenomenological Consequences of "Low Energy" Supersymmetry

Among new ($R \neq 0$) colorless particles one has the goldstino (the goldstone fermion consequent to spontaneous supersymmetry breaking), the photino, and spin-0 leptons. The Goldstino in the $\tilde{U}(1)$ model[7)] mentioned above is a superposition of the fermionic members of the $\tilde{U}(1)$ gauge and the X chiral super-multiplets. As long as gravity can be ignored, so that supersymmetry is a global symmetry, the goldstino must be strictly massless. This is acceptable experimentally, as is the low mass of the photino, because the effective interactions of these two with matter is weak, neutral-current-like, as shown in Fig. 3.[16)] The amplitude is $\sim e^2/m_s^2$. From PETRA we know that the spin-0 e, μ, and τ (and q's) have masses $\gtrsim$ 16 GeV. If not, a striking signal in $e^+e^- \rightarrow$ non-coplanar (e^+e^-) + missing energy would have been seen, coming from diagrams such as Fig. 4.[17)]

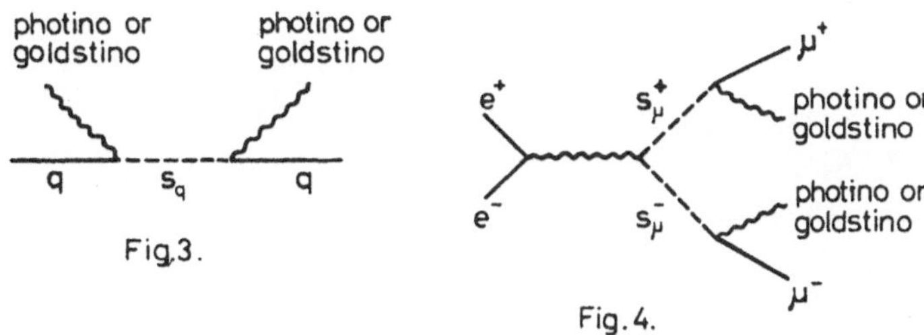

Fig.3.

Fig.4.

Colored R $\neq$ 0 particles are bound into hadrons. The lightest of these is probably a gluino-gluon or gluino-$q\bar{q}$ bound state. From estimates of glueball masses one would expect these to have masses $\sim 1\frac{1}{2}$ GeV if the gluino is massless. In that case their lifetimes would be $O(10^{-13})$ sec decaying into ordinary hadrons + a goldstino or photino which escapes. Beam dump experiments exclude such particles with masses less than $1\frac{1}{2}$ or 2 GeV[14)] and these limits will be improved in the next several years. Although a massless gluino is not completely excluded, it seems an improbable possibility

References

1. Like M. Grisaru's "Lectures on Supergraphs" on this point.
2. J. Wess, IAS preprint, to be published in Princeton University Press.
3. M. Grisaru, M. Rocek and W. Siegel, Nucl. Phys. B159, 429 (1979).
4. E. Witten, Nucl. Phys. B185, 513 (1981).
5. S. Dimopoulos and S. Raby, LASL preprint 1982; M. Dine and W. Fischler, IAS preprint 1982.
6. M. Suzuki, LBL preprint UCB-PTH-82/7.
7. P. Fayet, Phys. Lett. 69B, 489 (1977).
8. S. Weinberg, Phys. Rev. D, to be published; L. Alvarez-Gaume, M. Claudson and M. Wise, Harvard preprint 1981.
9. E. Witten, Phys. Lett. 105B, 267 (1981).
10. G. R. Farrar and S. Weinberg, "Supersymmetry at Ordinary Energies II: R-Invariance, Goldstone Bosons, and Gauge Fermion Masses", preprint in preparation, RU-82-38.
11. G. R. Farrar and E. Maina, Rutgers preprint RU-82-34.
12. P. Fayet, Phys. Lett. 78B, 417 (1978).
13. G. R. Farrar, "Mass Splitting in Supersymmetric Theories From An Effective Lagrangian Point of View", RU-82-12, to be published in Nucl. Phys. B.
14. Hadrons containing a gluino must have masses $\lesssim 1\frac{1}{2}$-2 GeV (See G. R. Farrar and P. Fayet, Phys. Lett. 76B, 575 (1978); ibid 79B, 442 (1978)); since glueballs with "valence" massless gluons are thought to have masses $\sim 1\frac{1}{2}$ GeV we cannot yet exclude $m_{gl} = 0$ with certainty. G. Kane and J. Leveille, Phys. Lett. 112B, 227 (1982) have used perturbative QCD to exclude the range 1 GeV $\lesssim m_{gl} \lesssim 3$ GeV.

15. N. Cabibbo, G. Farrar, L. Maiani, "Massive Photinos: Unstable and Interesting", Phys. Letts. 105B, 155 (1981).

16. G. R. Farrar and P. Fayet, ref. 14 and P. Fayet, Phys. Lett. 86B, 272 (1979).

17. G. R. Farrar and P. Fayet, Phys. Lett. 89B, 191 (1980).

Deep Inelastic Scattering and Jets

H.E. Montgomery
CERN
Geneva, Switzerland

Abstract

An attempt is made to describe some of the interest and fascination of Deep Inelastic Lepton Scattering. The basic kinematics is outlined and various experimental aspects are discussed. The data, both on Structure Functions and on Hadron Production, are discussed first within the framework of the Quark Parton Model and then within that of perturbative QCD.

Contents

I) Introduction

In general the aim of the physicist and especially the high energy physicist is to understand the most fundamental forces in matter. This implies that a study is made of matter in its most particular form. The conventional wisdom that arose from studies of the spectrum of hadrons [1] was that hadrons were made of quarks. This scenario even had the success of incorporating into the scheme the new quark flavour, charm. However searches for free quarks have met with a singular lack of success [2] which in turn has meant that experimental work with quarks has been to a greater or lesser extent indirect.

One of the most direct of these indirect methods has been provided by Deep Inelastic Lepton Scattering where, as we shall see, the data can be extremely well described by assuming that the quarks in whatever nucleon, proton or neutron, or nucleus are completely free and independent of their environment. Measurements of electron scattering and nuclei show the clear signature of quasi-elastic scattering from the individual nucleons. Electron, muon and neutrino scattering from nucleons is described by the incoherent sum of quasi-elastic scatters from the partons. This scenario originally proposed by Bjorken [3] and rapidly supported by experiment [4] has formed the basis of $\sim$ 15 years of experimental and theoretical effort.

II) Kinematics and Cross-sections

The basic deep inelastic lepton scattering process is illustrated in fig. 1 in which are also given the kinematic notation and definitions which will be used later.

The derivation of the expression for the cross-section has been given many times in many different ways [5] and only the results will be recalled here [6]. The cross-section for charged lepton (e or μ) scattering within the 1-photon exchange approximation is given by

$$\frac{d\sigma}{d\Omega dE'} = \frac{4\alpha^2(E'^2)}{Q^4} \left(\cos^2\frac{\theta}{2} W_2(\nu,Q^2) + 2\sin^2\frac{\theta}{2} W_1(\nu,Q^2)\right) \qquad (1)$$

$$= \Gamma_t \left(\sigma_T + \varepsilon\sigma_S\right) \qquad (2)$$

W_1 and W_2 are the structure functions of the nuclear target defined in a way analogous to the Electromagnetic Form Factors G_E and G_M of the nucleon but in

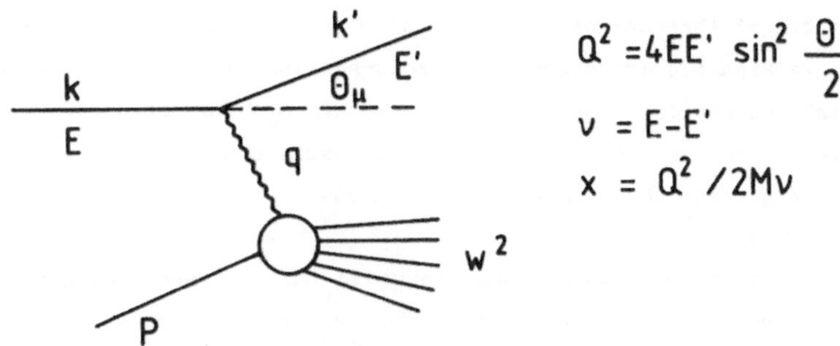

Fig. 1 - Deep Inelastic Scattering process and kinematics.

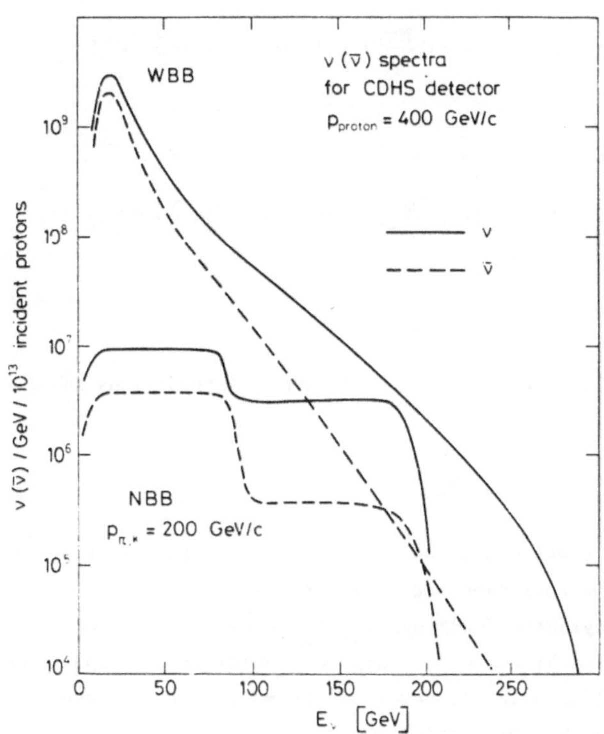

Fig. 2 - CERN neutrino beam flux.

general functions of the two variables Q^2 and ν rather than just Q^2. The entirely equivalent expression (2) expresses the cross-section in terms of a photon flux Γ_t and the total absorption cross-sections σ_T and σ_S for transverse and scalar photons (since the photon has $Q^2 \neq 0$ it was one extra degree of polarisation compared to the purely transverse real photon with $Q^2 \equiv 0$).

Recently it has become conventional to write the cross-section for both charged lepton and neutrino scattering in a common form so as to emphasise their similarity.

Charged Leptons

$$\frac{d\sigma^{e,\mu}}{dxdy} = \frac{4\pi\alpha^2 ME}{Q^4}\left[y^2 xF_1(x,Q^2) + \left(1-y-\frac{Mxy}{2E}\right)F_2(x,Q^2)\right] \qquad (3)$$

Neutrinos

$$\frac{d\sigma^{\nu}}{dxdy} = \frac{G^2 ME}{\pi}\left[xy^2F_1(x,Q^2) + \left(1-y-\frac{Mxy}{2E}\right)F_2(x,Q^2) + y\left(1-\frac{y}{2}\right)xF_3(x,Q^2)\right] \qquad (4)$$

Antineutrinos

$$\frac{d\sigma^{\bar{\nu}}}{dxdy} = \frac{G^2 ME}{\pi}\left[xy^2F_1(x,Q^2) + \left(1-y-\frac{Mxy}{2E}\right)F_2(x,Q^2) - y\left(1-\frac{y}{2}\right)xF_3(x,Q^2)\right] \qquad (5)$$

Some remarks are in order.

1) The change from electromagnetic to weak interactions involves the simple substitution

$$\frac{4\pi\alpha^2}{Q^2} \leftrightarrow \frac{G^2}{\pi}$$

2) Since parity is not conserved one extra structure function $xF_3(x,Q^2)$ is required in the weak case and this changes sign as $\nu \leftrightarrow \bar{\nu}$.

3) Although the symbolic functions F_1, F_2 used are the same in each case, there is no general reason to suppose a physical correspondence, this only comes either from data (a powerful argument) or through specific models such as the Quark Parton Model.

4) The expressions are approximations assuming such reasonable things as negligible lepton mass, but also more arguably single boson exchange and for neutrinos an infinite boson mass.

Contact with the Parton Model

Consider the electromagnetic scattering of leptons on quarks with charge q_i of different helicity

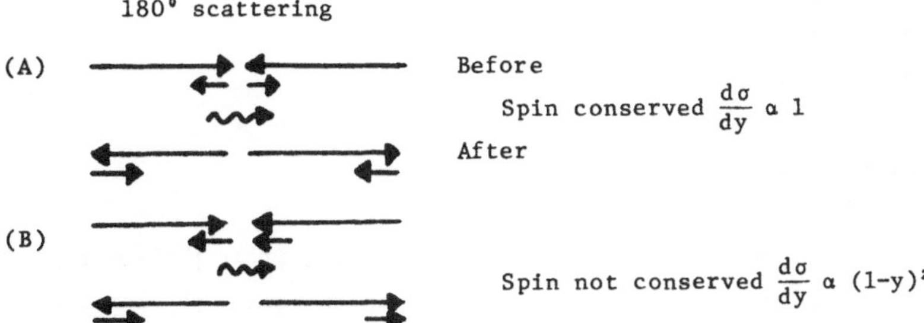

Now the electromagnetic coupling γ_μ can be written

$$\gamma_\mu = \frac{1}{2}\,[\gamma_\mu(1+\gamma_5) + (\gamma_\mu(1-\gamma_5)].$$

Each of the two terms is equivalent to the spin projections illustrated above so that the overall cross-section for the scattering has the form

$$\frac{d\sigma}{dy} \propto \frac{1}{2}\,[1 + (1-y)^2],$$

more precisely for a quark i with charge q_i

$$\frac{d\sigma}{dy} = \frac{4\pi\alpha^2}{Q^4}\,ME\,q_i^2\,\frac{1}{2}\,[1 + (1-y)^2].$$

Now, considering several quarks i each carrying a momentum fraction x of the proton momentum such that the distribution in x is $f_i(x)$, then

$$\frac{d\sigma}{dxdy} = \frac{4\pi\alpha^2}{Q^4}\,ME\,(\sum q_i^2 x f_i(x))\,\frac{1}{2}\,[1 + (1-y)^2]$$

$$= \frac{4\pi^2\alpha^2}{Q^4}\,ME\,(\sum q_i^2 x f_i(x))\,[\frac{y^2}{2} + (1-y)] \qquad (7)$$

A comparison of terms between equations (3) and (7) then yields the identification

$$2xF_1 \leftrightarrow \sum q_i^2\,x\,f(x)$$

ie. the structure functions are directly measures of the momentum distribution of quarks in the target nucleon. This identification made most explicitly $\gtrsim$ 10 years ago [7] will be the start point for comparison of data with models.

III) Experiments

The results discussed later have the following properties. They come from:

DIFFERENT EXPERIMENTS
with DIFFERENT BEAMS
DIFFERENT APPARATUS
DIFFERENT AIMS
and DIFFERENT CAPABILITIES

Since the audience is composed mainly of young theorists, it is perhaps appropriate to discuss some aspects of the daily work of the experimental physicist [8] in a modest attempt to reduce (in the future) comments of the type:

"We can calculate x, y, α with high precision, but (uncomprehending eyes to heaven) apparently the experimentalists find it hard to measure"

or (tone of outrage)

"the experimentalists must make every effort to understand their errors".
The truth is that often a number for the answer is very quickly obtained, but the experimentalist spends the longer time from that point to publication in understanding the systematic error without which the measurement has no value.

As pointed out in the previous section, the cross-section is a function of three variables (assuming the target to be at rest)

$$\frac{d\sigma}{dxdQ^2} = f\ (E,\ E',\ \theta)$$

therefore experiments measure all three of these or equivalent linear combinations.

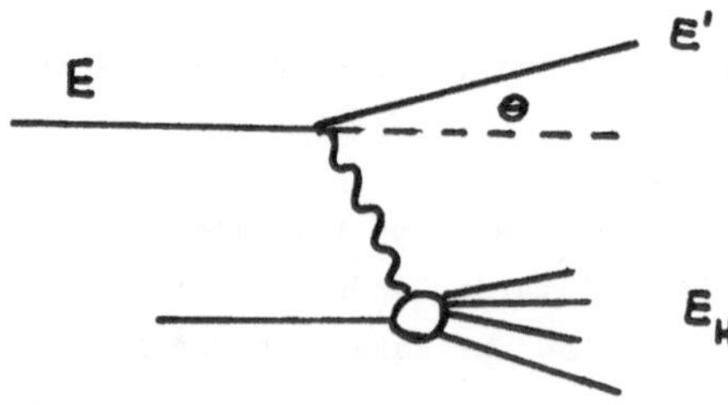

In general charged lepton experiments measure the incident and outgoing lepton energies and the scattering angle more or less directly. The neutrino experiments on the other hand cannot measure with any precision the incident neutrino energy and so measure the total hadronic energy E_h. Furthermore the mode of measurement of

E_h differs drastically between the electronic neutrino experiments which use calorimetric methods and the bubble chambers which measure and sum all tracks. This situation is summarised in Table I, what needs to be considered is the reflection on the structure functions $F(x,Q^2)$ of

a) Resolution smearing

b) Systematic measurement errors

on these quantities. My remarks, for lack of time, will be confined to some aspects of the latter in the hope that an incomplete story nevertheless prompts interest.

1) Electronic Neutrino Experiments, eg. CDHS

The beam used by the CDHS experiment is the narrow band beam at the CERN SPS. The flux in this beam is sketched as a function of neutrino energy in fig. 2, the two visible components arise from decay of pions and kaons respectively. In order to determine the absolute flux at a given energy it is necessary to have detailed knowledge of the π and K production spectra, to measure the relative π, K content in the hadron beam with a Cerenkov counter and also to monitor the muon yields at various points in the shielding. The commonly estimated combined systematic error from these various measurements is $\sim 5\%$ however from the spread of world measurements of $\frac{\sigma}{E}$ [9] this appears to be an underestimate.

Within a narrow band beam there is a well defined relationship between the radius of impact of the neutrino on the detector and its energy. This relationship allows events at a given energy to be separated into "pion" and "kaon" samples, but while it provides a useful check it's precision is low. This difficulty to measure the neutrino energy implies that the experiment has to measure E_h extremely well.

The CDHS apparatus [10] sketched in fig. 3 is a magnetised iron toroid with calorimeters which had (for the measurements discussed later) scintillators interleaved between 5 cm iron plates for a part of the detector and 15 cm iron in the major part. With 5 cm Fe spacing the resolution is given by $\frac{\Delta E}{E} \sim \frac{100\%}{\sqrt{E}}$ which clearly gives important smearing corrections at low ν. The absolute calibration which determines the resultant systematic error depends on five separate operations.

i) The response of a small (baby) calorimeter with the same characteristics as a large module is measured in a hadron test beam. This gives the response in terms of GeV/peak of muon pulse height. It also permits the development of an algorithm for taking proper account of the electromagnetic component of the shower.

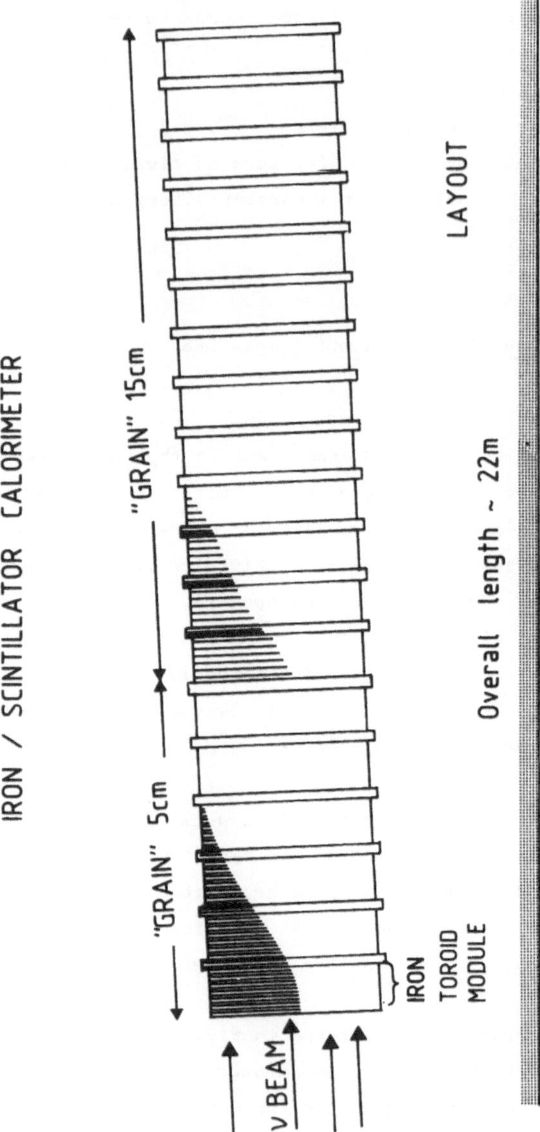

Fig. 3 - Sketch of CDHS experiment apparatus.

ii) The response of a few full size modules is measured in a hadron beam with respect to a distributed laser pulse height (GeV/LASER).

iii) Cosmic muons are measured in all modules between SPS spills giving a measure of LASER/COSMIC MEAN.

iv) A combination of ii) and iii) gives GeV/COSMIC MEAN.

v) The energy deposition in the neutrino beam is checked to be the same in all modules.

The final systematic error quoted is of the order of 1%! The importance is demonstrated in fig. 4 where, for a series of discrete neutrino beam energies, the ratio F_2^{True}/F_2^{Meas} for two different supposed systematic errors. The error is assumed to lead to a non linearity and even when, as in fig. 4, the deviation is 0.00025/GeV, it generates a non scaling of $\sim$ 5%.

2. Muon Scattering Experiments, eg. BCDMS, EMC.

Sketches of the apparatus for these experiments are shown in fig. 5. In the SPS muon beam the transmitted momentum bite is $\sim$ ±10% and typical beam fluxes are $5 \cdot 10^6 - 5 \cdot 10^7$ muons/SPS pulse (1.5-2.5 secs).

Since muons are charged particles the situation is vastly different from the neutrino case, the incident flux is in general measured in one of two ways:

i) The incident beam is counted directly and corrections for counter and electronics dead time are made. Corrections have also to be applied to account for the extra beam requirements which may be made in the physics analysis.

ii) A random trigger [11] is used to take beam data in a well defined time window. These data are then passed through exactly the same analysis as physics events and a measure of the useful flux (with whatever analysis requirements) is directly obtained. The typical normalisation error estimated for a given data set is ≲3%

The momentum for individual beam tracks is also measured and the systematic error on this measurement is controlled by:

i) The absolute field measurement and surveying of the beam spectrometer;

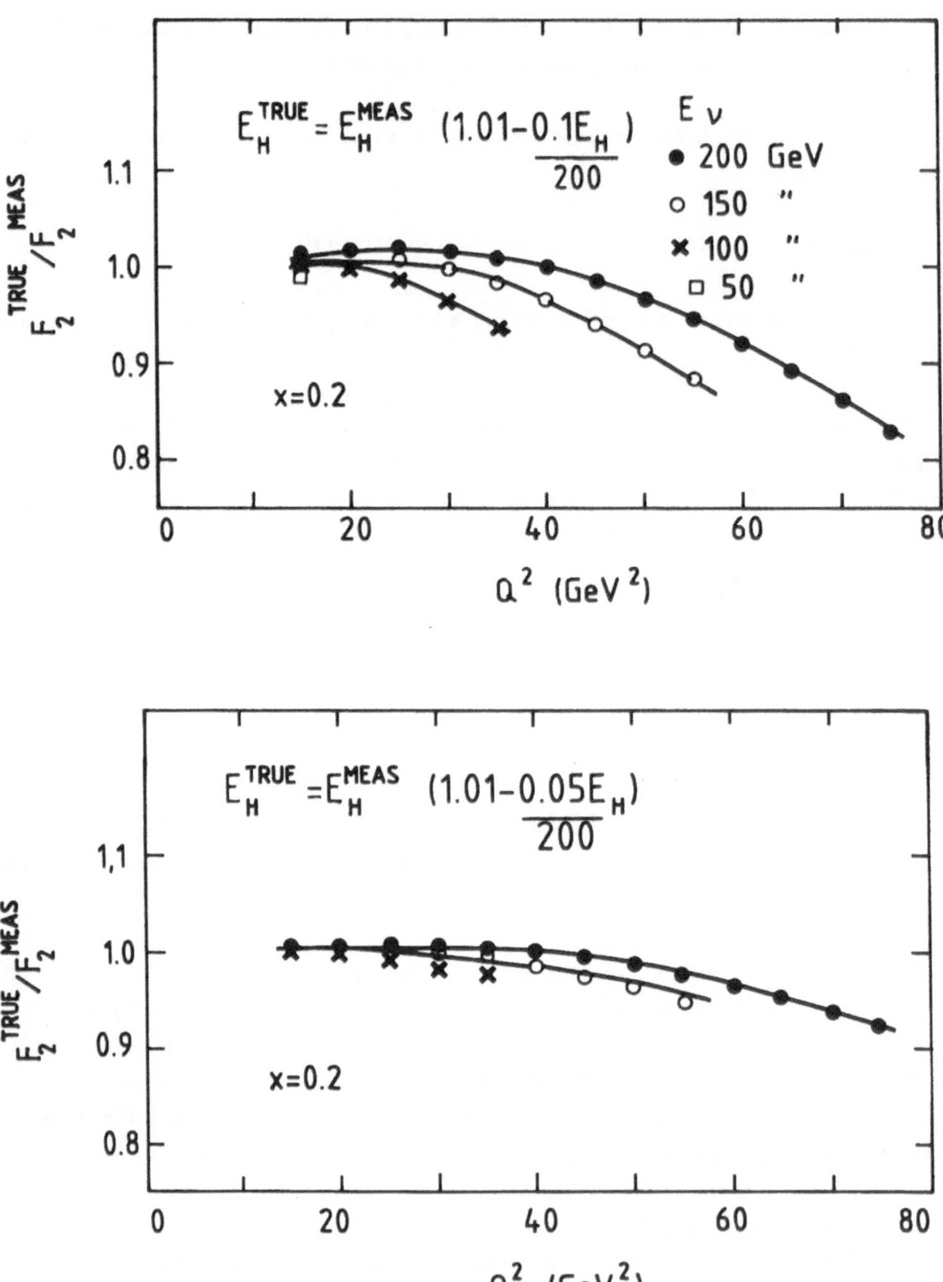

Fig. 4 - Ratio between F_2^{True} and F_2^{Meas} assuming a systematic error in the measurement of E_H the total hadronic energy.

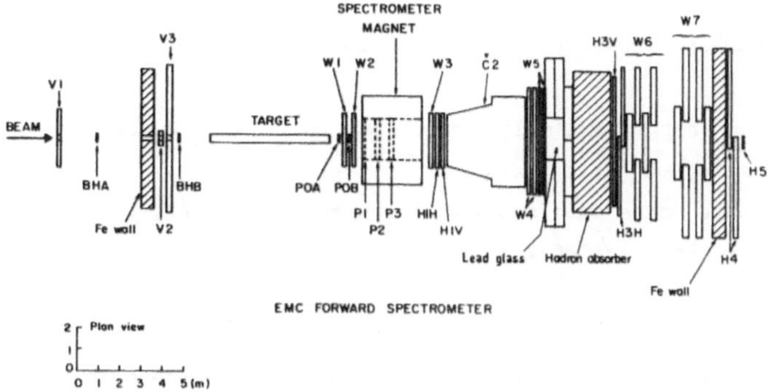

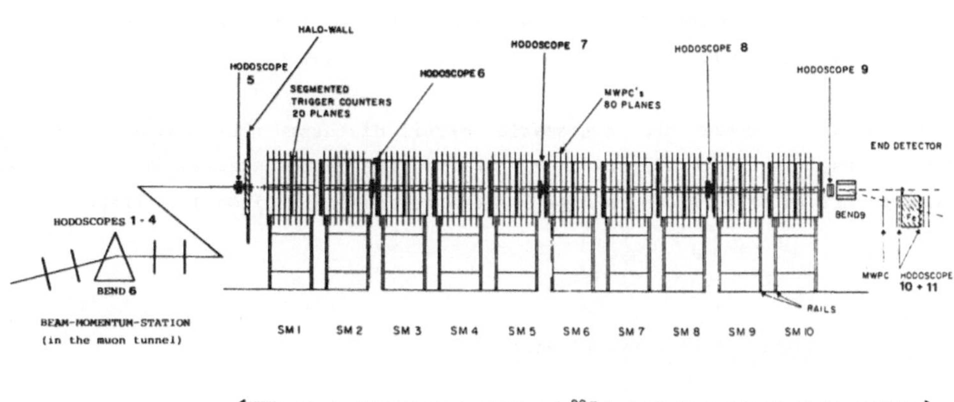

Fig. 5 - a) EMC experiment apparatus
b) BCDMS experiment apparatus.

ii) A cross-check of the beam spectrometer with the experiment's scattered muon spectrometer. The absolute error obtained is $\sim$ 0.3-0.4%.

The scattered muon energy similarly depends on the absolute measurement of the experiment's spectrometer magnetic field and detector survey. A good check is provided by the absolute mass obtained for known resonances K^0, ψ. An example dimuon spectrum showing the ψ in EMC liquid target data is shown in fig. 6. Again the precision is $\sim$ 0.3-0.4%.

The effect on the measurement of F_2 is shown in fig. 7 where we have exagerated the effect by considering a 1% error on incident beam momentum and we show only a single muon energy, in general the systematic errors from different beam energies are large in a different regions of phase space. It can also be shown [12] that if the relative error between incident and scattered muon can be reduced then the sensitivity to the absolute scale is much diminished at least for a measurement of the slopes with Q^2 of F_2.

Taking into account the systematic errors discussed above and all others and combining their separate effects on F_2 yields the representation of the final systematic errors on F_2 shown in fig. 8 which refers to EMC iron data [13].

3. Neutrino Bubble Chamber Experiments

Because of their lack of statistics bubble chamber measurements of scale breaking of F_2 are of little significance however they are still used to make measurements of the final state hadrons and for this aim there are no competitive neutrino experiments. However the weakness engendered by the lack of knowledge of the neutrino energy still appears, and is perhaps increased, because often the wide band neutrino beam is used for which there is no relationship between impact radius and energy.

The total hadronic energy is obtained by adding up the momenta of all event associated tracks observed. For a hydrogen filling this implies charged hadron tracks, but for a neon fill can also include e^+e^- pairs from the photons of π^0 decay (i.e. the systematic errors are different in the two cases).

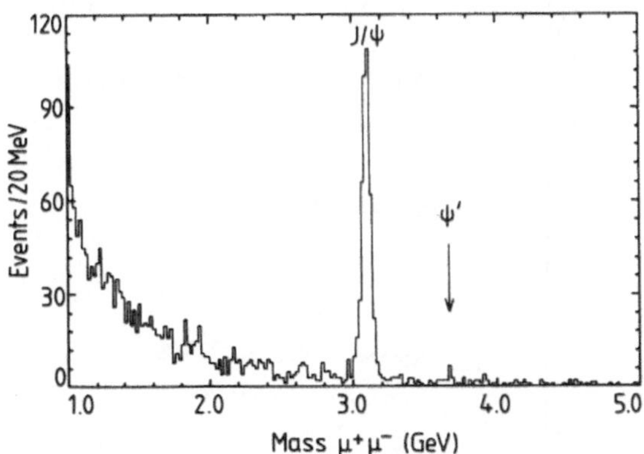

Fig. 6 - Dimuon spectrum showing J/ψ peak as measured with H_2/D_2 targets in EMC experiment.

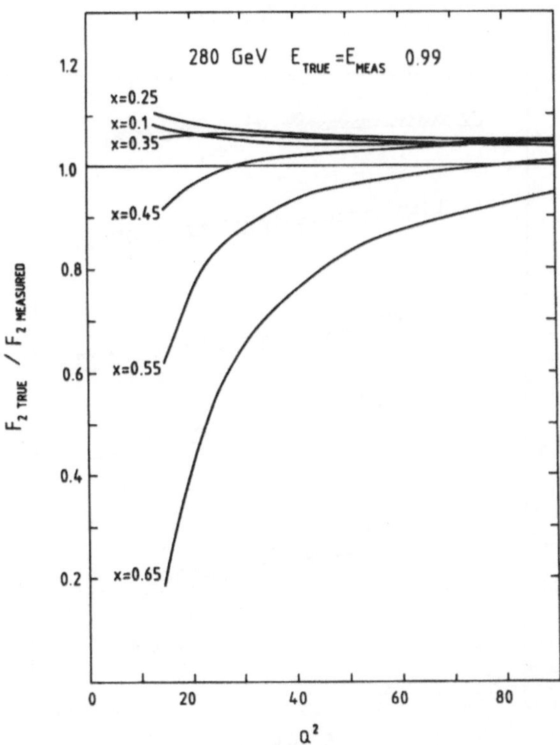

Fig. 7 - Effect of 1% systematic error in beam energy on measurement of F_2 with 280 GeV muon beam.

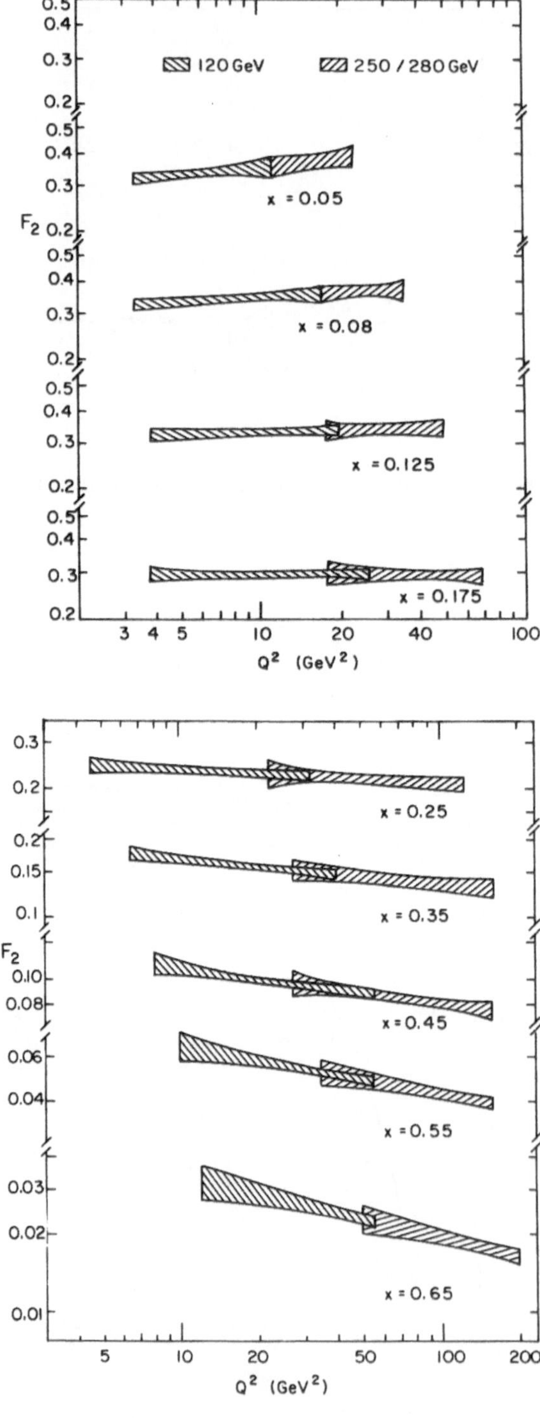

Fig. 8 - Overall systematic errors on F_2 for EMC iron data.

The techniques used are ways to make the transition from the transverse momentum conservation constraint

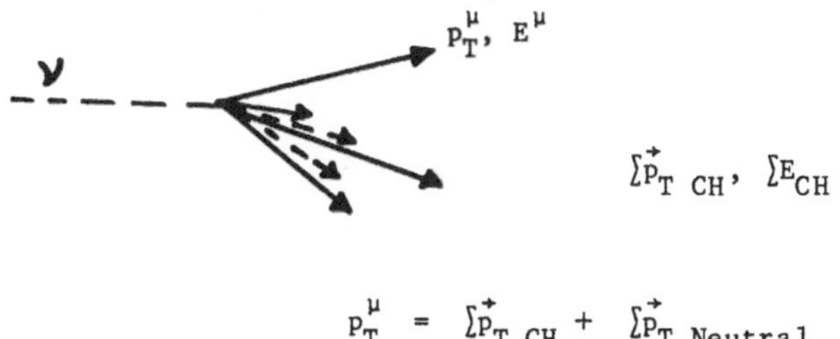

$\sum\vec{p}_{T\ CH}$, $\sum E_{CH}$

$$p_T^\mu = \sum\vec{p}_{T\ CH} + \sum\vec{p}_{T\ Neutral}$$

to the determination

$$E_H = \sum E_{H\ CH} + \sum E_{Neutral}.$$

This jump is made using different models and the reliability is studied using Monte Carlo data with some physics models included. The kind of variation possible is illustrated in fig. 9 which shows the estimated bias from three different methods, the figure is taken from the doctorate thesis of R. Giles [14] and refers to a hydrogen experiment in BEBC.

The importance of such errors for the measurement of a typical quantity, eg. $\frac{1}{N_{ev}}\frac{dN}{dz}$ can be estimated.

Typically

$$\frac{1}{\sigma}\frac{d\sigma}{dz} = \frac{1}{N}\frac{dN}{dz} \equiv f \sim e^{-6z} \qquad \text{where} \quad z = \frac{E_h}{\nu} = \frac{E_h}{E_H}$$

$$\text{if} \quad \left(\frac{\Delta\nu}{\nu}\right)^{Syst} = a \qquad \text{then } \left(\frac{\Delta z}{z}\right)^{Syst} = a, \qquad \text{consequently} \quad \left(\frac{\Delta f}{f}\right)^{Syst} = -6az$$

If one takes the case of METHOD 2 of fig. 9 then with $E_H > 20$ GeV and high z we obtain $\left(\frac{\Delta f}{f}\right)^{Syst} \sim 30\text{–}40\%$ as shown in fig. 10a). For muon scattering with a 280 GeV beam systematic effects of the order of 1% on E_μ^{Scatt} coupled with a desire to limit systematic effects to less than 10% gave [15] a low ν cut of 70 GeV (fig. 10b) and again emphasised the importance of a second energy (120 GeV) which allowed scale breaking effects to be measured down to ν = 30 GeV.

These examples by no means give a complete treatment of systematic errors but are meant to illustrate the level of thought which should go into whatever statement is made about systematic effects. A further practical point is that often the subtleties are only discussed in theses, a serious treatment is never possible in a letter publication.

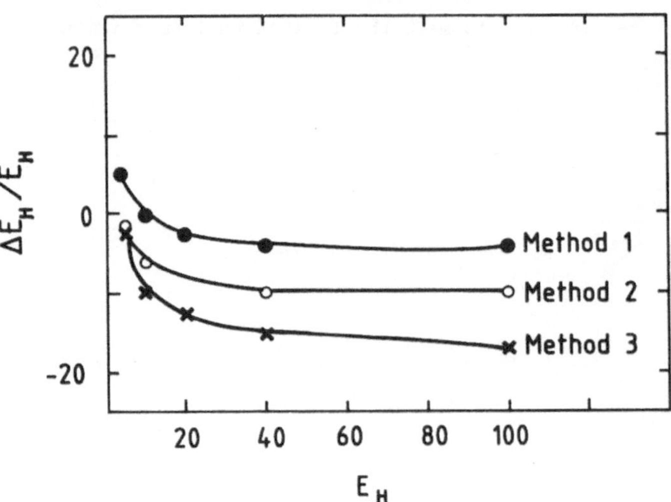

Fig. 9 - Bias resulting from using different methods to estimate the total hadronic energy in a hydrogen bubble chamber experiment [14].

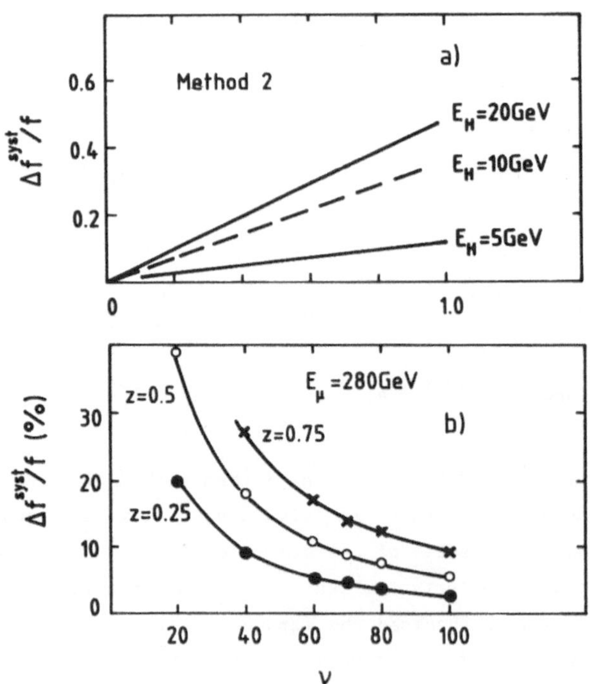

Fig. 10 - a) Systematic error in $\frac{1}{N}\frac{dN}{dz}$ resulting from bias of method 2, fig. 9.
b) Systematic error as a function of ν in $\frac{1}{N}\frac{dN}{dz}$ resulting from 1% systematic error in scattered muon momentum.

IV. Structure Functions in the Quark Parton Model

As was seen in section II for a world with only spin $\frac{1}{2}$ partons we have the identification

$$2xF_1(x) = F_2(x) \equiv \sum \varepsilon_i^2 x f_i(x)$$

for charged lepton scattering. Specifically for a proton target

$$F_2^{\mu p}(x) = \frac{4}{9} x\,[u^p(x) + \bar{u}^p(x)] + \frac{1}{9} x\,[d^p(x) + \bar{d}^p(x)] + \frac{1}{9} x\,[s^p(x) + \bar{s}^p(x)]$$

If we define $u^p(x) = u$

$d^p(x) = d$ where the argument x is dropped for brevity, then isospin invariance yields $u^n = d$

$d^n = u.$

It is possible to perform a series of manipulations and obtain Quark Parton Model (QPM) predictions which can be compared directly with the data.

Example I - Proton vs Neutron

$$\frac{1}{x} F_2^{\mu p} = \frac{4}{9}(u+\bar{u}) + \frac{1}{9}(d+\bar{d}) + \frac{1}{9}(s+\bar{s}) + \ldots.$$

$$\frac{1}{x} F_2^{\mu n} = \frac{1}{9}(u+\bar{u}) + \frac{4}{9}(d+\bar{d}) + \frac{1}{9}(s+\bar{s}) + \ldots.$$

$$F_2^{\mu p} - F_2^{\mu n} = \frac{x}{3}(u-d)$$

This quantity contains no contribution from sea quarks (since they should be the same in proton and neutron) and we could expect that it should show a quasi-elastic quark scattering peak, the data are shown in fig. 11 and indeed a broad peak at $x \sim 0.3$ is observed. If alternatively the ratio F_2^n/F_2^p is constructed then there are QPM limits

$$\frac{1}{4} < \frac{F_2^{\mu n}}{F_2^{\mu p}} < 4$$

the value 1 would be expected if the sea dominated and $\frac{1}{4}$ if a single u quark dominated. The data are shown as a function of x in fig. 12 and we see that they push close to the lower QPM limit.

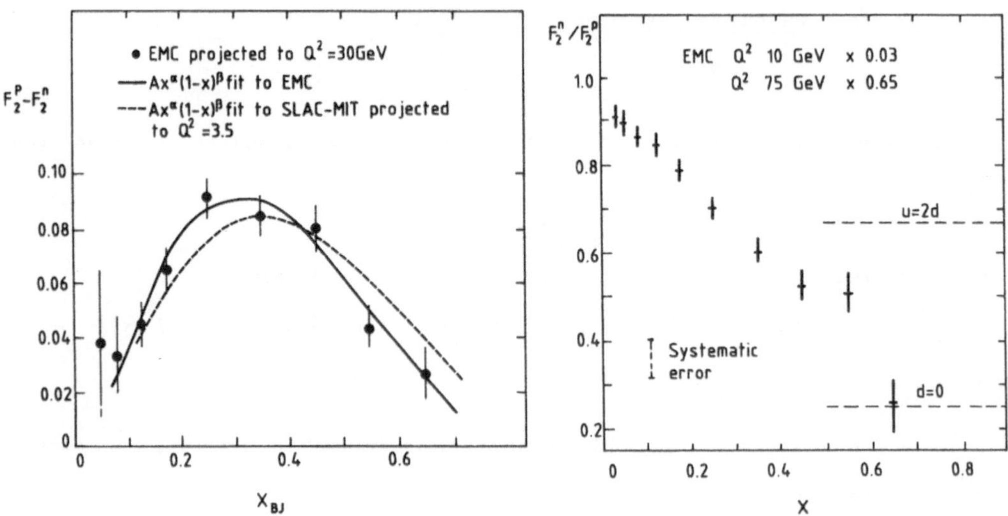

Fig. 11 - Difference $F_2^{p}-F_2^{n}$ from EMC.

Fig. 12 - Ratio F_2^{n}/F_2^{p} as a function of x.

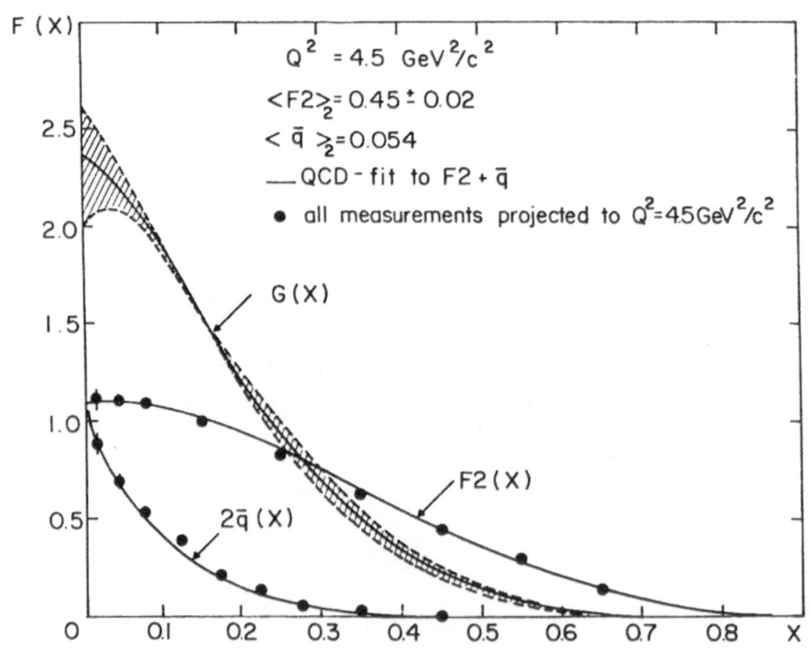

Fig. 13 - Decomposition of $F_2^{\nu N}$, $xF_3^{\nu N}$ into F_2, $\bar{q}$ and gluon distributions at $Q_0^2 = 5$ GeV² (CDHS).

Example II - Measurement of Quark Sea with neutrinos and antineutrinos

$$d\sigma^{\nu} + d\sigma^{\bar{\nu}} \rightarrow F_2^{\nu N}$$

$$d\sigma^{\nu} - d\sigma^{\bar{\nu}} \rightarrow F_3^{\nu N}$$

$$F_2^{\nu m} = 2x\,[u + d + \bar{u} + \bar{d} + \ldots.$$

$$xF_3^{\nu N} = 2x\,[u - \bar{u} + d - \bar{d} + \ldots.$$

$$\frac{F_2^{\nu N} + xF_3^{\nu N}}{2} = 2x\,[u + d + \ldots.] \rightarrow q(x)$$

$$\frac{F_2^{\nu N} - xF_3^{\nu N}}{2} = 2x\,[\bar{u} + \bar{d} + \ldots.] \rightarrow \bar{q}(x)$$

Recent data from CDHS are shown in fig. 13, the $\bar{q}(x)$ distribution clearly peaks towards $x = 0$ as we expect.

Example III - Total Quark Momentum

$$\int F_2^{\nu N}(x)dx = \int 2x\,[u(x) + d(x) + \bar{u}(x) + \bar{d}(x) + \ldots\,]\,dx$$

$$\int F_2^{\nu N}(x)dx \quad 2(1-\varepsilon)$$

where ε is the fraction of the proton momentum which is carried by partons which have no coupling to the exchanged boson. Alternatively

$$\int F_2^{\mu N}(x)dx = \int \tfrac{5}{9}x\,[u(x) + \bar{u}(x) + d(x) + \bar{d}(x) + \ldots\,]\,dx$$

$$\frac{9}{5}\int F_2^{\mu N}(x)dx = 1-\varepsilon$$

The data since 1970 have indicated that $\varepsilon \sim 50\%$ and this fact was at the origin of the term "gluon", the theory of QCD came later.

Example IV - Relation μN to νN

As implied in the previous example there is a clear QPM prediction for the relationship between $F_2^{\mu N}$ and $F_2^{\nu N}$

$$F_2^{\nu N} = 2x\,[u + \bar{u} + d + \bar{d} + \ldots\,]$$

$$F_2^{\mu N} = \tfrac{5}{9}x\,[u + \bar{u} + d + \bar{d} + \ldots\,]$$

$$\frac{F_2^{\mu N}}{F_2^{\nu N}} \simeq \frac{5}{18}$$

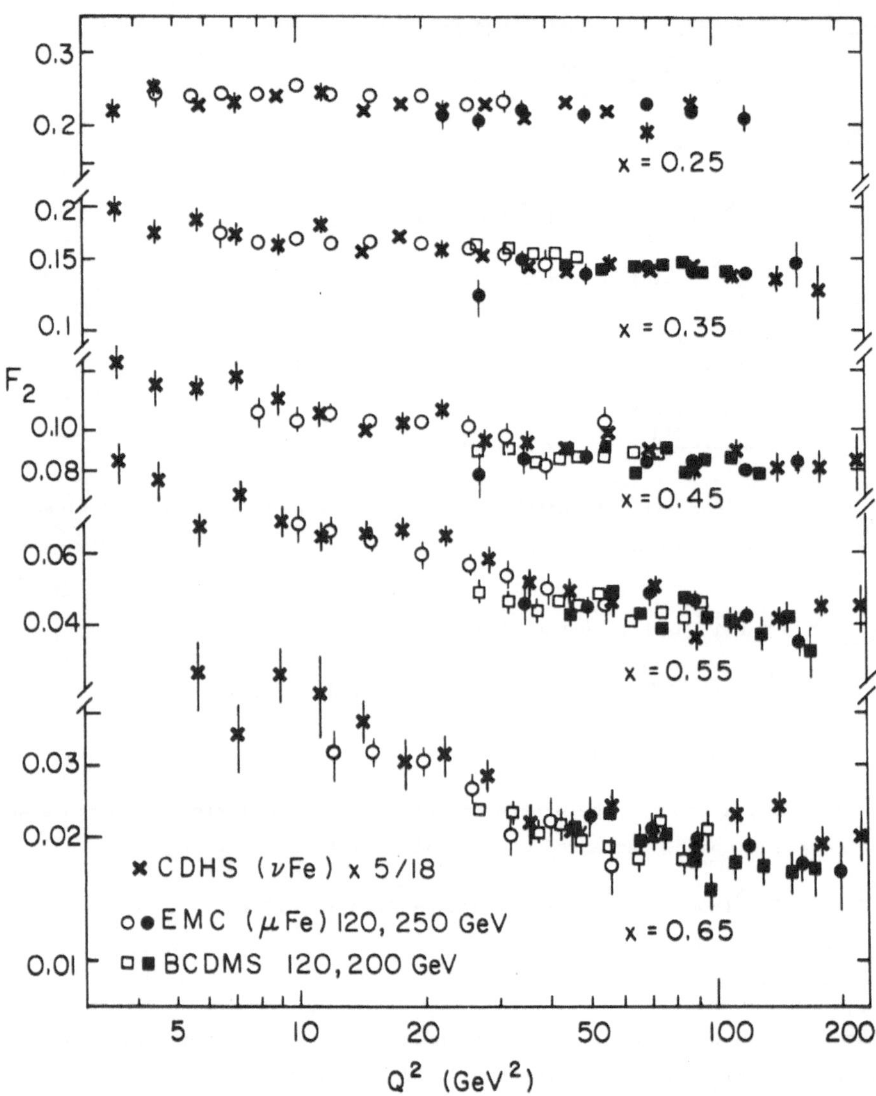

Fig. 14 - Comparison between F_2^N measured in νN interactions by CDHS and in muon interactions by BCDMS and EMC assuming ratio 5/18.

A recent data comparison between EMC, BCDMS and CDHS is shown in fig. 14 and in the high x region there is remarkable agreement. At low x the presence of strange and charm quarks gives a correction term

$$F_2^{\mu N} - \frac{5}{18} F_2^{\nu N} = \frac{1}{3} (C-S)$$

The problem with using this relation is that the fact that $M_{charm} \neq 0$ gives very significant threshold corrections to the contribution of s, $\bar{s}$, c and $\bar{c}$ to the neutrino data and to the contribution of $c\bar{c}$ to the muon data.

As has already been to some extent shown in these previous examples, the QPM gives a wealth of predictions and relationships with which the data agree very well. Most of the predictions however are sensitive only to the u, $\bar{u}$, d, $\bar{d}$, i.e. light quark distributions. We can also see what the data can say about the $s\bar{s}$ and $c\bar{c}$ distributions.

Minority Quark Distributions I - Determination of the Strange Sea

When a neutrino interacts with the various quarks there are two ways to produce charm quarks

$\nu + d \rightarrow \mu^- + c$ valence, Cabibbo Suppressed

$c \rightarrow s + \mu^+ + \nu$

$\nu + s \rightarrow \mu^- + c$ sea, Cabibbo Favoured

and for antiquarks

$\bar{\nu} + \bar{d} \rightarrow \mu^+ + \bar{c}$ sea, Cabibbo Suppressed

$\bar{c} \rightarrow \bar{s} + \bar{\mu} + \nu$

$\bar{\nu} + \bar{s} \rightarrow \mu^+ + \bar{c}$ sea, Cabibbo Favoured

In all these processes the charm hadron produced decays $\sim 10\%$ of the time semileptonically giving a second muon in the final state and also a significant missing energy corresponding to the neutrino. This clear experimental signature allows the s quark distribution in the nucleon to be determined from the following expressions

$$d\sigma^{\nu} = \frac{G^2 M E_{\nu} x}{\pi} [U_{cd}^2 [u+d] + |U_{cs}^2| 2s]$$

$$d\sigma^{\bar{\nu}} = \frac{G^2 M E_{\nu} x}{\pi} [U_{cd}^2 [\overline{u+d}] + |U_{cs}^2| 2\bar{s}]$$

the charm quark mass necessitates the use of

$\xi = x + \frac{M_c^2}{2M\nu}$ in place of x and gives an extra factor $(1 - y + \frac{xy}{\xi})$

in the cross-section. The results of measurements by CDHS [16] are shown in fig. 15, it is clear that the antineutrino case with a negligible light quark contribution is very clean. The result for the relative SU (3) breaking is given by

$$\frac{\int x\,(s(x) + \bar{s}(x))\,dx}{\int x\,(\bar{u}(x) + \bar{d}(x))\,dx} = 0.52 \pm 0.09$$

Minority Quark Distribution II - Determination of Charm Sea

In terms of quark interactions the charm quark should contribute to muon interactions through

$$\mu^+ + c \rightarrow \mu^+ + c$$
$$\qquad\qquad \rightarrow c \rightarrow \mu^+ + \nu + s$$

$$\mu^+ + \bar{c} \rightarrow \mu^+ + \bar{c}$$
$$\qquad\qquad \rightarrow \bar{c} \rightarrow \mu^- + \nu + s$$

and should give the dimuon + missing energy signature as for the strange sea in neutrino interactions. In fact because the data in general are not far above threshold for $c\bar{c}$ production, the data seem to be better described if the c and $\bar{c}$ are correlated as for example in the photon gluon fusion model

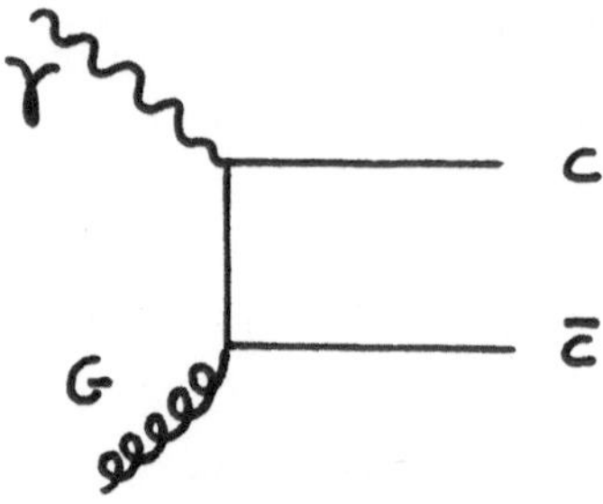

The data [17] show (fig. 16) very strong scale breaking compared to the full F_2 and, although their total contribution is small, it must be subtracted from the F_2 data before detailed analysis can be made. There have also been proposals [18] that there could be a significant intrinsic charm component in the proton wave function. The muon data put a rather stringent limit [19] on this, the result is illustrated in fig. 17 where the data clearly do not accomodate the 1% intrinsic charm suggested.

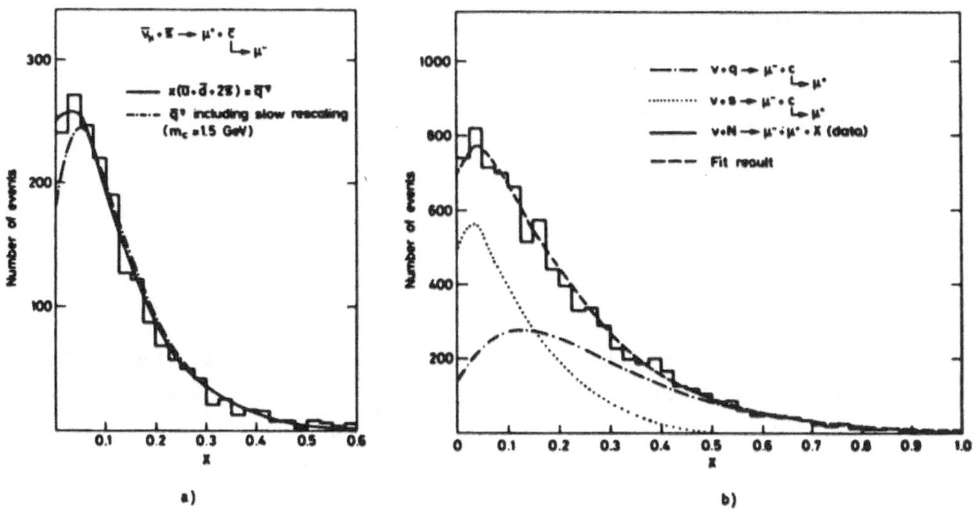

Fig. 15 - Measurements of strange sea distribution using dimuon production [16].

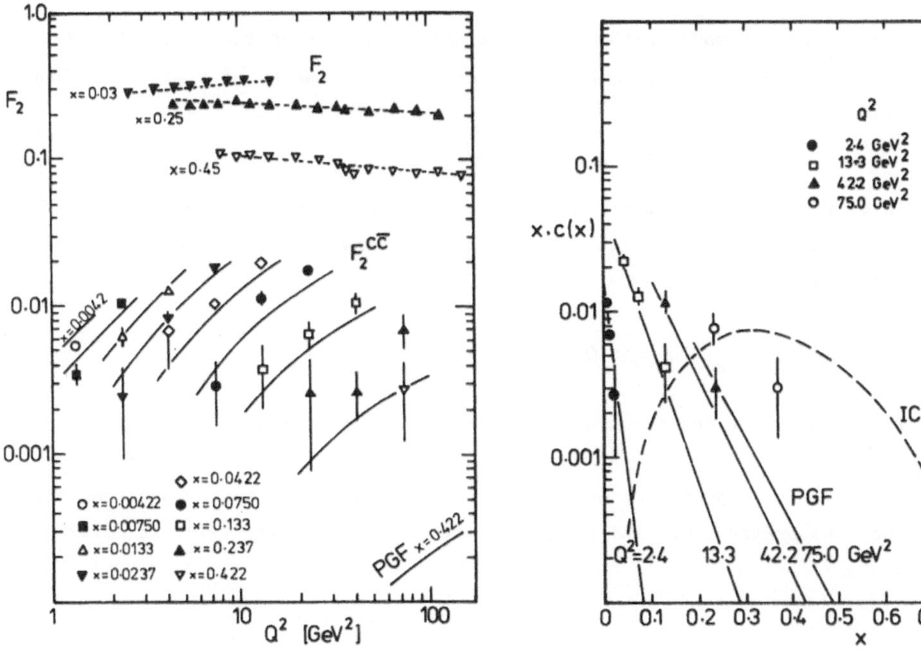

Fig. 16 - Structure function $F_2^{c\bar{c}}$ from EMC.

Fig. 17 - x distribution of charm sea compared to intrinsic charm predictions.

V. Structure Functions: Perturbative QCD Effects

i) Measurement of R

The quantity R is defined by

$$R \equiv \frac{\sigma_L}{\sigma_T} = \frac{F_2(1+Q^2/\nu^2) - 2xF_2}{2xF_1}$$

and is the ratio between the absorption cross-sections for longitudinal and transverse polarised photons. A similar quantity often used by neutrino experiment is

$$R' \equiv \frac{F_L}{F_2} = \frac{F_2 - 2xF_1}{F_2} \qquad (R' < R)$$

The importance of R appears at several levels. Basically it is impossible to deduce F_2 from the cross-section without having a measurement of R and it is this fact which allows R to be measured.

Within the parton model with massless quarks and no transverse momenta, $R = 0$. In general, if there is a p_T from no matter what source and a quark mass M then

$$R = 4\,\frac{p_T^2 + M^2}{Q^2}$$

One source of p_T^2 could be the intrinsic "Fermi motion" k_T of the quarks within the nucleon which can only be guessed. On the other hand perturbative QCD gives the explicit expression

$$F_L = \frac{\alpha_s(Q^2)}{2\pi}\,x^2 \int_x^1 \frac{dz}{z^3}\left[\frac{8}{3}F_2(z,Q^2) + 4a\left(1-\frac{x}{z}\right) zG(z,Q^2)\right]$$

so that in principle at least a measurement of F_L is directly related to α_s.

In practice the experiments are only slightly sensitive to R over the major part of their kinematic range so that the determination of its value is extremely difficult.

Neutrinos: Method I

$$\frac{d\sigma^\nu}{dxdy} + \frac{d\sigma^{\bar\nu}}{dxdy} = \frac{G^2ME_\nu}{\pi}\left(F_2\,[1 + (1-y)^2 - R'y^2] + [\delta(q-\bar q) + 2(s-c)]\,(2y-y^2)\right)$$

so that a determination of the y^2 term in the y dependence gives a measure of R. This method [20] required that the neutrino and antineutrino ratio be renormalised to 1 at $y = 0$ and the difficulty of the measurement is demonstrated by fig. 18 where the difference between the $R = 0$ curve and the best fit is barely visible. The method is also sensitive to the relative yield of π and K neutrinos in order to have a y dependence at fixed ν. Nevertheless systematic errors of ~ 0.1 are claimed and the results as a function of ν and then x are shown in fig. 19. There is no evidence within errors of any dependence on these variables.

Neutrinos: Method II

If one considers the cross-section for antineutrinos as x and y tend to 1 then one finds

$$\frac{d\sigma^{\bar{\nu}}}{dxdy} \begin{matrix} x \to 1 \\ y \to 1 \end{matrix} \sim \frac{G^2 M E_\nu}{\pi} x F_3 \left(\frac{1}{1-R'} - y\right)(1 - y)$$

ie. if $R' = 0$ $\sim (1 - y)^2$

if $R' \neq 0$ expect more events at high x, y

This method has been used to give an upper limit [21] which is extraordinarily restrictive ($R = 0.02 \pm 0.025$). The small error comes basically from the confidence that the event reconstruction is highly efficient. It is not yet clear however what the final published systematic error will be.

Charged Leptons, Electrons, Muons

Since these experiments employ discrete energy beams then the separation of σ_L and σ_T is based on the cross-section

$$\frac{d\sigma}{dQ^2 d\nu} \sim \Gamma_t (\sigma_T + \varepsilon\sigma_L) = \Gamma_t \sigma_t (1 + \varepsilon R)$$

being linear in ε at fixed Q^2 and ν. The different ε values come from different energies. The results from a recent measurement [22] for three different Q^2 ranges are shown in fig. 20. It is clear that the determination of R from such fits is sensitive to the relative normalisations and it is useful to have regions where the data are completely insensitive to R: R as a function of x is shown for hydrogen in fig. 21. Table II, which is an updated version of that prepared by Drees [23], shows that the high energy data with small systematic errors (CDHS, EMC) tend to favour low (zero) values of R in contrast to the classical low energy measurements from SLAC which stubbornly [24] insist on remaining at $R \sim 0.2 \pm 0.1$. Unfortunately SLAC measurements are at high x and EMC at lowish x so that no

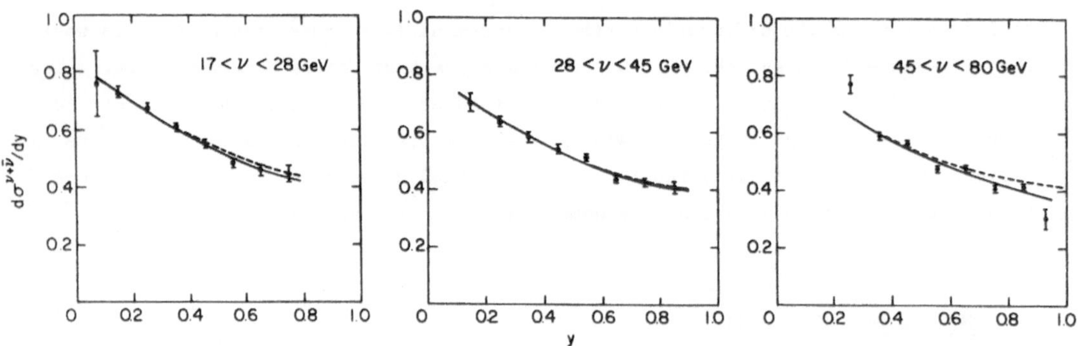

Fig. 18 - Basic data for CDHS determination of R
----- R' = 0, ——— best fit.

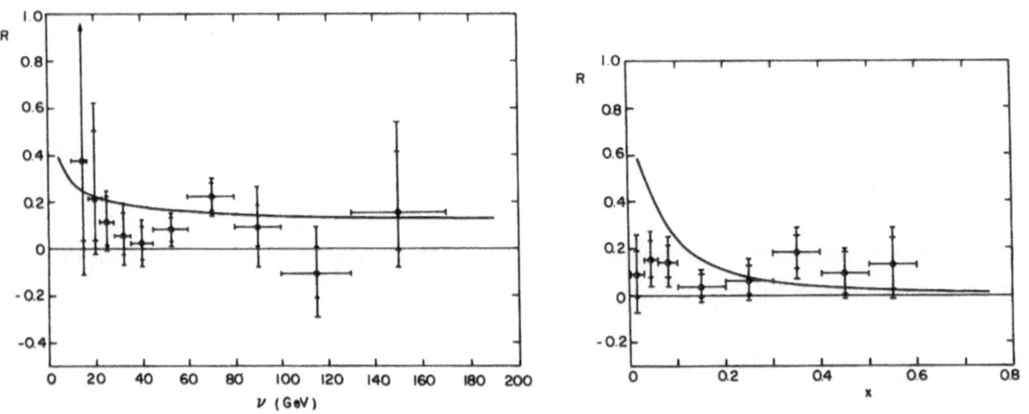

Fig. 19 - CDHS R measurement as a function of ν and x.

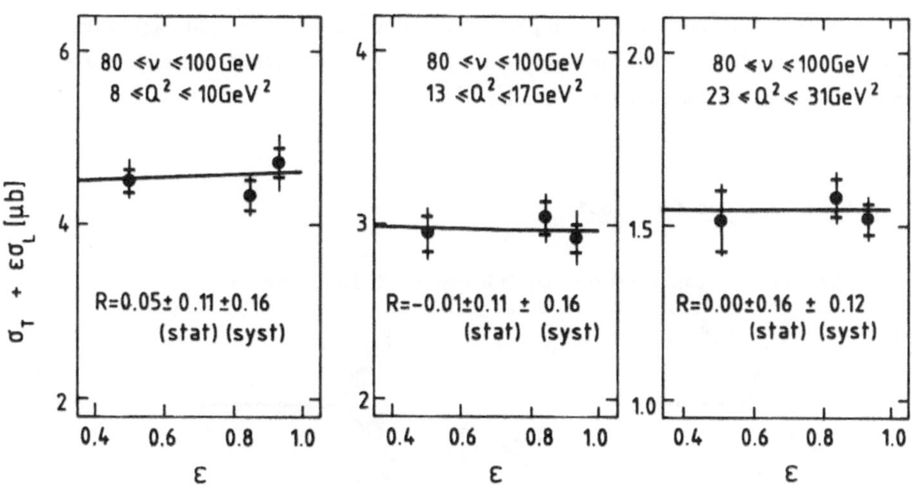

Fig. 20 - Basic data for EMC determination of R.

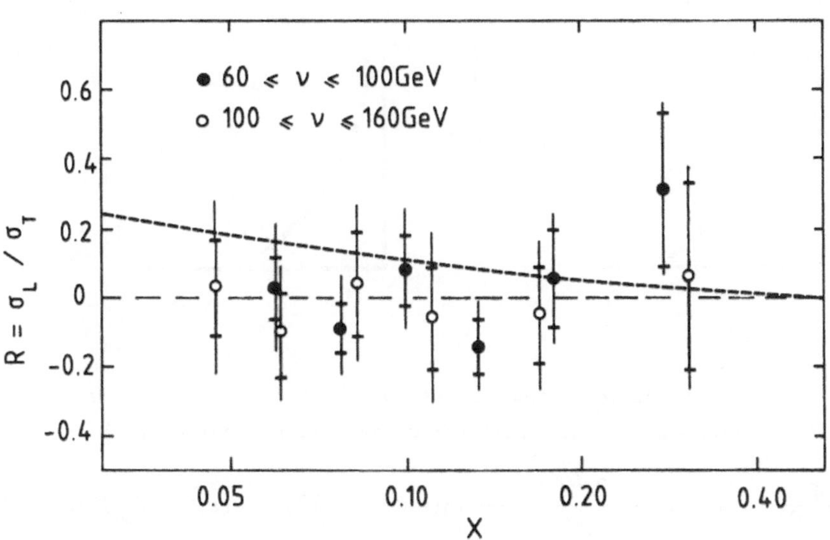

Fig. 21 - EMC results for R as a function of x.

conflict exists but nor is a Q^2 dependence detectable. The QCD calculations would prefer rather higher values at low x than are being obtained altough the errors are not small enough to allow contradiction.

ii) <u>What is expected from QCD?</u>

The transition from QPM to QCD can be illustrated by

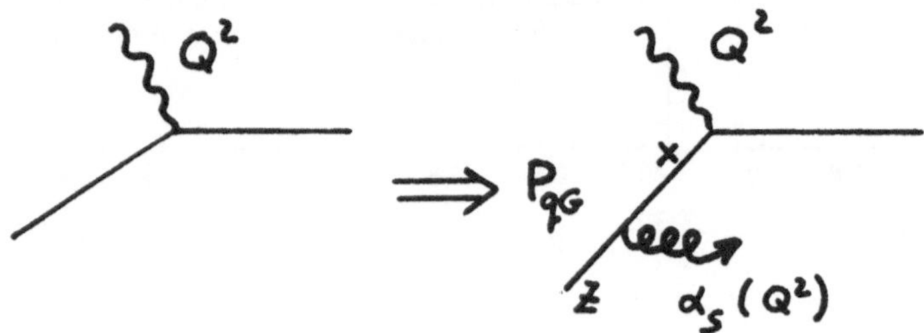

and the result of the presence of the gluon emission is that the quark distributions change as a function of Q^2.

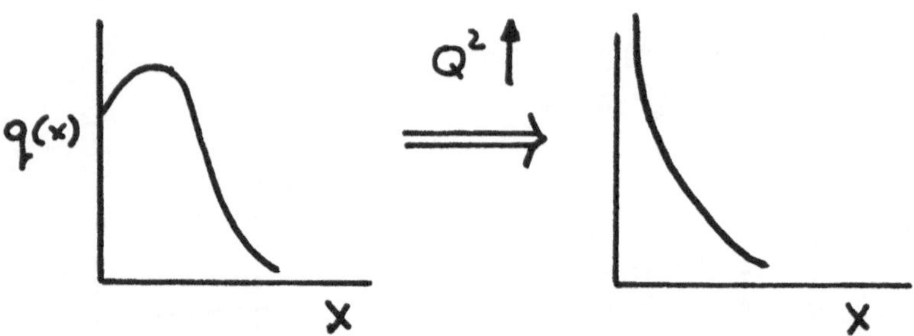

How they change is governed to some extent by how they are to begin with and this is explicitly expressed by the Altarelli-Parisi evolution equation [25]

$$\frac{d}{d\ \ln Q^2}\ q(x,Q^2) = \frac{\alpha_s(Q^2)}{2\pi}\int_x^1 \frac{dz}{z}\ [q(z,Q^2)\ P_{qq}(\tfrac{x}{z}) + G(z,Q^2)\ P_{Gq}(\tfrac{x}{z})]$$

$$\frac{d}{d\ \ln Q^2}\ G(x,Q2) = \frac{\alpha_s(Q^2)}{2\pi}\int_x^1 \frac{dz}{z}\ [q(z,Q^2)\ P_{qG}(\tfrac{x}{z}) + G(z,Q^2)\ P_{GG}(\tfrac{x}{z})]$$

These equations prompt two remarks

1) When CHIO, CDHS, CFRR, EMC, BFP, BCDMS et al. were designed the designers did not realise they were looking for logarithms (if they had, many would have stayed at home).

2) The coupled system makes for a difficult life, so the first move is usually to simplify things which reduces to finding a situation where $G(x,Q^2) \sim 0$.

Charged Lepton Data Analysis

This analysis is based on the procedures proposed by Gonzales-Arroyo, Lopez and Yndurain [26] and also by Abbott and Barnett [27]. For $x > 0.25$ the gluon and sea quark distributions are expected to be negligible, in which case the evolution equation for F_2 can be reduced to the form

$$F_2(x,Q^2) = \int_x dz\, F_2(z,Q_0^2)\, b(x,z;\, Q_0^2,Q^2)$$

the validity of ignoring the glue and sea must of course be checked. A parametric form for the structure function at Q_0^2 is chosen

$$F_2(x,Q_0^2) = Ax^\alpha\,(1-x)^\beta\,(1-\gamma x)$$

which is sufficiently general. The actual use of Q_0^2 is symbolic since the data at all Q^2 are then fitted simultaneously to yield A, α, β, γ and also Λ_{QCD}. The procedure is also available at next to leading order [28]. The fits, as shown in figs 22, 23, 24, give a reasonably good description of the data with

BCDMS	Carbon	Λ_{LO} =	85^{+60}_{-40} MeV
EMC	Iron	=	125^{+22}_{-20} MeV
	Hydrogen	=	110^{+58}_{-46} MeV

The errors quoted are statistical only, there is a systematic error of approximately $^{+100}_{-70}$ MeV. At next to leading order the fits are of similar quality and in general $\Lambda_{\overline{MS}} \sim 1.3\,\Lambda_{LO}$ is obtained. The result survives a wide range of checks on the procedure [29], in particular it can be shown that to significantly change the result one needs $G(x,Q_0^2) \sim \frac{(1-x)^n}{x}$ with $n < 4$ at $Q^2 = 5\ \mathrm{GeV}_0^2$. As discussed later the CDHS neutrino data indicate that this is very unlikely.

The analysis can be extended to all x values provided some information on the proton-neutron difference is included, in this case the full coupled evolution form is used with the parametric forms

$$F_2^{\,\mathrm{Singlet}}(x,Q_0^2) = Ax^\alpha(1-x)^\beta + B(1-x)^\gamma$$

$$F_2^{\,\mathrm{Non\ singlet}}(x,Q_0^2) = Cx^{1/2}\,(1-x)^\delta\,(1-\varepsilon x) \qquad (\text{SLAC-EMC } F_2^{\,p}/F_2^{\,n})$$

$$xG(x,Q_0^2) = D(1-x)^{5.9}\,(1+3.5x) \qquad (\text{CDHS})$$

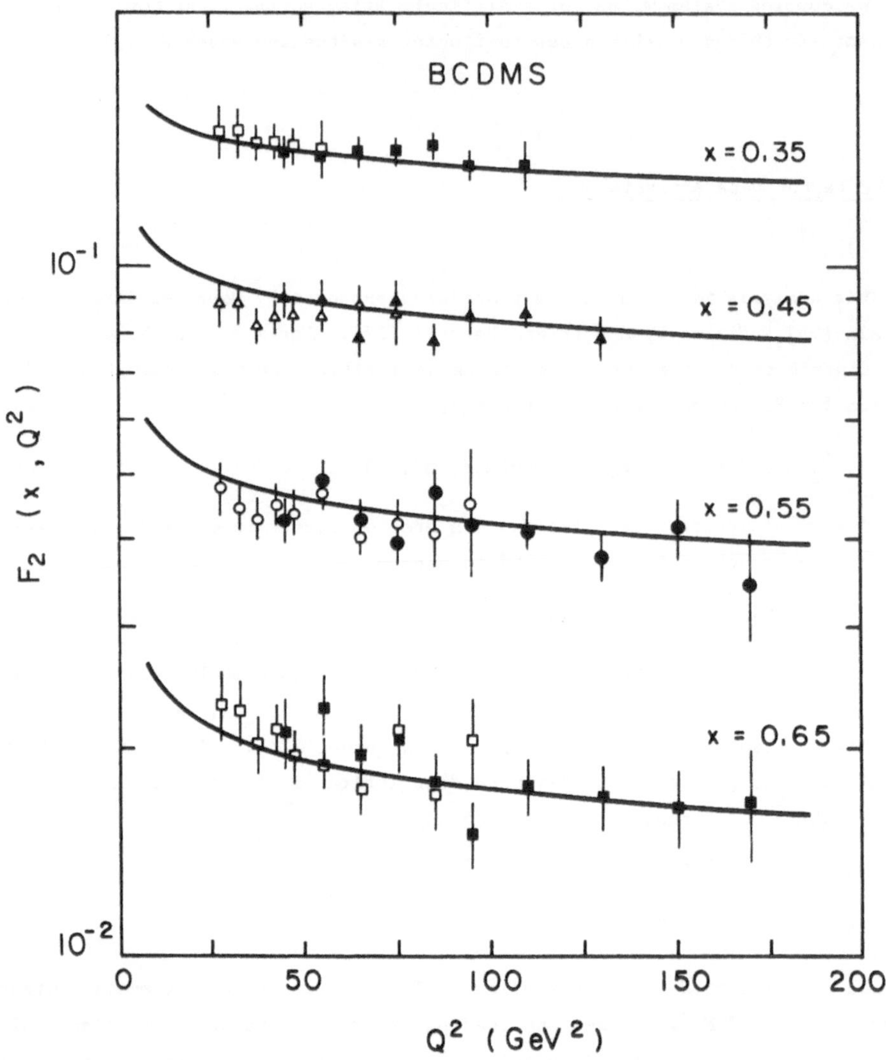

Fig. 22 - BCDMS (D. Bollini et al., Phys. Lett. 104B (1981) 403) measurement of F_2 and QCD fit.

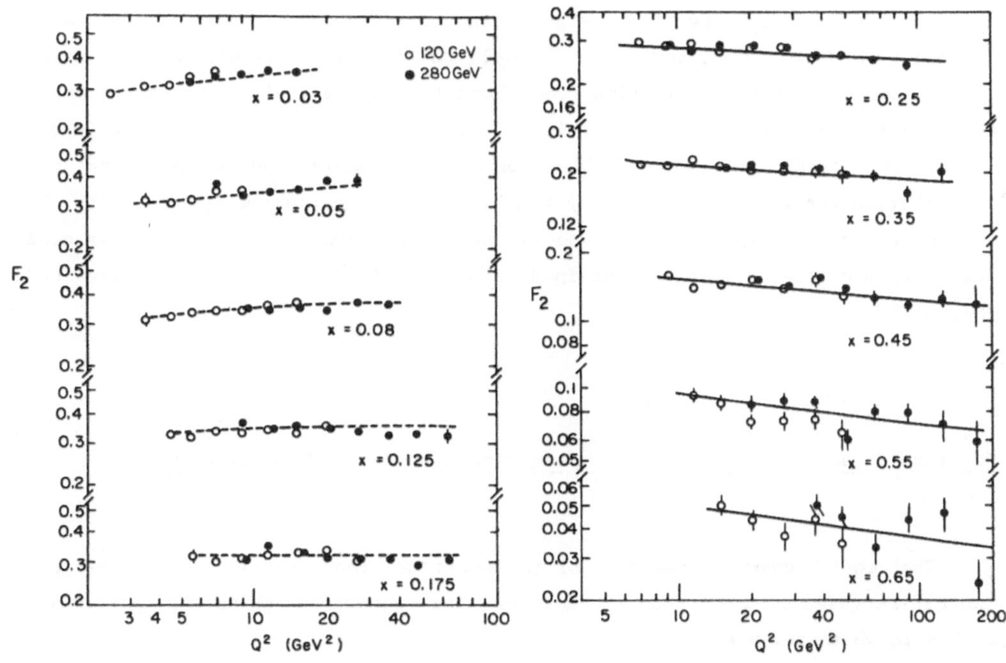

Fig. 23 - EMC measurement of F_2^p and QCD fit.

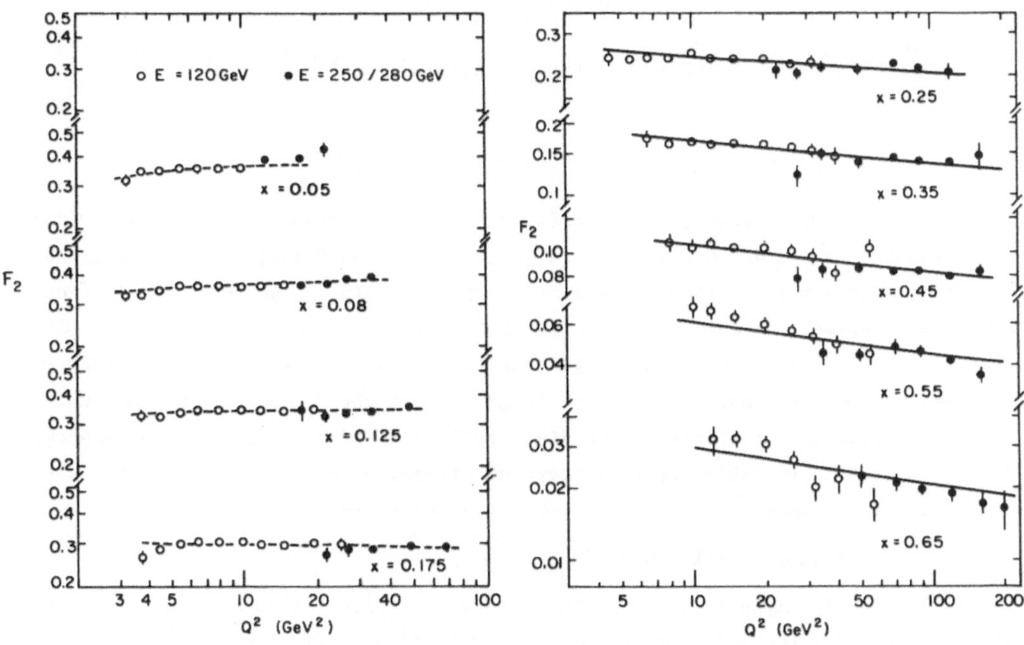

Fig. 24 -EMC measurement of F_2^N (Iron) and QCD fit.

at $Q_0^2 = 5\ \mathrm{GeV}^2$. Since there is a significant contribution of charm at low x a first attempt is made in which, for $x < 0.08$, only the lowest Q^2 points are retained. In a second stage the contribution from charm [17] is explicitly subtracted and all data points are illustrated for the low x region in fig. 25. There is therefore the satisfactory conclusion that a combination of perturbative QCD and measured charm contribution provides a good description of the muon scattering F_2 data sets. This comment is not however exclusive, and it has been shown [30] that just as good a description of the data can be obtained within the framework of the Massive Quark Model [31].

Neutrino Data Analysis, CDHS [32], Determination of the Gluon Distribution

The most advanced analysis of the neutrino (antineutrino) data employs yet another representation of the Altarelli-Parisi equations which is particularly suited to what they measure

$$\frac{d}{d\ \ln Q^2} F_2(x,Q^2) = \frac{\alpha_s}{2\pi}(Q^2) \int_x^1 [P_{qq}(\tfrac{x}{z})\ F_2(z,Q^2) + 2N_{F_2} P_{Gq}(\tfrac{x}{z})\ G(z,Q^2)]\ \frac{dz}{z^2}$$

$$\frac{d}{d\ \ln Q^2} F_2(x,Q^2) = \frac{\alpha_s}{2\pi}(Q^2) \int_x^1 [P_{qq}(\tfrac{x}{z})\ \bar{q}(z,Q^2) + N_{\bar{q}}\ P_{Gq}(\tfrac{x}{z})\ G(z,Q^2)]\ \frac{dz}{z^2}$$

$$\frac{d}{d\ \ln Q^2} F_2(x,Q^2) = \frac{\alpha_s}{2\pi}(Q^2) \int_x^1 [P_{Gq}(\tfrac{x}{z})\ F_2(z,Q^2) + P_{GG}(\tfrac{x}{z})\ G(z,Q^2)]\ \frac{dz}{z^2}$$

with

$$N_{F_2} = N_{\bar{q}} = 4 \qquad \text{and} \qquad \alpha_s(Q^2) = \frac{12\pi}{25\ \ln Q^2/\Lambda^2}$$

The "measured" input is $\bar{q}(x,Q^2)$ and $F_2(x,Q^2)$ and the quantities extracted from the analysis are Λ_{LO} and $G(x,Q_0^2)$. $F_2(x,Q_0^2)$, $\bar{q}(x,Q_0^2)$ and $G(x,Q_0^2)$ are parameterised and the choice for the latter is extremely important since, as $G(x,Q^2)$ is not directly measured, there is no direct check form for the gluon. The description obtained is shown in figs 26 and 27 and the trends in the data are well reproduced although at low x an extension in Q^2 range of both the F_2 and $\bar{q}$ measurements would be welcome. The shapes of the distributions at Q_0^2 have already been shown in fig. 13. This probably the most thorough determination of the form of the gluon distribution yet made. The parameterisation obtained is

$$G(x,Q_0^2) = a(1 + bx)\ (1-x)^c$$

although it is emphasised that this form is valid for no other Q^2. In the standard analysis the normalisation of the gluon distribution is constrained by the momentum sum rule, if its normalisation is free the value 0.55 ± 0.11 is obtained for the fraction of the nucleon momentum carried by the gluons. The value for

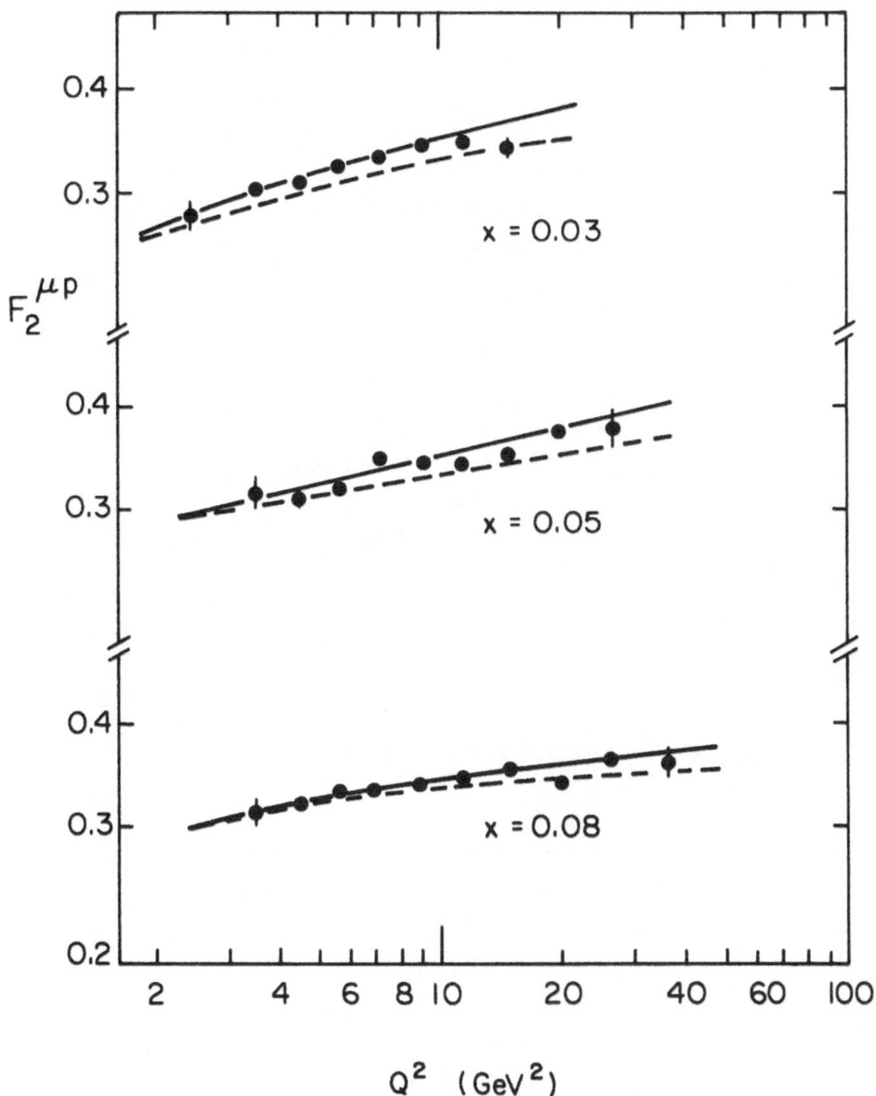

Fig. 25 - QCD fit to low x EMC F_2^p data showing the contribution of charm (see text).

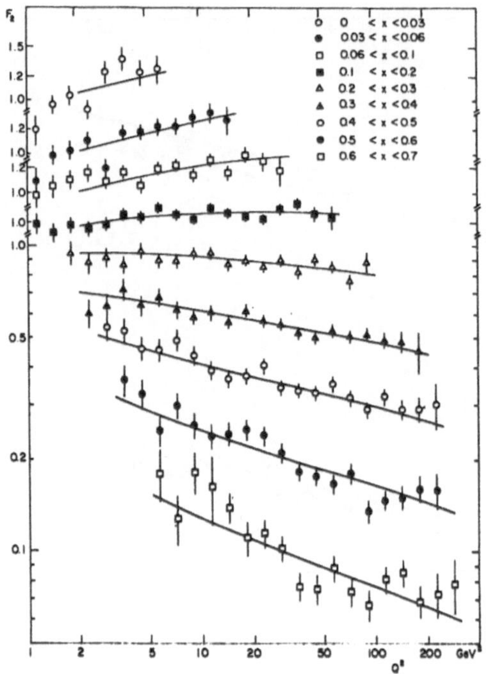

Fig. 26 - Measurement of $F_2^{\nu N}$ and QCD fit from CDHS.

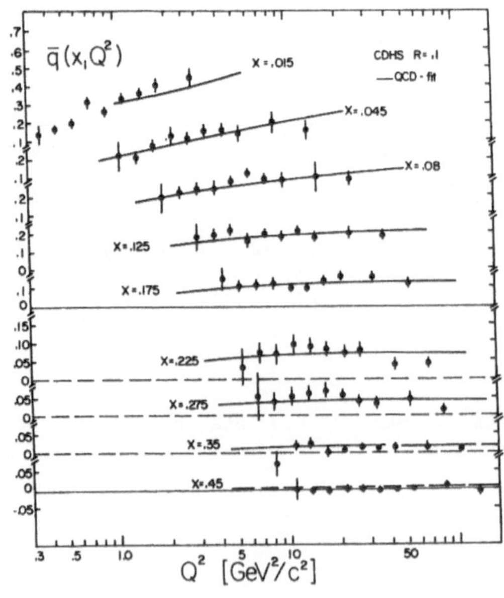

Fig. 27 - Measurement of $\bar{q}^{\nu N}$ and QCD fit from CDHS.

$\Lambda_{LO} = 0.18 \pm 0.02$ (stat.) is slightly higher than that obtained by analysis of the muon data however given the systematic errors (not quoted in the neutrino case) there is no disagreement.

Alternative Determination of the Gluon Distribution [17]

Although it is very nice to have an internally consistent determination of the gluon distribution as in the above neutrino analysis, it is nevertheless interesting to search elsewhere for a determination. Such a possibility is offered by the muon charm production data which, as remarked in section 4, is well described by the photon gluon fusion model.

The data can be fitted using the form $\eta\, G(\eta) = A\,(1-\eta)^{m(Q^2)}$ and the power m obtained for each bin in Q^2. A supplementary point is obtained by including the J/ψ production data. The result is compared with various gluon models in fig. 28 with the two choices Q^2 and $Q^2 + \hat{s}$ for the argument of the evolution.

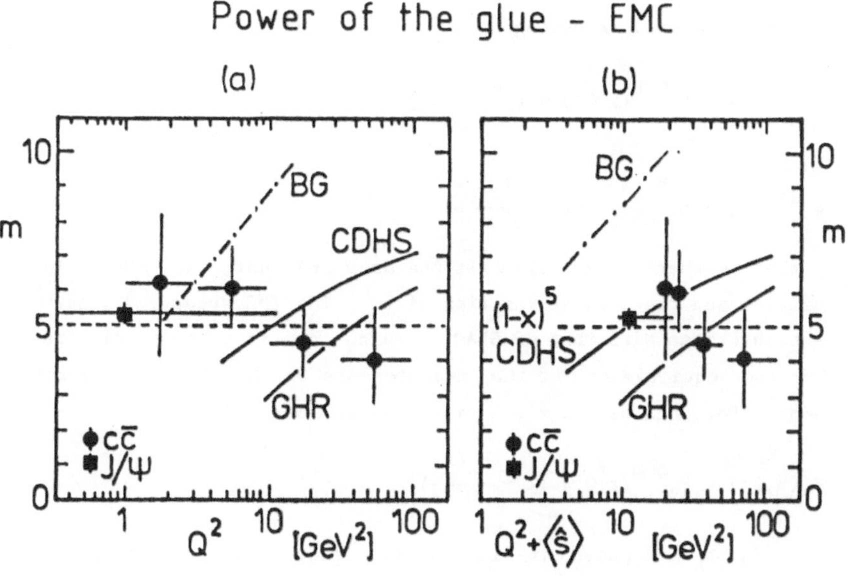

Fig. 28 - Power n of $(1-x)^n$ in gluon distribution from EMC.

The data give $m \simeq 5$ and are in agreement with the neutrino evaluation and it can be further remarked that comparatively little evolution with Q^2 is observed.

Higher Twist

The analyses of the structure function data discussed so far contemplate the interaction of the exchanged boson with a single quark only, commonly known as the Leading Twist terms. There are in addition also terms which involve more than one parton and possibly parton composites (eg. diquarks). These terms can be thought of as analogous to the quasi-elastic α-particle contribution to electron scattering from nuclei and their main property is that they are expected to decrease in importance as inverse powers of Q^2. There is little theoretical constraint on their behaviour, perhaps the complete series can be written

$$F_i(x,Q^2) = F_i^{\text{Pert QCD}}(x,Q^2)\left(1 + \frac{h_4(x)}{Q^2} + \frac{h_6(x)}{Q^4} + \ldots\right)$$

but unfortunately not even the signs of h_4, h_6 ... have yet been predicted. The phenomenological analyses which look for these terms must necessarily use the maximum Q^2 range available. The CDHS group [33] combine their neutrino data with SLAC measurements of eD. This may or may not be a valid procedure since both beam and target are different. Assuming

$$F_2(x,Q^2) = F_2^{QCD}(x,Q^2) + F_2^{HT}(x,Q^2)$$

with

$$F_2^{HT}(x,Q^2) = \frac{h_4(x)}{Q^2} \quad \text{or} \quad \frac{h_6(x)}{Q^4}$$

they find that either form will fit the data and that, as shown in fig. 29, F_2^{HT} is positive and rises as a function of x. The EMC group [29] have combined their data on hydrogen with that of SLAC in which case there is no dispute possible about either the equivalence of the measurements or the effect of Fermi motion, there is none. For each bin in x, $h_4(x)$ is determined using

$$F_2(x,Q^2) = F_2^{QCD}(x,Q^2)\left(1 + \frac{h_4(x)}{Q^2}\right)$$

The results for h_4 are shown in fig. 30 and they can be described by the functional form $x^2/(1-x)^2$. The value of $\Lambda_{\overline{MS}} = 130^{+50}_{-40}$ MeV is compatible with that determined by a fit to the high energy data alone.

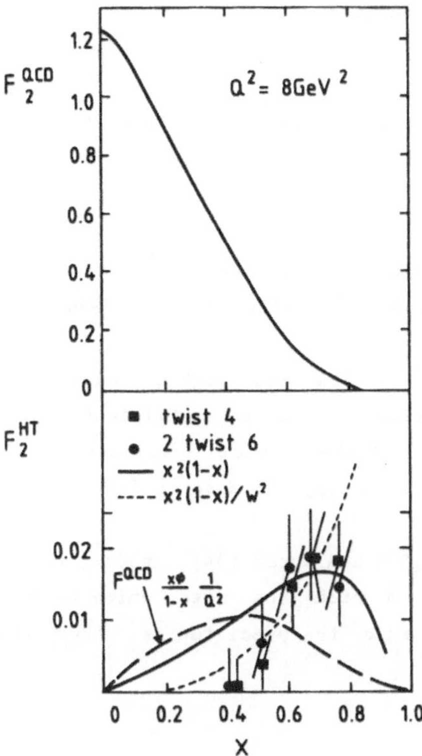

Fig. 29 - Higher twist contribution to F_2 from CDHS (νN) + SLAC (eD) data.

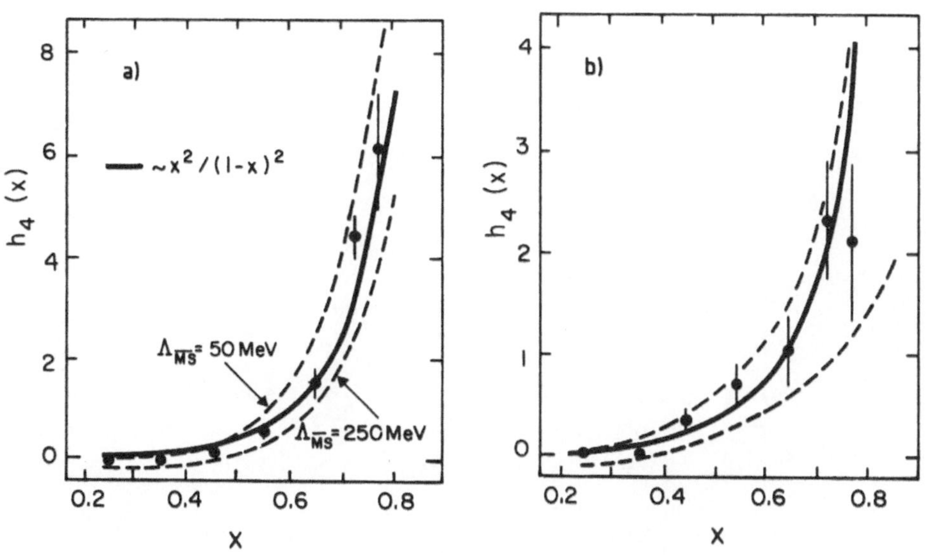

Fig. 30 - $h_4(x)$ determined from EMC (μp) and SLAC (ep) data.

Structure Functions and QCD: Summary

i) For the experiments performed in the period 1969-1978, mainly but not exclusively low energy, the data could be described if $\Lambda \sim 400$-700 MeV. The data were also compatible with $1/Q^2$.

ii) For the recent experiments 1978-1982, the data are described by $\Lambda \sim 100$ MeV, by themselves they are also compatible with power law behaviour.

iii) If a combined analysis of low and high Q^2 data is made, then neither logarithmic nor $1/Q^2$ behaviour alone is sufficient and this allows some preliminary determination of higher twist behaviour to be made. However an infinite series $\sum_{i=1}^{\infty} a_i/(Q^2)^i$ can always be made to fit anything.

iv) As summarised by Drees [23] and Buras [34], there is remarkable consistency between the determination of Λ from the data which is even more impressive if expressed in terms of α_s for which the precision is better than 10% from Deep Inelastic Scattering.

VI. The Quark Parton Model and Hadron Production Data

The Quark Parton Model describes hadron production as indicated in fig. 31 in which the kinematic definitions are also given. The production cross-section then has the form

$$\frac{1}{N_\mu}\frac{dN^h}{dz} = \frac{\sum_i \varepsilon_i(x)\, D_i^h(z)}{\sum \varepsilon_i(x)}$$

$$\varepsilon_i = e_i^2\, q_i(x) \qquad\qquad i = u, d, d, \bar{u} \ldots$$

$$h = \pi^{\pm,0},\ K^{\pm 0,\bar{0}}$$

in which we see explicitly the product form which goes by the name "factorisation" and the independence of Q^2 which is called scaling. In addition to these two properties the QPM allows rather straightforward predictions based on isospin symmetry. In addition various models are more or less specific about the form of $D_i^h(z)$, but in general the treatment is mainly based on experimental data.

An example of a particular model is the Feynman-Field [35] model which envisages the fragmentation process as a series of independent emissions of mesons by the original quark. At each stage the meson takes a fraction z of the remaining quark momentum such that $f(z) = 1-a_F+3a_F(1-z)^2$. At each branch point the necessary $q\bar{q}$ pair receives a transverse momentum controlled by σ_q which results in the hadrons picking up a fragmentation p_T, σ_{Frag}, each pair internally conserves p_T. Normally only light quark pair production is admitted with relative probabilities

$$u:d:s = \gamma:\gamma:\gamma_s$$

such that $\gamma_s = 1-2\gamma$. Pseudoscalar and vector mesons are assumed to be produced with probabilities α_s and α_v respectively, $\alpha_s + \alpha_v = 1$. So there are 4 parameters α_F, σ_q, γ_s and α_v with which to describe the data.

For $e^+ e^-$ interactions

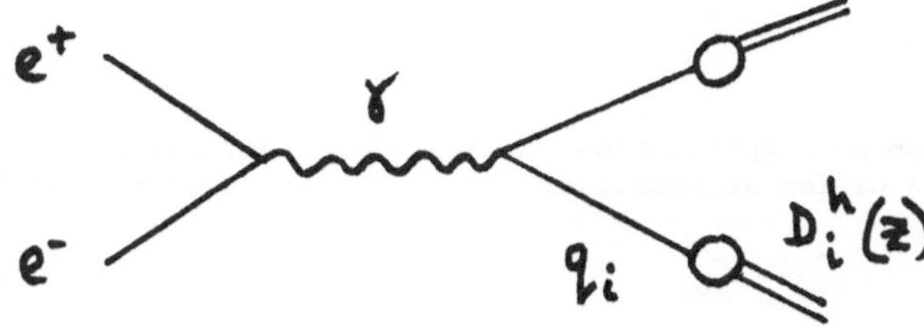

we have

$$\frac{d\sigma}{dz}(e^+e^- \to q\bar{q} \to h) = \sigma_{q\bar{q}}(D_q^h(z) + D_{\bar{q}}^h(z))$$

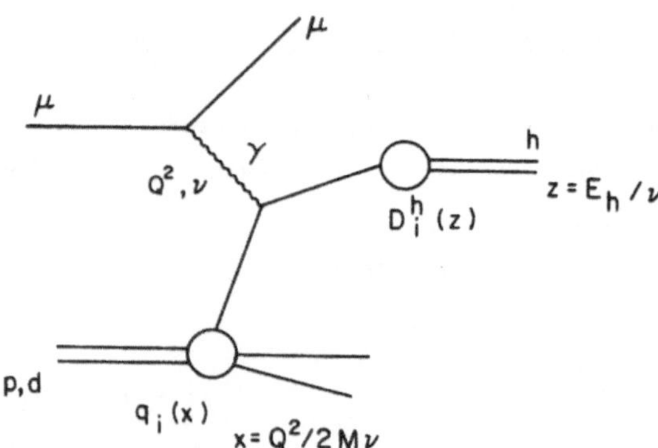

Fig. 31 - Deep inelastic hadron production and kinematics in QPM.

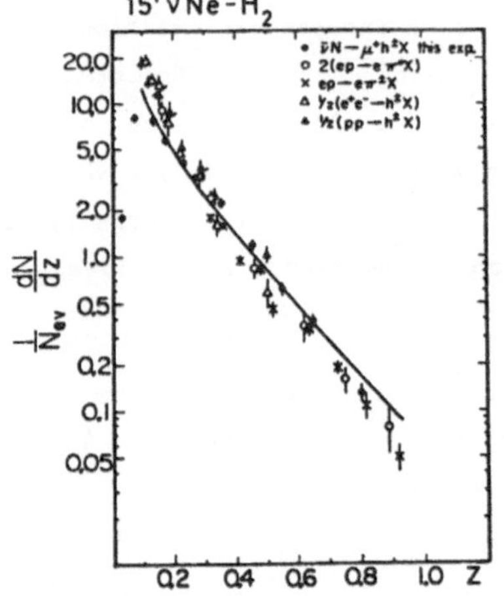

Fig. 32 - Comparison of fragmentation data from different processes.

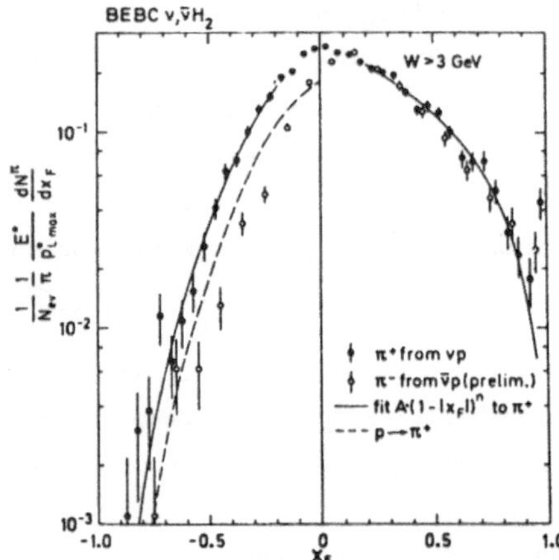

Fig. 33 - Comparison of π^+ production by neutrinos and π^- production by antineutrinos as a function of x_F.

Summing over all quarks i we then obtain

$$\frac{d\sigma}{dz}(e^+e^- \to h) = \sum_i \sigma_i [D_i^h(z) + D_{\bar{i}}^h(z)]$$

$$= \sum e_i \, \sigma_{\mu\mu}[D_i^h(z) + D_{\bar{i}}^h(z)]$$

Now, if there are only u and d quarks and pions

$$\frac{d\sigma}{dz}(e^+e^- \to \pi) = \sigma_{\mu\mu}\,[e_u^2(D_u^{\pi^+} + D_{\bar{u}}^{\pi^+} + D_u^{\pi^-} + D_{\bar{u}}^{\pi^-}) + e_d^2(D_d^{\pi^+} + D_{\bar{d}}^{\pi^+} + D_d^{\pi^-} + D_{\bar{d}}^{\pi^-})]$$

$$D_u^{\pi^+} = D_d^{\pi^-} = D_{\bar{u}}^{\pi^-} = D_{\bar{d}}^{\pi^+}$$

$$D_u^{\pi^-} = D_d^{\pi^+} = D_{\bar{d}}^{\pi^-} = D_{\bar{u}}^{\pi^+}$$

$$= \sigma_{\mu\mu}\,[e_u^2\, 2(D_u^{\pi^+} + D_u^{\pi^-}) + e_d^2\, 2(D_u^{\pi^+} + D_u^{\pi^-})]$$

$$= [\sigma_{\mu\mu}\,(e_u^2 + e_d^2)]\; 2[D_u^{\pi^+} + D_u^{\pi^-}]$$

$$\boxed{\frac{1}{\sigma}\frac{d\sigma}{dz}(e^+e^- \to h) = 2\,D_u^{\pi}}$$

A similar derivation yields in the three different deep inelastic cases

Neutrino

$$\frac{1}{N_\nu}\frac{dN}{dz} = D_u^{\pi}$$

Antineutrino

$$\frac{1}{N_{\bar{\nu}}}\frac{dN}{dz} = D_u^{\pi}$$

Muons, Electrons

$$\frac{1}{N_\mu}\frac{dN}{dz} = D_u^{\pi}$$

These simple results allow a direct test of the basic QPM conjecture and a typical data comparison [36] is shown in fig. 32. A comparable compilation [37] of data from many different experiments shows also good agreement. The data can be compared in further detail with the QPM by considering the charges separately. In neutrino interactions we have

$$\nu + d \to u + x$$

$$\to D_u^{\pi^+} \text{ favoured}$$

$$D_u^{\pi^-} \text{ unfavoured}$$

and for antineutrinos

$$\bar{\nu} + u \rightarrow d + x$$
$$\rightarrow D_d^{\pi^+} \text{ unfavoured}$$
$$D_d^{\pi^-} \text{ favoured}$$

from which we expect

$$\frac{1}{N}\frac{dN^{\nu\to\pi^+}}{dz} \sim \frac{1}{N}\frac{dN^{\bar{\nu}\to\pi^-}}{dz}$$

Such behaviour is nicely shown in fig. 33 where we see that the data agree in the forward ($x_F > 0$) region where the quark fragmentation should dominate, but not in the backward, target fragmentation region. In muon scattering the quark composition can be varied by varying the x_{Bj} and it can be shown that

$$\frac{\pi^+}{\pi^-} \sim \frac{[\frac{4}{9}u + \frac{1}{9}\bar{d}]\ D_u^{\pi^+} + [\frac{1}{9}d + \frac{4}{9}\bar{u}]\ D_u^{\pi^-}}{[\frac{1}{9}d + \frac{4}{9}\bar{u}]\ D_u^{\pi^+} + [\frac{4}{9}u + \frac{1}{9}\bar{d}]\ D_u^{\pi^-}},$$

if $x_{Bj} \sim 0$ then the quark sea dominates and $u \sim d \sim \bar{u} \sim \bar{d}$ gives $\frac{\pi^+}{\pi^-} \sim 1$, in contrast for $x_{Bj} \to 1$ we expect $\frac{\pi^+}{\pi^-} >> 1$. The data [38] are shown in fig. 34 and they demonstrate the expected features mainly for $z \to 1$ indicating that the particles in this region of phase space are more faithful indicators of the parent quark.

The isospin invariance test is shown for muons in fig. 35 in which π^0 data are compared with identified pions. A fraction 0.27 ± 0.02 (stat.) ± 0.05 (syst.) of the energy of the virtual photon is carried by the π^0s [39]. Equivalent data from Tasso (e^+e^-) [40] are shown in fig. 36. These and data from other groups, eg. the fraction of energy in γ rays is measured by the Cello group [41] to be 0.251 ± 0.003 ± 0.04, suggest that in high energy e^+e^- the situation may be more complicated.

As alluded to with respect to fig. 33 the QPM envisages a fairly well defined "quark fragmentation region" and at high energies one might expect a clear separation from the target fragmentation. Possible evidence for this [42] is shown in fig. 37 where positive hadron, negative hadron yields and their difference are plotted as a function of centre of mass rapidity y. For W > 8 GeV there is indeed an indication of a separation at y = 0. Preliminary high energy muon data with the EMC vertex system [43], fig. 38, also suggest that something is happening in the region of $y_{cms} = 0$.

The idea of separating a quark fragmentation region was pushed to its limit with the idea [44] that the quark charge could be measured if the mean charge of resultant

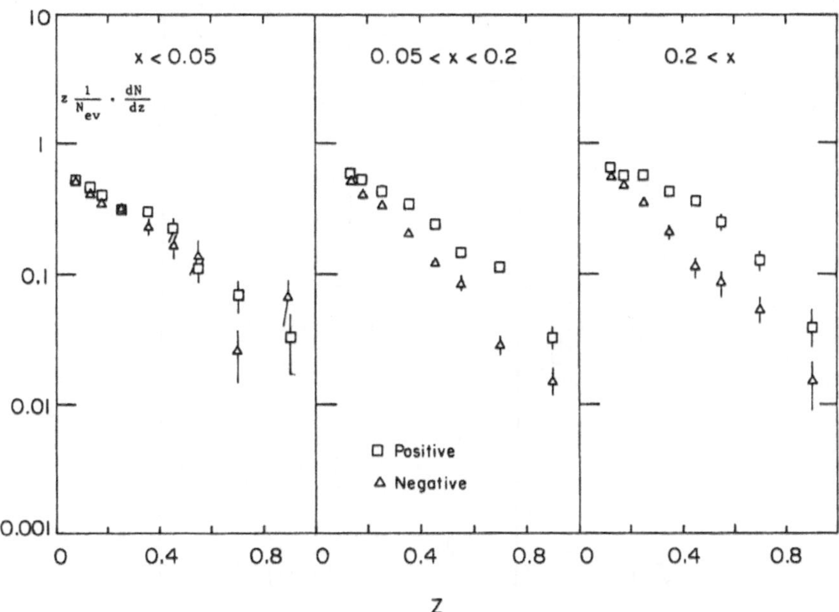

Fig. 34 – Dependence of h^+ and h^- production, as a function of z, on x.

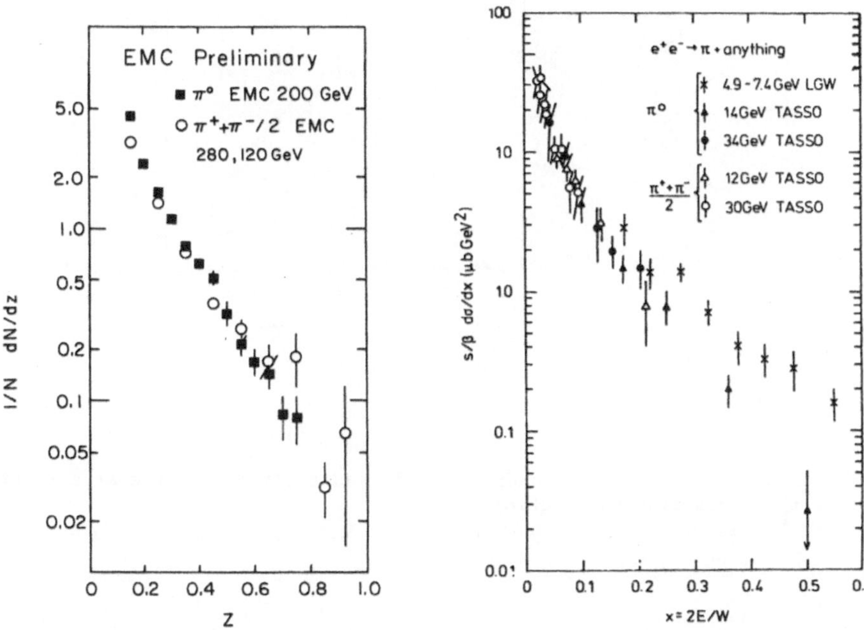

Fig. 35 – Comparison of π^0 production with $(\pi^+ + \pi^-)/2$ from EMC.

Fig. 36 – Comparison of π^0 and charged pion production in e^+e^- interactions.

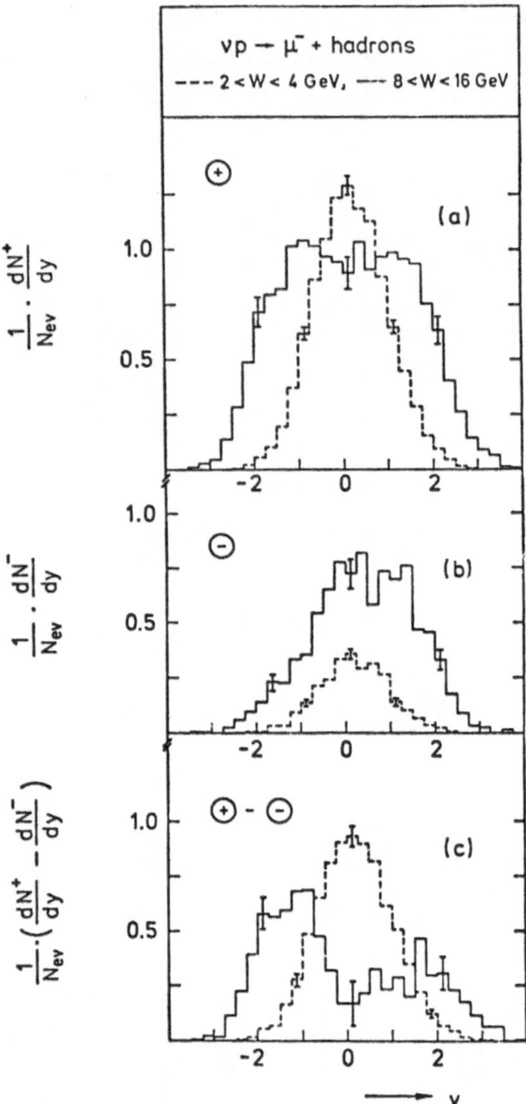

Fig. 37 - Positive and negative hadron yields and their difference as a function of x_F in neutrino interactions.

hadrons were measured. However any cut in the chain of quark antiquark pairs is is made at the boundary of a hadron in which case

$$<Q_q> = e_q - \sum_i \gamma_i e_i$$

$$<Q_u> = 1 - \gamma_u \qquad\qquad <Q_d> = <Q_s> = -\gamma_u$$

so it becomes clear that the quark charge cannot be measured in this way. Furthermore the resultant measure of $\gamma_u = 0.45 \pm 0.04$ [45] yields $\gamma_s/\gamma = 0.22^{+0.22}_{-0.18}$ which is not very discriminative either. The attempts have become more sophisticated [46] and, as shown in fig. 39 where the weighted charge distribution

$$Q_w = \sum_i z_i^r Q_i \qquad\qquad \text{with} \quad r = 0.5$$

is plotted, the data are very consistent with model predictions which of course have $Q_u = \frac{2}{3}$, $Q_d = \frac{1}{3}$. It cannot unfortunately be called a measurement of the quark charge.

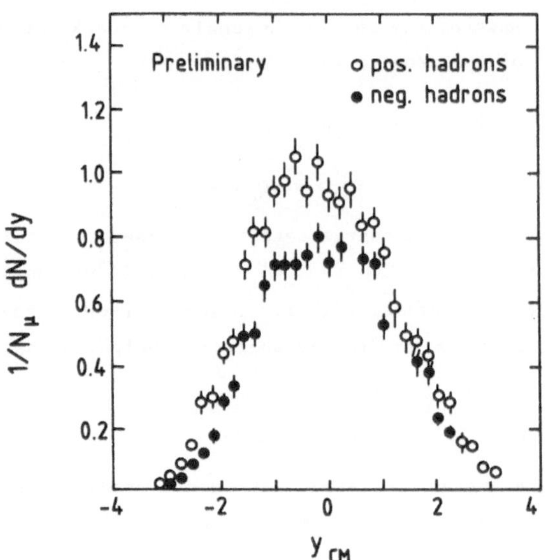

Fig. 38 - Positive and negative hadron yields as a function of x_F in muon scattering.

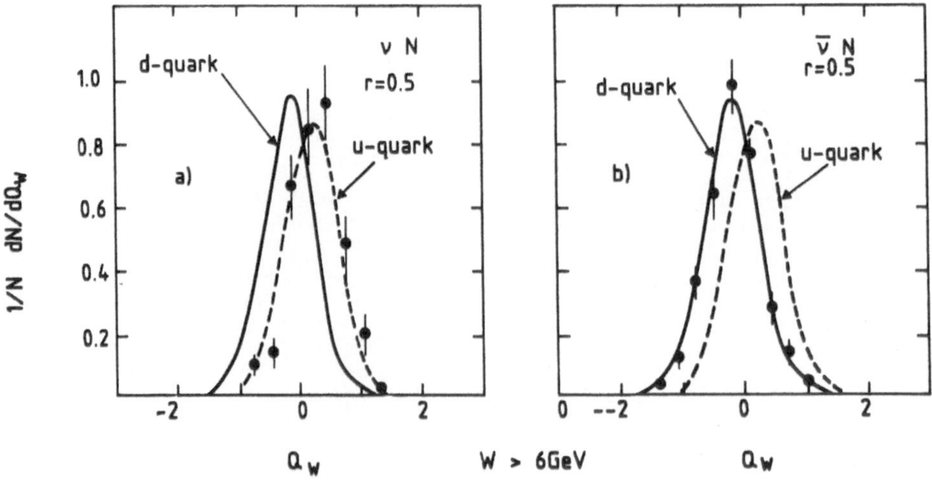

Fig. 39 - Weighted charge distribution in neutrino and antineutrino interactions.

VII) Fragmentation Properties

As mentioned in the previous section most fragmentation models have at least four parameters characterising the longitudinal fragmentation, the transverse fragmentation, the strange particle production and the ratio of vector to pseudoscalar particles. The latter two have always been the least well constrained by experiment but there is some progress which is worth discussing.

Strange Mesons - γ_s/γ

From muon scattering the strange particle system is quite involved but with some assumption can be reduced to a manageable problem

$$\frac{1}{N_\mu}\frac{dN^{K^+}}{dz} \simeq \varepsilon_u D_u^{K^+} + \varepsilon_{\bar{u}} D_u^{K^-} + \varepsilon_{\bar{s}} D_u^{\pi^+}$$

$$\frac{1}{N_\mu}\frac{dN^{K^-}}{dz} \simeq \varepsilon_u D_u^{K^-} + \varepsilon_{\bar{u}} D_u^{K^+} + \varepsilon_s D_u^{\pi^+}$$

$$\frac{1}{N_\mu}\frac{dN^{K^0}}{dz} \simeq 2\varepsilon_u D_u^{K^-} + (\varepsilon_d + \varepsilon_{\bar{d}})D_u^{K^+} + (\varepsilon_s + \varepsilon_{\bar{s}})D_u^{\pi^+}$$

The original Feynman-Field guess of $\gamma_s/\gamma = 0.5$ was based on "high p_T" $\frac{K}{\pi}$ data, the electron and muon data [47], fig. 40, favour something like $\gamma_s/\gamma = 0.25$ and a recent neutrino measurement of the K^0/π^0 ratio [48] yields 0.27 ± 0.04. For e^+e^- data the plethora of strange particle sources makes life difficult (see fig. 41), but a recent determination [49] also gives a value consistent with the deep inelastic data.

Vector Mesons - α_v/α_s

The standard FF models of fragmentation consider the production of either pseudoscalar mesons or vector mesons and assign the probabilities α_s, α_v respectively such that $\alpha_s + \alpha_v = 1$. The measurement of the relative magnitude α_v/α_s is quite difficult since it involves the determination of the relative yields of a direct pseudoscalar and vector production in the presence of decay products. Perhaps the most direct measurement has been made in muon scattering in

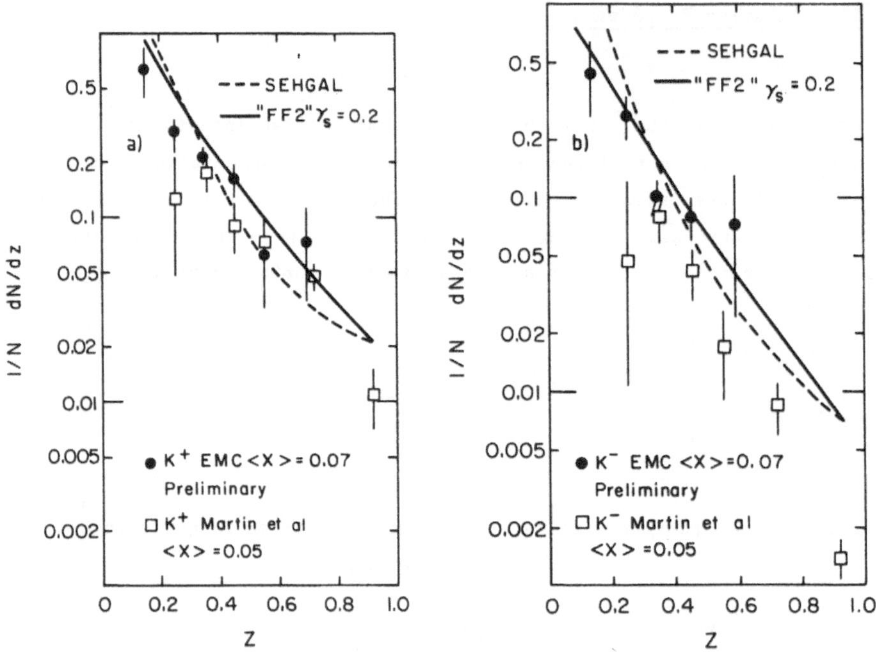

Fig. 40 – K^+ production data as function of z;
K^- production data as function of z, by charged leptons.

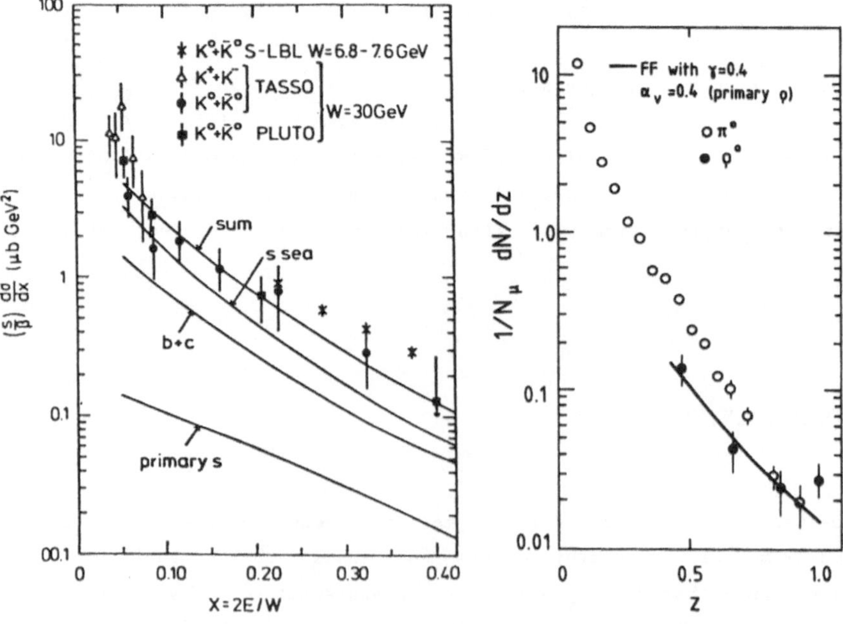

Fig. 41 – Strange particles in e^+e^- production.

Fig. 42 – Yields of π^0 and ρ^0 mesons as a function of z in muon interactions. The z = 1 point for the ρ^0 contains "elastic production".

which the yields of ρ^0 and π^0 (fig. 42) mesons is measured for $z \to 1$. At this point only direct production contributes, the result is

$$\frac{\rho^0}{\pi^0}(z = 1) = \alpha_V/\alpha_S = 1 \pm 0.3 \pm 0.4 \text{ (syst.)}.$$

While this is the standard value assumed by Feynman and Field it is not clear why the ratio should not be 3 corresponding to the number of possible spin states.

Baryon Production

Until recently the primary source of baryons was assumed to be target fragmentation and the presence of protons in the forward region was assumed to be due to the comparatively low energies which did not allow separation. Within the last years however there have been measurements of baryon and antibaryon yields in both Deep Inelastic [50] and e^+e^- production [51]. These data as a function of the fragmentation variable are shown in figs 43 and 44. One conceivable source of these baryons is diquark-antidiquark production in the fragmentation chain [52] and a consistent description of the present proton/antiproton production can certainly be obtained. In addition, the deep inelastic lambda production also fits in the model as shown in fig. 45. An exciting possibility is that baryon (antibaryon) production is enhanced for gluons [53] and indeed there is some indication [54] that baryon production is enhanced on the Upsilon resonances which are expected to have a significant three gluon decay mode. Some further hints in the same direction are provided by some new muon data [43] which, as illustrated in fig. 46, show a rise of the antiproton/proton yield as a function of p_T. (As we shall see later, a high p_T with respect to the photon axis is a good indicator for events with the characteristics associated with gluon emission.)

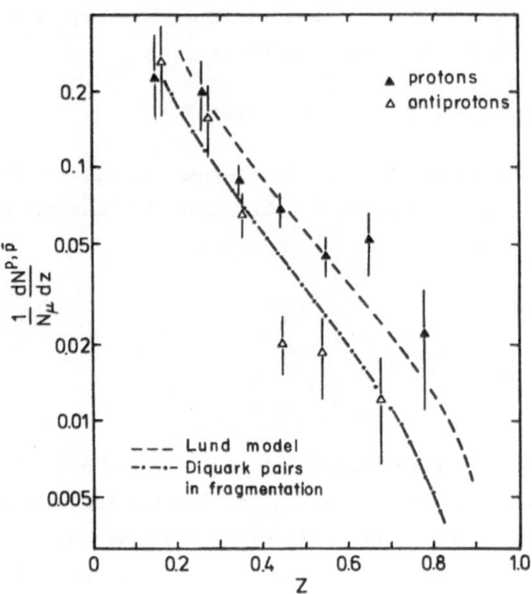

Fig. 43 - z dependence of p and $\bar{p}$ yields in high energy muon scattering.

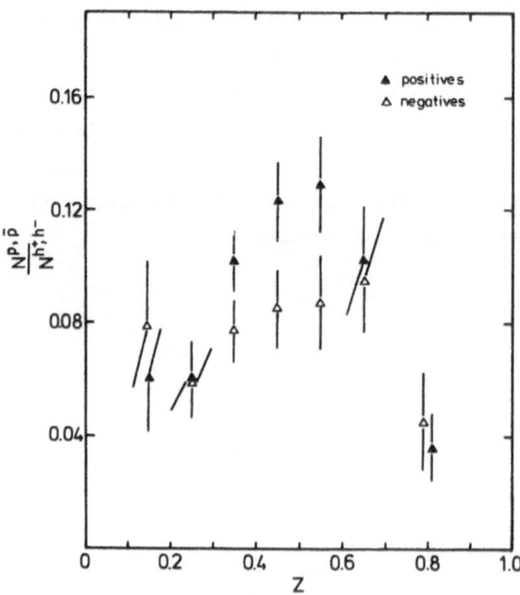

Fig. 44 - Ratio of p to all positive charged hadrons and $\bar{p}$ to all negative

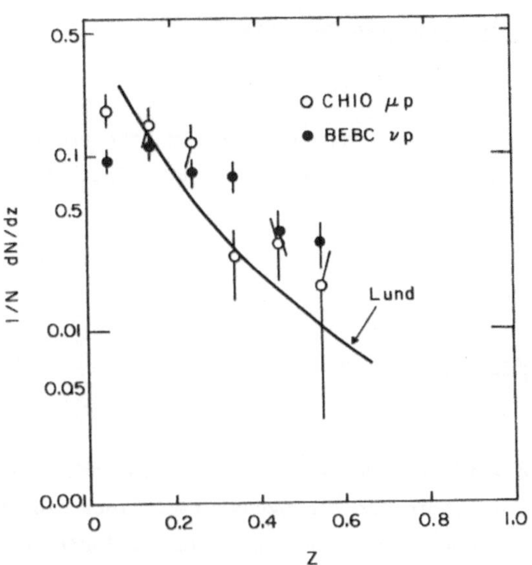

Fig. 45 - High energy lambda production in deep inelastic scattering.

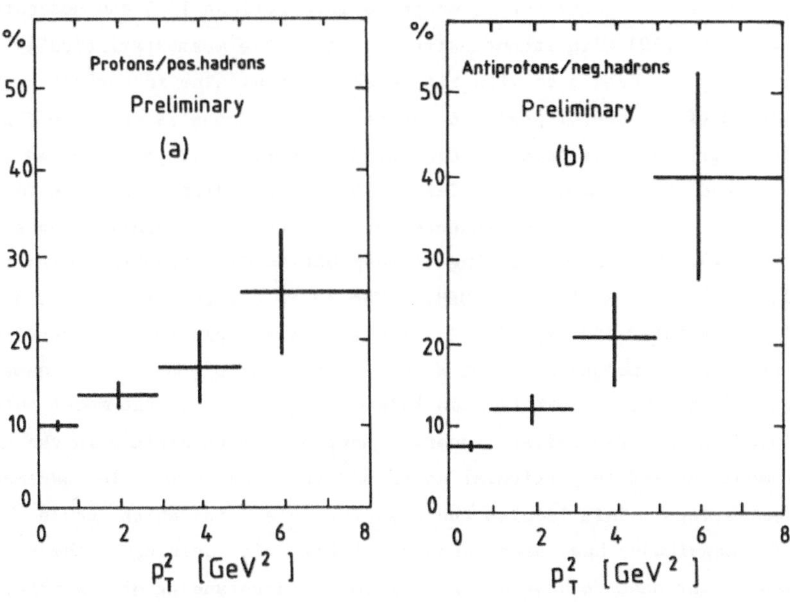

Fig. 46 - Proton and antiproton production compared to all hadrons as a function of p_T^2 in muon scattering.

VIII) Hadron Production: Perturbative QCD Effects

Given the remarkable success, demonstrated in the previous section, it is clear (at least in retrospect) that a search for QCD effects implies high statistics data.

Scale Breaking, Factorisation Breaking

The acquisition of a Q^2 dependence is taken as the basic sign of QCD in the structure function data and there are analogous predictions for the fragmentation function [55]. The QPM is modified in leading order as indicated in fig. 47a). In next to leading order (fig. 47b) the explicit terms of order α_s in the parton cross-section [56,57] are supplemented [58] by order α_s terms from the quark distribution functions and fragmentation functions which further modify the expected Q^2 dependence.

The characteristic scale breaking and, in next to leading order, factorisation breaking, have been searched for in neutrino interactions [59] and moderate energy electroproduction [60] with inconclusive results. The most statistically powerful measurement [61] has been made with high energy muons. The factorisation breaking (x dependence) of the sum of positive and negative hadrons is shown in fig. 48. The systematic errors on these data (as discussed in section II) are limited to $\leq 10\%$ and the observed effect is of the order of 25-30% depending on z. The factorisation breaking demands that the Q^2 dependence be displayed for small ranges in x as shown in fig. 49. The scale breaking is weak but hardly compatible with the scaling (dotted lines) expected from the QPM. The solid curves are a phenomenological calculation from Baier and Fey [57]. If the data are plotted as a function of W^2 there is then no dependence at fixed W^2 on either x or Q^2, but the W^2 dependence is quite marked (fig. 50). What is also interesting is that the model curves behave qualitatively in a very similar manner. None of the parameters in the calculation have been modified and in particular no value for the QCD scale parameter Λ has yet been extracted. This is also the case for e^+e^- in which scale breaking similar in magnitude has been observed [62]. In this case the situation is complicated by the need to have a very complete understanding of the effect of heavy quark (c and b) production and decay which may well account for a large part of the observed effect [63].

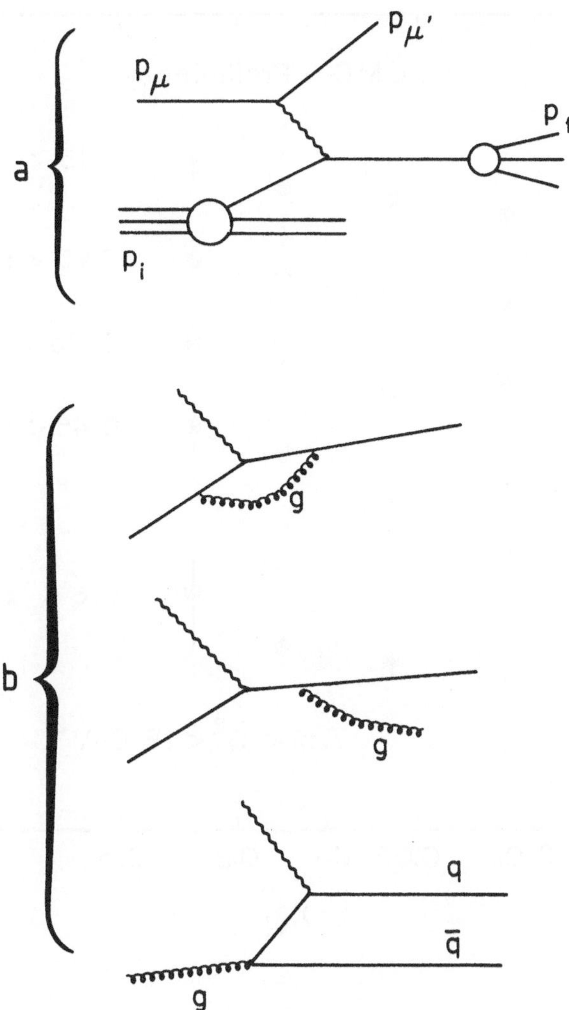

Fig. 47 - a) QPM or QCD leading order;
b) Next to leading order QCD corrections.

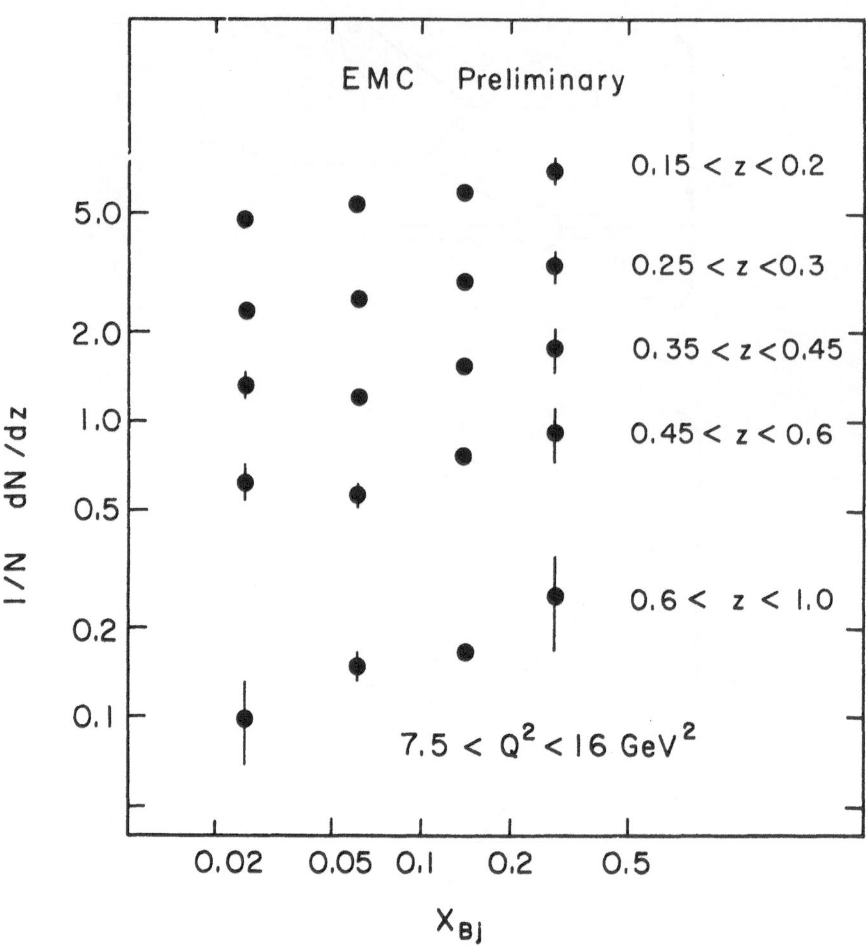

Fig. 48 – x-dependence of $\frac{1}{N}\frac{dN}{dz}$ in muon interactions.

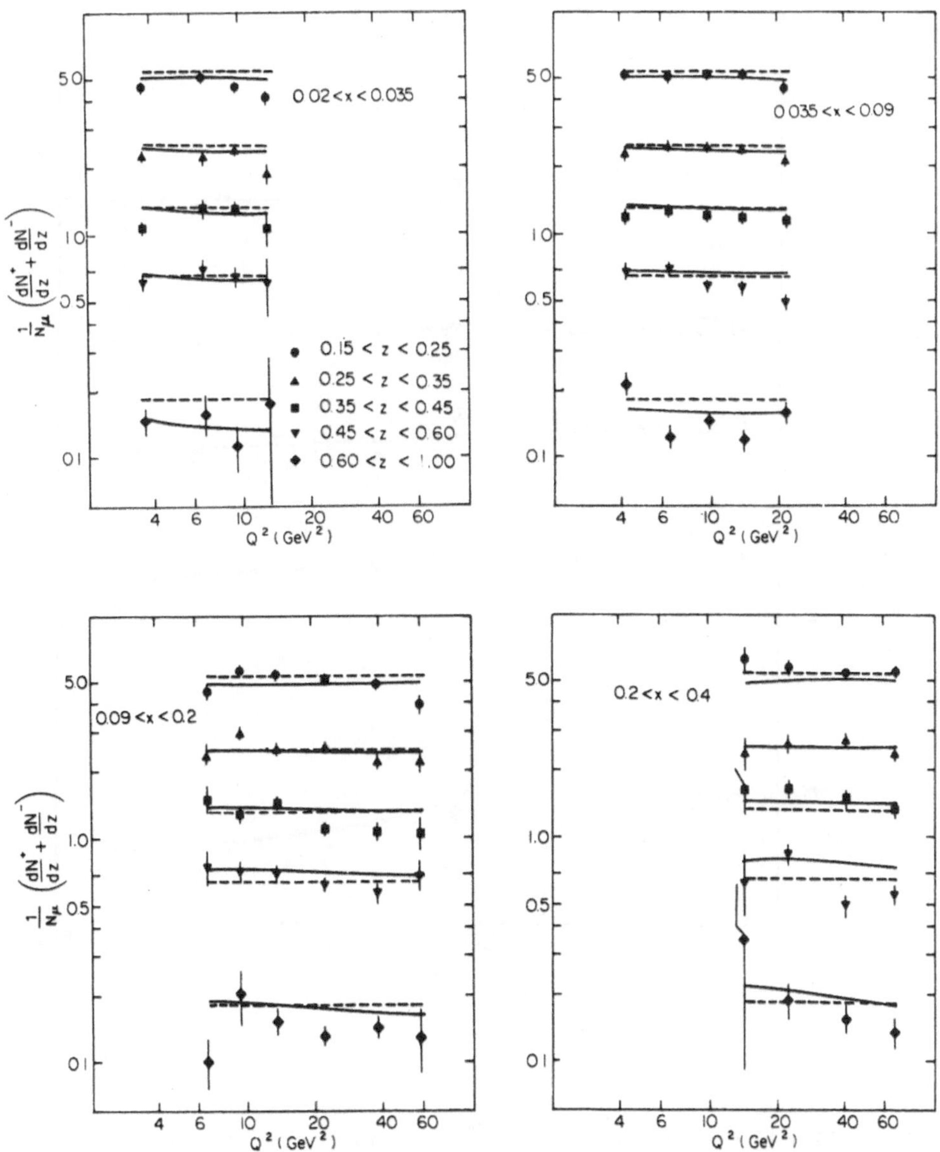

Fig. 49 – Q^2 dependence of $\frac{1}{N}\frac{dN}{dz}$ in muon interactions for different x ranges, the solid line is the QCD model.

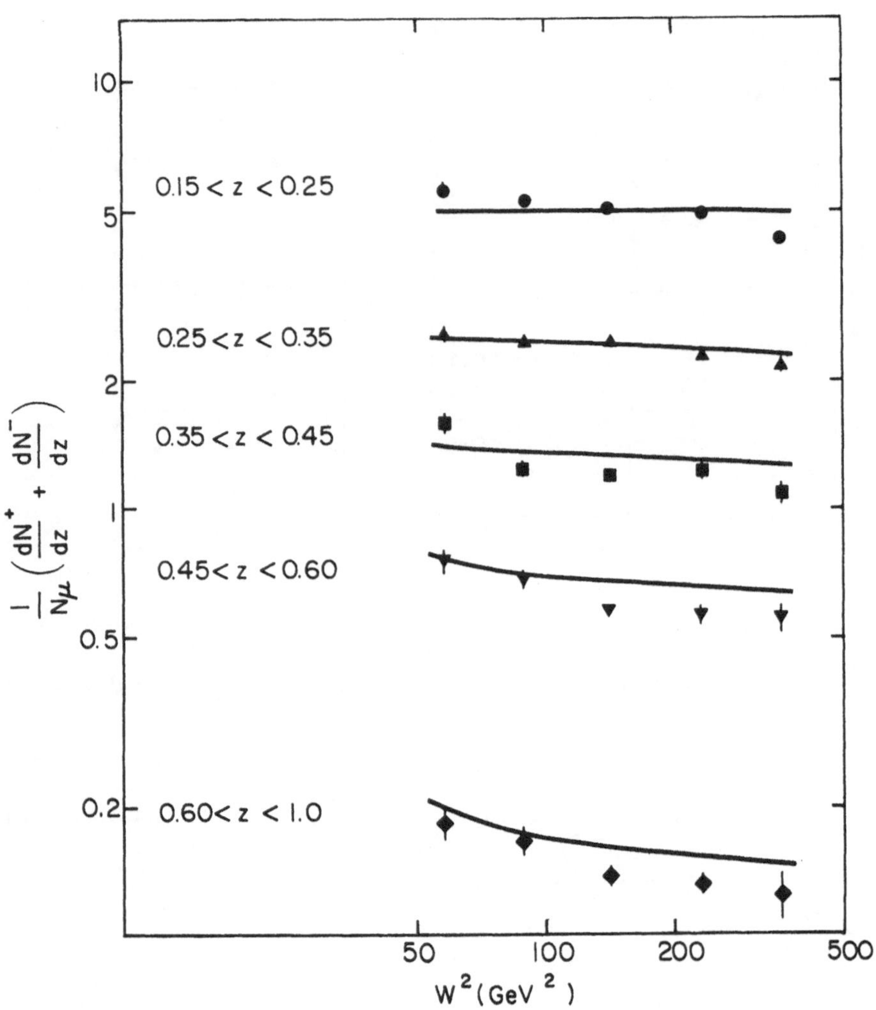

Fig. 50 - W² dependence of $\frac{1}{N}\frac{dN}{dz}$, the line is the QCD model.

QCD Effects in the Transverse Momentum Distributions

Before considering the data it is necessary to enumerate the sources of transverse momentum available in Deep Inelastic Scattering.

- Fragmentation p_T
  - generated by the quark antiquark pair production and locally balanced, denoted by σ_q or $\langle p_T^2\rangle_{Frag}$
- Intrinsic k_T
  - quark motion within the target nucleon analogous to the classical Fermi motion of a nucleon in a nucleus. A hadron receiving a fraction z of the quark momentum would expect on average to receive $z^2 \langle k_T^2\rangle$.
- QCD Effects
  Non-collinear gluon emission and the photon gluon fusion graphs lead to additional p_T with respect to the photon direction.

The predictions are therefore: i) extra p_T beyond σ_q should be observed, and eventually separated jets should appear; ii) the $\langle p_T^2\rangle$ should vary as the kinematics of the current are changed [64]

$$\langle p_T^2\rangle \sim \alpha_s(Q^2) \cdot Q^2 \cdot f(x,y)$$
$$\sim \alpha\,(Q^2) \cdot W^2,$$

the proportionality to W^2 is understood as coming from the integration over the phase space available for gluon emission; iii) the current interacts with a single quark which then suffers gluon emission, therefore there should be an asymmetry between the forward quark system and the backward diquark system

$$\langle p_T^2\rangle|_{x_F>0} > \langle p_T^2\rangle|_{x_F<0},$$

following ii) this asymmetry should rise as a function of W^2.

The p_T^2 distributions of charged hadrons observed in two neutrino experiments is shown in fig. 51 [59]. The tail for $W^2 > 50$ GeV^2 is clearly more marked in each case. Fig. 52 shows the p_T^2 distribution of charged hadrons observed in high energy muon interactions in comparison with a model [65] with and without the QCD effects included. If the QCD effects are not present then a very big modification of the fragmentation model is necessary. The qualitative behaviour of the data is very similar to e^+e^- data [66]. The behaviour of $\langle p_T^2\rangle$ with W^2 is illustrated in fig. 53 in which charged and neutral hadron data from muon scattering are collected together with neutrino data. Within their limited W range the latter are in agreement and a clear rise, in agreement with the QCD model is seen.

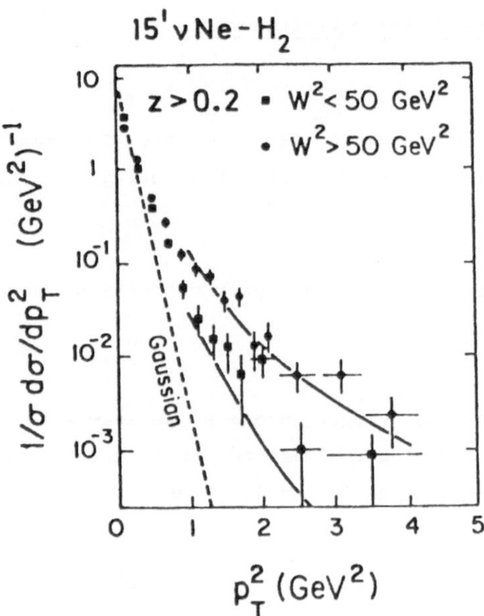

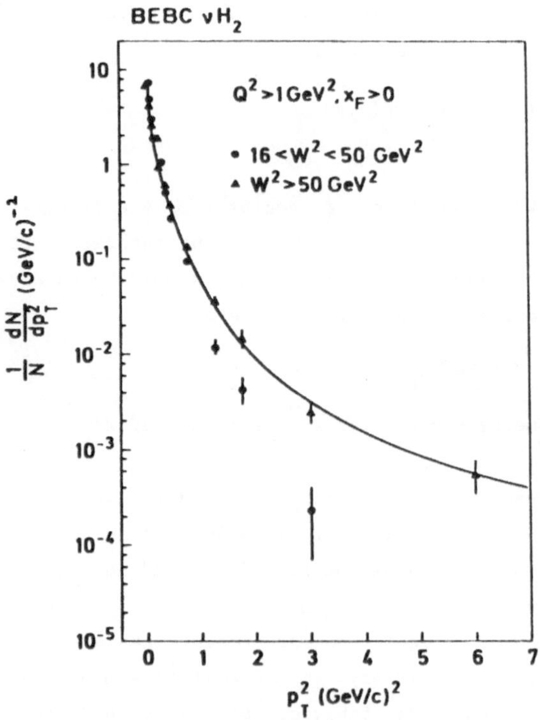

Fig. 51 - Observed p_T^2 spectra in high energy neutrino experiments.

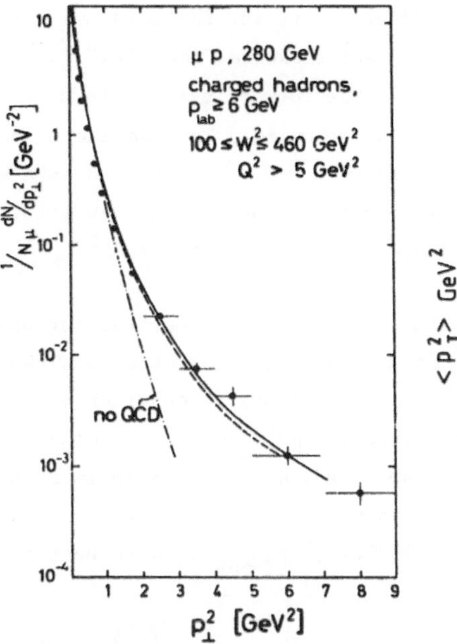

Fig. 52 - Observed p_T^2 spectra in high energy muon scattering compared with model predictions.

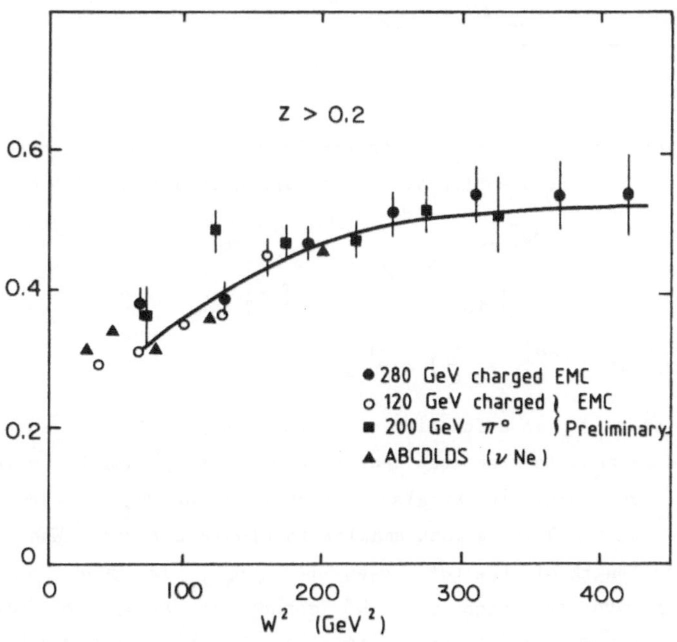

Fig. 53 - Dependence of $<p_T^2>$ on W^2, solid line is QCD prediction.

That the behaviour is different in the backward cms region is shown in fig. 54. Fig. 54a) shows the forward and backward "seagulls" for two different W ranges from a neutrino experiment [67]. The forward seagull is more marked at high W whereas the backward seagull is hardly changed. This asymmetry has been exploited by a Fermilab neutrino group [68] who use the backward as a measure of non QCD effects and take the forward-backward difference in $\langle p_T^2 \rangle$ as a function of W^2 and compare directly with a QCD calculation as shown in fig. 55.

Remembering the list of predictions, the item for which evidence has not yet been demonstrated is "and eventually separated jets should appear". In order to look for radiative jets it is necessary to consider event shapes. Provided that a natural axis exists, as given by the exchanged boson direction in deep inelastic scattering, and is well measured, as is the case in muon interactions, then the procedure is very straightforward. As indicated in the sketch the event is rotated about the photon axis such that $\sum p_{T\,out}^2$ with respect to a plane is minimised.

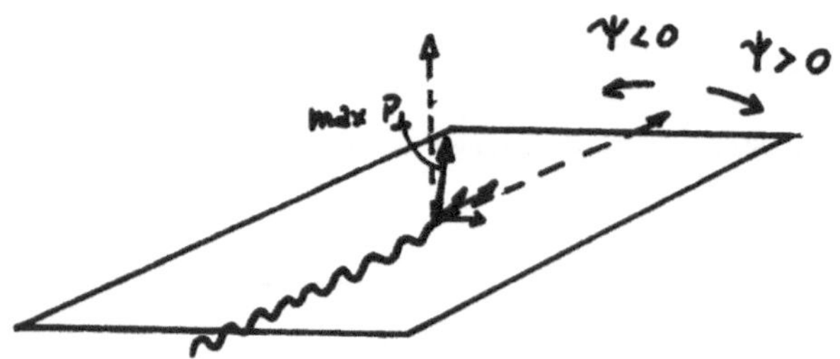

The ambiguity in this procedure is resolved by ordering the highest p_T particle to $\psi < 0$. There is then a series of interesting quantities which can be studied.

eg. $\langle p_T^2 \rangle_{in}$ $\quad$ $\langle p_T^2 \rangle_{out}$

$\sum \vec{p}_{T\,in}^{\,2}$ $\quad$ $\sum \vec{p}_{T\,out}^{\,2}$

Energy flow $$\frac{d\Sigma}{d\psi} = \frac{1}{\sigma} \int z \frac{d\sigma}{dz\,d\psi} dz$$

The behaviour of $\langle p_T^2 \rangle_{in}$ as a function of W^2 for neutrino interactions [69] is plotted in fig. 56 and is seen to rise such that a comparatively smooth continuation to e^+e^- measurements is possible. It can also be seen that the $\langle p_T^2 \rangle_{out}$ also rises with W^2 even though (as must be) it is much smaller in absolute terms. What is perhaps more striking is the length of the tail seen in $\sum p_{T\,in}^2$ in muon interactions [70], fig. 57, the growth of which with W^2 cannot be easily reproduced by modified fragmentation. The energy flow in the corresponding high energy data is shown with and without a p_T^2 cut in fig. 58.

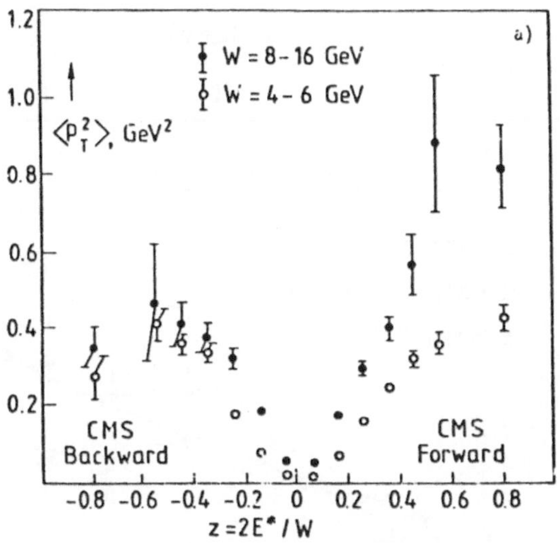

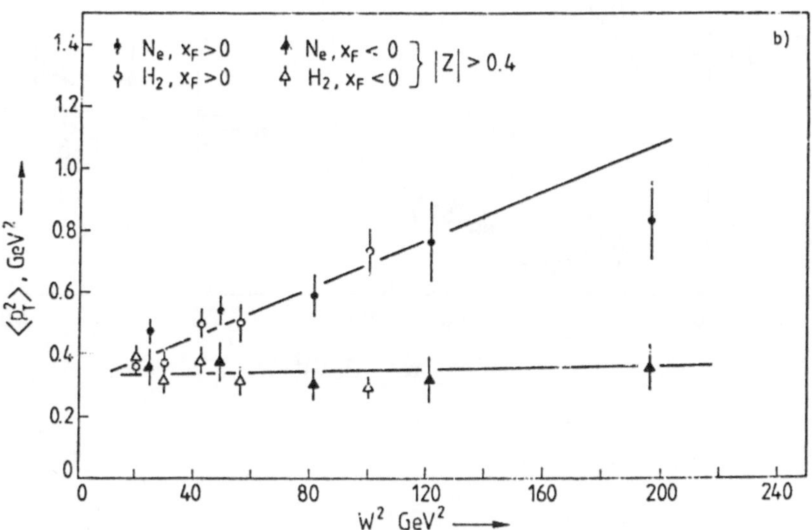

Fig. 54 - a) Dependence of $<p_T^2>$ on x_F in neutrino interactions.
b) Dependence of forward and backward $<p_T^2>$ on W^2.

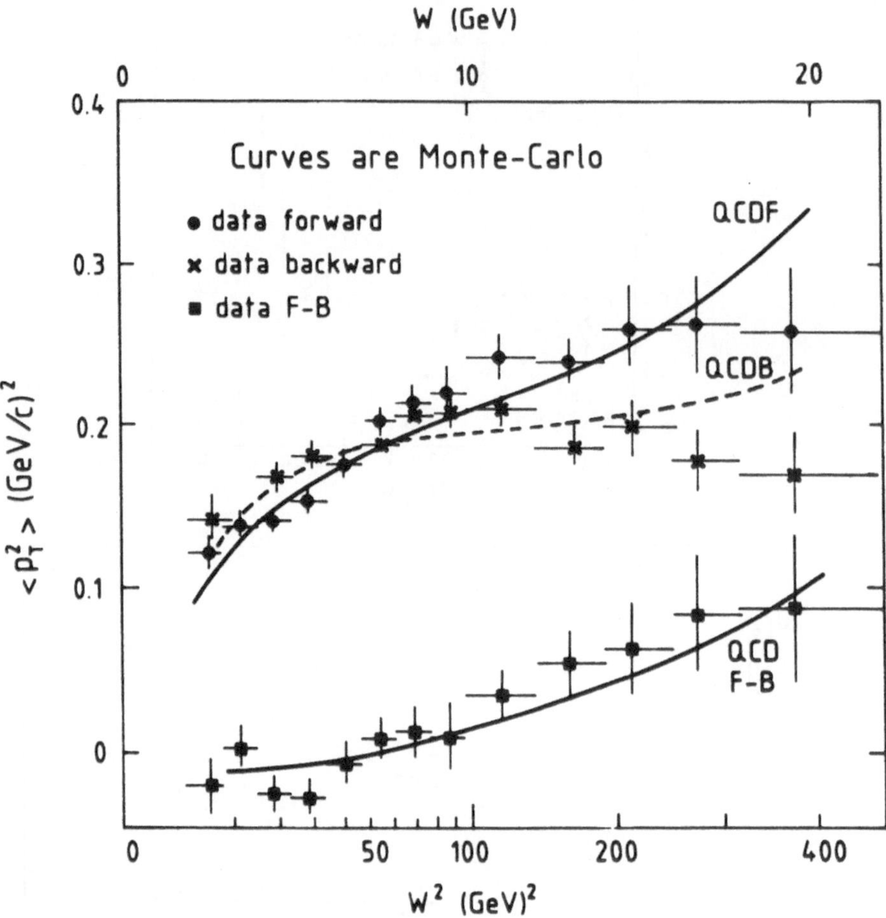

Fig. 55 - W^2 dependence of forward and backward p_T^2 and the difference compared to QCD model in neutrino interactions.

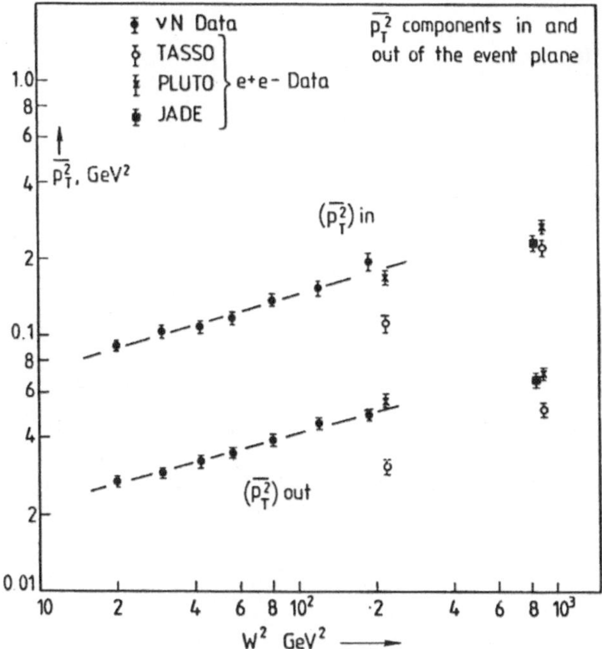

Fig. 56 – $\langle p_T^2 \rangle_{in,\ out}$ as a function of W² and comparison with e^+e^- results.

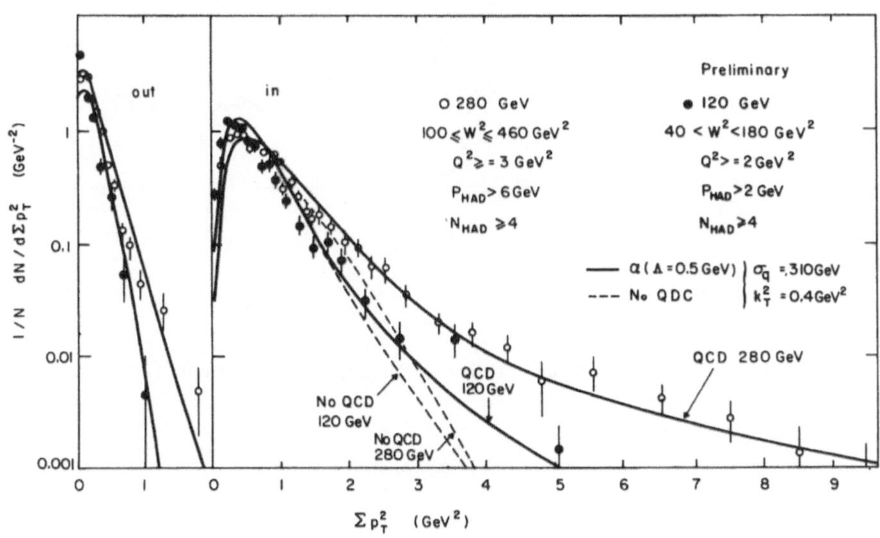

Fig. 57 – $\Sigma p_{T\ out}^2$ and $\Sigma p_{T\ in}^2$ for muon interactions in different W² ranges, comparison with models.

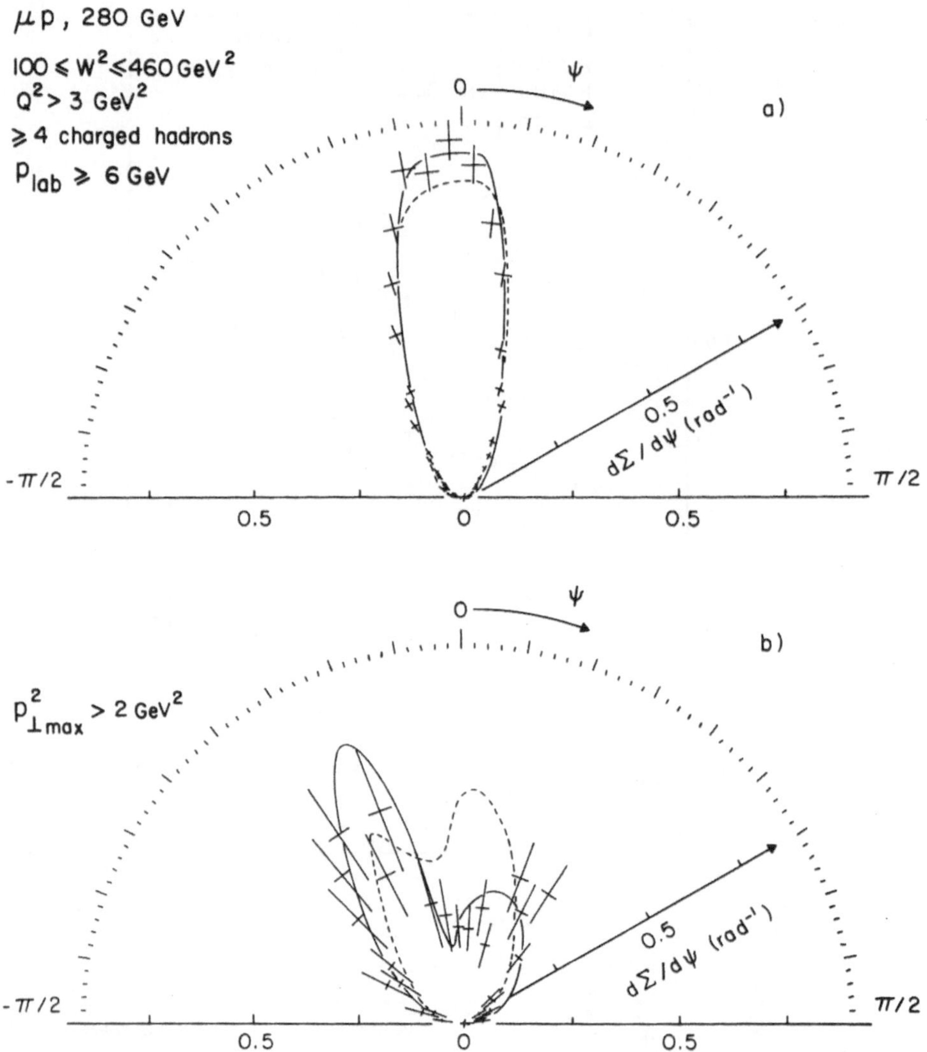

Fig. 58 - Energy flow of hadrons produced in high energy muon interactions.

A less marked structure is obtained with a similar but lower p_T^2 cut in the 120 GeV data although in this case and for similar W neutrino data [67,68] the gymnastics required of "No QCD" fragmentation models to reproduce the data is not great and the major part of the effect can be attributed to the analysis procedure and momentum conservation. The interpretation of the effect seen in fig. 58 as being due to the presence of two forward jets is supported by the azimuthal correlation observed between other tracks and the high p_T track (not included in fig. 59) [71]. If no cuts are applied an expected "away side" enhancement is seen. If a p_T cut is applied this enhancement is accentuated and a positive correlation in the region of the "trigger particle" is also seen which cannot be reproduced by resonances.

Measurement of k_T?

A source of hadronic p_T which in any general discussion must be present is the intrinsic k_T of the quarks in the nucleon corresponding to their confinement. This is analogous to the classical Fermi motion of nucleons in a nucleus. However the inability of theorists to solve the confinement problem has lead to the use of the term intrinsic k_T as a useful repository for all not understood but measured p_T values. An example is the Drell-Yan process for lepton pair production, the observed values of apparent k_T [72] exceeded what was reasonable for some time and it is only recently that at least some part of this large p_T has been explicitly calculated. In deep inelastic scattering the simplest observable indication of a k_T is, as indicated earlier, that a hadron carrying a fraction z of the quark momentum should also carry a fraction z of its k_T, hence at high z

$$\langle p_T^2 \rangle \sim z^2 \, k_T^2 .$$

As shown in fig. 60 such a term is necessary to describe the data.

A second method which is more difficult to measure experimentally is related also to the R measurement (see section V). The general expression for the azimuthal distribution of hadrons with respect to the <u>lepton scattering plane</u> is given by

$$\frac{dN}{d\phi} = a + b\, f_1(y) \cos\phi + c\, f_2(y) \cos 2\phi .$$

The term in $\cos\phi$ arises from an interference between transverse and scalar photons and the $\cos 2\phi$ term from an interference between photons in the two transverse polarisation states. These effects are extremely difficult to measure because the error on the direction of the virtual photon must be very small. This measurement was once advertised as a gold plated QCD test [73] however it was pointed out [74]

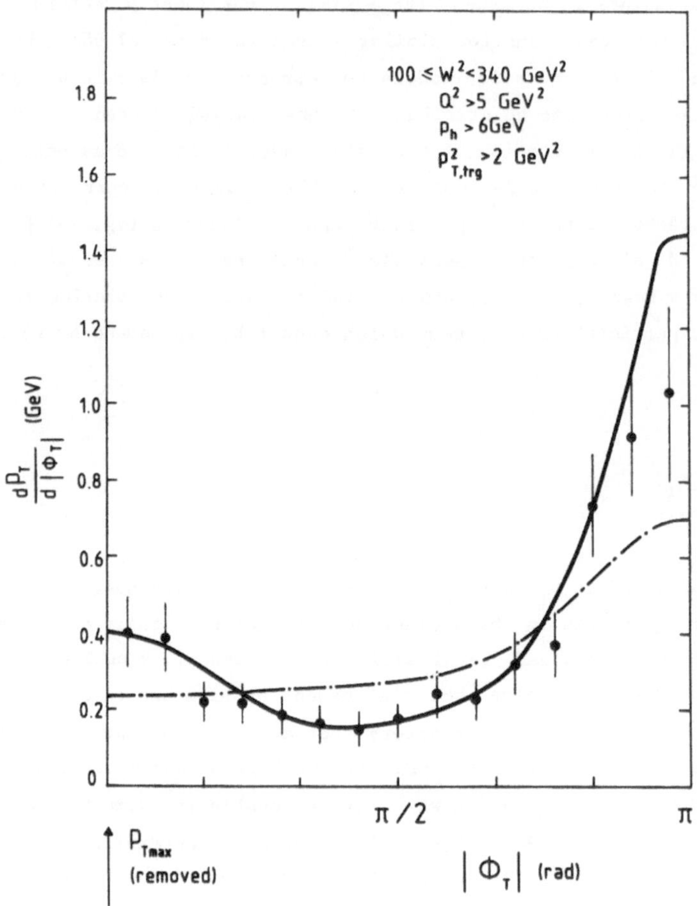

Fig. 59 - Azimuthal correlation between "trigger" hadron and other hadrons in the event, —— QCD model, - · - · - QPM broad fragmentation.

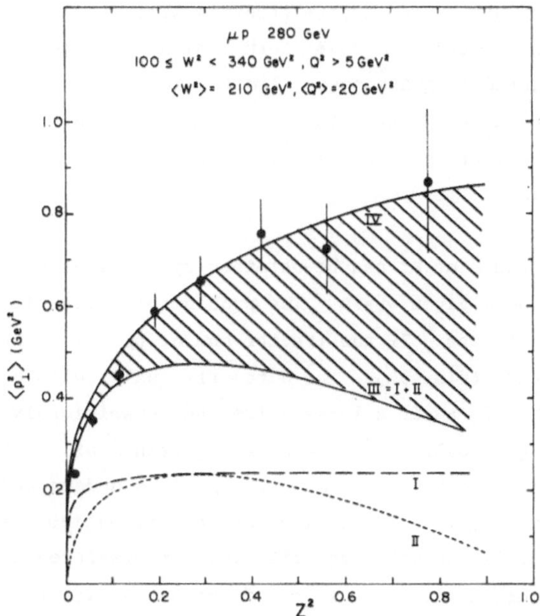

Fig. 60 - Dependence of $<p_T^2>$ on z^2 in deep inelastic muon scattering. The curves represent the various contributions, the shaded part is the contribution of k_T^2.

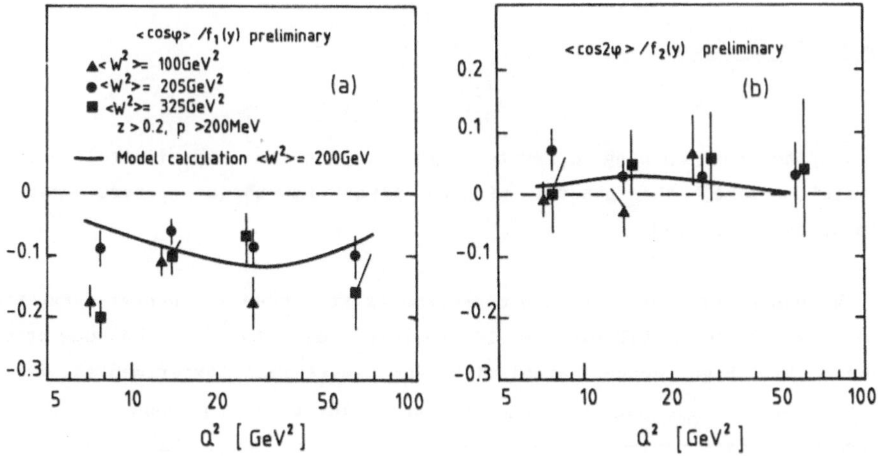

Fig. 61 - $<\cos\phi>$ and $<\cos 2\phi>$ as measured in high energy muon scattering.

that the presence of any k_T could also generate similar effects. Recent data [75] (fig. 61) shows an appreciable $\cos\phi$ term which compares in magnitude with a calculation [76] including phase space limitations and fragmentation. The major contribution is from k_T and the value required, as for the high z^2 single particles, is of the order of 0.7–0.8 GeV². Like the Drell-Yan measurements this is an "unreal" value.

Yet another measure of k_T^2 can be obtained by taking the vector sum of all the hadron fragments from a given quark. The problem of course lies in defining <u>all the products</u> not more not less. If the (vector sum of p_T)² for those events for which $\sum z_h > 0.7$ is plotted (fig. 62), the data fit well with a Monte Carlo using $k_T^2 = 0.4$ GeV². This is again a large value but suspiciously different to the previous values. Eventually the repeated quoting of such large values had the desired effects and some theoreticians attempted to understand the effect in terms of soft gluon emission [77]. This alternative explanation gave a perfectly good description of the existing data, eg. fig. 56, but predicted that rather than having the p_T balanced almost entirely by target fragments as for k_T, there should be some balancing in the central region. The high energy muon scattering data are compared with these two models in fig. 63 and indeed the data favour the soft gluon explanation.

The residual k_T^2 required by the data is $\langle k_T^2 \rangle \sim 0.2$ GeV² which is also compatible within errors with the null measurement of $R = \frac{\sigma_L}{\sigma_T}$.

Summary, Conclusions

Despite the warning comments with respect to experimental effects and systematic errors, the general picture given in these lecture was, I think, optimistically slanted.

The basic structure function measurements from muon and neutrino experiments faithfully follow the QPM picture to the level of $\sim 10\%$*). The deviations are measured with high enough statistics and sufficient experimental care for a "measurement" of Λ_{QCD} such that the relevant value of α_s is known to about 10%. The present understanding of the corrections suggests that this is one of the more

---

*) There is perhaps an interesting blip on the horizon with the presentation by EMC of a comparison between iron and deuterium data which indicates a surprising difference in the behaviour as a function of x.

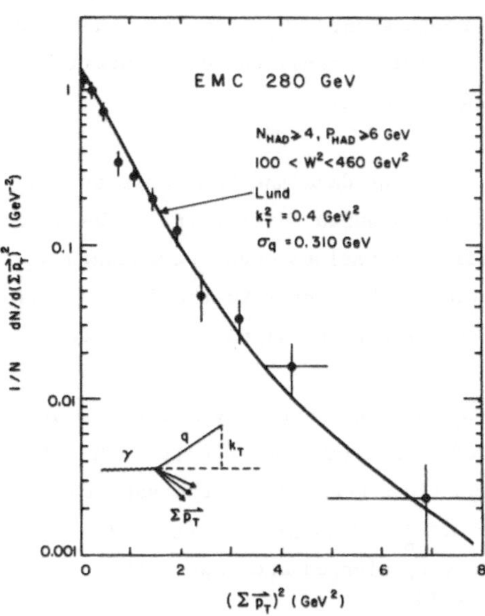

Fig. 62 - Distribution of $\sum(\vec{p}_T)^2$ for $\sum z_h > 0.7$, curve corresponds to $\langle k_T^2 \rangle = 0.4$ GeV^2.

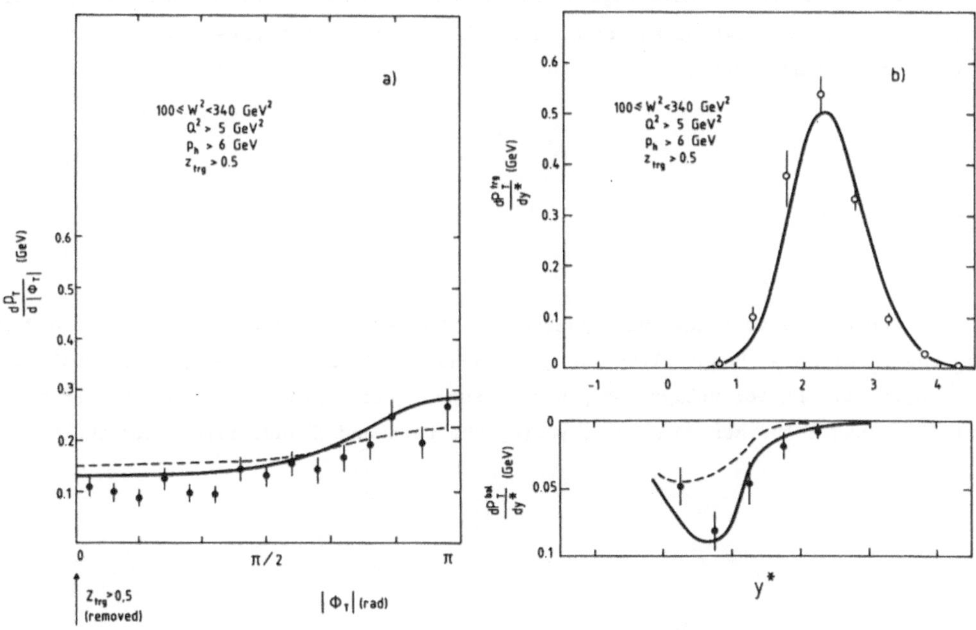

Fig. 63 - Distribution of the p_T which balances the "trigger" particle as a function of azimuth and as a function of y^*. The dotted line is from model with no soft gluons and $\langle k_T^2 \rangle = 0.7$ GeV^2, the solid line from model with soft gluons but $\langle k_T^2 \rangle = 0.2$ GeV^2.

reliable sources of such a measurement. Nevertheless there are higher twist effects expected and indeed within a limited sense the data provide an indication of their size and behaviour as a function of x.

As with the structure function data the QPM works so well that the measurement of deviations from the model is quite difficult. The recent deep inelastic data have refined the determination of various model parameters γ_s, α_v and have caused the major step of introduction of baryons into the fragmentation models. It remains to be seen whether these baryons are the signal of more fundamental effects.

The highest energy data pass qualitatively all the tests required for them to be advanced as data showing QCD bremsstrahlung effects. This is of primary importance, given that the structure functions measured in the same events provide one of the best quantitative measured of α_s. However the transition from fragmentation tail to out and out QCD effect is very blurred and makes the progression to quantitative comparison with QCD difficult*).

Overall, given that deep inelastic scattering is, as stated in the introduction, an indirect means to study parton interactions, the results of the study are so transparent that the disadvantage at times seems minimal. Furthermore the progress made with improved high energy data in the last two years suggests that it is a field worthy of further effort.

Acknowledgements

I would like to thank the organisers of the school and the director Risto Raitio for the formidable hospitality at the school. I am, as always, grateful to my colleagues in EMC for enlightening discussions. Yet again I would like to thank Ariella Mazzari for her work in preparing the text and Claude Rigoni for the figures.

---

*) This has also been retrospectively appreciated in the case of e^+e^- [54].

References

[1] eg. A. Donnachie, Proc. 1970 CERN School of Physics, CERN 71-7 (1971).

[2] L.W. Jones, Rev. Mod. Phys. 49 (1977) 717;
G. Susinno, Contr. to Conf. on Physics in Collision, Blacksburg, 1981.

[3] J.D. Bjorken, Phys. Rev. 179 (969) 1547.

[4] W.K.H. Panofsky, Proc. 14th Int. Conf. on High Energy Physics, Ed. J. Prentki, J. Steinberger, Vienna 1968.

[5] eg. M. Gourdin, CERN Summer School 1972, Grado, Italy.
J. Drees, CERN Summer School 1980, Malenta, Germany.

[6] F.E. Close, An Introduction to Quarks and Partons, Academic Press, 1979.

[7] R.P. Feynman, Photon-Hadron Interactions, W.A. Benjamin Inc. 1972.

[8] A similar attempt at the appraisal of the systematic effects in Deep Inelastic Scattering was made in:
W.B. Atwood, SLAC-PUB-2428 (1979), Published in Proceedings of the SLAC Summer Institute 19798, Ed. Anne Moshe.

[9] M.J. Murtagh, Proc. 1981 Int. Conf. on Neutrino Physics and Astrophysics, Maui, 1981, Vol. I, p. 388, Ed. R.J. Cence, E. Ma, A. Roberts.

[10] H. Abramowicz et al., Nucl. Instr. and Meth. 180 (1981) 429;
I would like to thank F. Dydak for discussion on the calibration procedure.

[11] R.P. Mount, Nucl. Instr. and Meth. 187 (1981) 401.

[12] C. Zupancic, NA4 Experiment Internal Note 1980.

[13] EMC, J.J. Aubert et al., Phys. Lett. 105B (1981) 322.

[14] R.T. Giles, D.Phil. Thesis 1981, Oxford University, Nuclear Physics Laboratory.

[15] EMC, J.J. Aubert et al., Phys. Lett. 114B (1982) 373.

[16] H. Abramowicz et al., Preprint CERN-EP/82-77 (1982).

[17] EMC, J.J. Aubert et al., Preprint CERN-EP-82/153 (1982); submitted to Nuclear Physics B.

[18] S.J. Brodsky et al., Phys. Lett. 93B (1980) 451;
S.J. Brodsky et al., Phys. Rev. D23 (1981) 2745;
D.P. Roy, Phys. Rev. Lett. 47 (1981) 213.

[19] EMC, J.J. Aubert et al., Phys. Lett. 110B (1982) 73.

[20] H. Abramowicz et al., Phys. Lett. 107B (1981) 141.

[21] F. Eisele, Invited talk at the X Int. Winter Meeting on Fundamental Physics and XIII GIFT Int. Seminar on Theoretical Physics; DO-EXP-82/5 (1982).

[22] EMC, J.J. Aubert et al., Preprint CERN-EP-82/159 (1982); submitted to Physics Letters.

[23] J. Drees, Proc. 1981 Int. Symposium on Lepton and Photon Interactions at High Energies, Ed. W. Pfeil, Bonn 1981.

[24] M.D. Mestayer et al., SLAC-PUB-2933, 1982; submitted to Phys. Rev. D.

[25] G. Altarelli and G. Parisi, Nucl. Phys. B126 (1977) 298.

[26] A. Gonzales-Arroyo, C. Lopez and F.J. Yndurain, Nucl. Phys. B153 (1979) 161; B159 (1979) 512.

[27] L.F. Abbott, W.B. Atwood and R.M. Barnett, Phys. Rev. D22 (1980) 582.

[28] A. Gonzales-Arroyo and C. Lopez, Nucl. Phys. B166 (1980) 429.

[29] EMC, J.J. Aubert et al., Phys. Lett. 114B (1982) 291.

[30] S.J. Wimpenny, Ph.D. Thesis, Sheffield University 1981, RAL HEP/T/90 1981.

[31] P. Castorina et al., Phys. Rev. Lett. 47 (1981) 468, Bari University Report No BA-GT-81/03 (1981).

[32] H. Abramowicz et al., Z. Phys. C12 (1982) 289;
Z. Phys. C13 (1982) 199.

[33] F. Eisele, Proc. 1981 Int. Conf. on Neutrino Physics and Astrophysics, Maui, 1981, Vol. I, p. 297, Ed. R.J. Cence, E. Ma, A. Roberts.

[34] A. Buras, Proc. 1981 Int. Symposium on Lepton and Photon Interactions at High Energies, Ed. W. Pfeil, Bonn 1981.

[35] R. Field and R.P. Feynman, Nucl. Phys. B136 (1978) 1.

[36] J.P. Berge et al., Nucl. Phys. B203 (1982) 1.

[37] P. Renton and W.S.C. Williams, Ann. Rev. Sci. 31 (1981) 193.

[38] EMC, J.J. Aubert et al., Preprint CERN-EP/80-137 (1980).

[39] EMC, J.J. Aubert et al., in preparation, see also ref. [43].

[40] R. Brandelik et al., Phys. Lett. 108B (1982) 71.

[41] H.J. Behrend et al., Z. Phys. C14 (1982) 189.

[42] eg. N. Schmitz, MPI-PAE/Exp. El. 88 (1980), published in Acta Physica Polonica.

[43] EMC, K.H. Becks, presented at the XXI Int. Conf. on High Energy Physics, Paris 1982, also Wuppertal Preprint WU-B 82-13.

[44] R.P. Feynman, Proc. Int. Conf. on Neutrino Physics and Astrophysics, Balaton 1972.

[45] eg. S. Barlag, Paper # 189 contributed to 1981 Int. Symp. on Lepton and Photon Interactions at High Energies, Ed. W. Pfeil, Bonn 1981; see ref. [54] for compilation.

[46] J.P. Berge et al., Nucl. Phys. B184 (1981) 30.

[47] M. Henckes, Ph.D. Thesis University of Wuppertal 1981.

[48] V.V. Ammosov et al., Phys. Lett. 93B (1980) 210.

[49] Ch. Berger et al., Phys. Lett. 104B (1981) 79; see also ref. [51].

[50] EMC, J.J. Aubert et al., Phys. Lett. 104B (1981) 388.

[51] R. Felst, Proc. 1981 Int. Symp. on Lepton and Photon Interactions at High Energies, Ed. W. Pfeil, Bonn 1981.

[52] B. Andersson et al., Z. Phys. C13 (1982) 361.

[53] G. Schierholz and M. Teper, Z. Phys. C13 (1982) 53.

[54] G. Wolf, Rapporteur's Talk XXI Int. Conf. on High Energy Physics, Paris 1982.

[55] J.F. Owens, Phys. Lett. 76B (1978) 85;
T. Uematsu, Phys. Lett. 79B (1978) 97.

[56] G. Altarelli, R.K. Ellis, G. Martinelli and So Young Pi, Nucl. Phys. B160 (1979) 301.

[57] R. Baier and K. Fey, Z. Phys. C2 (1979) 339.

[58] W. Furmanski and R. Petronzio, Z. Phys. C11 (1982) 293.

[59] eg. N. Schmitz, Proc. 1981 Int. Symp. on Lepton and Photon Interactions at High Energies, Ed. W. Pfeil, Bonn 1981.

[60] C. Drews et al., Phys. Rev. Lett. 41 (1978) 1433.

[61] EMC, J.J. Aubert et al., Phys. Lett. 114B (1982) 373.

[62] R. Brandelik et al., DESY 81-013 (1982);
Phys. Lett. 114B (1982) 65.

[63] C. Peterson et al., SLAC-PUB-2912 (1982).

[64] G. Altarelli and G. Martinelli, Phys. Lett. 76B (1978) 89.

[65] B. Andersson et al., Z. Phys. C1 (1979) 105;
B. Andersson and G. Gustaffson, Z. Phys. C3 (1980) 22;
B. Andersson and G. Gustaffson, Lund Univ. Preprint LU-TP-80-1 (1980);
B. Andersson et al., Z. Phys. C9 (1981) 233.

[66] eg. G. Wolf, DESY 81-086;
K.H. Mess and B.H. Wiik, DESY 82-011.

[67] H. Deden et al., Nucl. Phys. B181 (1981) 375.

[68] H.C. Ballagh et al., Phys. Rev. Lett. 47 (1981) 556.

[69] D.H. Perkins, Proc. 1981 CERN Summer School, CERN 82-04 (1982).

[70] EMC, J.J. Aubert et al., Phys. Lett. 100B (1981) 433;
see also J. Gayler, DESY 81-063;
H.E. Montgomery, Proc. 1981 Int. Symp. on Lepton and Photon Interactions at High Energies, Ed. W. Pfeil, Bonn 1981.

[71] EMC, J.J. Aubert et al., Preprint CERN-EP/82-146 (1982), to be published in Phys. Lett.

[72] A. Michelini, Preprint CERN-EP/81-128 (1981);
Proc. 1981 European Conf. on High Energy Physics, Lisbon 1981.

[73] H. Georgi and H.D. Politzer, Phys. Rev. Lett. 40 (1978) 3.

[74] R.W. Cahn, Phys. Lett. 78B (1978) 269.

[75] EMC, Paper #750 Contributed to XXI Int. Conf. on High Energy Physics, Paris 1982;
see also ref. [43].

[76] A. König and P. Kroll, Univ. of Wuppertal Preprint WU-B 82-8 (1982).

[77] B. Andersson, G. Gustaffson and T. Sjöstrand, Z. Phys. C12 (1982) 49;
G. Ingelman et al., Lund Preprint LU-TP 81-8 (1981).

TABLE I

Experiment	Quantities Measured
Muons	
BCDMS	E_μ E'_μ θ_μ
EMC	E_μ E'_μ θ_μ (also E_H for Iron expt)
Neutrinos	
CDHS, CFRR	E_H (calorimetry) E'_μ θ_μ
BEBC, 15'	$E_{H\ VIS}$ (measured track momenta) E'_μ θ_μ

Quantities measured in different types of experiments.

TABLE II

Experiment	Reaction	R	R'
SLAC	ep	0.21±0.10	
SLAC-MIT	ed	0.17±0.07	
CHIO (x>0.01)	μp	0.38±0.38	
EMC	μp	0.0 ±0.10	
	μFe	0.03±0.13	
HWPFOR	νN		0.18±0.07
CDHS I	νN	0.10±0.07	
CDHS II (High x)	νN	<0.02±0.025	
BEBC	νN		0.04±0.16
FIIM	νN		0.03±0.12
CHARM	NC		0.10±0.10

Table of R measurements based on [23].

Antiproton Collisions*

Richard Lednický
JINR, Dubna, USSR

1. Introduction

Special interest to antinucleon-nucleon interactions is in particular connected with the unique possibility of studying processes of complete baryon disintegration in annihilation reactions $\bar{N}N \rightarrow$ mesons. Characteristic feature of these processes is a large multiplicity of produced particles allowing one to study multiparticle production even at relatively low energy. This interest is reflected by the fact that a whole series of conferences, starting in 1972, is devoted to $\bar{N}N$-interactions. These conferences embrased also other interesting topics of antiproton physics such as $\bar{p}$-atoms, baryonium spectroscopy etc.,which will certainly achieve in near future a new qualitative level with the help of antiproton cooling devices due to enormous increase in beam intensity by factor of 10^3 and the amazing mass resolutions of a few keV. The exciting data at very high energies are starting to appear from $\bar{p}p$ collider, e.g. clear hard parton jets have been observed allowing for a check of quantum chromodynamics /QCD/.

In this lecture, I concern mainly the questions of multiparticle production in $\bar{p}p$-interactions and also in other processes and discuss the experimental data in terms of Regge phenomenology and quark-parton models /QM/. In Section 2, the question to what extent are differences between $\bar{p}p$ and pp interactions caused by $\bar{p}p$ annihilations is discussed from experimental as well as from theoretical point of view. Comparison of the moments of multiplicity distributions in various processes with the predictions of dual unitarization /DU/ scheme and the corresponding QM is made in Section 3. Threshhold effects, jet multiplication, KNO scaling, similarities between $e^+e^- \rightarrow$ hadrons and hadron interactions are also discussed. Some questions connected with spin effects in resonance production are considered in Section 4. Spin alignment of vector mesons recently observed in $\bar{p}p$ interactions and its relation to similar effects in

*The full text with references, tables and all figures has been submitted to Czech.J.Phys.B.

other processes is discussed in terms of Regge phenomenology. Quark models for spin effects in hadron production are discussed in Section 5.

2. $\bar{p}p$-pp differences and $\bar{p}p$ annihilations.

The question to what extent are differences between $\bar{p}p$ and pp interactions caused by $\bar{p}p$ annihilations, i.e.

$$\bar{p}p_A \overset{?}{=} \bar{p}p-pp \ , \qquad /1/$$

is important not only for theory but also in practice since almost all information on $\bar{p}p_A$ at high primary momenta $/\gtrsim 10$ GeV/c/ has been extracted from $\bar{p}p$-pp differences. The conclusion is made that $\bar{p}p$-pp differences /with some limitations, e.g. for low multiplicities/ give reasonable estimates for $\bar{p}p_A$ at least up to several tens of GeV/c. Approximate validity of Eq. /1/ for total cross sections and combined $\pi^{\mp}$ inclusive spectra can be explained by Regge phenomenology assuming that the annihilation channels build up the cross channel ω Regge pole via unitarity. Note in this context that the Mueller-Regge predictions for inclusive cross section differences in proton fragmentation region

$$\Delta f_{\pi^-}(\pi^{\mp}p) = \Delta f_{\pi^-}(K^{\mp}p) = \tfrac{1}{2}\Delta f_{\pi^-}(p^{\mp}p) = A_- s^{-1/2}, \qquad /2/$$
$$\Delta f_{\pi^+}(K^{\mp}p) = 0,\ \Delta f_{\pi^+}(\pi^{\mp}p) = -\Delta f_{\pi^+}(p^{\mp}p) = -A_+ s^{-1/2},$$

are in reasonable agreement with experimental data /Fig. 1/, except for low energy $\bar{p}p$-pp differences. In the latter case we see a strong nonleading contribution $\sim s^n$, $n \simeq -2$, presumably connected with low-lying trajectories /baryonia/ strongly coupled to baryon-antibaryon systems. In DU scheme, such contribution corresponds to the 1- or 2-jet $\bar{p}p_A$ /Fig. 2a,b/ and to the diquark annihilation diagram for $\bar{p}p_{NA}$ /Fig. 3a/, as compared with the leading 3-jet $\bar{p}p_A$ /Fig. 2c/ and 1- and 2-jet $\bar{p}p_{NA}$ /Fig. 3b,c/. The predictions of DU scheme are obtained in terms of simple fragmentation QM and found in qualitative agreement with the data on meson x-spectra in proton fragmentation region. In particular, production of more energetic pions in $\bar{p}p_A$ in comparison with pp is explained due to contribution of quark and diquark annihilation diagrams in the former case. Since these contributions rapidly vanish with the increasing energy, we expect an universal shape of such spectra at high enough energies. It is important that x-distribution of a constituent quark /dressed valence quark, valon/ in a nucleon is unambiguously determined in DU approach by the intercepts of effective $q\bar{q}$- and $2q2\bar{q}$-trajectories

$$p(x) = Ax^{-\alpha_{q\bar{q}}}(1-x)^{-\alpha_{2q2\bar{q}}} \simeq \frac{1}{2\sqrt{x}} \; . \qquad /3/$$

3. Multiplicities in $\bar{p}p$-annihilations and other processes.

Characteristic feature of low energy $\bar{p}p_A$ is a large mean charged particle multiplicity $\langle n_{ch}\rangle$ and small relative dispersion

$$d = D/\langle n\rangle = (\langle n^2\rangle - \langle n\rangle^2)^{1/2}/\langle n\rangle \qquad /4/$$

of the multiplicity distribution as compared with other reactions, see Figs. 4,5. Both these features are naturally explained in DU scheme due to larger number of jets in $\bar{p}p_A$ in comparison with other processes. Note that in the multiperipheral approach we expect asymptotically the linear lns-dependence of mean particle multiplicity $\langle n\rangle = Ng\langle k\rangle \ln s$ with the slope proportional to the number N of jets /g is quark-meson/cluster/ coupling constant squared, $\langle k\rangle$ is particle multiplicity in an average cluster/. Instead, $\langle n_{ch}\rangle$ for $\bar{p}p_A$ /3 jets/ and pp /2 jets/ show essentially nonlinear and approximately the same lns dependence /Fig. 4/. We connect this with threshold effects and jet multiplication /branching/. We account for threshold effects by

$$\langle n_{ch}\rangle_{(1)} = \begin{cases} a & , \; s_j \le s_o \\ a + b\ln(s_j/s_o), & s_j > s_o \end{cases} \qquad /5/$$

where $s_o > m_{qT}^2$, $m_{\pi T}^2$. For $s > s_o$, the averaging over jet effective masses squared $s_j \simeq s x_q x_{\bar{q}}$ leads to /see Eq. /3//

$$\langle n_{ch}\rangle_{\bar{p}p_A} = 3a + 3b\left[4\left(\sqrt{\frac{s_o}{s}} - 1\right) + \left(\sqrt{\frac{s_o}{s}} + 1\right)\ln\frac{s}{s_o}\right] . \qquad /6/$$

The data on ratios of $\bar{p}p_A$ and pp charged particle topological cross sections yield $\langle k_{ch}\rangle \simeq 1.88$, i.e. $b = g\langle k_{ch}\rangle \simeq 0.94$ close to the lns-slope of $\langle n_{ch}\rangle_{e^+e^-}$ at $\sqrt{s} < 7$ GeV. Choosing $\sqrt{s_o} = 0.62$ and $a = 0.78$ to agree with the low energy $\bar{p}p_A$ data, formula /6/ describes well the experimental points on Fig. 4. Comparison of Eq. /6/ with its asymptotic form indicates /Fig. 4/ that the threshold effects in $\bar{p}p_A$ are important up to $\sqrt{s} \sim 10$ GeV. However, the lns-dependence of $\langle n_{ch}\rangle_{pp}$ is not linear even at $\sqrt{s} > 10$ GeV. The parametrization of pp-data at s=8-65 GeV

$$\langle n_{ch}\rangle_{pp} = 0.88 + 0.44\cdot\ln s + 0.118\cdot\ln^2 s \qquad /7/$$

suggests that the effective number of jets increases with energy approximately linear in lns. At $\sqrt{s} = 540$ GeV, Eq. /7/ predicts $\langle n_{ch}\rangle_{pp} = 25.1$ close to the value $\langle n_{ch}\rangle_{\bar{p}p} = 24.6 \pm 1.6$ obtained recently on CERN $\bar{p}p$-collider. It should be stressed that the effect of jet multiplication and the related fast increase of $\langle n\rangle$ and the rising height of rapidity plateau with energy /violation of Feynman scaling/ has been predicted by Regge phenomenology due to multiple Pomeron exchange contributions in correspondence with AGK cutting rules. Interpreting the multiple Pomeron sheets as jets in DU scheme and generalizing the quark distribution /3/ for the multiquark case, a good description of the data on $d\sigma/dy$ and $\langle n_{ch}\rangle$ at $\sqrt{s} \gtrsim 20$ GeV has been obtained. The effect of jet multiplication-branching is also in qualitative agreement with the positive correlation between $\langle p_T\rangle$ and n_{ch} recently observed in collider data.

We thus see that the conclusive check of DU predictions with the help of data on $\langle n\rangle$ requires very high primary energies /to determine contribution of the leading term in the energy dependence of $\langle n\rangle$ /. It may therefore be useful to check these predictions using the quantities which are known to be only weakly dependent on energy, such as relative dispersion $d=D/\langle n\rangle$. In fact, naive DU prediction /dispersions squared and mean particle multiplicities in independent jets are added/

$$d_{(N)} = \frac{1}{\sqrt{N}}\, d_{(1)} \qquad /8/$$

is at least in qualitative agreement with the data on 3-jet /$\bar{p}p_A$/, 2-jet /pp_{ND}/ and 1-jet /K^-p-K^+p/ processes, see Fig. 5. The naive DU prediction is however essentially modified if we account for fluctuations of jet effective masses. Thus even in the case when $d_{(1)}$=const, we have

$$d_{(N)} = \frac{1}{\sqrt{N}}\left(d^2_{(1)} + \Delta^2\right)^{1/2} . \qquad /9/$$

Besides the dispersion due to energy fluctuations, Δ^2 contains negative contribution connected with the energy-momentum conservation. We have found $\Delta^2_{\bar{p}p_A}$ roughly constant at $\sqrt{s} \lesssim 30$ GeV and close to $d^2_{(1)}=d^2_{e^+e^-}$ - 0.18. The corresponding curve shown in Fig. 5 is in reasonable agreement with the data on $d^2_{\bar{p}p_A}$. Analogous calculations in the case of non-diffractive pp-interactions yield again $\Delta^2_{pp_{ND}} \simeq \Delta^2_{e^+e^-}$ which explains the success of naive DU prediction /8/ and also the approximate equality $d_{pp_{ND}} \simeq d_{e^+e^-}$. On the other hand, the naive relation /8/ fails for K^-p-K^+p difference /see Fig. 5/ since $\Delta^2_{K^-p-K^+p} > \Delta^2_{\bar{p}p_A,pp_{ND}}$ /there is no negative correlation due to energy-momentum conservation in 1-jet process K^-p-K^+p/.

4. Spin effects in vector meson production.

The resonance decay angular distributions can be obtained by fitting the effective mass distributions of decay particles in various intervals of decay angles. The distributions of cosine of polar angles Θ_J and Θ_T of ρ^o-decay π^+-momentum in Jackson and transversity frames for the reactions $\bar{p}p \to \rho^o\pi^+\pi^-$+neutrals at 5.7 GeV/c and $\bar{p}p \to \rho^o X$ at 12 and 22.4 GeV/c are shown in Figs. 6,7. We see a strong deviation from uniform distribution, i.e. ρ^o-spin alignment. Absence of ρ^o-spin alignment in pp-interactions and also the dominant or large contribution of annihilation channels to the considered $\bar{p}p$ reactions, make it possible to connect the observed effect mainly with annihilation channels. The analogus analysis of ω-decay distributions yields an indication for different character of ρ^o- and ω-spin alignment in $\bar{p}p$-interactions. Namely, while ρ^o is produced predominantly in a state with zero spin projection on the production plane normal /the corresponding probability ρ^T_{oo} is large/, ω has a large probability of zero spin projection on the c.m.s. reaction axis. It is interesting that the similar difference of ρ^o- and ω - spin characteristics has been observed in a number of binary meson-nucleon reactions, with the vector mesons produced near the backward direction. Such an analogy is, in fact, not unexpected since in multiperipheral approach, $\bar{p}p_A$ as well as backward meson production are described by baryon exchange. As is well-known, $\bar{B}B\omega$-coupling predominantly doesn't flip the baryon helicity, while $\bar{B}B\rho$-coupling is mostly helicity flip. This explains a large /small/ probability of zero ω/ρ/-spin projection on the reaction axis, but it is not sufficient to explain the large ρ^T_{oo} value for ρ-meson. Probably, absorption or cuts should be taken into account.

5. Quark polarization.

Interesting semi-classical model for Λ-polarization in $p \to \Lambda$ fragmentation has been recently suggested. In this model, Λ is produced by recombination of fast valence spin-zero diquark ud with slow sea s-quark, the latter being accelerated by a force $\vec{F}$ along the reaction axis. If the corresponding confining potential is scalar, the effective Hamiltonian contains the spin-dependent term $U=\vec{s}\vec{\omega}_T$ completely determined by the Thomas frequency

$$\vec{\omega}_T = \frac{\gamma}{\gamma+1}\,\frac{\vec{F}\times\vec{\beta}}{m}\,, \qquad /10/$$

where m and $\vec{\beta}$ are the s-quark mass and velocity, $\gamma=(1-\beta^2)^{-1/2}$. For large enough $p_{\Lambda T}$, the transverse momenta of s quark and Λ will be roughly parallel, i.e. $\vec{\omega}_T$ roughly parallel to the production normal $\hat{n}=\hat{p}_p\times\hat{p}_\Lambda$. Thus the minimal energy configuration corresponds to the s-quark spin /i.e. Λ-spin/ directed opposite to $\hat{n}$, in agreement with experiment. The corresponding polarization mechanism could be an analog of spontaneous radiation polarization of electrons in an external magnetic or electric field /Sokolov-Ternov effect/. In modern electron-positron storage rings, the polarization $P=P_o(1-e^{-t/t_P})$ reaches its maximal value $P_o=0.92$ during $t\sim$ several minutes- electrons /positrons/ are polarized antiparallel /parallel/ to the magnetic field direction. Simple estimates "per analogiam" show that the polarization time in the rest frame of a quark moving in confining field is of reasonable order of 10^{-23}sec,

$$t_P^{*-1}\sim \alpha_s p_{qT}^3 F^3/m_q^8 , \qquad /11/$$

where $\alpha_s\sim 1$, $m_q \lesssim 0.3$ GeV/c^2 and the confining force is $F\sim \Lambda^2\sim 0.1$ GeV^2. To explain the negative sign of Λ-polarization, we should require either a large anomal color-magnetic moment of the s-quark compensating the Dirac contribution, or the dominance of the scalar component of the confining field. In the latter case, however, the energy independent reversed time of radiation polarization in Eq. /11/ should be multiplied by γ^3 thus leading to vanishing Λ-polarization with increasing energy, in contradiction with experiment. The discussed model can be applied to explain also the data on polarization asymmetry $A_c=\frac{1}{P}(N_+-N_-)/(N_++N_-)$, where $N_{+(-)}$ is the production rate of the particle c in the case when the incoming proton polarization vector $\vec{P}$ is parallel /antiparallel/ to the production plane normal $\hat{n}=\hat{p}_p x\hat{p}_c$. The asymmetry for the production of positive mesons in pp and $\bar{p}p$ reactions has been found positive and quite large at $p_T\gtrsim 1$ GeV/c, see also Fig. 8. It is obvious from Fig. 9 that the positive asymmetry corresponds to the most favourable configuration $\vec{s}_u \parallel \vec{\beta}_u x\vec{F}$. Note that according to SU(6) proton wave function $p\uparrow=2u\uparrow u\uparrow d\downarrow - u\uparrow u\downarrow d\uparrow$, almost all the proton polarization is carried by the valence u-quarks. This model explains also the spin alignment /large ρ_{oo}^T/ of ρ^o's in $\bar{p}p$-interactions and in meson-nucleon interactions with large momentum transfer. Note that the different spin characteristics of ρ^o and ω produced near the backward direction in binary meson-nucleon reactions are naturally explained due to coherent superposition of valence-valence and valence-sea recombination amplitudes, see Fig. 10. It should be stressed that the similar effect observed in $\bar{p}p$-interactions requires the account for coherence of the quark recombination amplitudes even in the case of multiparticle production.

6. Concluding remarks.

The approximate KNO-scaling seen in a number of multiparticle processes indicates an universal /presumably jet/ mechanism of multiparticle production. However, the different relative dispersions of multiplicity distributions in various processes indicate, contrary to some suggestions, that this universal mechanism is not simply determined by the energy available for particle production. In fact, multiplicity distributions and one-particle inclusive spectra are successfully described in DU scheme, unifying the approach of quark fragmentation models with Regge phenomenology. The latter turnes out to be highly usefull for discussion of antiparticle-particle cross-section differences and their relation to annihilations. The corresponding multiperipheral approach, accounting for absorption, pretends to describe the spin effects in hadron interactions as well. On the other hand, the observation of substantial spin effects at high energies suggests that the underlying processes involved might be simple at the level of constituents. Indeed, the interesting interpretations of these effects based on color and electric charge analogy have been given in terms of simple quark models. There are also several attempts to estimate spin effects in the framework of perturbative QCD. The expected polarization is about one order of magnitude smaller than that required by experiment. On the other hand, the simple models for quark polarization based on analogy with QED can give only hints for future theory of soft processes- presumably nonperturbative QCD. Hopefully, data from new polarization facilities in Batavia, Brookhaven and Dubna will help in clarifying the nature of these effects. It is unlikely that the nonperturbative QCD will be elucidated without the guidence of experiment. With the gradual exhaustion of the possibilities offered by hadron spectroscopy, the data on multiparticle production will be increasingly exploited. In particular, we may expect that the available data on antiproton collisions, as well as the new data coming from $\bar{p}p$-collider and Batavia, CERN and Stanford hybrid systems /the latter allowing for separation of $\bar{p}p$ annihilations and nonannihilations/ will considerably contribute to understanding of the complicated confinement problem.

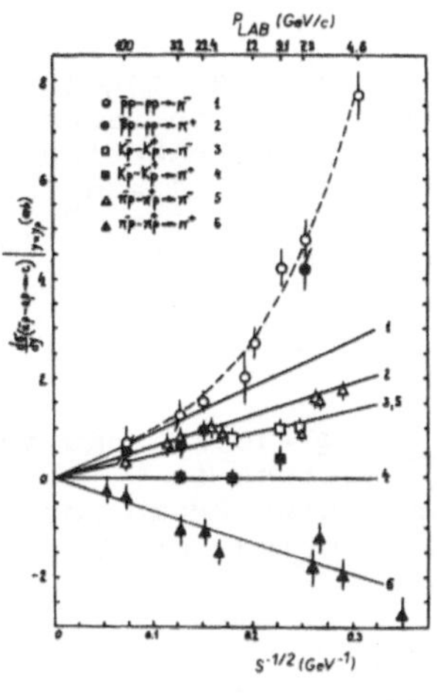

Fig. 1. Energy dependence of invariant cross section differences for the reactions $a^{\pm}p \to cX$ at rapidity $y_c = y_p$. The lines 1-6 correspond to Eqs. /2/. The dashed curve is the line 1 with the added contribution As^{-2}.

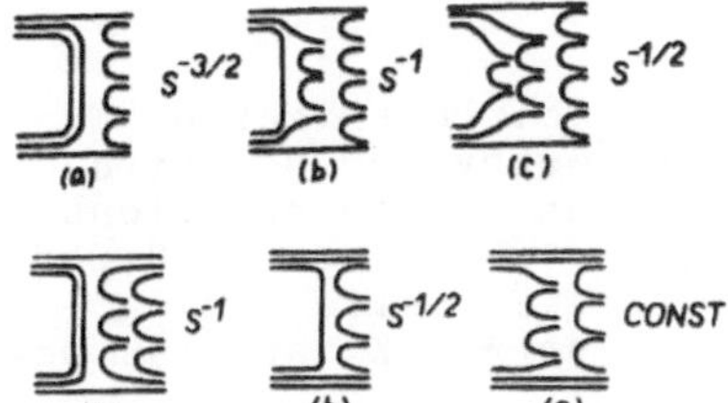

Fig. 2. Dual quark-line diagrams for $\bar{p}p$-annihilations.

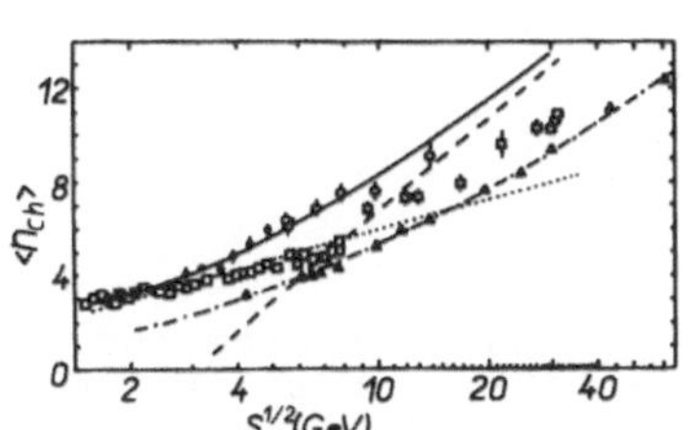

Fig. 3. Dual quark-line diagrams for $\bar{p}p$-nonannihilations.

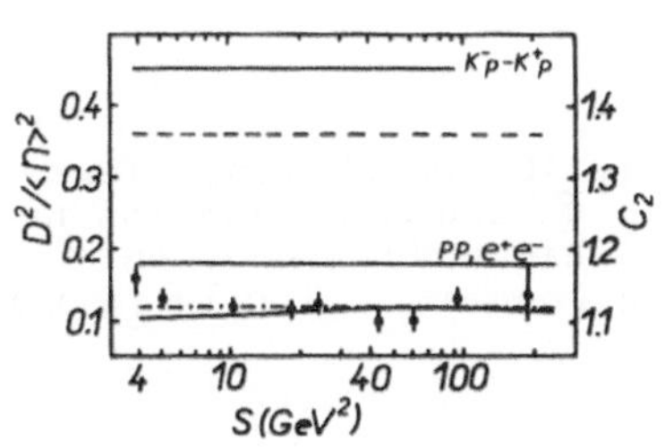

Fig. 4. Energy dependence of $<n_{ch}>$ in $\bar{p}p$-annihilations /•/, $\bar{p}p$-pp /○/, pp /△/ and e^+e^- /□/. The dotted, full, dashed-dotted and dashed curves correspond to Egs. /5-7/ and to the asymptotic form of Eg. /6/, respectively.

Fig. 5. Energy dependence of $D_{ch}^2/<n_{ch}>^2$ for $\bar{p}p$-annihilations /•/ and $\bar{p}p$-pp /○/. The full lines describe the data on nondiffractive pp-interactions, e^+e^--annihilations and K^-p-K^+p differences. The dashed and dashed-dotted lines are naive DU predictions /Eq. /8// for 1- and 3-jet processes based on pp_{ND} data. The curve corresponds to Eg. /9/.

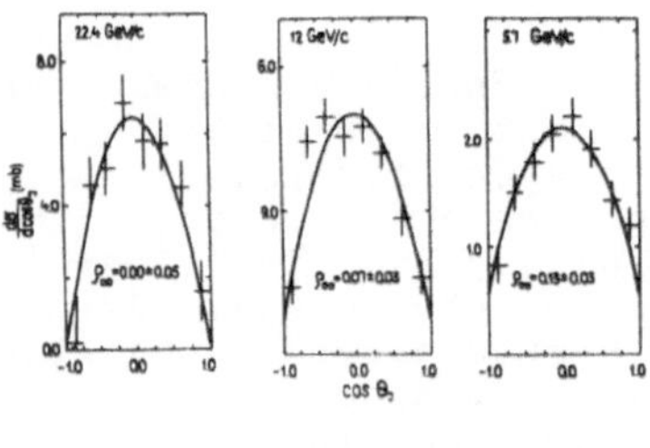

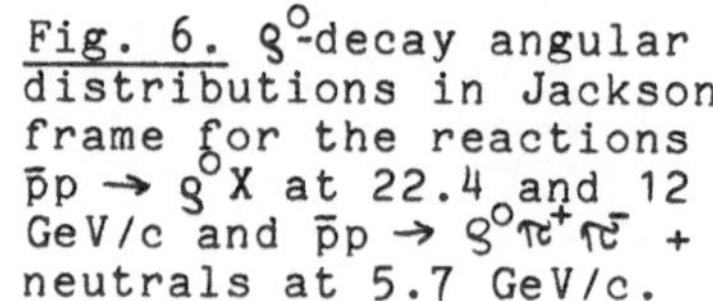

Fig. 6. ρ^0-decay angular distributions in Jackson frame for the reactions $\bar{p}p \to \rho^0 X$ at 22.4 and 12 GeV/c and $\bar{p}p \to \rho^0\pi^+\pi^-$ + neutrals at 5.7 GeV/c.

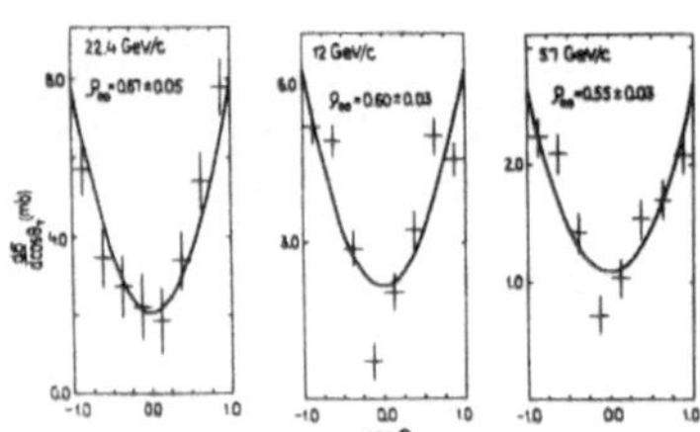

Fig. 7. ρ^0-decay angular distributions in transversity frame for the same reactions as in Fig. 6.

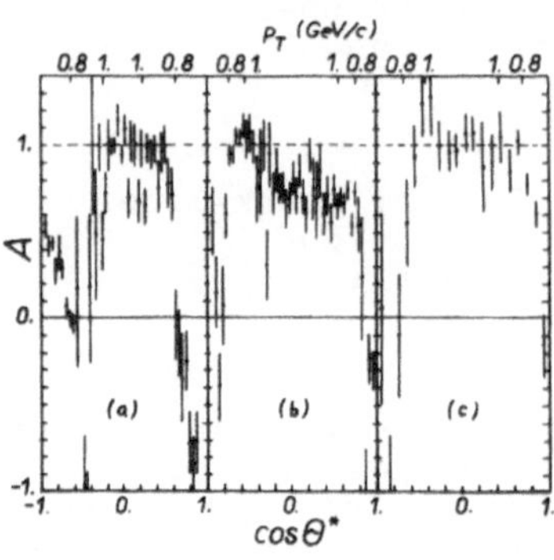

Fig. 8. Polarization asymmetry for positive mesons in the reactions $\bar{p}p \to \pi^-\pi^+$ at /a/ 1.0 GeV/c, /b/ 1.99 GeV/c and /c/ $\bar{p}p \to K^-K^+$ at 1.99 GeV/c.

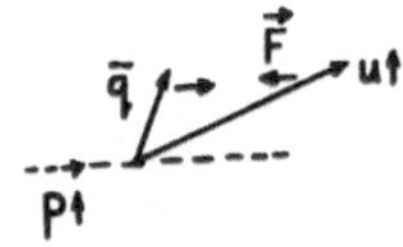

Fig. 9. Quark diagram for fragmentation of polarized proton $p\uparrow \to \pi^+(K^+)$.

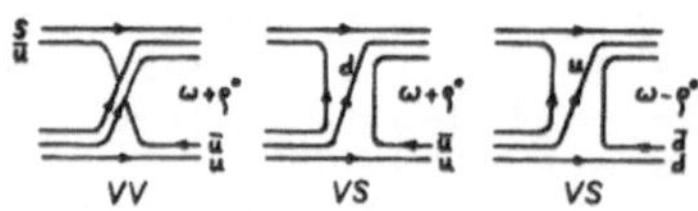

Fig. 10. Quark-line diagrams for ρ^0 and ω production near the backward direction in the reactions $K^-p \to \Lambda\rho^0, \Lambda\omega$.

ARCTIC SCHOOL OF PHYSICS 1982

LIST OF PARTICIPANTS

Achiman, Yoav
Universität Wuppertal
Fachbereich 8 - Physik
Postfach 100127
D-5600 Wuppertal
BRD

Bengtsson, Anders
Institute for Theor.Physics
Chalmers Univ. of Technology
S-41296 Göteborg
SWEDEN

Bengtsson, Ingemar
Vattugatan 2 B
S-41316 Göteborg
SWEDEN

Bennett, Donald
Department of Physics
Universitetsparken 2
DK-2100 Copenhagen Ø
DENMARK

Cronström, Christofer
University of Helsinki
Dept. of Theoretical Physics
Siltavuorenpenger 20 C
SF-00170 Helsinki 17
FINLAND

DeTar, Carleton
University of Utah
Department of Physics
Salt Lake City, UT 84112
USA

Diamandis, George
University of Athenes
Physics Department
Ano Ilisia - Kouponia 621
Athens
GREECE

Dyakonov, D.
Leningrad Nuclear Physics Institute
Gatchina
Leningrad 188350
USSR

Ehtamo, Harri
University of Helsinki
Research Institute for Theoretical Physics
Siltavuorenpenger 20 C
SF-00170 Helsinki 17
FINLAND

Enqvist, Kari
University of Helsinki
Siltavuorenpenger 20 C
SF-00170 Helsinki 17
FINLAND

Farrar, Glennys
Rutgers University
Physiscs Department
New Brunswick, NJ 08903
USA

Fritzsch, Harald
Max Planck - Institut für Physik
Föhringer Ring 6
D-8 München 40
BRD

Gatheral, James
DAMTP, Cambridge University
Silver Street
Cambridge CB3 9EW
GREAT BRITAIN

Gavai, Rajiv V.
Universität Bielefeld
Fakultät für Physik
Postfach 8640
D-4800 Bielefeld 1
BRD

Generalis, S. C.
The Open University, Physics Dept.
Walton Hall
Milton Keynes, MK7 6AA
GREAT BRITAIN

Gepner, Doron
Weizmann Institute
Rehovot
ISRAEL

Glass, Henry
Fermilab/MS 221
P.O. Box 500
Batavia, IL 60510
USA

Goossens, Michel
CERN/EP
CH-1211 Geneva 23
SWITZERLAND

Hwang, Stephen
Institute for Theoretical Physics
Chalmers University
S-41296 Göteborg
SWEDEN

Häkkinen, Lasse
University of Helsinki
Research Institute for Theoretical Physics
Siltavuorenpenger 20 C
SF-00170 Helsinki 17
FINLAND

Immonen, Olli
Koroistentie 8 B 11
SF-00280 Helsinki 28
FINLAND

Jackiw, Roman
MIT, Department of Physics
Cambridge, MA 02139
USA

Jacobs, Laurence
Institute for Theoretical Physics
University of California
Santa Barbara, CA 93106
USA

Jensen, Henrik
Institute of Physics
University of Aarhus
DK-8000 Aarhus C
DENMARK

Johansson, Lars
Institute of Theoretical Physics
Chalmers University
S-41296 Göteborg
SWEDEN

Karsch, Frithjof
Fakultät für Physik
Universität Bielefeld
Universitätsstrasse
D-4800 Bielefeld 1
BRD

Kleppe, Astri
Fysisk Institutt
Bergen Universitetet
Allégt. 55
NORWAY

Kupiainen, Antti
University of Helsinki
Research Institute for Theoretical Physics
Siltavuorenpenger 20 C
SF-00170 Helsinki 17
FINLAND

Kurki-Suonio, Hannu
University of Helsinki
Department of Theoretical Physics
SF-00170 Helsinki 17
FINLAND

Launer, Guy
CNRS, Centre de Physique Theorique
Case 907 - Luminy
F-13288 Marseille Cedex 2
FRANCE

Leung, C. N.
School of Physics, University of Minnesota
116 Church St. SE
Minneapolis, MN 55 455
USA

Lednicky, R.
JINR, Dubna, Head Post Office
P.O. Box 79
101 000 Moscow
USSR

Lindfors, Juha
University of Helsinki
Research Institute for Theoretical Physics
Siltavuorenpenger 20 C
SF-00170 Helsinki 17
FINLAND

Lütken, Carsten
Fysisk Institutt, University of Oslo
Boks 1048
Blindern, Oslo 3
NORWAY

Mack, Gehard
II Institut für Theor. Physik
Universität Hamburg
Notkestrasse 85
2 Hamburg 52
BRD

Makeenko, Yu M.
ITEP
B.Cheremushkinskaya 25
117259 Moscow
USSR

McLerran, Larry
University of Washington
Physics Department
Seattle, WA 98195
USA

Meyer, Hildegard
II Institut für Theor. Physik
Universität Hamburg
Luruper Chaussee 149
2 Hamburg
BRD

Mickelsson, Jouko
Dept. of Mathematics
University of Jyväskylä
Seminaarinkatu 15, Jyväskylä 15
FINLAND

Miettinen, Hannu I.
University of Helsinki
Research Institute for Theoretical Physics
Siltavuorenpenger 20 C
SF-00170 Helsinki 17
FINLAND

Montonen, Claus
University of Helsinki
Research Institute for Theoretical Physics
Siltavuorenpenger 20 C
SF-00170 Helsinki 17
FINLAND

Montgomery, Hugh
CERN/EP
CH-1211 Geneva 23
SWITZERLAND

Mursula, Kalevi
University of Helsinki
Research Institute for Theoretical Physics
Siltavuorenpenger 20 C
SF-00170 Helsinki 17
FINLAND

Mäkinen, Markku

Ida Aalbergintie 7A8
SF-00400 Helsinki 40
FINLAND

Nadeau, Hélène
Sloane Labs., Yale University
217 Prospect St.
New Haven, CT 06511
USA

Nielsen, Holger B.
Niels Bohr Institute
Blegdamsvej 17
DK-2100 Copenhagen Ø
DENMARK

Niemi, Antti
MIT, Dept. of Physics
Cambridge, MA 02139
USA

Nordling, Johan
Lilla Tavastgatan 14 A 5
SF-20500 Turku 50
FINLAND

Peccei, Roberto
Max Planck-Institut für Physik
Föhringen Ring 6
D-8 München 40
BRD

Pellinen, Asta
University of Helsinki
Dept. of High Energy Physics
Siltavuorenpenger 20 C
SF-00170 Helsinki 17
FINLAND

Petcov, Sergei
Institute of Nuclear Research
Boul. Lenin 72
Sofia 1184
BULGARIA

Pi, So-Young
Harvard University
Physics Department
Cambridge, Ma 02138
USA

Pitkänen, Matti
Framnäsintie 2 C 13
02430 Masala
FINLAND

Raitio, Risto
University of Helsinki
Dept. of Theoretical Physics
Siltavuorenpenger 20 C
SF-00170 Helsinki 17
FINLAND

Yahalom, Ram
Physics Department, Technion
3200 Haifa
ISRAEL

Roos, Matts
University of Helsinki
Dept. of High Energy Physics
Siltavuorenpenger 20 C
SF-00170 Helsinki 17
FINLAND

Rubbia, Carlo
CERN/EP
CH-1211 Geneva 23
SWITZERLAND

Ruuskanen, Vesa
University of Jyväskylä
Physics Department
Nisulankatu 78
SF-40720 Jyväskylä 72
FINLAND

Satz, Helmut
Universität Bielefeld
Fakultät für Physik
Postfach 8640
D-4800 Bielefeld 1
BRD

Schepkin, M.
Institute for Theoretical and Experimental Physics
B. Cheremushkinskaya 89
117 259 Moscow
USSR

Sinha, Amarendra
Ohio State University, Physics Dept.
174 West 18th Ave.
Columbus, OH 43210
USA

Smulter, Leif
Åbo Akademi, Physics Department
SF-20500 Åbo 50
FINLAND

Sokachev, Emery
JINR, Dubna, Head Post Office
P.O. Box 79
101 000 Moscow
USSR

Stelle, Kellogg
Imperial College, Theoretical Physics
London SW7 2AZ
GREAT BRITAIN

Davies, Christine
Cavendish Laboratory
Madingley Road
Cambridge CB3 OHE
GREAT BRITAIN

Streng, Karl-Heinz
Institut der Theor. Physik
Universität München
Theresienstr. 37
D-8000 München
BRD

Sutela, Tytti
University of Helsinki
Dept. of Theoretical Physics
Siltavuorenpenger 20 C
SF-00170 Helsinki 17
FINLAND

Sørensen, Paul
DAMTP, Cambridge University
Silver Street , Cambridge CB3 9EW
Great Britain

Tangen, Kjell
Theory Group, Fysisk Institutt
Boks 1049
Blidern, Oslo 3
NORWAY

Thodberg, Hans Henrik
Tagensvej 15,
DK-2200 København N
DENMARK

Toimela, Tuomo
University of Helsinki
Research Institute for Theoretical Physics
Siltavuorenpenger 20 C
SF-00170 Helsinki 17
FINLAND

Törnqvist, Nils
University of Helsinki
Research Institute for Theoretical Physics
Siltavuorenpenger 20 C
SF-00170 Helsinki 17
FINLAND

Visnjić, Vladimir
University of Minnesota
Department of Physics
Minneapolis, MN 55455
USA

Vrba, Vaclav
JINR , Dubna, Head Post Office
P.O. Box 79
101 000 Moscow
USSR

Lecture Notes in Physics

Vol. 144: Topics in Nuclear Physics I. A Comprehensive Review of Recent Developments. Edited by T.T.S. Kuo and S.S.M. Wong. XX, 567 pages. 1981.

Vol. 145: Topics in Nuclear Physics II. A Comprehensive Review of Recent Developments. Proceedings 1980/81. Edited by T. T. S. Kuo and S. S. M. Wong. VIII, 571-1.082 pages. 1981.

Vol. 146: B. J. West, On the Simpler Aspects of Nonlinear Fluctuating. Deep Gravity Waves. VI, 341 pages. 1981.

Vol. 147: J. Messer, Temperature Dependent Thomas-Fermi Theory. IX, 131 pages. 1981.

Vol. 148: Advances in Fluid Mechanics. Proceedings, 1980. Edited by E. Krause. VII, 361 pages. 1981.

Vol. 149: Disordered Systems and Localization. Proceedings, 1981. Edited by C. Castellani, C. Castro, and L. Peliti. XII, 308 pages. 1981.

Vol. 150: N. Straumann, Allgemeine Relativitätstheorie und relativistische Astrophysik. VII, 418 Seiten. 1981.

Vol. 151: Integrable Quantum Field Theory. Proceedings, 1981. Edited by J. Hietarinta and C. Montonen. V, 251 pages. 1982.

Vol. 152: Physics of Narrow Gap Semiconductors. Proceedings, 1981. Edited by E. Gornik, H. Heinrich and L. Palmetshofer. XIII, 485 pages. 1982.

Vol. 153: Mathematical Problems in Theoretical Physics. Proceedings, 1981. Edited by R. Schrader, R. Seiler, and D.A. Uhlenbrock. XII, 429 pages. 1982.

Vol. 154: Macroscopic Properties of Disordered Media. Proceedings, 1981. Edited by R. Burridge, S. Childress, and G. Papanicolaou. VII, 307 pages. 1982.

Vol. 155: Quantum Optics. Proceedings, 1981. Edited by C.A. Engelbrecht. VIII, 329 pages. 1982.

Vol. 156: Resonances in Heavy Ion Reactions. Proceedings, 1981. Edited by K.A. Eberhard. XII, 448 pages. 1982.

Vol. 157: P. Niyogi, Integral Equation Method in Transonic Flow. XI, 189 pages. 1982.

Vol. 158: Dynamics of Nuclear Fission and Related Collective Phenomena. Proceedings, 1981. Edited by P. David, T. Mayer-Kuckuk, and A. van der Woude. X, 462 pages. 1982.

Vol. 159: E. Seiler, Gauge Theories as a Problem of Constructive Quantum Field Theory and Statistical Mechanics. V, 192 pages. 1982.

Vol. 160: Unified Theories of Elementary Particles. Critical Assessment and Prospects. Proceedings, 1981. Edited by P. Breitenlohner and H.P. Dürr. VI, 217 pages. 1982.

Vol. 161: Interacting Bosons in Nuclei. Proceedings, 1981. Edited by J.S. Dehesa, J.M.G. Gomez, and J. Ros. V, 209 pages. 1982.

Vol. 162: Relativistic Action at a Distance: Classical and Quantum Aspects. Proceedings, 1981. Edited by J. Llosa. X, 263 pages. 1982.

Vol. 163: J. S. Darrozes, C. Francois, Mécanique des Fluides Incompressibles. XIX, 459 pages. 1982.

Vol. 164: Stability of Thermodynamic Systems. Proceedings, 1981. Edited by J. Casas-Vázquez and G. Lebon. VII, 321 pages. 1982.

Vol. 165: N. Mukunda, H. van Dam, L.C. Biedenharn, Relativistic Models of Extended Hadrons Obeying a Mass-Spin Trajectory Constraint. Edited by A. Böhm and J.D. Dollard. VI, 163 pages. 1982.

Vol. 166: Computer Simulation of Solids. Edited by C.R.A. Catlow and W.C. Mackrodt. XII, 320 pages. 1982.

Vol. 167: G. Fieck, Symmetry of Polycentric Systems. VI, 137 pages, 1982.

Vol. 168: Heavy-Ion Collisions. Proceedings, 1982. Edited by G. Madurga and M. Lozano. VI, 429 pages. 1982.

Vol. 169: K. Sundermeyer, Constrained Dynamics. IV, 318 pages. 1982.

Vol. 170: Eighth International Conference on Numerical Methods in Fluid Dynamics. Proceedings, 1982. Edited by E. Krause. X, 569 pages. 1982.

Vol. 171: Time-Dependent Hartree-Fock and Beyond. Proceedings, 1982. Edited by K. Goeke and P.-G. Reinhard. VIII, 426 pages. 1982.

Vol. 172: Ionic Liquids, Molten Salts and Polyelectrolytes. Proceedings, 1982. Edited by K.-H. Bennemann, F. Brouers, and D. Quitmann. VII, 253 pages. 1982.

Vol. 173: Stochastic Processes in Quantum Theory and Statistical Physics. Proceedings, 1981. Edited by S. Albeverio, Ph. Combe, and M. Sirugue-Collin. VIII, 337 pages. 1982.

Vol. 174: A. Kadić, D.G.B. Edelen, A Gauge Theory of Dislocations and Disclinations. VII, 290 pages. 1983.

Vol. 175: Defect Complexes in Semiconductor Structures. Proceedings, 1982. Edited by J. Giber, F. Beleznay, J. C. Szép, and J. László. VI, 308 pages. 1983.

Vol. 176: Gauge Theory and Gravitation. Proceedings, 1982. Edited by K. Kikkawa, N. Nakanishi, and H. Nariai. X, 316 pages. 1983.

Vol. 177: Application of High Magnetic Fields in Semiconductor Physics. Proceedings, 1982. Edited by G. Landwehr. XII, 552 pages. 1983.

Vol. 178: Detectors in Heavy-Ion Reactions. Proceedings, 1982. Edited by W. von Oertzen. VIII, 258 pages. 1983.

Vol. 179: Dynamical Systems and Chaos. Proceedings, 1982. Edited by L. Garrido. XIV, 298 pages. 1983.

Vol. 180: Group Theoretical Methods in Physics. Proceedings, 1982. Edited by M. Serdaroğlu and E. İnönü. XI, 569 pages. 1983.

Vol. 181: Gauge Theories of the Eighties. Proceedings, 1982. Edited by R. Raitio and J. Lindfors. V, 644 pages. 1983.

Selected Issues from

Lecture Notes in Mathematics

Vol. 836: Differential Geometrical Methods in Mathematical Physics. Proceedings, 1979. Edited by P. L. García, A. Pérez-Rendón, and J. M. Souriau. XII, 538 pages. 1980.

Vol. 837: J. Meixner, F. W. Schäfke and G. Wolf, Mathieu Functions and Spheroidal Functions and their Mathematical Foundations Further Studies. VII, 126 pages. 1980.

Vol. 838: Global Differential Geometry and Global Analysis. Proceedings 1979. Edited by D. Ferus et al. XI, 299 pages. 1981.

Vol. 840: D. Henry, Geometric Theory of Semilinear Parabolic Equations. IV, 348 pages. 1981.

Vol. 841: A. Haraux, Nonlinear Evolution Equations- Global Behaviour of Solutions. XII, 313 pages. 1981.

Vol. 842: Séminaire Bourbaki vol. 1979/80. Exposés 543–560. IV, 317 pages. 1981.

Vol. 843: Functional Analysis, Holomorphy, and Approximation Theory. Proceedings. Edited by S. Machado. VI, 636 pages. 1981.

Vol. 845: A. Tannenbaum, Invariance and System Theory: Algebraic and Geometric Aspects. X, 161 pages. 1981.

Vol. 849: P. Major, Multiple Wiener-Itô Integrals. VII, 127 pages. 1981.

Vol. 851: Stochastic Integrals. Proceedings, 1980. Edited by D. Williams. IX, 540 pages. 1981.

Vol. 852: L. Schwartz, Geometry and Probability in Banach Spaces. X, 101 pages. 1981.

Vol. 856: R. Lascar, Propagation des Singularités des Solutions d'Equations Pseudo-Différentielles à Caractéristiques de Multiplicités Variables. VIII, 237 pages. 1981.

Vol. 858: E. A. Coddington, H. S. V. de Snoo: Regular Boundary Value Problems Associated with Pairs of Ordinary Differential Expressions. V, 225 pages. 1981.

Vol. 861: Analytical Methods in Probability Theory. Proceedings 1980. Edited by D. Dugué, E. Lukacs, V. K. Rohatgi. X, 183 pages. 1981.

Vol. 866: J.-M. Bismut, Mécanique Aléatoire. XVI, 563 pages. 1981.

Vol. 878: Numerical Solution of Nonlinear Equations. Proceedings, 1980. Edited by E. L. Allgower, K. Glashoff, and H.-O. Peitgen. XIV, 440 pages. 1981.

Vol. 881: R. Lutz, M. Goze, Nonstandard Analysis. XIV, 261 pages. 1981.

Vol. 888: Padé Approximation and its Applications. Proceedings, 1980. Edited by M. G. de Bruin and H. van Rossum. VI, 383 pages. 1981.

Vol. 898: Dynamical Systems and Turbulence, Warwick, 1980. Proceedings. Edited by D. Rand and L.-S. Young. VI, 390 pages. 1981.

Vol. 901: Séminaire Bourbaki vol. 1980/81 Exposés 561–578. III, 299 pages. 1981.

Vol. 904: K. Donner, Extension of Positive Operators and Korovkin Theorems. XII, 182 pages. 1982.

Vol. 905: Differential Geometric Methods in Mathematical Physics. Proceedings, 1980. Edited by S.J. Andersson, H.-D. Doebner, and H.R. Petry. VI, 309 pages. 1982.

Vol. 909: Numerical Analysis. Proceedings, 1981. Edited by J.P. Hennart. VII, 247 pages. 1982.

Vol. 912: Numerical Analysis. Proceedings, 1981. Edited by G. A. Watson. XIII, 245 pages. 1982.

Vol. 920: Séminaire de Probabilités XVI, 1980/81. Proceedings. Edité par J. Azéma et M. Yor. V, 622 pages. 1982.

Vol. 921: Séminaire de Probabilités XVI, 1980–81 Supplément: Géométrie Différentielle Stochastique. Proceedings. Edité par J. Azéma et M. Yor. III, 285 pages. 1982.

Vol. 922: B. Dacorogna, Weak Continuity and Weak Lower Semicontinuity of Non-Linear Functionals. V, 120 pages. 1982.

Vol. 923: Functional Analysis in Markov Processes. Proceedings, 1981. Edited by M. Fukushima. V, 307 pages. 1982.

Vol. 926: Geometric Techniques in Gauge Theories. Proceedings, 1981. Edited by R. Martini and E.M.de Jager. IX 219 pages. 1982.

Vol. 927: Y. Z. Flicker, The Trace Formula and Base Change for GL (3). XII, 204 pages. 1982.

Vol. 928: Probability Measures on Groups. Proceedings 1981. Edited by H. Heyer. X, 477 pages. 1982.

Vol. 929: Ecole d'Eté de Probabilités de Saint-Flour X - 1980. Proceedings, 1980. Edited by P.L. Hennequin. X, 313 pages. 1982.

Vol. 930: P. Berthelot, L. Breen, et W. Messing, Théorie de Dieudonné Cristalline II. XI, 261 pages. 1982.

Vol. 931: D.M. Arnold, Finite Rank Torsion Free Abelian Groups and Rings. VII, 191 pages. 1982.

Vol. 932: Analytic Theory of Continued Fractions. Proceedings, 1981. Edited by W.B. Jones, W.J. Thron, and H. Waadeland. VI, 240 pages. 1982.

Vol. 934: M. Sakai, Quadrature Domains. IV, 133 pages. 1982.

Vol. 935: R. Sot, Simple Morphisms in Algebraic Geometry. IV, 146 pages. 1982.

Vol. 936: S.M. Khaleelulla, Counterexamples in Topological Vector Spaces. XXI, 179 pages. 1982.

Vol. 937: E. Combet, Intégrales Exponentielles. VIII, 114 pages. 1982.

Vol. 938: Number Theory. Proceedings, 1981. Edited by K. Alladi. IX, 177 pages. 1982.

Vol. 942: Theory and Applications of Singular Perturbations. Proceedings, 1981. Edited by W. Eckhaus and E.M. de Jager. V, 363 pages. 1982.

Vol. 953: Iterative Solution of Nonlinear Systems of Equations. Proceedings, 1982. Edited by R. Ansorge, Th. Meis, and W. Törnig. VII, 202 pages. 1982.

Vol. 956: Group Actions and Vector Fields. Proceedings, 1981. Edited by J.B. Carrell. V, 144 pages. 1982.

Vol. 957: Differential Equations. Proceedings, 1981. Edited by D.G. de Figueiredo. VIII, 301 pages. 1982.

Vol. 963: R. Nottrot, Optimal Processes on Manifolds. VI, 124 pages. 1982.

Vol. 964: Ordinary and Partial Differential Equations. Proceedings, 1982. Edited by W.N. Everitt and B.D. Sleeman. XVIII, 726 pages. 1982.

Vol. 968: Numerical Integration of Differential Equations and Large Linear Systems. Proceedings, 1980. Edited by J. Hinze. VI, 412 pages. 1982.

Vol. 970: Twistor Geometry and Non-Linear Systems. Proceedings, 1980. Edited by H.-D. Doebner and T.D. Palev. V, 216 pages. 1982.

Vol. 972: Nonlinear Filtering and Stochastic Control. Proceedings, 1981. Edited by S.K. Mitter and A. Moro. VIII, 297 pages. 1983.

Vol. 978: J. Ławrynowicz, J. Krzyż, Quasiconformal Mappings in the Plane. VI, 177 pages. 1983.

Vol. 979: Mathematical Theories of Optimization. Proceedings, 1981. Edited by J.P. Cecconi and T. Zolezzi. V, 268 pages. 1983.